BIOLOGY

Foundation for Class X

BIOLOGY

Foundation for Class X

Abhimanyu Kumar Jha
Ph.D. (Biotechnology)
Professor and Head, Department of Biotechnology
School of Engineering and Technology (SET), Sharda University
Greater Noida, Uttar Pradesh

Balasubramanian Velramar
Ph.D. (Environmental Microbiology)
Assistant Professor, Amity Institute of Biotechnology
Amity University Chhattisgarh, Raipur, Chhattisgarh

Archana Awasthi
M.Phil.
Biology Teacher, Delhi Public School
Ghaziabad, Uttar Pradesh

Saurabh Gupta
M.Sc. (Biochemistry)
Research Scholar in Environmental Biotechnology
Amity University Chhattisgarh, Raipur, Chhattisgarh

Universities Press

BIOLOGY FOUNDATION FOR CLASS X

UNIVERSITIES PRESS (INDIA) PRIVATE LIMITED

Registered Office
3-6-747/1/A & 3-6-754/1, Himayatnagar, Hyderabad 500 029, Telangana, India
info@universitiespress.com; www.universitiespress.com

Distributed by
Orient Blackswan Private Limited

Registered Office
3-6-752 Himayatnagar, Hyderabad 500 029, Telangana, India

Other Offices
Bengaluru / Chennai / Guwahati / Hyderabad / Kolkata
Mumbai / New Delhi / Noida / Patna / Visakhapatnam

First published by Universities Press (India) Private Ltd 2024

Cover and book design

0501960

ISBN: 978-93-93330-54-3

Typeset in Times New Roman PS Std 10 *by*
SRS Publishing Services, Puducherry

Printed in India by
Rajiv Book Binding House, Delhi

Published by
Universities Press (India) Private Limited
3-6-747/1/A & 3-6-754/1, Himayatnagar, Hyderabad 500 029, Telangana, India

About the Authors

Abhimanyu Kumar Jha is a molecular biologist with vast experience in teaching NEET UG (previously AIPMT), NEET PG and medical students from around the world with a very high success ratio. He is currently working as Professor and Head of the Department of Biotechnology at Sharda University, Greater Noida. Previously he worked as Dean, Faculty of Life Sciences, and Head of the Department of Biotechnology at the Institute of Applied Medicines and Research, Ghaziabad. He also worked as Head of the Department of Biotechnology at IMS Engineering College, Ghaziabad. He obtained his Ph.D. in Biotechnology from Panjab University, Chandigarh. He has published more than 125 papers in national and international journals and contributed 28 chapters in books published by Elsevier, Springer and Apple Academic Press. He has authored four books and worked on an edited book. He was awarded the honorary degree of Doctor of Science (Honorary D.Sc.) with specialisation in Biological Science by the American University, Washington D.C., USA. He has won many awards such as Distinguished Faculty, Best Indian Research and Young Scientist. His current area of research is on the epidemiology and epigenetics of oral squamous cell carcinoma and liver cancer.

Balasubramanian Velramar is an Assistant Professor at Amity Institute of Biotechnology, Amity University Chhattisgarh, Raipur. He is well-recognised internationally in the field of biodegradation of plastic wastes, more specifically using genes and enzymes. He is also working on the development of a band-aid or dressing material coated with nanoparticles for incurable wounds associated with diabetes and leprosy. He is the recipient of awards, honours and fellowships such as DST-SERB Young Scientist and UGC-Dr. D.S. Kothari Post-Doctoral Fellowship. He has written and co-authored more than 50 publications.

Archana Awasthi has been associated with Delhi Public School Ghaziabad for 17 years. She obtained her M.Sc. in Zoology and M.Phil. from Chaudhary Charan Singh University, Meerut. She was one of the top ten educators from India selected for Microsoft E2 Education Exchange 2019 held in Paris, France. She is a Microsoft Certified Educator, MIE Expert and MIE Master Trainer. She secured national rank 70, subject and city rank 2 and a cash prize of Rs. 5000 in CENTA Teaching Professionals' Olympiad 2018.

Saurabh Gupta is a Ph.D. Scholar at Amity University in Raipur. He has worked for more than 15 years in Environmental Biotechnology and with various MNCs. He has a PG degree in Biochemistry and an MBA.

List of Contributors

Runjhun Mathur is currently working for her Ph.D. from Dr. A.P.J Abdul Kalam Technical University, Lucknow, Uttar Pradesh. She has published more than 18 papers in national and international journals and contributed chapters to books published by Springer. She worked as a lecturer at Mahatma Jyoti Rao Phoole Women's College in Jaipur for three years. She obtained her M.Sc. in Biotechnology from the University of Rajasthan, Jaipur. She has also authored one book.

Hrithika Panday is a Ph.D. Scholar in Biotechnology at the School of Engineering and Technology, Sharda University, Greater Noida, with a keen interest in Bioinformatics (Drug Discovery). She has a Master's degree in Biotechnology and a Bachelor's degree in Basic Sciences from Deen Dayal Upadhyaya Gorakhpur University. She has published research papers, review articles and book chapters.

Sakshi Chauhan is a Research Scholar at the Department of Biotechnology, School of Engineering and Technology, Sharda University, Greater Noida. She obtained her M.Sc. in Biotechnology from Hemwati Nandan Bahuguna Garhwal University, Uttarakhand, and has contributed to a book published by Springer.

Preface

Welcome to the world of Biology! This textbook has been carefully designed to cater to students studying in Class X under both the ICSE (Indian Certificate of Secondary Education) and CBSE (Central Board of Secondary Education) syllabi. With comprehensive coverage of topics and concepts, this book aims to provide a solid foundation in Biology and equip students with the knowledge and skills necessary for their academic success.

This textbook follows a student-friendly approach, presenting information in a clear and concise manner. Each chapter begins with an introduction that provides an overview of the topic and its relevance. The content is organised systematically, building upon previous knowledge and gradually introducing more complex concepts. Care has been taken to present the subject matter in a logical sequence, making it easier for students to grasp and retain the information.

To facilitate effective learning, the book incorporates a variety of learning aids. These include labelled diagrams, illustrations, tables and hierarchy charts that visually enhance understanding. Key terms and definitions are highlighted for quick reference, and important points are summarised at the end of each chapter.

The exercise section is composed of two parts: school exam-based and competitive exam-based. The school exam-based Practice Questions consist of MCQs, Fill in the Blanks, True or False, Match the Following, Very Short Answer, Short Answer, Long Answer, Assertion–Reason, One-word Answer, Odd One Out, Define the Term, Groups of Terms, Image-Based, Case Study-Based and NCERT questions. The Competition Window presents MCQs for exams such as NEET and Olympiads to test comprehension and reinforce learning. Answers or hints to all questions are provided.

It is important to note that this textbook aligns with the syllabi prescribed by both ICSE and CBSE, ensuring that students are well-prepared for their respective board examinations. While the content covers the core concepts required by both boards, it also offers additional information to broaden the students' understanding of the subject.

Abhimanyu Kumar Jha
Balasubramanian Velramar
Archana Awasthi
Saurabh Gupta

Acknowledgements

We are grateful to many people for their support throughout our lives. For the sake of brevity, we would like to thank and recognise those who have had the greatest impact on this book's success.

First, we thank our student audience over the past few years, and our publisher, Universities Press, for inviting us to serve this audience.

We would like to express our heartfelt gratitude to the management of Sharda University, Greater Noida, especially Sh. Pradeep Kumar Gupta, Chancellor; Sh. Yatendra Kumar Gupta, Pro-Chancellor; and Prof. (Dr.) Sibaram Khara, Vice Chancellor, for providing constant motivation and support.

We are especially indebted to Prof. (Dr) Parma Nand, Dean Academics, Sharda University, and Prof. (Dr.) Shankar Ramamoorthy, Dean, School of Engineering and Technology, Sharda University, who have been a constant support and source of strength.

This book would not exist without the continual support and vision of the team at Universities Press: Thomas Mathew Rajesh, Kallol Das and Madhavi Sethupathi, who edited this book with enormous skill, warmth and precision.

Further, we would like to take this opportunity to express our gratitude to our parents and to our children for their endurance, love, wishes and support, which helped us to complete our work. In particular, Abhimanyu Kumar Jha would like to thank his parents, Sh. Arjun Kumar Jha and Smt. Subhadra Jha, his wife Mrs. Meenakshi Jha and his son, Master Aayushman Jha.

We would also like to acknowledge the blessings and support of our in-laws. We also want to thank and appreciate all our family members who offered their help willingly.

We want to convey our admiration and gratitude to the scientific community for everything they have done and are now doing to provide us with the best scientific advancements.

Above all, we want to express our gratitude to the Supreme Power, who has blessed us abundantly.

Abhimanyu Kumar Jha
Balasubramanian Velramar
Archana Awasthi
Saurabh Gupta

Contents

To

Lord Shiva, Mata Parvati and their family

and

My parents, Sh. Arjuṇ Kumar Jha and Smt. Subhadra Jha

—*Abhimanyu Kumar Jha*

CHAPTER 1

Cells: The Structural and Functional Units of Life

PREREQUISITES

You must have knowledge of:

- Cell biology and cell structure
- Cell membrane and its functions
- Basic biochemistry, such as the structure and functions of macromolecules like proteins, carbohydrates, lipids and nucleic acids
- Basic genetics, including DNA structure, replication, transcription and translation

LEARNING OBJECTIVES

At the end of this chapter, you will be able to:

- Identify and describe the functions of the different organelles found within cells, such as the nucleus, mitochondria, ribosomes, endoplasmic reticulum, Golgi apparatus and lysosomes
- List the differences between plant and animal cells
- Explain the importance of cells and organelles in the overall health and functioning of organisms

INTRODUCTION

All living beings are made up of cells and they develop from pre-existing cells. The number of cells in an organism's body increases with its size. Single-celled organisms are those that are composed of one cell. Examples are bacteria, yeast and *Amoeba*. Some organisms are composed of a few cells, such as *Spirogyra* and *Volvox*. Multi-celled organisms are made up of millions and billions of cells, such as humans and mango.

Cells are very small and can be seen only with a microscope. Some bacteria and human red blood cells are among the smallest cells. Nerve cells are the longest. The eggs of birds are the largest cells (actually the central yellow sphere). The largest cell in the modern living world is the ostrich egg before development begins. The egg's shell and albumen are components that are added as the egg travels through the reproductive system.

An average-sized adult human has approximately:

- 1000 million million cells
- 10,000 million nerve cells in the brain cortex
- 25 million million red blood cells
- 30,000 million white blood cells

Cells carry out various functions that are frequently correlated with their shape:

- Human red blood cells can easily pass through blood capillaries and carry oxygen because of their biconcave shape.
- White blood cells can squeeze through capillary walls because they have pseudopodia and amoeboid movement.
- Due to the length of nerve cells, the brain can receive 'impulses' from distant parts of the body and vice versa.

- Guard cells in the leaves of plants have a bean-like shape that allows them to open and close the stomatal pore.

GROSS STRUCTURE OF A CELL

The cell is surrounded by a **plasma membrane or cell membrane**. It is semi-permeable (or selectively or differentially permeable), which means that only some substances can pass through while others are blocked. It is a living membrane with fine pores.

The cell membrane is encircled by a **cell wall**, an additional outermost non-living layer, in a plant cell. The majority of the cell wall is composed of cellulose. It gives the cell its rigidity and structure. It can move freely through (allowing substances in solutions to enter and leave the cell without hindrance). Enzymes catalyse numerous chemical reactions in the semi-liquid substance known as the **cytoplasm**, which is found inside the cell membrane and outside the nucleus. Different organelles having different functions are present in the cytoplasm. The cytosol, which mainly consists of the liquid medium, is the part of the cytoplasm that does not include organelles (such as mitochondria).

The large, roughly spherical **nucleus** takes up almost the entire cytoplasm. A nuclear membrane with two layers and nuclear pores encircles it (Fig. 1.1). The ground substance of the nucleus (nucleoplasm) is a semi-solid one containing one or more rounded nucleoli.

A network of chromatin fibres, which are dark coloured, can be found in the nucleoplasm. In the course of cell division, these chromatin fibres group together to form short, thick chromosomes

In Fig. 1.2, an animal cell at the onset of cell division, two pairs of chromosomes can be seen condensed from the chromatin network; they are situated inside the nucleus. Each organism has a different number of chromosomes.

Cell Organelles

Organelles are specialised membrane-bound structures with predetermined roles. The organelles that make up a cell resemble the organs of a body. Each organelle performs one or more specific functions, much like how each organ in the body does.

The term **protoplasm or protoplast** refers to the entire living portion of a cell, including the cytoplasm, nucleus and other living bodies. The latter term, protoplast, is more appropriate to describe a plant cell that no longer has a cell wall. The living components, known as organelles, are the same in both plant and animal cells, with the exception of cell walls in plants and centrosomes with centrioles in animals. The various organelles are the cell membrane, nucleus, nucleolus, mitochondria, endoplasmic reticulum, ribosomes, Golgi bodies, lysosomes, chloroplasts and vacuoles (Table 1.1).

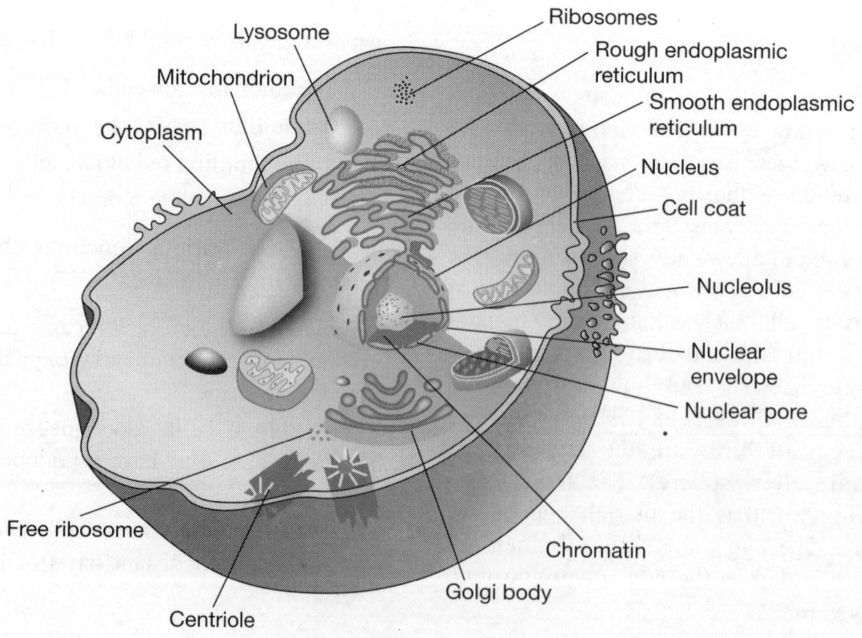

Fig. 1.1 Cell structure

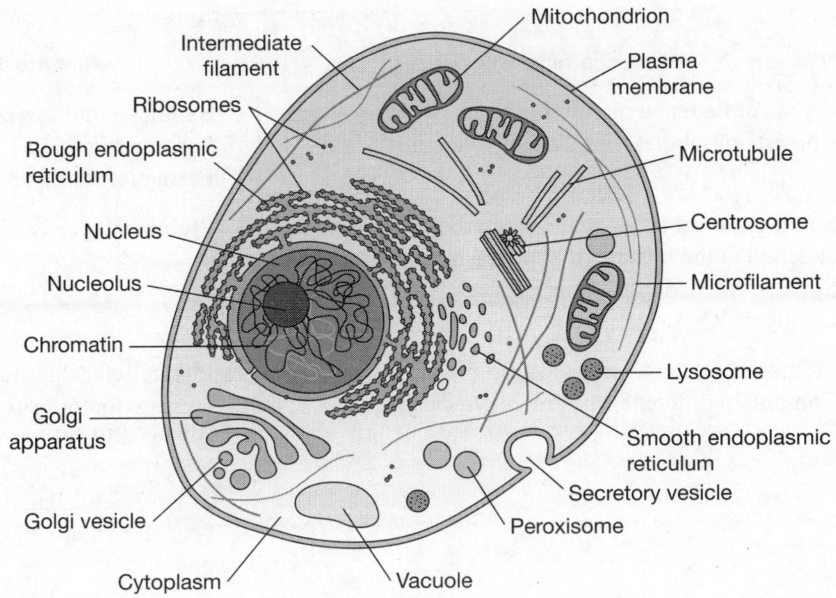

Fig. 1.2 An animal cell

Table 1.1 Cell organelles and their functions

Cell organelle	Main characteristics	Chief function(s)
Cell membrane	1. Outermost in animal cells. 2. Lies next to the cell wall in plant cells. 3. Very thin, flexible, living membrane. 4. Possesses fine pores. 5. Semi-permeable. 6. Made up of lipoproteins.	1. Separates contents of cell from its surroundings. 2. Regulates the entry of certain solutes and ions. 3. Maintains the shape of the cell (in animal cells).
Cell wall	1. Non-living rigid layer surrounding the plasma membrane. 2. Mainly composed of cellulose. 3. Freely permeable.	1. Gives rigidity and shape to plant cells. 2. Allows substances in solution to enter and leave the cell without hindrance. 3. Provides protection.
Cytoplasm	1. Comprises all the parts inside the plasma membrane excluding the nucleus. 2. Contains a mixture of water and soluble inorganic and organic compounds, and various organelles.	1. Different organelles contained in it perform different functions. 2. All metabolic activities occur in it. 3. Medium of earlier steps of respiration (production of pyruvic acid) (anaerobic respiration).
Endoplasmic reticulum	1. Irregular network of tubular double membrane. 2. It is continuous with the plasma membrane on the outside and the nuclear membrane on the inside. 3. May be smooth or rough (attached ribosomes).	1. Supportive framework for the cell. 2. Synthesis and transport of proteins and fat.
Mitochondria	1. Various shapes but usually sausage-like. 2. Double walled; inner wall thrown into folds (cristae). 3. Have their own DNA (containing several genes) 4. Contain their own ribosomes.	1. Release of energy from pyruvic acid produced in the cytoplasm in the form of ATP. (Seat of cellular aerobic respiration and stores energy). 2. Synthesis of respiratory enzymes.

(Continued)

Cell organelle	Main characteristics	Chief function(s)
Golgi body	1. Stacks of flattened membrane sacs. 2. Consists of tubules, vesicles and vacuoles.	1. Synthesis and secretion of enzymes, hormones, etc. 2. Formation of acrosome of sperm.
Ribosomes	1. Small granules either scattered in the cytoplasm or attached to the outside of endoplasmic reticulum. 2. Single-walled, dense, spherical bodies composed mainly of RNA.	1. Protein synthesis.
Lysosomes	1. Membranous sacs budded off from Golgi body. 2. Contain 40 different types of enzymes.	1. Intracellular digestion. 2. Destroy foreign substances. 3. When cell is old or injured, these rapidly destroy the organelles (hence called suicide bags). 4. Digest cartilage during formation of bones.
Centriole	1. A region surrounding the centrioles, located near nucleus. 2. Contains one or two centrioles 3. Centrioles are surrounded by microtubules.	1. Initiates and regulates cell division. 2. Forms spindle fibres, with the help of asters.
Plastids	1. Several types; most common ones are chloroplasts containing chlorophyll. 2. Double-membraned, proteinaceous matrix, containing DNA. 3. Disc-like structures called thylakoids contain chlorophyll.	1. Chloroplasts (green) trap solar energy for photosynthesis, contain the pigment chlorophyll. 2. Leucoplasts store starch. 3. Chromoplasts impart colour to flowers and fruits; contain the pigments xanthophyll (yellow) and carotene (orange-red). 4. Anthocyanin (blue-violet) is dissolved in the cytoplasm.
Nucleus	1. Largest cell organelle. 2. Mostly spherical and dense. 3. Nuclear membrane with pores to allow substances to enter and leave. 4. Contains network of thread-like structures called chromatin fibres which contain DNA.	1. Regulates cell functions. 2. If removed, the cell dies. 3. Contains chromosomes (bearers of genes that control hereditary characters).
Nucleolus	1. One or more round-shaped nucleoli inside the nucleus.	1. Produces ribosomes 2. Participates in protein synthesis by forming and storing RNA. 3. Directs ribosomes to synthesise proteins.
Chromosome	1. The chromatin network in resting stage of the nucleus condenses into chromosomes during cell division. 2. Made up of DNA threads.	1. Chromosomes carry hereditary information or genes.

(*Continued*)

Cell organelle	Main characteristics	Chief function(s)
The next two organelles, vacuoles and granules, are non-living.		
Vacuoles	1. Clear spaces with water or other substances in solution. 2. Plant cells have larger vacuoles, while animal cells have fewer and smaller ones. 3. Covered by tonoplast.	1. Storage of water and other substances, food, pigments and waste products. 2. Give turgidity to plant cells by pressing against the cell wall.
Granules	1. Small particles, crystals or droplets.	1. Starch (in plant cells), glycogen (in animal cells) and fat-containing granules serve as food for the cell.

Table 1.2 Differences between plant and animal cells

Feature	Plant cell	Animal cell
Cell wall	A definite cell wall, made up of cellulose.	No cell wall.
Centrosome	Absent.	Present.
Vacuoles	Prominent, one or more.	Vacuoles, if any, are small and temporary; concerned with excretion or secretion.
Plastids	Usually present..	No plastids.
Size	Usually larger.	Usually smaller.
Cytoplasm	Cytoplasm not so dense.	Cytoplasm denser and more granular.
Arrangement of cytoplasm	Only a thin lining of cytoplasm, mostly pushed to the periphery.	Cytoplasm fills almost the entire cell.

Table 1.2 lists the main differences between plant and animal cells.

Microscopic Examination of Onion Peel to View Cell Organelles

1. Cut an onion bulb lengthwise into four equal pieces (quarters).
2. Take one thick scale leaf out of one of these quarters. Grab an onion's fleshy scale in your hand, and tear it from the interior (concave) side to create a thin, transparent strip.
3. Remove this strip (peel) with a pair of forceps, and place it in a water-filled watch glass.
4. Slice a square piece of this peel (approximately 5 × 5 mm) and mount it on a slide in a drop of water.
5. Carefully place a cover slip over the peel to prevent the tissue from wrinkling.
6. Take a low-power microscope and examine the peel. The material can also be stained with an eosin or iodine solution to highlight the nucleus.
7. Under a high-power microscope, the cell structure will be more clearly visible (Figs 1.3 and 1.4).

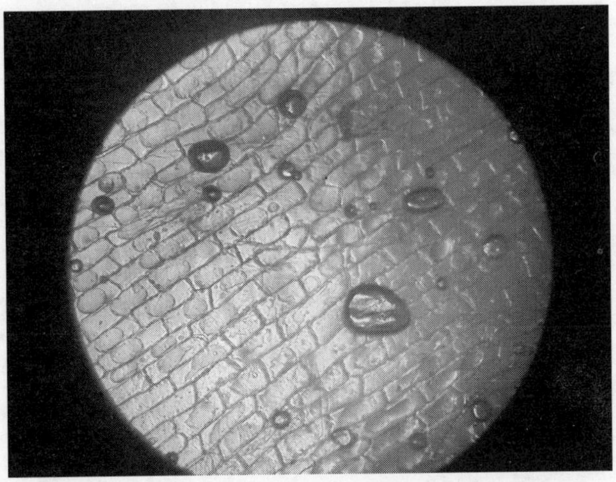

Fig. 1.3 Microscopic structure of onion peel
(Deplasmolysis of the onion peel…; Kristina Tsimlyakowa; 2016; https://commons.wikimedia.org/wiki/File:Onion_peel_under_a_microscope(restoration_of_the_norm).jpg. Licensed under: CC-BY-4.0)

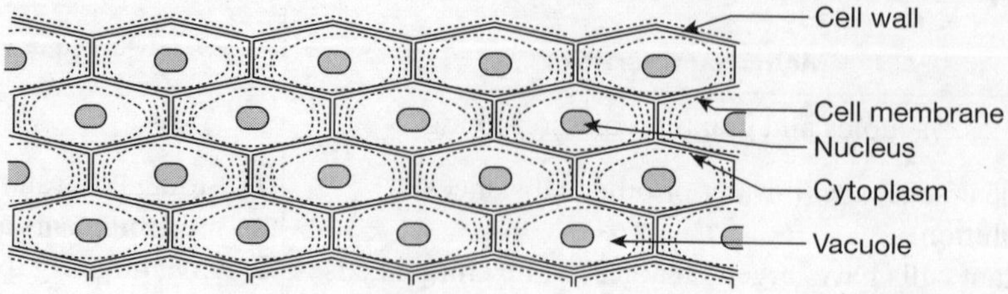

Fig. 1.4 Cell structure of an onion peel

The linear or rectangular shape of the onion peel's cells makes them appear somewhat regular. A prominent cell wall, nucleus and cytoplasm encircling one or two sizable vacuoles are all present in every cell. High magnification makes the details, particularly the thick cell wall, more visible. The nucleus is positioned to one side, as is typical in almost all plant cells, and the cells are bound tightly together.

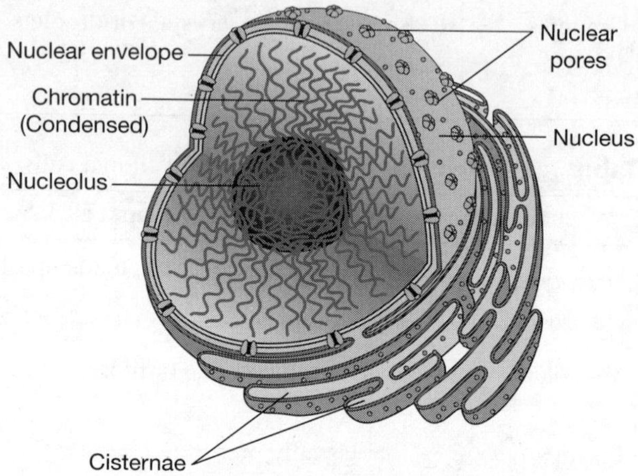

Fig. 1.5 Nucleus of a cell

Nucleus

The nucleus (Fig. 1.5) is the primary actor or master of the cell, even though each component of a cell contributes to the life of the cell as a whole. An experiment carried out on the single-celled *Amoeba* makes this clear. When an enucleated *Amoeba* receives the nucleus of another, the donor *Amoeba* dies, but the recipient survives and divides.

Cells carry out all the bodily functions of an organism. However, each type of cell is tailored for a specific purpose.

SUMMARY

- The cell is the structural and functional unit of all living organisms. Every cell in the body has a specific function. For example, an alveolar cell.
- A group of cells that have similar structure and function constitute a tissue. For example, a group of alveolar cells form alveolar tissue.
- Tissues aggregate to form an organ. For example, alveolar tissues aggregate to form lungs.
- Many organs together constitute an organ system. For example, nostrils, nasal cavity, trachea, bronchi, bronchioles and lungs together form the respiratory system.
- Organ systems aggregate to form an organism. A human being is a complete organism made up of 11 organ systems: digestive system, respiratory system, circulatory system, immune system, muscular system, skeletal system, nervous system, endocrine system, integumentary system, excretory system and reproductive system.
- The cell is surrounded by a plasma membrane or cell membrane. Enzymes catalyse numerous chemical reactions in the semi-liquid substance known as the cytoplasm, which is found inside the cell membrane and outside the nucleus. The large, roughly spherical nucleus takes up almost the entire cytoplasm.
- Organelles are specialised membrane-bound structures with predetermined roles. The organelles that make up a cell resemble bodily organs. Each organelle performs one or more specific functions, much like how each organ in the body does.

PRACTICE QUESTIONS

Analyse Your Concepts (School Exam Based)

Multiple Choice Questions

1. Chromatin is formed of:
(a) DNA only (b) DNA and histones
(c) Histones only (d) Nucleotides

2. The term chromosome means:
(a) Inherited body (b) Twisted thread
(c) Coloured body (d) Shining thread

3. The number of chromosomes in a certain type of cell division is halved. This type of cell division occurs in:
(a) Only testis (b) Only ovary
(c) Both ovary and testis (d) All body cells

4. In which one of the following options have the stages of mitosis been given in correct sequence?
(a) Prophase, metaphase, telophase, anaphase
(b) Prophase, metaphase, anaphase, telophase
(c) Anaphase, telophase, prophase, metaphase
(d) Telophase, anaphase, prophase, metaphase

5. Synthesis phase in the cell cycle is so-named because of the synthesis of:
(a) RNA (b) RNA and proteins
(c) DNA (d) Glucose

Match the Following

Each question contains statements given in two columns which have to be matched. Statements in column I have to be matched with statements in column II.

Column I	Column II
(a) Cristae	(i) Flat membranous sacs in stroma
(b) Cisternae	(ii) Infoldings in mitochondria
(c) Thylakoids	(iii) Disc-shaped sacs in Golgi apparatus
(d) Endoplasmic reticulum	(iv) Kitchen of the cell
(e) Chloroplast	(v) Protein synthesis

True or False

Read the following statements and write your answer as true or false.

1. Lysosomes arise from the Golgi apparatus.
2. Lipid molecules are synthesized by rough endoplasmic reticulum.
3. The unit of measurement used for expressing the dimension (size) of cells is micrometre.
4. The nucleus of a cell contains genetic material in the form of DNA.
5. Plant cells have a cell wall made up of cellulose.

Fill in the Blanks

Complete the following statements with an appropriate word/term to be filled in the blank spaces.

1. A substance like a jelly present inside the plasma membrane in which all the cell organelles are floating is referred to as ________.
2. Digestive enzymes are present in ________.
3. The glycocalyx is associated with ________.
4. ________ organelles contain DNA.
5. The green colour of leaves is due to the presence of a pigment called ________.

Assertion–Reason Questions

The following questions consist of two statements: Assertion (A) and Reason (R). Answer these questions selecting the appropriate option given below:
(a) Both A and R are true and R is the correct explanation of A
(b) Both A and R are true but R is not the correct explanation of A
(c) A is true but R is false
(d) A is false but R is true

1. Assertion (A): Organisms are made up of cells.
Reason (R): Cells are the structural units of living organisms.

2. Assertion (A): Specialisation of cells is useful for an organism.

Reason (A): It increases the operational efficiency of an organism.

3. Assertion (A): The number of cells in a multicellular organism is inversely proportional to the size of its body.
Reason (R): All the cells of the biological world are of the same size.

4. Assertion (A): Living organisms possess specific individuality with definite shape and size.
Reason (R): Both living and non-living entities resemble each other at the lower level of organisation.

5. Assertion (A): It is important that organisms have cells.
Reason (R): A cell keeps its chemical composition steady within its boundary.

One-word Answers

Identify the cellular structures with the help of the following clues:

1. It synthesizes the respiratory enzymes.
2. It is made up of lipoprotein.
3. It is a non-living rigid layer surrounding the plasma membrane.
4. It is the supportive framework of the cell.
5. It consists of cisternae, vesicles and vacuoles.
6. It destroys foreign substances.
7. It gives turgidity to plant cells.
8. It is made up of DNA threads.
9. It contains chromatin fibres.
10. It initiates and regulates cell division.

Define the Term

1. Cell
2. Organelles
3. Cytoplasm
4. Protoplasm
5. Nucleus

Groups of Terms

Given below are certain groups of terms. In each group, the first pair indicates a relationship between the two terms. Rewrite and complete the second pair on a similar basis.

(a) Cell membrane: Selective permeability :: Cell wall: ________

(b) Mitochondria: Cellular respiration :: Chloroplasts: ________

(c) Prokaryotic cell: No nucleus :: Eukaryotic cell: ________

(d) Animal cell: Centrioles :: Plant cell: ________

(e) Chromosomes: DNA and protein :: Genes: ________

Identify the Odd One Out

1. Which of the following is the odd one out?
(a) Prokaryotic cells (b) Eukaryotic cells
(c) Bacterial cells (d) Fungal cells

2. Which of the following is the odd one out?
(a) Diffusion
(b) Osmosis
(c) Active transport
(d) Facilitated diffusion

3. Which of the following is the odd one out?
(a) DNA (b) RNA
(c) Lipids (d) Proteins

Image-Based Questions

1. Identify and label the cell organelles.

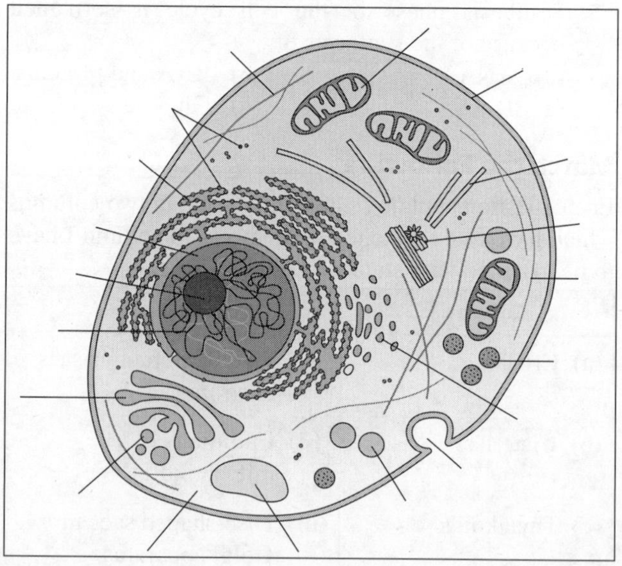

2. Identify and label this structure.

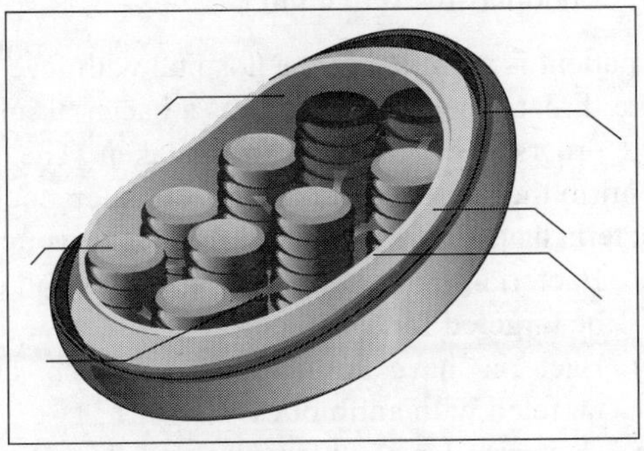

3. Label this structure.

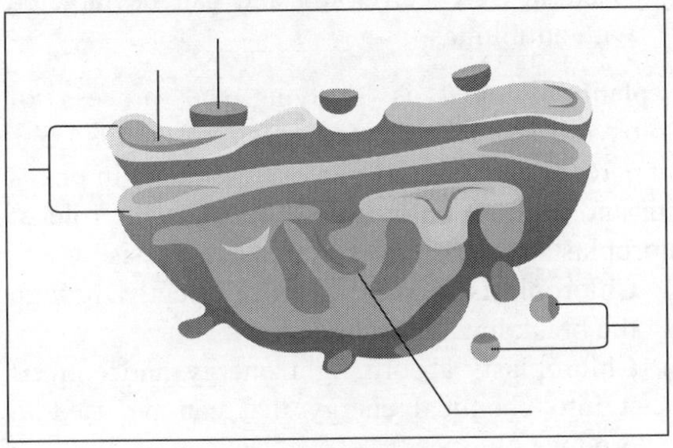

4. Identify and label this structure.

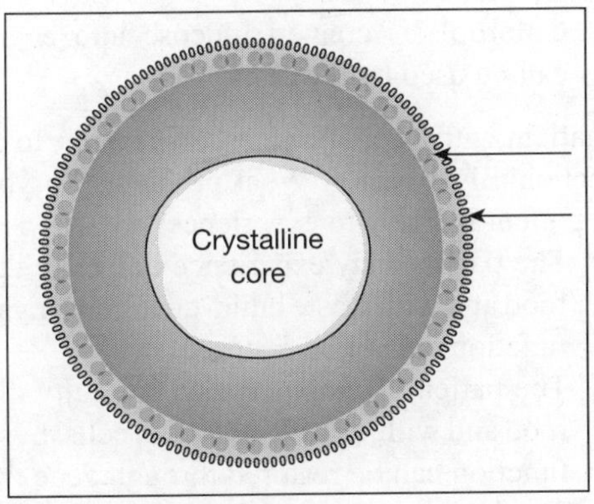

5. Identify the movement and label the structure of cilia and flagella.

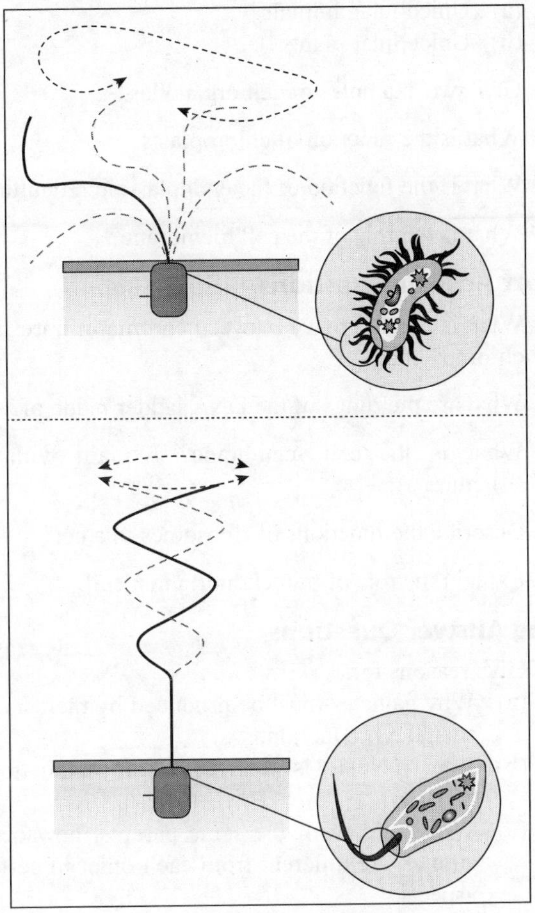

Very Short Answer Questions

1. Give two examples of each:
   (a) Unicellular animals
   (b) Unicellular plants
2. Give two examples of cell organelles.
3. What is the function of chloroplasts.
4. What is the function of the endoplasmic reticulum?
5. What is the role of the cell membrane?

Short Answer Questions

1. What is the difference between chromatin fibres and chromosomes?
2. What are the rungs of the DNA ladder made of?
3. What is the cell membrane? Explain with its structure.
4. Describe the functions of ribosomes in a cell.
5. Explain the role of mitochondria in a cell.

Long Answer Questions

1. Give reasons for:
   (a) Why gametes must be produced by meiosis for sexual reproduction.
   (b) Why meiosis is referred to as 'reductional division'.
   (c) Why children of the same parents, howsoever similar, are different from each other in certain aspects.
2. Define the term cell. Describe the basic structure of a typical animal cell.
3. What are the functions of the nucleus in a cell? Describe the structure of a typical nucleus.
4. Enumerate the various changes that occur in the nucleus of the cell during (a) prophase (b) anaphase of mitotic division.
5. Name and explain the various stages of the cell cycle.

Case Study-Based Questions

1. A research team is studying the effects of radiation on cells. They expose two groups of cells to different levels of radiation and observe their behaviour. One group is found to have a higher rate of mutation and cell death compared to the other. Which organelle do you think is the most affected by radiation?
   (a) Mitochondria
   (b) Golgi apparatus
   (c) Nucleus
   (d) Endoplasmic reticulum
2. A patient is admitted to the hospital with fever and a sore throat. The doctor suspects a bacterial infection and orders a throat swab to be taken. The results confirm the presence of bacteria. What organelles do bacteria have that are different from eukaryotic cells?
   (a) Bacteria do not have any organelles and cannot be targeted for treatment.
   (b) Bacteria have mitochondria, which can be targeted with antibiotics.
   (c) Bacteria have chloroplasts, which can be targeted with antivirals.
   (d) Bacteria have cell walls and can be targeted with antibiotics.
3. A plant biologist is studying the process of photosynthesis in different types of plants. She compares the amount of glucose produced in plants with and without chloroplasts. What is the role of chloroplasts in the process of photosynthesis?
   (a) Chloroplasts produce glucose directly through the breakdown of light energy.
   (b) Chloroplasts absorb light energy and convert it into chemical energy that can be used to produce glucose.
   (c) Chloroplasts store glucose until it is needed by the plant.
   (d) Chloroplasts convert glucose into energy that can be used by the plant.
4. A patient with a genetic disorder is unable to produce functional lysosomes. What problems do you think the patient is likely to experience?
   (a) The patient may experience difficulty digesting food and with waste build-up in cells. Lysosomal function cannot be restored.
   (b) The patient may experience difficulty digesting food and with waste build-up in cells. Lysosomal function can be restored through gene therapy.
   (c) The patient may experience difficulty breathing and have abnormal heart function. Lysosomal function cannot be restored.
   (d) The patient may experience difficulty breathing and have abnormal heart function. Lysosomal function can be restored through stem cell therapy.
5. A scientist is studying the effects of a new drug on protein synthesis. She observed that cells treated with the drug produce less protein than untreated cells. Which organelle do you think is most affected by the drug?

(a) Ribosomes, as they are the site of protein synthesis in cells.
(b) Nucleus, as it contains the DNA that codes for protein synthesis.
(c) Golgi apparatus, as it packages and transports proteins.
(d) Mitochondria, as they provide energy for protein synthesis.

NCERT Corner

1. Name two cell organelles that are double-membrane-bound. What are the characteristics of these two organelles?
2. What are the differences between a prokaryotic cell and a eukaryotic cell?
3. State the differences between rough endoplasmic reticulum and smooth endoplasmic reticulum.
4. What is the function of the cytoskeleton in the cell?
5. Name two types of plastids found in plant cells. What is the function of each type of plastid?

COMPETITION WINDOW (OBJECTIVE TYPE)

1. The ________ is the structural and functional unit of the body.
   (a) Tissue (b) Cell
   (c) Organs (d) Bacteria
2. Which cells are circular and biconcave?
   (a) Nerve cells
   (b) Human red blood cells
   (c) Guard cells
   (d) WBCs
3. Tissues combine to form:
   (a) Nucleus (b) Cells
   (c) Organism (d) Organs
4. The coloured organ cells which are found only in plants are:
   (a) WBC (b) Plastids
   (c) Cell wall (d) Chlorophyll
5. The powerhouse of the cell is:
   (a) Cell wall (b) Mitochondria
   (c) Ribosome (d) Nucleus
6. Single-celled organisms are called:
   (a) Multi-cellular (b) Unicellular
   (c) Both of these (d) None of these
7. The cell organelle that contains chromosomes is:
   (a) Ribosome (b) Nucleus
   (c) Mitochondria (d) Plastids
8. This cell is the longest in the human body:
   (a) Gland cells (b) Bone cells
   (c) Nerve cells (d) Muscle cells
9. ________ is mostly made up of cellulose:
   (a) Cell membrane (b) Cytoplasm
   (c) Cell wall (d) Nucleus
10. Which is a thick fluid in the cell?
    (a) Cytoplasm (b) Chromosomes
    (c) Golgi body (d) Mitochondria
11. Which among the following sentences is incorrect?
    (a) Centrosome is absent in plant cells
    (b) Chromatin fibre is made up of DNA threads
    (c) Cell wall is absent in animal cells
    (d) Ribosomes destroy foreign substances
12. Which is an example of a unicellular organism?
    (a) *Chlamydomonas* (b) *Volvox*
    (c) Mushroom (d) *Chiton*
13. Which is the biggest animal cell?
    (a) Ostrich egg (b) *Mycoplasma*
    (c) Hen's egg (d) Lizard's egg
14. Which plastid has chlorophyll pigments?
    (a) Xanthophyllous (b) Anthrocynin
    (c) Carotene (d) Chloroplast
15. Which of the following is not a function of vacuoles?
    (a) Storage
    (b) Providing turgidity to the cell
    (c) Locomotion
    (d) Waste excretion
16. The cytoplasmic connections from cell to cell are known as:

(a) Middle lamella
(b) Plasmodesmata
(c) Cell membrane system
(d) Endoplasmic reticulum

17. Bacterial flagella is made up of:
(a) Tubulin (b) Flagellin
(c) Chitin (d) None of these

18. Plasmolysis occurs due to:
(a) Absorption (b) Endosmosis
(c) Osmosis (d) Exosmosis

19. Which is called suicide bag?
(a) Centrosome (b) Lysosome
(c) Mesosome (d) Chromosome

20. Cell secretion is carried out by:
(a) Plastids
(b) ER
(c) Golgi apparatus
(d) Nucleolus

Answer Keys

Multiple Choice Questions

1. (b) 2. (c) 3. (c) 4. (b) 5. (c)

Match the Following

Column I	Column II
(a) Cristae	(ii) Infoldings in mitochondria
(b) Cisternae	(iii) Disc-shaped sacs in Golgi apparatus
(c) Thylakoids	(i) Flat membranous sac
(d) Endoplasmic reticulum	(v) Protein synthesis
(e) Chloroplasts	(iv) Kitchen of the cell

True or False

1. T 2. F 3. T 4. T 5. T

Fill in the Blanks

1. Cytoplasm
2. Lysosomes
3. Plasma membrane
4. Mitochondria
5. Chlorophyll

Assertion–Reason Questions

1. (a) 2. (a) 3. (d) 4. (b) 5. (a)

One-word Answers

1. Mitochondria
2. Cell membrane/Plasma membrane
3. Cell wall
4. Endoplasmic reticulum
5. Golgi apparatus
6. Lysosomes
7. Vacuoles
8. Chromatin fibres
9. Nucleus
10. Centrosome

Define the Term

1. The cell is the structural and functional unit of life capable of independent existence. All cells are alike in chemical composition and metabolic processes and arise from a pre-existing cell.
2. Organelles are specialised and membrane-bound living structures in a cell that perform specific functions.
3. The cytoplasm is the part of the cell which is inside the cell membrane and outside the nucleus. It is a semi-liquid substance and contains several organelles, each concerned with a specific function.
4. The living parts of the cell which consist of the cytoplasm, nucleus and other living bodies collectively constitute the protoplasm or protoplast.
5. The nucleus is a large spherical body found close to the centre of the cytoplasm. It is surrounded by a double-layered nuclear membrane with nuclear pores.

Groups of Terms

(a) Rigidity and support
(b) Photosynthesis
(c) True nucleus
(d) Plastids
(e) Segments of DNA

Identify the Odd One Out

1. (a) 2. (c) 3. (c)

Image-Based Questions

1.

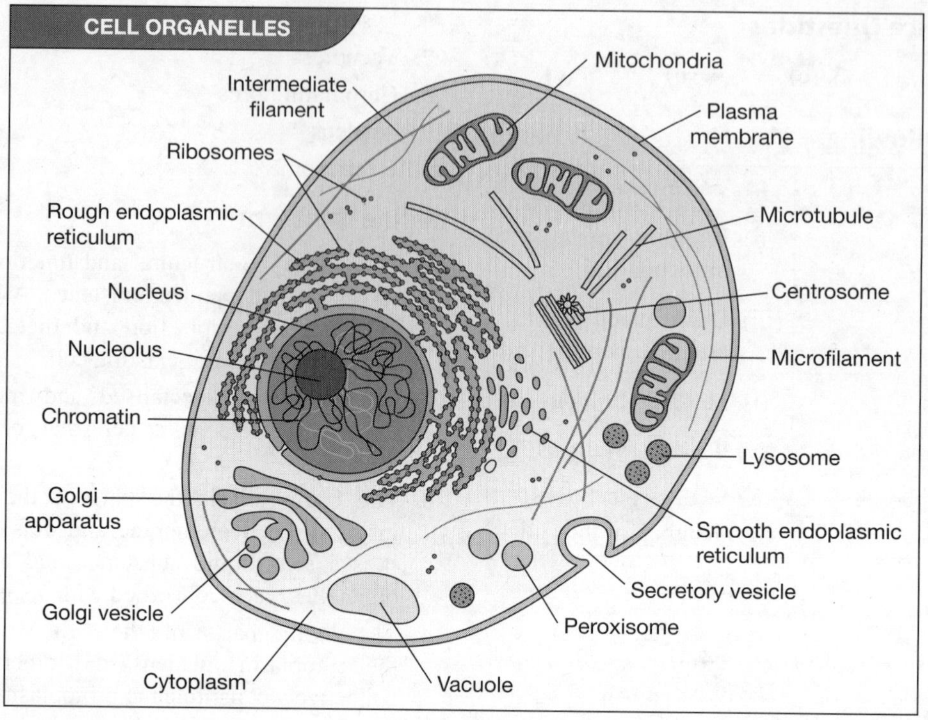

2.

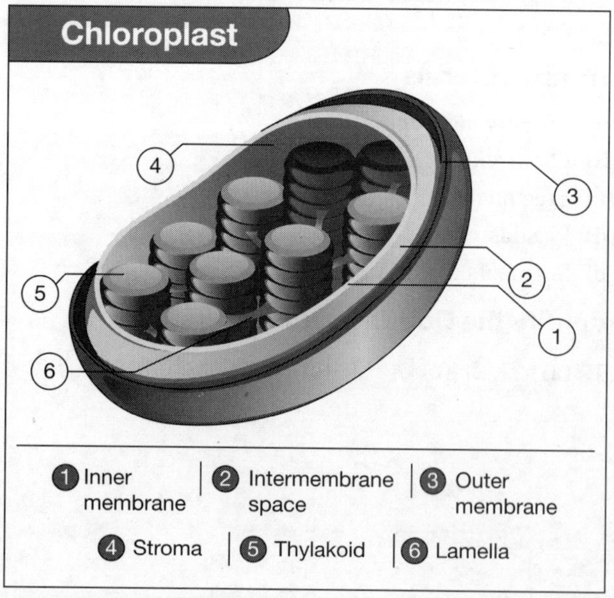

3.

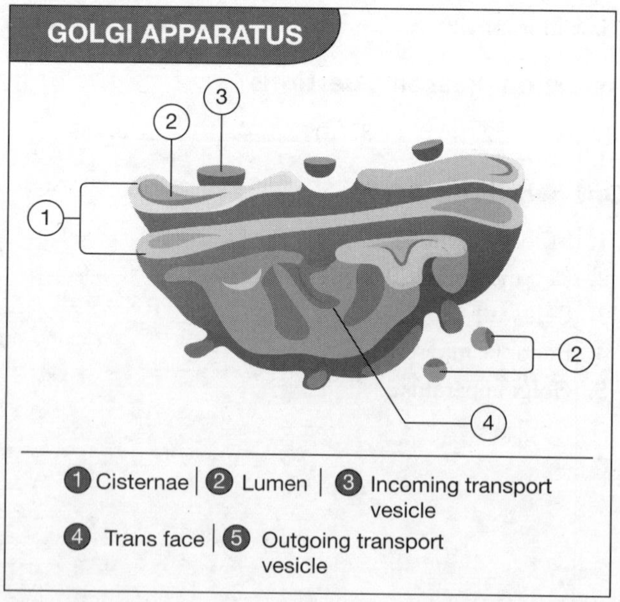

4.

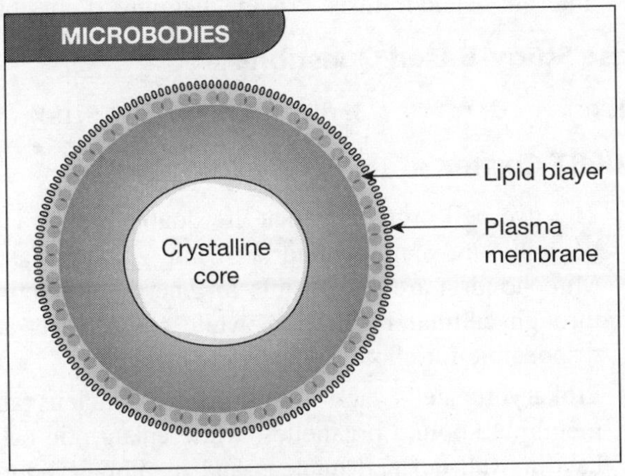

5.

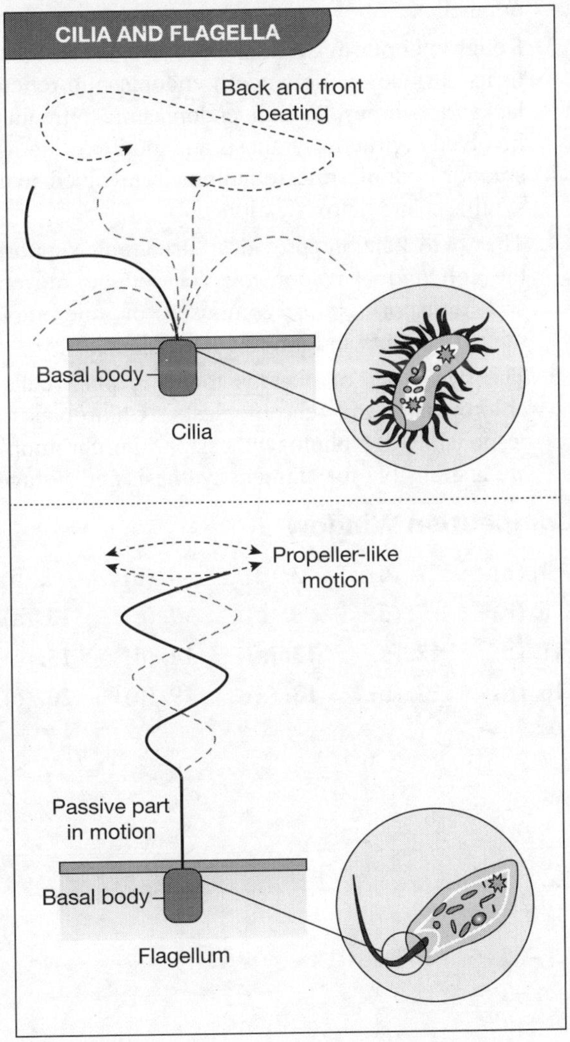

Very Short Answer Questions

1. (a) *Amoeba*, *Paramecium*
   (b) *Chlamydomonas*, Diatoms
   (c) Mitochondria, Nucleus
   (d) Vacuoles, Granules
   (e) Iodine, Eosin
2. **Mitochondria:** They are the sites of cellular respiration and produce energy for the cell in the form of ATP.

   Chloroplasts: They are the sites of photosynthesis and produce organic molecules for the cell using light energy.
3. Chloroplasts are the site of photosynthesis in plant cells, producing glucose from sunlight, water and carbon dioxide.
4. The endoplasmic reticulum is involved in the synthesis, folding and transport of proteins and lipids within the cell.
5. The cell membrane controls what enters and leaves the cell and provides structural support.

Short Answer Questions

1.

Chromatin Fibre	Chromosome
Composed of nucleosomes.	Condensed chromatin fibres.
Thin, long, uncoiled structure.	Thick, compact, ribbon-like structure.
Single, unpaired fibres.	Exists as pairs.
Appears in interphase of cell cycle.	Appears during metaphase and exists in the anaphase of nuclear division.
Allows genetic material to be packed into the nucleus.	Ensures proper arrangement of genetic material in the cell equator to allow equal separation of genetic material between the two cells.

2. The rungs of the DNA ladder are made of four types of nitrogenous bases. These are:
   - Adenine (A)
   - Guanine (G)
   - Cytosine (C)
   - Thymine (T)

3. The cell membrane is the thin layer that surrounds the cell and which separates it from the external environment. It is made up of a double layer of phospholipids and embedded proteins. It is selectively permeable, which means it allows only certain substances to pass through it. It also helps in maintaining the shape and stability of the cell.
4. Ribosomes are structures that make proteins in cells. They can be free or attached to the ER. They use mRNA to link amino acids in order. They have two subunits of rRNA and proteins. They are vital for cell function and structure.
5. Mitochondria are energy-producing organelles with two membranes and their own DNA. They make ATP from oxygen and food. They also store calcium, produce heat and control cell growth and death. They are found in most eukaryotic cells and vary in number and size. They are crucial for cell and organism survival and function.

Long Answer Questions

1. Hints: (a) Explain how meiosis reduces the chromosome number by half and how fertilisation restores the normal diploid number.
   (b) Define reductional division and give an example of diploid and haploid numbers in humans.
   (c) Describe how meiosis leads to genetic variation and diversity in the offspring by recombination of genes.
2. Hint: Mention that the cell is the basic unit of life. List the main components of an animal cell, such as cell membrane, cytoplasm, nucleus and organelles. Provide a diagram if possible.
3. Hint: Explain the role of the nucleus in controlling the activities of the cell and storing genetic material.
4. Hint: List the main events that occur in the nucleus of the cell during prophase and anaphase of mitotic division.
5. Hint: Define cell cycle and explain its two phases: interphase and mitosis. Provide diagrams if possible.

Case Study-Based Questions

1. (c)	**2.** (d)	**3.** (b)	**4.** (b)	**5.** (a)

NCERT Corner

1. The two cell organelles that are double-membrane-bound are mitochondria and chloroplasts. Mitochondria are responsible for energy production through cellular respiration, while chloroplasts are responsible for photosynthesis in plant cells.
2. Prokaryotic cells lack a well-defined nucleus and membrane-bound organelles, while eukaryotic cells have a well-defined nucleus and membrane-bound organelles.
3. Rough endoplasmic reticulum has ribosomes attached to its surface, while smooth endoplasmic reticulum lacks ribosomes. Rough endoplasmic reticulum is involved in protein synthesis and modification, while smooth endoplasmic reticulum is involved in lipid synthesis and detoxification.
4. The cytoskeleton provides structural support to the cell and is responsible for cellular movement and transport. It is composed of microtubules, microfilaments and intermediate filaments.
5. The two types of plastids found in plant cells are chloroplasts and chromoplasts. Chloroplasts are responsible for photosynthesis, while chromoplasts are responsible for pigment synthesis and storage.

Competition Window

1. (b)	**2.** (b)	**3.** (d)	**4.** (b)	**5.** (b)
6. (b)	**7.** (d)	**8.** (c)	**9.** (c)	**10.** (a)
11. (d)	**12.** (a)	**13.** (a)	**14.** (d)	**15.** (c)
16. (b)	**17.** (b)	**18.** (d)	**19.** (b)	**20.** (c)

CHAPTER 2

Structure of Chromosomes, Cell Cycle and Cell Division

PREREQUISITES

You must have knowledge of:

- Cell theory
- DNA structure and functions
- Mitosis
- Meiosis
- Protein synthesis

LEARNING OBJECTIVES

At the end of this chapter, you will be able to:

- Describe the structure of chromosomes
- Explain the cell cycle and cell division
- Describe mitosis and meiosis in detail

INTRODUCTION

One of life's most fundamental processes is cell division. This process makes it possible for life to continue for many generations. This holds true for the most basic organisms, like amoebas, as well as the most complex ones, like humans, elephants and coconut trees.

In this chapter, we will learn about cell division.

CHROMOSOMES

You already know that chromosomes play a role in all the notable events that take place during cell division. During mitosis, the duplicated chromosomes are evenly distributed throughout the daughter cells. As a result, all the body cells that are the outcome of a protracted series of successive mitotic divisions share the same type and quantity of chromosomes. This guarantees that the cells will function normally and that the organism will live as a result.

A normal, unstained living cell's nucleus might not seem to house anything specific when viewed under a light microscope. However, when the same cell is stained with the right dyes, different structures in the nucleus can be seen. Most likely, you will view a stage of the cell that is not dividing (interphase), though it may also display some division stages, depending on the location in the body from where the cell was taken.

Chromatin fibres are a network of long, thin, dark-staining fibres that can be seen in the nucleus during interphase. They condense to form chromosomes as the cell enters the first phase (prophase) of cell division (Fig. 2.1). The name chromosome is derived from the fact that certain dyes easily colour these structures (in Greek, '*chroma*' means 'coloured' and '*soma*' means 'body').

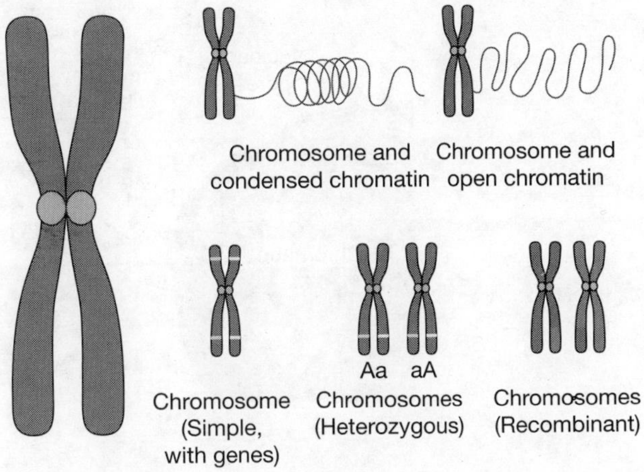

Fig. 2.1 Structure of chromosome

The first chromosome was identified in 1882 by Walther Flemming, a German scientist. He discovered them in a salamander larva's rapidly dividing cells. Using a microscope, he observed tiny threads that appeared to be dividing lengthwise. Flemming referred to this division as mitosis (literally meaning 'thread'). The chromosomes and their components have since been thoroughly studied with the aid of more potent microscopes and unique techniques.

Structure of Chromosomes

Each chromosome is made up of two chromatids that are connected along its length when it is condensed, as seen at the beginning of cell division. The centromere, which is the name of this point of attachment, resembles a small, constrained area (Fig. 2.2). During cell division, the centromere also serves as a point of attachment to the spindle fibre. The sister chromatids are divided at the centromere as the spindle fibre contracts, and they are then pulled apart from one another and towards the two poles of the dividing cell (Fig. 2.3).

After cell division is complete, the chromatids (now referred to as chromosomes) decondense and transform

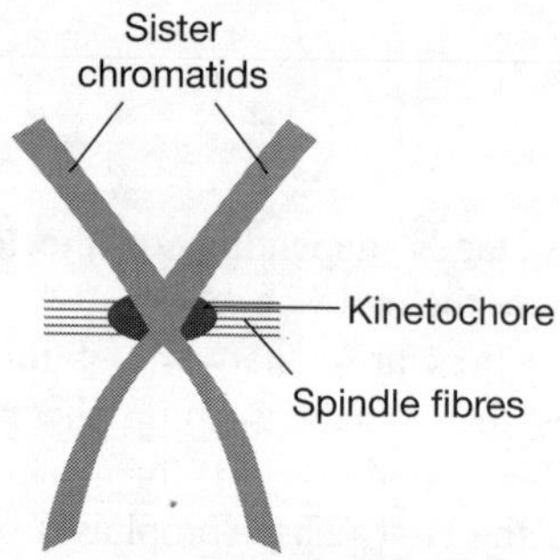

Fig. 2.2 Centromere

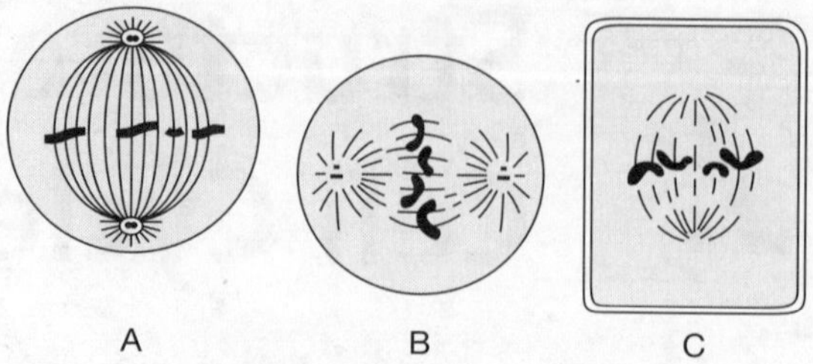

Fig. 2.3 Cell division (metaphase). A and B represent an animal cell (no cell wall), which undergoes furrow formation to produce two daughter cells. C is a plant cell (cell wall is present), where cell plate formation takes place to produce daughter cells.

back into their long and delicate chromatin fibres. The number of chromosomes that are produced during cell division corresponds to the number of chromatin fibres present inside the nucleus.

The chromatin material that constitutes the fibre is formed of two substances:

- DNA (deoxyribonucleic acid): about 40%
- Histones (a type of protein): about 60%

Figure 2.3 is a diagrammatic representation of the structure of chromatin fibres showing the sequentially enlarged view of the DNA strand. A central core of eight histone molecules is encircled by the DNA strand. A long rope with one or two loops is wound around this core, which is similar to a football. Each such complex is called a nucleosome. A single human chromosome may have about a million nucleosomes! The entire chromatin fibre is coiled and super coiled like in a telephone cord (Fig. 2.4).

Structure of DNA

The shape of the DNA molecule was studied by Rosalind Franklin in 1953 and its structure was finally worked out

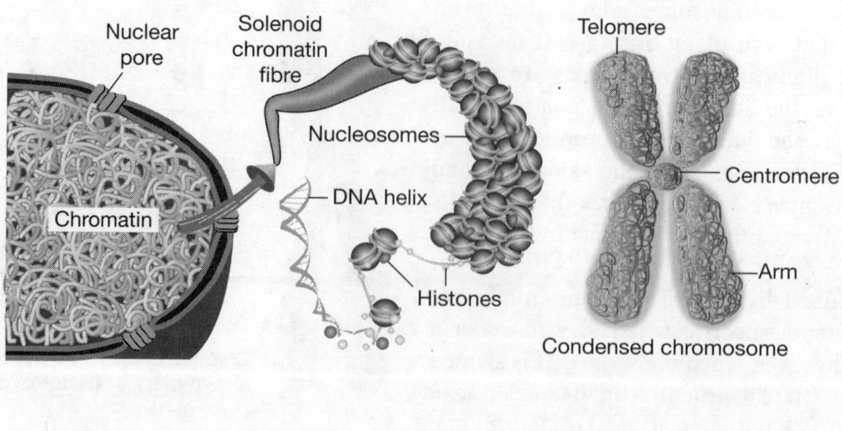

Fig. 2.4 Coiled chromatin fibre

by Watson and Crick in the same year. DNA is a very large molecule; it is referred to as a macromolecule. It consists of a double-helix composed of two complementary strands that are wound around one another (Fig. 2.5).

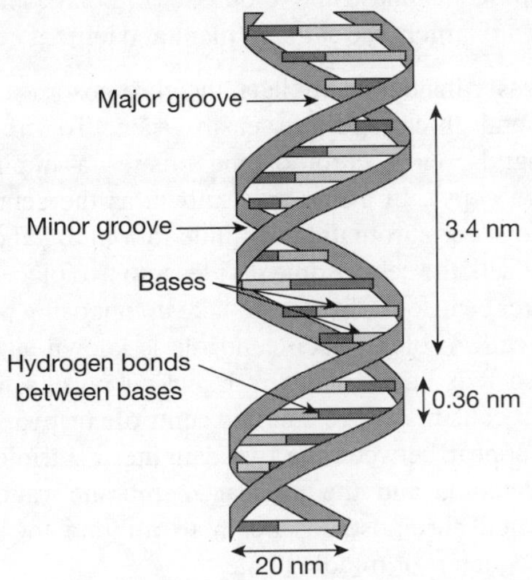

Fig. 2.5 DNA structure

The three components of repeating nucleotides – phosphate, sugar (pentose) arranged lengthwise, and a nitrogenous base attached to the sugar inwards – make up each DNA strand. The nitrogenous base extends to join (through a hydrogen bond) the complementary nitrogenous base from the other strand. The nitrogenous base serves as the 'rungs' of the ladder-like structure formed by the two strands when combined. Adenine (A), guanine (G), cytosine (C) and thymine are the bases. Two hydrogen bonds connect thymine and adenine. Guanine has three hydrogen bonds with cytosine (Fig. 2.6).

Each DNA molecule (that is, chromosome) replicates during the cell cycle's interphase in preparation for their even distribution in the two daughter cells during mitosis. For replication, the DNA double-helix opens at one end, making the two strands free; new strands begin to form and the process continues in sequence for the whole length of the DNA (Fig. 2.7).

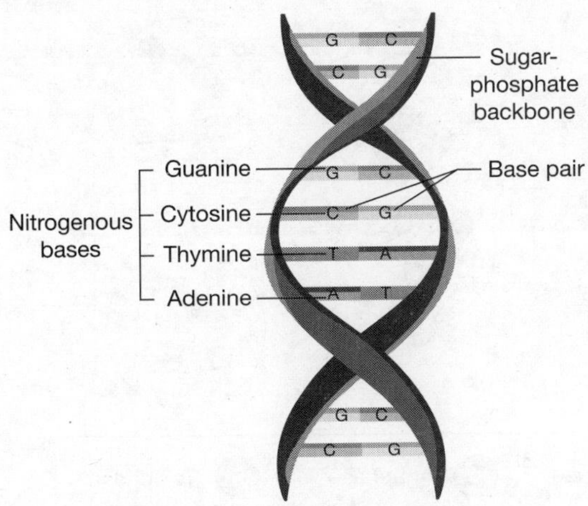

Fig. 2.6 Base pairs in DNA

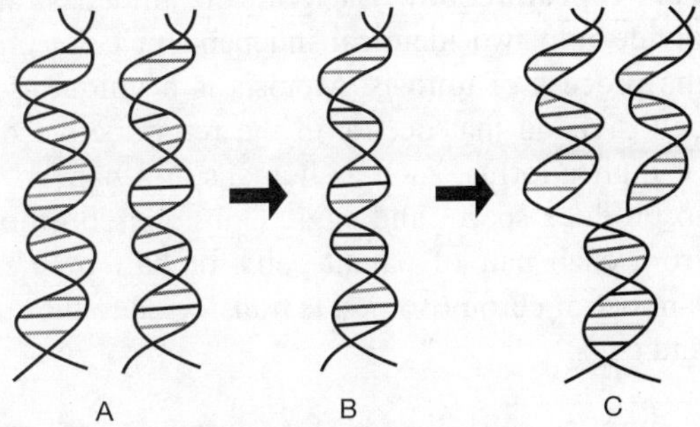

Fig. 2.7 DNA strands

What are Genes?

Genes are particular nucleotide sequences on a chromosome that are responsible for encoding specific proteins that are expressed as distinct bodily characteristics.

DNA Fingerprinting

Many intergenic regions, which are stretches of DNA sequences that occur between genes, do not encode proteins or other functional products. These regions (which make up nearly all of the DNA) exhibit enormous variation from person to person and aid in identifying the individual through a process better known as DNA profiling than DNA fingerprinting. In criminal cases, DNA fingerprinting can be used to prove paternity and maternity. In DNA fingerprinting, blood or semen stains, hair or articles of clothing discovered at the crime scene can be used to provide biological evidence.

CELL DIVISION

Cell division results in the creation of new cell. New cells need to be produced for the following **reasons**:

- **For growth:** All living things, whether plants or animals, start out as a single cell (the fertilised egg). This cell repeatedly divides to create a cluster of cells that form tissues and organs. Cell division is therefore necessary for growth.
- **For replacement:** In routine bodily processes, cells are constantly subject to wear and tear. For instance, our body loses 20 million red blood cells per minute. These are replaced by fresh cells produced in the bone marrow by the division of parent cells. New leaves emerge from plants as the old, dried ones fall off.

- **For repair:** Accidental injuries can occur in addition to normal ageing and deterioration of body tissues. Skin wounds or bone fractures are possible. The same cells that divide, fill in the holes, and join the broken ends are used to repair such wounds.
- **For reproduction:** Bacteria or amoebas simply divide into two identical independent cells through the process of mitosis. Meiosis is a unique type of cell division that occurs in the reproductive organs of higher forms, such as humans and banyan trees, to produce sperm and eggs. Only one chromosome from each pair of parent cells, or half of the total number of chromosomes, is transferred to the sperms and eggs.

Each time a cell divides for purposes of growth, replacement or repair, the number of chromosomes stays constant. In the daughter cells, the chromosomes multiply and are distributed equally. Mitosis is the term for this type of cell division, which takes place in all body cells with the exception of the spermatocytes and oocytes that produce eggs and sperm, respectively, in the ovaries and testes.

Cells of the eye lens, nerve cells of the cerebral cortex and most muscle cells last a lifetime, but once dead are not replaced. Skin cells are replaced every two weeks or so. Red blood cells last for about 120 days and are replaced. Bone cells are replaced every 10 years or so in adults. Epithelial cells lining the gut last only 5 days. The average life of other gut cells is about 15 years. Liver cells are replaced every 300–500 days.

Cell division is of two types: mitosis and meiosis.

Mitosis

Mitosis is cell division that leads to growth and development. Here, one parent cell divides into two identical daughter cells. The main function of mitosis is to ensure that each cell division maintains the same number of normal chromosomes.

The four main phases of mitosis are prophase, metaphase, anaphase and telophase (Fig. 2.8). These four stages are largely the same in plant and animal cells.

Prophase: Inside the nucleus, the chromosomes become short and thick and are easily seen. To create two chromatids, each chromosome creates a duplicate of itself (a copy). In a tiny area known as the centromere, the two sister chromatids continue to join together. In an animal cell, the centrosome divides into two pieces. These daughter centrioles divide and take up opposing positions at the cell's 'poles'. Each centriole is known as an aster because it is surrounded by radiating rays (aster star). Animal cells may have a single centriole or two. Spindle fibres appear between the two daughter centrioles. Both the nucleolus and the nuclear membrane vanish. The duplicated chromosomes begin to migrate towards the cell's 'equator' or middle plane.

Metaphase: The duplicated chromosomes are organised on the equatorial plane. By way of its centromere, each chromosome is joined to a spindle fibre.

Anaphase: The centromere that connects the two chromatids splits, and each chromosome's two sister chromatids separate and move in opposite directions. This results from contraction of the spindle fibres.

Telophase: Each daughter chromosome or chromatid lengthens, thins and transforms into a network of chromatin threads. The nuclear membrane is seen again. Each daughter nucleus contains the same nucleolus.

During karyokinesis, all of the cytoplasm's components, including the mitochondria and chloroplasts in plant cells, are randomly distributed in the daughter cells.

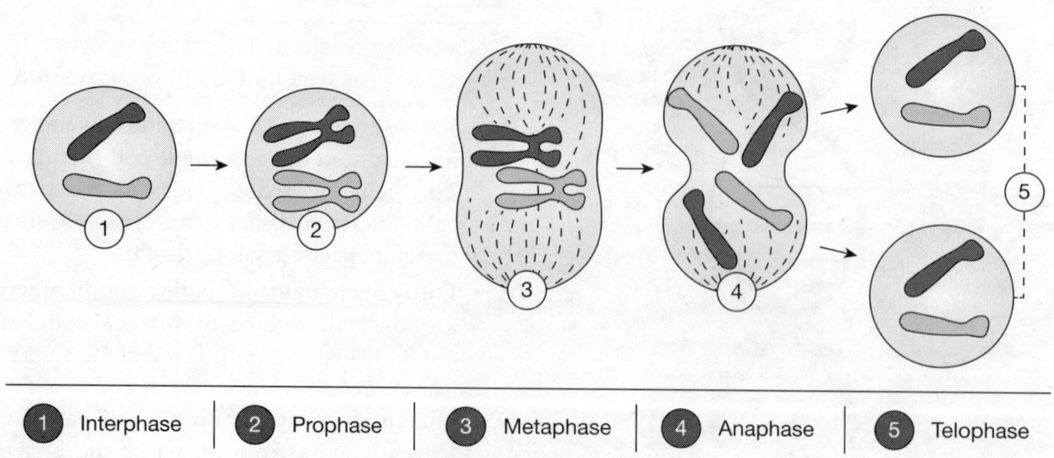

Fig. 2.8 Mitosis

Cytokinesis (division of cytoplasm): At the conclusion of telophase, a furrow forms in the middle of the cell membrane. It deepens and eventually splits the cytoplasm in half, resulting in the formation of two new cells (Fig. 2.9).

Interphase (resting phase): No discernible chromosome change is seen, but active DNA synthesis takes place.

Differences Between Animal and Plant Cells

The nuclear and chromosomal events of mitosis in plant cells are the same as those in animal cells. However, there are some differences (Table 2.1).

Table 2.1 Differences in mitosis between animal and plant cells

Animal cell	Plant cell
1. Asters are formed. 2. Cytokinesis by furrowing of cytoplasm.	1. Asters are not formed. 2. Cytokinesis is by cell plate formation.
3. Occurs in most tissues throughout the body (for growth and replacement).	3. Occurs mainly at the growing tips (for lengthening) and sides (for increase in girth).

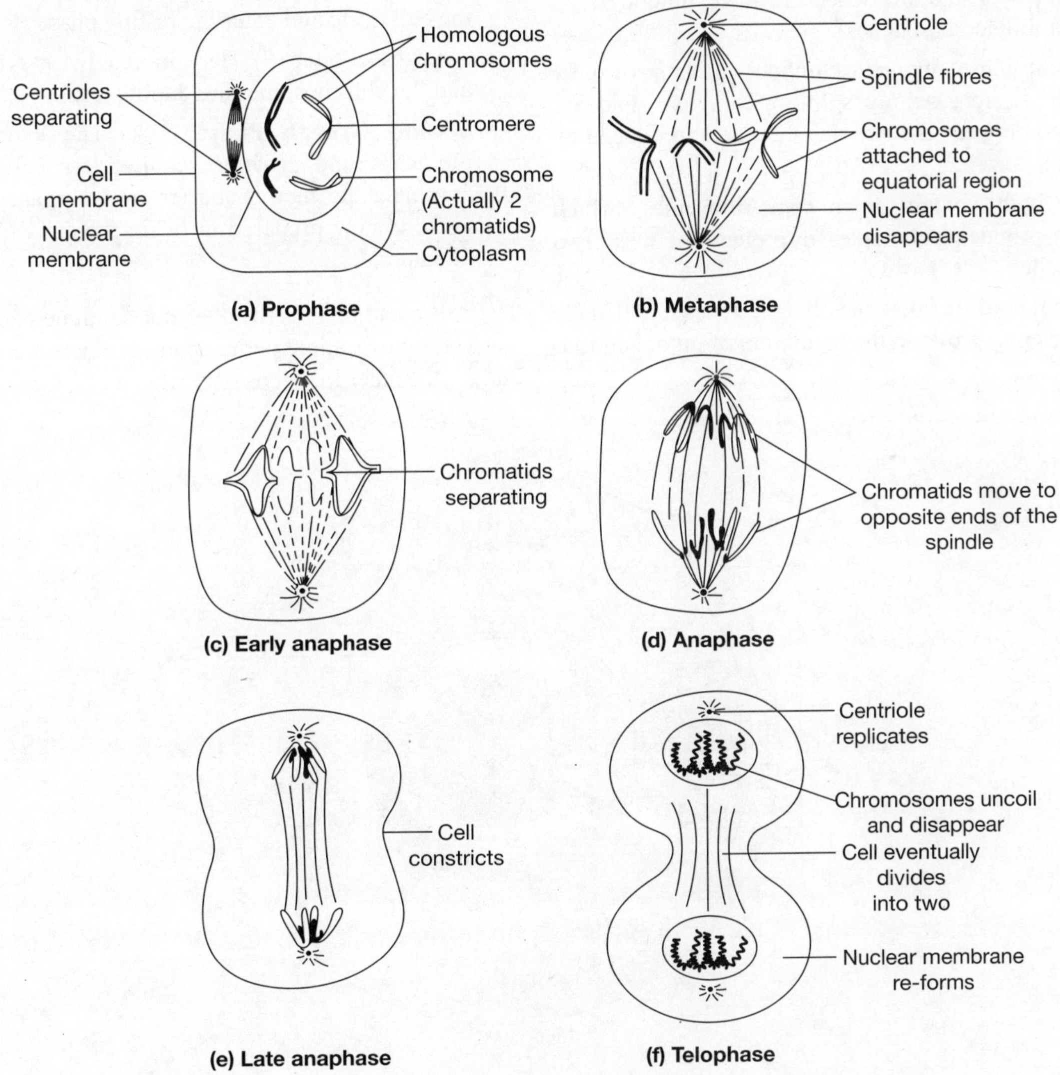

Fig. 2.9 Stages in mitosis

Significance of Mitosis

- Increases body size through the production of new cells in tissues.
- Replaces lost cells resulting in healing of injured and damaged tissues.
- Replaces old and dead cells, such as that of the skin's epidermal cells and blood cells.
- A type of asexual reproduction, in which yeast cells and other unicellular organisms like amoeba divide into two cells.
- Ensures that the same number of chromosomes is present in daughter cells.

Mitochondria and Chloroplasts in Cell Division

- Chloroplasts and mitochondria have unique DNA (containing certain genes).
- Both of these organelles divide on their own by simple fission, simply splitting into two, and are divided between the two daughter cells produced by mitosis.
- They both contain their own ribosomes, which aid in producing the specific proteins of these two organelles.
- The cytoplasmic ribosomes and the genes found in the nucleus play a role in the regulation of mitochondrial division.

Cell Cycle

At the end of mitosis, the newly formed cells have a full-sized nucleus but only a small amount of cytoplasm. They are now in interphase, where they enlarge to the size of their mother cell and prepare for the next cell division. Three phases make up interphase:

- **First Growth Phase (G_1):** Here, the production of proteins and RNA results in expansion of the cytoplasm. The two organelles that divide, mitochondria (found in all cells) and chloroplasts (found in plant cells), each have their own DNA. All the cells in the late G phase must take one of two routes: they can start preparing for subsequent division by entering the synthesis phase (S) or exit the cell cycle and enter the resting phase (R).
- **Synthesis Phase (S):** Here, more DNA is synthesized and the chromosomes are duplicated.
- **Second Growth Phase (G_2):** The synthesis of the RNA and proteins required for cell division continues during this shorter growth phase. The cell cycle is now prepared to begin the subsequent cell division.

The cell cycle (Fig. 2.10) does not continue endlessly. It stops at some locations permanently, at others momentarily, and at still other locations only when necessary.

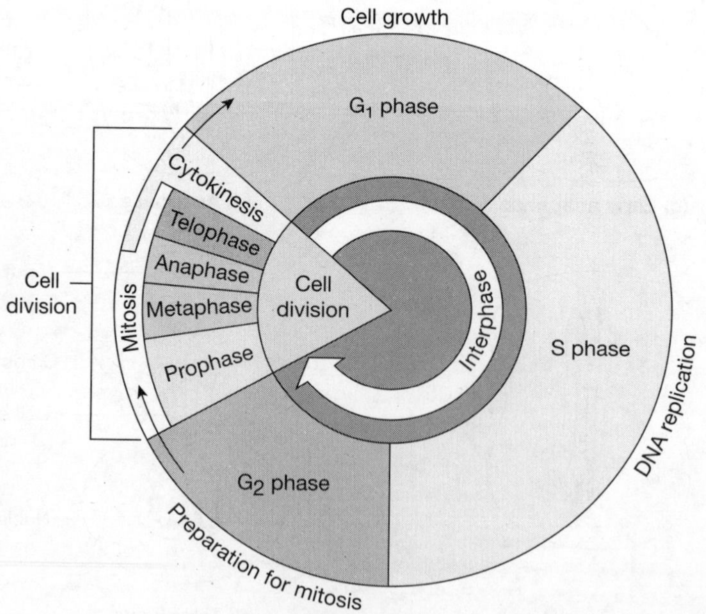

Fig. 2.10 Cell cycle

- After developing in the embryo, brain and other nerve cells stop dividing. They are not replaced after death.
- Only once every one to two years, liver cells may divide, replacing any damaged or dead cells.
- The underlying cells of the skin gradually replace the surface cells. Human skin cells are present in large quantities in household dust.
- In plants, the cells at the growth points (meristems) rapidly divide to create new leaves, buds and flowers, among other things.
- The other type of cell division, known as meiosis, creates sex cells in the ovary and testis of animals, as well as in the ovary and anthers of plants (Table 2.2).
- Cell cycles that are unchecked and continuous can result in tumours that may or may not be cancerous.

Children develop because they constantly produce more new cells than dying ones. Our cell population remains constant as we age. The number of dying cells and the number of newly formed cells are equal. The number of newly formed cells falls short of the number of dying cells as we age.

Meiosis

This type of cell division leads to the production of gametes (sex cells). The human reproductive organs (the testis and ovary) are where the sperm and ova are produced. Pollen grains and ovules are produced in the anthers and ovary, respectively, of flowering plants.

The fact that the number of chromosomes in the sex cells is halved during meiosis is its most important feature. For instance, only single chromosomes, or one member of each pair (haploid), is transferred to the sex cells out of the 23 pairs of chromosomes in humans. This is crucial because during fertilisation, the male and female gametes fuse, reacquiring the typical double (diploid) number of chromosomes (in pairs). In general, the haploid number is written as n, and the diploid number is written as 2n.

Meiosis is split into two main phases: meiosis I and meiosis II. The first phase is the reduction phase and the second is the mitotic phase (Fig. 2.11).

Significance of Meiosis

- The number of chromosomes is cut in half in gametes (sex cells), restoring the normal number (2n) upon fertilisation.
- It allows for gene admixture, which can happen in one of two ways:
  - As the homologous pairs separate during the first (reduction) division, the maternal and paternal chromosomes become mixed up.
  - The chromatid material frequently exchanges between the two members of a homologous pair during the separation of the maternal and paternal chromosomes, leading to genetic recombination.

  These permutations and combinations allow for the progeny's countless variations. Because of this, despite their similarities, children of the same parents are distinct from one another in some ways. Variations frequently aid in evolution.

Table 2.2 Differences between mitosis and meiosis

	Mitosis	**Meiosis**
1. Where it occurs	In the somatic (body) cells	In the reproductive cells
2. What it is for	To provide for growth and replacement	To enable gamete formation
3. When it occurs	Throughout life	Only in reproductively active age
4. Number of daughter cells produced	Two	Four
5. Number of chromosomes passed on to each cell	Full set of chromosomes is passed on to each daughter cell; this is the diploid (2n) number of chromosomes	Only half the number of chromosomes is passed (only one member from each pair); this is the haploid (n) number of chromosomes
6. Number of nuclear divisions	A single nuclear division after chromosome duplication	Two nuclear divisions after chromosome duplication
7. Identity of chromosomes and genes in daughter cell	Identical	Randomly assorted between the gametes produced; this results in genetic variation

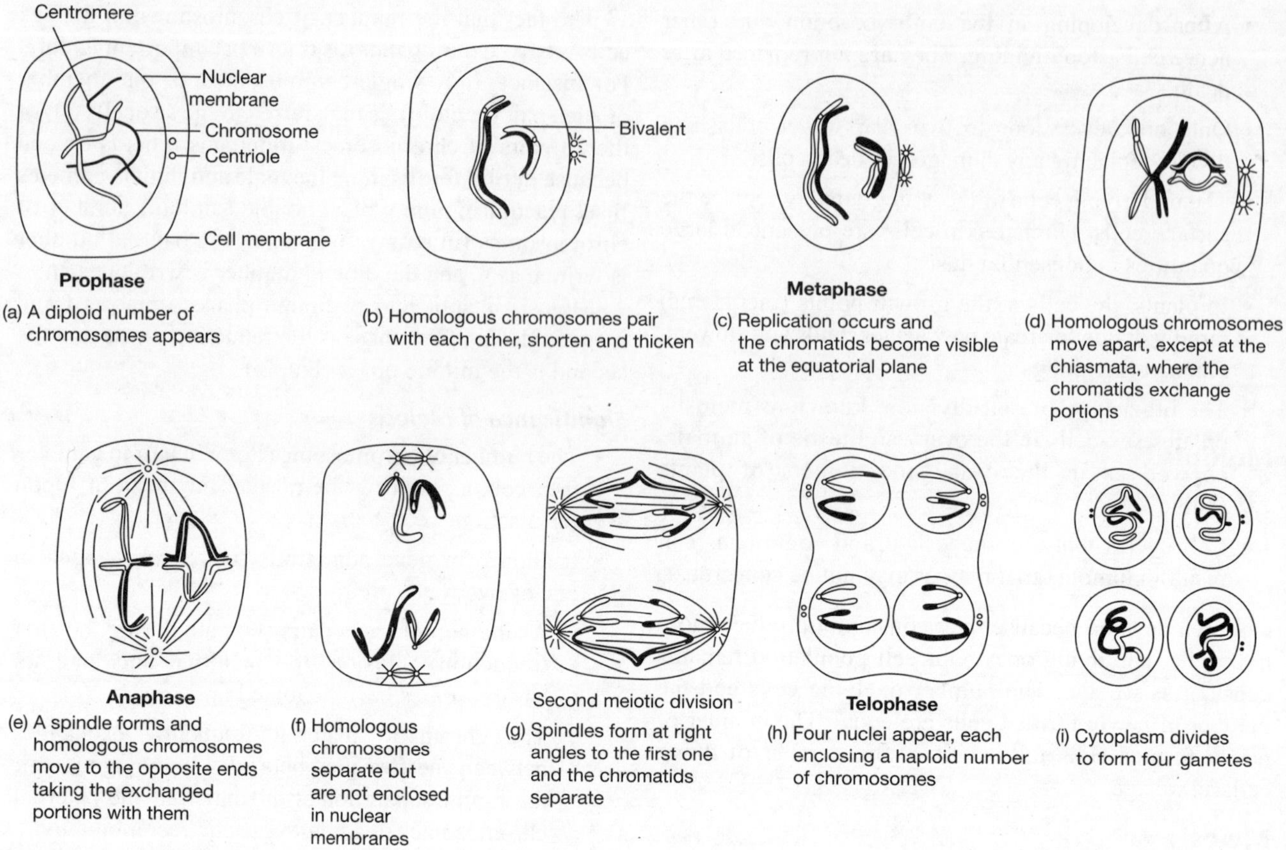

Fig. 2.11 Stages in meiosis

Stem Cells

Stem cells are undifferentiated cells that can undergo unlimited division and give rise to one or several different types of cells.

These days, the term is frequently mentioned in the media, particularly in relation to the treatment of specific human diseases. Although the word 'stem' is more commonly associated with plants, it has a wider meaning that means 'to branch off'. Unspecialised cells that can differentiate into leaves, flowers, fruits, seeds, etc. are present in the plant stem.

Animals have embryonic cells that develop into various tissues and organs. Embryonic stem cells develop in a variety of ways. Examples include those that specialise to produce blood cells, nerve stem cells, and so forth. Even in adults, these tissue-specific stem cells continue to exist.

SUMMARY

- Cell theory states that cells develop from pre-existing cells. Cell division is the process through which this takes place.
- Every sexually reproducing organism begins its life cycle as a zygote, which is a single cell. The development of the adult organism does not mark the end of cell division; rather, it continues throughout the organism's lifetime.
- The cell cycle refers to the phases that a cell goes through as it progresses from one division to the next.
- The cell cycle is divided into two phases: (i) Interphase, which is a period of cell division

preparation, and (ii) Mitosis (M phase), which is the actual cell division period.

- G_1, S and G_2 are further divisions of interphase. The G_1 phase is when a cell matures and engages in regular metabolism. During this period, the majority of organelle duplication also takes place. Chromosome duplication and DNA replication occur during the S phase. The G_2 phase is when the cytoplasm grows.
- Prophase, Metaphase, Anaphase and Telophase are the four stages of mitosis.

PRACTICE QUESTIONS

Analyse Your Concepts (School Exam Based)

Multiple Choice Questions

1. The cell division which occurs in somatic cells is:
(a) Mitosis (b) Meiosis
(c) Binary fission (d) Budding

2. The process in which cytoplasmic division occurs after nuclear division is called:
(a) Mitosis (b) Meiosis
(c) Cytokinesis (d) Interphase

3. Which of the following is not a stage of mitosis?
(a) Prophase (b) Metaphase
(c) Anaphase (d) Telophase

4. The point at which two chromatids are joined together is called:
(a) Kinetochore (b) Centromere
(c) Spindle (d) Chromatin

5. The number of chromosomes in a human somatic cell is:
(a) 23 (b) 46
(c) 69 (d) 92

6. The process of cell division in which the resulting daughter cells have half the number of chromosomes as compared to the parent cell is called:
(a) Mitosis (b) Meiosis
(c) Cytokinesis (d) Interphase

7. The process of DNA replication occurs during which phase of the cell cycle?
(a) G_1 phase (b) G_2 phase
(c) S phase (d) M phase

8. The structure responsible for pulling the chromatids apart during mitosis is called:
(a) Kinetochore (b) Centromere
(c) Spindle (d) Chromatin

9. Which of the following is not a function of mitosis?
(a) Growth and repair of tissues
(b) Asexual reproduction
(c) Formation of gametes
(d) Development of embryos

10. The period of the cell cycle during which the cell grows and carries out its metabolic activities is called:
(a) G_1 phase (b) G_2 phase
(c) S phase (d) M phase

Match the Following

Each question contains statements given in two columns which have to be matched. Statements in column I have to be matched with statements in column II.

Column I	Column II
(a) Chromosomes are arranged in a horizontal plane at the equator.	(i) Anaphase
(b) Daughter chromosomes move to the opposite poles of a spindle.	(ii) Prophase
(c) Chromosomes become visible as fine long threads.	(iii) Telophase
(d) Chromosomes lose their distinctiveness and gradually become transformed into a chromatin network.	(iv) Meiosis
(e) Results in the formation of four genetically diverse daughter cells.	(v) Metaphase

True or False

Read the following statements and write your answer as true or false.

1. As you grow from childhood to adulthood, your skin cells divide only to replace such cells as are lost from the surface.

2. The unfertilised human egg has half the number of chromosomes of the body cells.

3. The nuclear membrane in a mitotically dividing cell remains intact up to metaphase and disappears only in telophase.
4. Mitotic cell division can be a mode of reproduction.
5. Crossing-over between chromatids can occur only between homologous chromosomes.

Fill in the Blanks

Complete the following statements with an appropriate word/term to be filled in the blank spaces.

1. Mitosis is a type of cell division that takes place in _______ cells.
2. Meiosis is a type of cell division that takes place in _______ cells.
3. In mitosis, the chromosome number is _______.
4. In meiosis, the chromosome number is _______.
5. Mitosis occurs at the time of _______ reproduction
6. Meiosis occurs at the time of _______ reproduction.
7. The two non-sister chromatids of a paired chromosome are attached to each other at _______ during the process of crossing over.

Assertion–Reason Questions

The following questions consist of two statements: Assertion (A) and Reason (R). Answer these questions selecting the appropriate option given below:

(a) Both A and R are true and R is the correct explanation of A
(b) Both A and R are true but R is not the correct explanation of A
(c) A is true but R is false
(d) A is false but R is true

1. Assertion (A): Interphase is the stage of rest.
Reason (R): The cell in interphase is not metabolically active.
2. Assertion (A): Basic proteins called histones play a crucial role in the wrapping of eukaryotic DNA. The majority of a eukaryotic chromosome, chromatin, is made up of DNA and histones.
Reason (R): Histones are of 5 major types H1, H2A, H2B, H3 and H4.
3. Assertion (A): During the G_1 and G_2 phases of the cell cycle, DNA is synthesised.
Reason (R): The DNA concentration doubles during the G_1 and G_2 phases.
4. Assertion (A): Some cells enter the G_0 phase as a result of the cell cycle being inactive.
Reason (R): The G_0 phase arises as a result of the absence of mitogen and energy-rich substances.
5. Assertion (A): G_1 phase is the interval between mitosis and initiation of DNA replication.
Reason (R): During the G_1 phase, the cell is not metabolically active.

One-word Answers

1. Modern humans have 46 chromosomes. Their sperms and eggs will have how many chromosomes each?
2. In which division does the chromosome number in daughter cells remain the same as in the parent cell?
3. DNA replicates in which phase of the cell cycle?
4. What is the type of cell division that produces sex cells or gametes?
5. Which process usually takes place in a somatic cell?

Define the Term

1. Cytokinesis
2. Metaphase
3. Growth (G_1) phase
4. Centromere
5. Lamella

Groups of Terms

Given below are certain groups of terms. In each group the first pair indicates a relationship between the two terms. Rewrite and complete the second pair on a similar basis.

1. Mitosis : Cell division :: Photosynthesis : _________
2. Nucleus : DNA :: Chloroplast : _________
3. Chromosome : Gene :: Protein : _________
4. Meiosis : Gametes :: Fertilization : _________
5. Cell cycle : Interphase :: Cell division : _________

Identify the Odd One Out

1. Which of the following is the odd one out?
(a) Mitosis (b) Meiosis
(c) Cytokinesis (d) Interphase

2. Which of the following is the odd one out?
(a) Nucleus (b) Cell membrane
(c) Ribosome (d) Lysosome

3. Which of the following is the odd one out?
(a) Chromosome (b) Gene
(c) DNA (d) RNA

4. Which of the following is the odd one out?
(a) Epithelial tissue (b) Muscle tissue
(c) Connective tissue (d) Nervous tissue

5. Which of the following is the odd one out?
(a) Sickle cell anaemia (b) Down syndrome
(c) Haemophilia (d) Cystic fibrosis

Image-Based Questions

1. The diagram given below represents a certain phenomenon which occurs during meiosis. Name and explain the phenomenon by using the terms homologous chromosomes, chromatids and crossing-over.

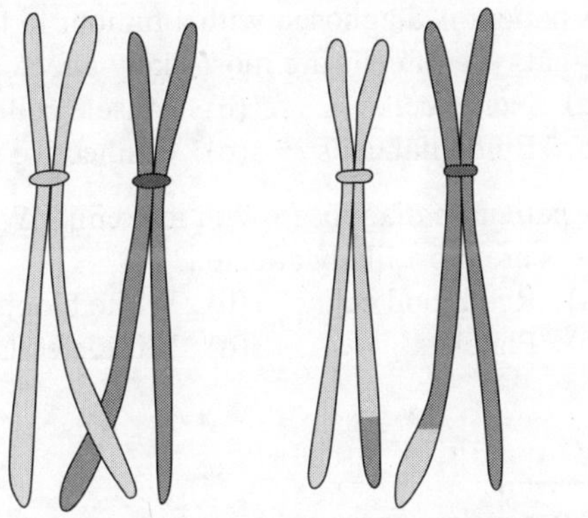

2. The three sketches given below (A, B and C) are intended to represent the replication of DNA. What should be their correct sequence starting with the first and ending with the last?

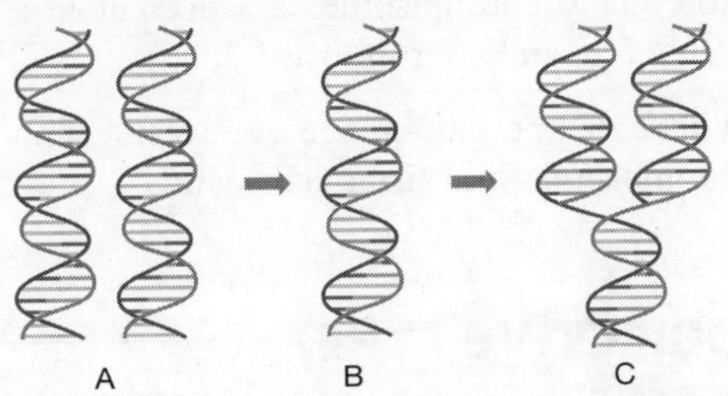

3. Identify the type of chromosomes shown in A, B, C, D based on the position of the centromere.

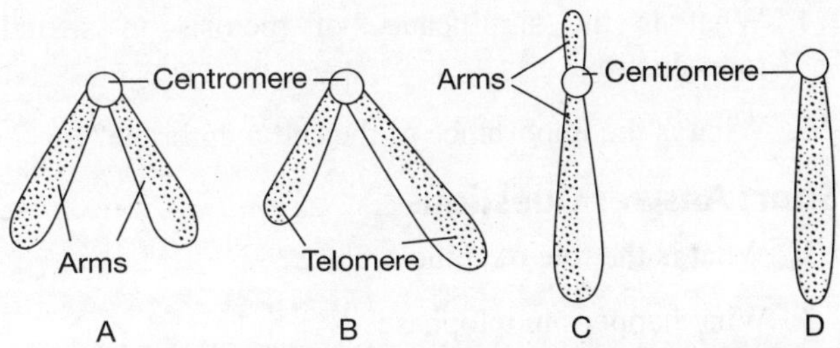

4. The diagram below represents a stage during cell division. Study the same and then answer the questions that follow:

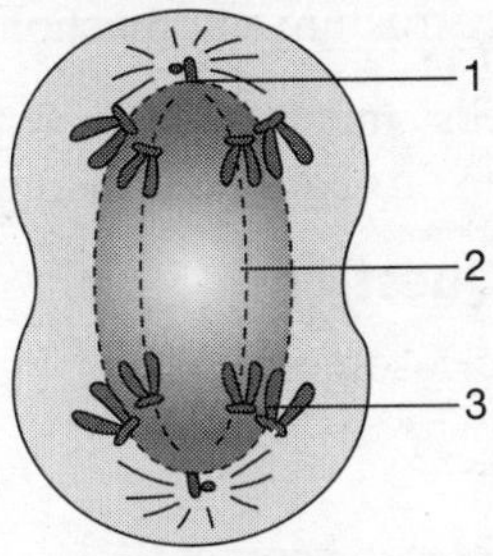

(a) Name the parts labelled 1, 2 and 3.
(b) Mention where in the body this type of cell division occurs.

5. Answer the following for the given diagram:
(a) The number of strands in the diagram is ____.
(b) Name the parts numbered 1, 2, 3, 4 and 5.
(c) 1, 2 and 3 are collectively termed ____.

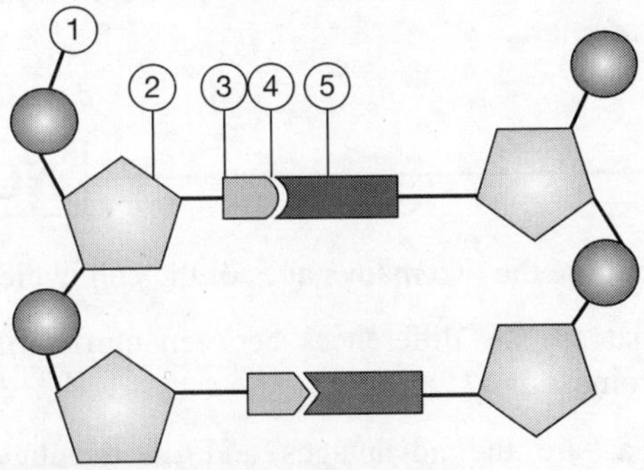

Very Short Answer Questions

1. What are the lengthwise repeating components of each DNA strand?

2. What is the complex structure consisting of a DNA strand and a core of histones?

3. Explain the type of bond which joins the complementary nitrogenous bases.

4. What is the significance of meiosis in sexual reproduction?
5. What is the importance of the cell membrane?

Short Answer Questions

1. What is the role of spindle fibres?
2. What happens in telophase?
3. What is the process of meiosis, and how does it contribute to genetic diversity?
4. What are the different types of cells, and how do they differ in their structure and function?
5. How do cells maintain homeostasis, and what happens when this balance is disrupted?

Long Answer Questions

1. What are nucleosomes?
2. What are genes?
3. State the difference between:
   (a) Chromosome and chromatid
   (b) Aster and spindle fibre
   (c) Haploid and diploid
4. "The first meiotic division is the reduction division". What does the word 'reduction' refer to in this statement?
5. Explain the difference between centrosome and centromere.

Case Study-Based Questions

1. A patient is diagnosed with sickle cell anaemia. What is the cause of this disease?
   (a) Mutation in the gene for haemoglobin
   (b) Mutation in the gene for insulin
   (c) Mutation in the gene for growth hormone
   (d) Mutation in the gene for thyroid hormone
2. A patient is diagnosed with Down syndrome. What is the cause of this disease?
   (a) An extra copy of chromosome 18
   (b) An extra copy of chromosome 21
   (c) A missing copy of chromosome 18
   (d) A missing copy of chromosome 21
3. A patient is diagnosed with cystic fibrosis. What is the cause of this disease?
   (a) Mutation in the gene for insulin
   (b) Mutation in the gene for haemoglobin
   (c) Mutation in the gene for CFTR protein
   (d) Mutation in the gene for thyroid hormone
4. A patient is diagnosed with a tumour in their breast. What type of cells are most likely affected?
   (a) Nerve cells (b) Muscle cells
   (c) Epithelial cells (d) Connective tissue cells
5. A patient is diagnosed with leukemia. What type of cells are most likely affected?
   (a) Red blood cells (b) White blood cells
   (c) Platelets (d) Muscle cells

NCERT Corner

1. What are the two main stages of the cell cycle?
2. What are the differences between chromatin and chromosomes?
3. What are the advantages and disadvantages of sexual reproduction over asexual reproduction?
4. Explain the process and significance of crossing over during meiosis.
5. How many chromosomes are present in a human cell before and after mitosis?
6. What is the difference between homologous chromosomes and sister chromatids?

COMPETITION WINDOW (OBJECTIVE TYPE)

1. Water is absorbed by plant cells through surface attraction called:
   (a) Diffusion (c) Imbibition
   (b) Osmosis (d) Endosmosis
2. Plasmolysis occurs when a plant cell is placed in a specific solution. What sort of a solution is it?
   (a) Isotonic sugar solution
   (b) Hypotonic salt solution

(c) Hypertonic salt solution
(d) Isotonic salt solution

3. When a cell is in this condition, the cell wall is hard and stretched due to increase in volume brought on by absorption of water:
(a) Flaccidity (b) Turgidity
(c) Capillarity (d) Tonicity

4. Which one of the following is a characteristic not related to the suitability of roots for absorbing water?
(a) Tremendous surface area
(b) Contain cell sap at a higher concentration than the surrounding soil water
(c) Root hairs have thin cell walls
(d) Grow downwards into the soil

5. Chromosome or (DNA) duplication takes place during:
(a) Interphase (b) Prophase
(c) Mitosis (d) Cytokinesis

6. The nuclear membrane and nucleolus reappear during:
(a) Prophase (b) Metaphase
(c) Anaphase (d) Mitotic Anaphase

7. Crossing over occurs during:
(a) Mitotic Metaphase (c) Meiosis Prophase
(b) Mitotic Anaphase (d) Meiosis Telophase

8. During chromosomal alignment, the chromosomes line up at the cell's centre during which stage?
(a) Metaphase (c) Prophase
(b) Anaphase (d) Telophase

9. The chromosomes are attached to the spindle fibres at their:
(a) Centriole (b) Centromere
(c) Chromomeres (d) Chromatids

10. In meiosis, duplication of chromosomes takes place during:
(a) Interphase (b) Prophase
(c) Both (a) and (b) (d) None of the above

11. How many cells will be produced if a cell divides mitotically four times?
(a) 24 (c) 32
(b) 16 (d) 64

12. A single cell gives rise to 32 cells by dividing mitotically. How many times has it undergone mitotic division?
(a) 16 (b) 8
(c) 4 (d) 5

13. If there are 32 pollen grains (male gametes) in total after one division, how many cells have undergone meiotic division?
(a) 16 (c) 4
(b) 8 (d) 5

14. The chromatids of a chromosome are attached at a point called:
(a) Centriole (b) Centromere
(c) Chromomere (d) Centrosome

15. The centromere divides into two during:
(a) Prophase (b) Metaphase
(c) Anaphase (d) Telophase

16. At the end of mitosis, two cells form that comprise:
(a) Same number of chromosomes as the original cell
(b) Half the number of chromosomes as the original cell
(c) Double the number of chromosomes of the original cell
(d) None of the above

17. Equatorial plate is formed during:
(a) Prophase (b) Anaphase
(c) Metaphase (d) Telophase

18. Among the following cells, which will be diploid?
(a) Pollen grain (b) Egg cell
(c) Skin cell (d) Red blood cell

19. Among the statements given below, select the most appropriate option associated with a cell:
(a) The smallest entity of the living organism
(b) The structural and functional unit of life
(c) The macroscopic unit of life
(d) Both (a) and (b)

20. The G_1, S and G_2 phases are present in which of these phases?
(a) Anaphase (b) Interphase
(c) Prophase (d) Metaphase

Answer Keys

Multiple Choice Questions

1. (a) 2. (c) 3. (d) 4. (b) 5. (b)
6. (b) 7. (c) 8. (c) 9. (c) 10. (a)

Match the Following

Column I	Column II
(a) Chromosomes are arranged in a horizontal plane at the equator.	(v) Metaphase
(b) Daughter chromosomes move to the opposite poles of a spindle.	(i) Anaphase
(c) Chromosomes become visible as fine long threads.	(ii) Prophase
(d) Chromosomes lose their distinctiveness and gradually become transformed into a chromatin network.	(iii) Telophase
(e) Results in the formation of four genetically diverse daughter cells	(iv) Meiosis

True or False

1. F 2. T 3. T 4. T 5. T

Fill in the Blanks

1. Somatic
2. Germinal
3. Remains the same
4. Reduces to half
5. Asexual
6. Sexual
7. Centromere

Assertion–Reason Questions

1. (c) 2. (b) 3. (d) 4. (a) 5. (c)

One-word Answers

1. 23
2. Mitosis
3. Synthesis
4. Meiosis
5. Mitosis

Define the Term

1. It is the division of the cytoplasm during cell division.
2. Chromosomes are arranged on the metaphase plate or equatorial plane.
3. In the growth (G_1) phase of the cell cycle, cells primarily undergo growth by increasing in size and producing the proteins and organelles required for normal cellular functions and DNA synthesis.
4. The region of a chromosome that holds the two sister chromatids together.
5. Thin, flat layers of connective tissue found between some cells in animal tissues.

Groups of Terms

1. Sugar production
2. Chlorophyll
3. Amino acid
4. Zygote
5. Mitosis or Meiosis

Identify the Odd One Out

1. (d) 2. (b) 3. (a) 4. (d) 5. (b)

Image-Based Questions

1. The exchange of chromatids between homologous chromosomes is known as crossing-over, which results in the halving of chromosome number in gametes. This allows for restoration of the normal number of chromosomes (2n) upon fertilisation. Cross-over facilitates gene mixing through two mechanisms:
   (i) The separation of maternal and paternal chromosomes during the first division results in their mixing.
   (ii) The exchange of chromatid material often occurs between the two members of a homologous pair during the separation of maternal and paternal chromosomes.
2. 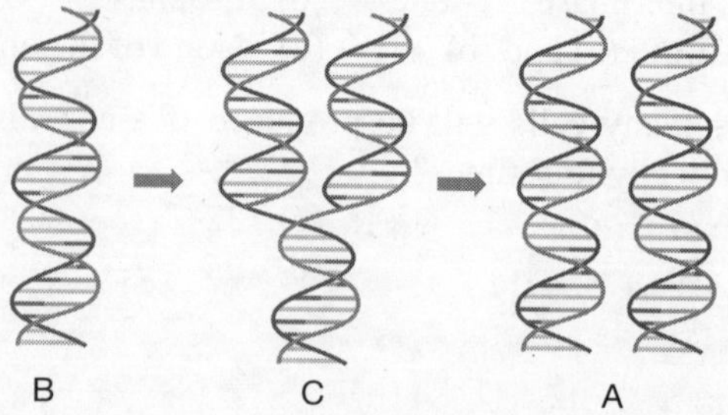

3. A. Metacentric: The centromere holds the chromatids at the centre with approximately equal p and q arms.

B. Sub metacentric: The centromere is slightly away from the centre. The p arm is slightly smaller than the q arm.

C. Acrocentric: The centromere is more towards one end and the p arm much shorter than the q arm.

D. Telocentric: The centromere occurs at one end of the chromosome.

4. (a) 1 = Centromere, 2 = Spindle fibres, 3 = Chromatids

(b) Mitotic division is the type of cell division that takes place in all the cells of the body, except the reproductive cells.

5. (a) There are two strands in the diagram.

(b) 1 = Phosphate, 2 = Sugar, 3 = Bases, 4 = Hydrogen bond, 5 = Base

(c) Nucleotide

Very Short Answer Questions

1. The repeating components of DNA strand lengthwise are nucleotides, which are made up of a sugar molecule, a phosphate group and a nitrogenous base. The nitrogenous base can be adenine (A), thymine (T), guanine (G), or cytosine (C), and it determines the genetic information carried by the DNA molecule.
2. The complex structure consisting of a DNA strand and a core of histones is called a nucleosome. Histones are proteins that DNA strands wrap around to form a compact structure. Nucleosomes are the building blocks of chromatin, which is the material that makes up chromosomes.
3. The type of bond which joins the complementary nitrogenous bases is hydrogen bond. Hydrogen bonds form between the nitrogenous bases of two complementary DNA strands, holding the strands together in a double-helix structure. The hydrogen bonds are weak, allowing the DNA strands to separate during DNA replication and transcription.
4. Meiosis results in the production of genetically diverse gametes, which leads to genetic variation in offspring during sexual reproduction.
5. The cell membrane protects the cell from its environment and regulates the movement of substances in and out of the cell.

Short Answer Questions

1. Spindle fibres help in the division of chromosomes equally in the daughter cells from the parent cell.
2. In this phase, the chromosomes are first uncoiled and then separate from the parent cell into two identical daughter cells.
3. Meiosis is a type of cell division that produces gametes (sex cells) with half the number of chromosomes as the parent cell. It consists of two rounds of cell division and results in the production of four haploid cells. Meiosis contributes to genetic diversity through the process of recombination, in which homologous chromosomes exchange genetic material. This results in unique combinations of genes in each of the four haploid cells. Additionally, meiosis can result in mutations, which can also contribute to genetic diversity.
4. There are two main types of cells: prokaryotic and eukaryotic. Prokaryotic cells are simple in structure and lack a nucleus and membrane-bound organelles. They are typically found in bacteria and archaea. Eukaryotic cells are more complex and have a nucleus and various membrane-bound organelles. They are found in all other living organisms, including animals, plants, and fungi. Different types of cells also have different functions. For example, muscle cells are specialised for contraction, while nerve cells are specialised for transmitting signals.
5. Cells maintain homeostasis by regulating their internal environment to maintain a stable and balanced state. This includes maintaining a constant pH, temperature, and concentration of ions, nutrients and waste products. To achieve this, cells have various mechanisms such as ion channels, transporters and pumps to move substances across their membrane.

Long Answer Questions

1. Hint: Explain that nucleosomes are the basic units of chromatin, the material that makes up chromosomes. Explain how they are formed by DNA coiling around a group of histone proteins. Describe how they help protect and regulate the expression of DNA.
2. Hint: Explain that genes are segments of DNA that code for proteins or other molecules. Explain how they are the units of heredity that determine the traits of an organism.
3. Hints: (a) Explain that a chromosome is a long, coiled molecule of DNA and protein that carries genetic information. A chromatid is one of two identical copies of a chromosome that are joined at the centromere before cell division.

   (b) Explain that an aster is a star-shaped structure of microtubules that radiates from the centrosome near the poles of a dividing cell. Spindle fibres are the microtubules that extend from the centrosomes and attach to the chromosomes during cell division.

(c) Describe a haploid cell as a cell that has one set of chromosomes, such as a gamete (sperm or egg). A diploid cell is a cell that has two sets of chromosomes, one from each parent, such as a somatic (body) cell.

4. Hint: Explain that the word 'reduction' refers to the halving of chromosome number in the first meiotic division. This means that the daughter cells produced by the first meiotic division have half as many chromosomes as the parent cell.

5. Hint: Explain how a centrosome is an organelle that contains a pair of centrioles and organises the microtubules of the spindle apparatus. A centromere is a region of DNA on a chromosome that holds the sister chromatids together and attaches to the spindle fibres.

Case Study-Based Questions

1. (a) **2.** (b) **3.** (c) **4.** (c) **5.** (b)

NCERT Corner

1. The two main stages of the cell cycle are interphase and mitotic (M) phase. Interphase is the phase when the cell grows and replicates its DNA. Mitotic (M) phase is the phase when the cell divides its nucleus and cytoplasm into two daughter cells.
2. Chromatin and chromosomes are both forms of DNA in the nucleus of a cell. Chromatin is a loosely coiled and dispersed form of DNA that is present during interphase. Chromosomes are tightly coiled and condensed forms of DNA that are visible during mitosis and meiosis.
3. Sexual reproduction combines gametes from two parents to form a new individual. Asexual reproduction produces offspring from one parent without gametes. Sexual reproduction has more genetic variation, fewer genetic defects and more evolution potential. Asexual reproduction has a lower time and energy cost, more offspring and smaller risk and complications.
4. Crossing over is the swapping of DNA segments between non-sister chromatids of homologous chromosomes in meiosis I. It makes recombinant chromosomes with mixed alleles from both parents. It increases genetic variation and ensures proper chromosome separation in gametes. It prevents errors like non-disjunction and aneuploidy.
5. A human cell has 46 chromosomes before mitosis, which are arranged in 23 pairs of homologous chromosomes. Each chromosome consists of two identical sister chromatids that are joined at the centromere. During mitosis, the sister chromatids separate and move to opposite poles of the cell. As a result, each daughter cell receives 46 chromosomes after mitosis, which are identical to those of the parent cell.
6. Homologous chromosomes are similar pairs of chromosomes from each parent. They have the same genes at the same locations. Sister chromatids are identical copies of one chromosome from DNA replication.

Competition Window

1. (c)	**2.** (c)	**3.** (b)	**4.** (d)	**5.** (a)
6. (d)	**7.** (c)	**8.** (a)	**9.** (b)	**10.** (a)
11. (b)	**12.** (d)	**13.** (b)	**14.** (b)	**15.** (c)
16. (a)	**17.** (c)	**18.** (c)	**19.** (d)	**20.** (b)

CHAPTER

3 Life Processes

PREREQUISITES

You must have knowledge of:

- Structures of plant and animal cells, their organelles and their functions
- Different types of nutrients required by living organisms and their sources
- Process of respiration
- Transport of materials in plants and animals
- Process of excretion
- Control and coordination

LEARNING OBJECTIVES

At the end of this chapter, you will be able to:

- Define life processes
- Explain nutrition and its types
- Explain the process of respiration
- Describe the transport of food and oxygen
- Describe the process of excretion of waste matter

INTRODUCTION

Living organisms are made up of cells, which in turn form tissues, organs and organ systems. This organisation is sustained by molecules moving about among the various parts. Life processes can thus be defined as all the functions that are carried out by the body and which keep organisms alive.

How do we know if an organism is alive or not? If we see a crow flying, a donkey running or a woman talking to her friends, we understand that these are all living beings. But what happens if the crow, donkey or woman are asleep? How do we know that they are alive? We see them breathing, and we know that they are alive because they continually perform various activities which differentiate them from the non-living. Similarly, how do we know if a plant is alive? It will grow over time and perform various processes, such as flowering and opening and closing its buds, that indicate it is alive.

The main life processes are nutrition, respiration, transportation and excretion (Fig. 3.1).

Nutrition

Nutrition is the process of ingesting food and breaking it down into smaller molecules that the body can absorb. As soon as we pedal a bike or walk, our energy levels begin to decline. Energy is required to maintain our body, even when we appear to be doing nothing. We frequently require elements from the outside world in order to grow, develop, synthesize macromolecules and create other compounds. Nutrition provides energy for carrying out any activity. The materials that make up the food consumed by organisms are referred to as nutrients. Carbohydrates and lipids are the primary sources of energy for the body.

How Do Organisms Obtain Food?

All organisms need energy and resources, but this need is satisfied in different ways. Mode refers to the method by which an organism obtains nutrition. Some organisms consume readily available food that comes from inorganic sources such as CO_2 and water. Others depend on other organisms for their food.

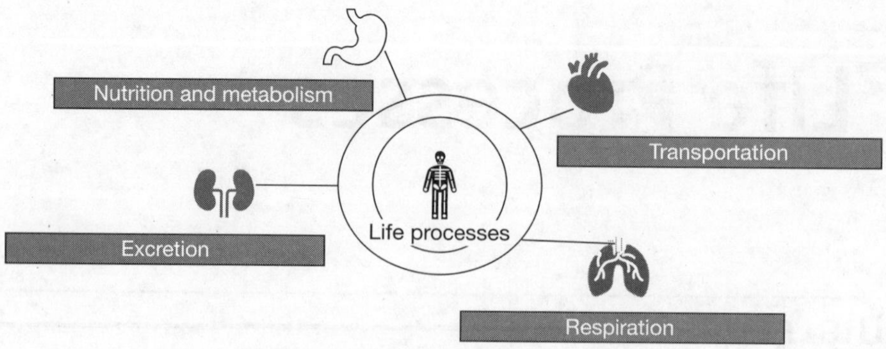

Fig. 3.1 Types of life processes

- **Autotrophs:** In the presence of sunshine and chlorophyll, these organisms prepare their own food from basic inorganic elements like carbon dioxide and water. Examples are some microorganisms and all green plants.
- **Heterotrophs:** These organisms directly or indirectly obtain their sustenance from plants. For example, all animals, some bacteria and fungi. Heterotrophs can be further classified as:
  - **Carnivores:** Organisms that eat only animals are known as carnivores. For example, tiger, lion, snake and frog.
  - **Herbivores:** Organisms that eat only plants are known as herbivores. For example, cow, deer, rabbit and elephant.
  - **Omnivores:** Organisms that eat plants and animals are known as omnivores. For example, crow, human, dog and sparrow.

The heterotrophic mode of nutrition is of three types:

- **Saprophytic nutrition:** This type of nutrition occurs when organisms eat dead and decomposing matter. They digest the food outside their bodies before absorbing it. For instance, mushroom, bread mould, yeast and some bacteria.
- **Parasitic feeding:** This type of nutrition occurs when an organism feeds on another living thing (the host) without causing harm to the host. For instance, *Cuscuta*, orchid, tick, lice, leech, roundworm, tapeworm and *Plasmodium*.
- **Holozoic nutrition:** This type of nutrition occurs when an organism consumes food without first digesting or assimilating it. Examples include *Amoeba*, *Paramecium*, birds, fish and humans.

Nutrition in Plants

Photosynthesis, which occurs when sunlight and chlorophyll are present, is the process by which plants manufacture food using carbon dioxide and water (Fig. 3.2). Carbohydrate, which is stored as starch, is present in the prepared food. In this process, oxygen is released.

There are three key steps in photosynthesis:

1. Absorption of light energy by chlorophyll.
2. The splitting of water molecules into hydrogen and oxygen and the conversion of light energy into chemical energy.
3. The reduction of carbon dioxide by hydrogen to create carbohydrates.

$$6CO_2 + 6H_2O \xrightarrow{\text{Sunlight + Chlorophyll}} C_6H_{12}O_6 + 6O_2$$

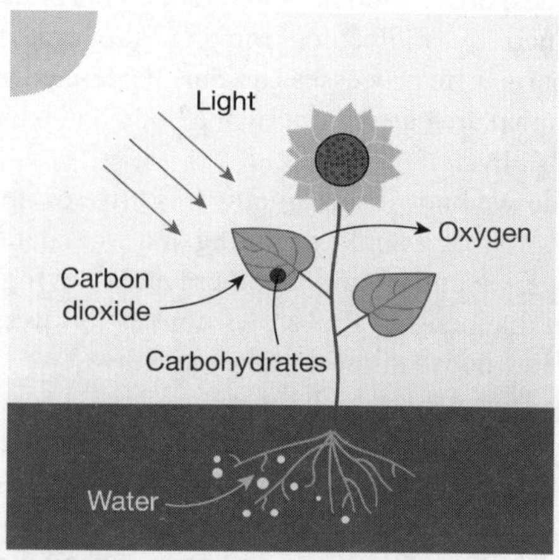

Fig. 3.2 Process of photosynthesis

The green pigment found in leaves is called **chlorophyll**. A cross-section of a leaf can be examined under a microscope to identify cells with chloroplasts, which are tiny green dots that contain chlorophyll (Fig. 3.3).

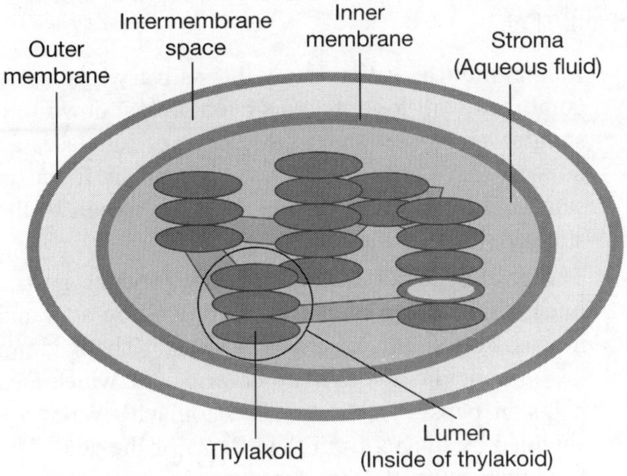

Fig. 3.3 Structure of the chloroplast

SCIENCE IN DAILY LIFE

1. Select a potted plant with iridescent leaves, like a croton or a money plant. For three days, keep the plant in a dark area to ensure that all of the starch is consumed.
2. Place the plant in sunlight for roughly six hours.
3. Pluck one of the leaves. On a piece of paper, highlight and trace the green regions of the leaf.
4. Submerge the leaf briefly in boiling water. Then, submerge it in an alcohol-filled beaker. Place this beaker in a water bath and heat it until the alcohol starts to boil. What happens to the leaf's colour? What colour is the solution?
5. Soak the leaf for a few minutes in a weak iodine solution. Remove the leaf, then thoroughly clean it with water. Examine the leaf's colour and contrast it with the initial trace you made. What can you deduce from the presence of starch in various areas of the leaf?

Small pores called **stomata** are found in the leaves; they are used for gas exchange (Fig. 3.4). The guard cells at each stoma regulate the opening and closing of the stomatal hole (Fig. 3.5). The guard cells expand when water enters, opening the pore, and contract when water leaves them, closing the hole.

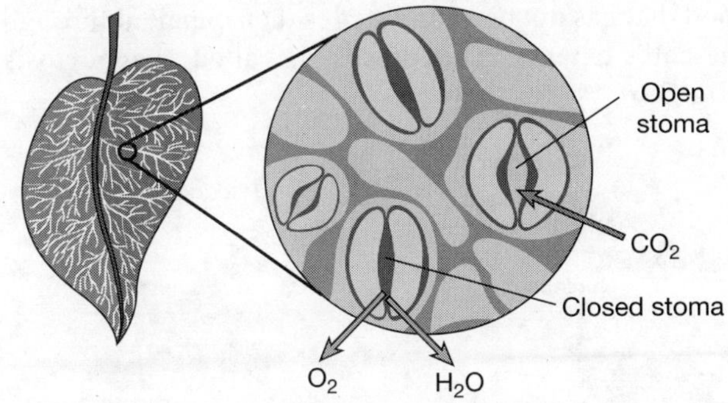

Fig. 3.4 Structure of stomata

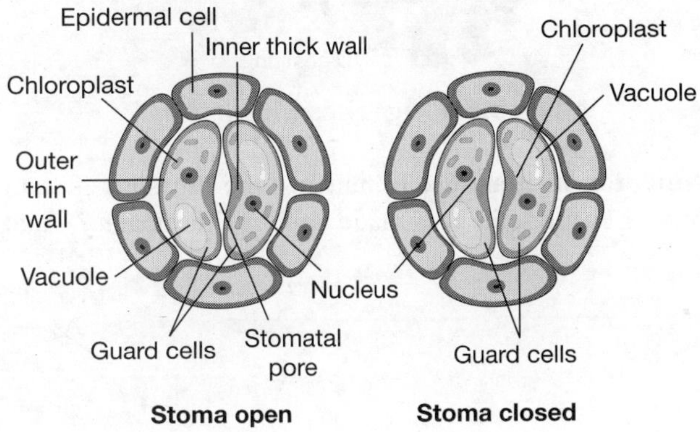

Fig. 3.5 Open and closed stomata

SCIENCE IN DAILY LIFE

1. Place two identical-sized potted plants in a dark area for three days. Afterwards, keep each plant on a different glass plate.
2. Keep a watch glass with potassium hydroxide close to one plant.
3. Cover both plants with bell jars. Use Vaseline on the bottom of the jars to create an airtight seal. Expose the plants to sunlight for three hours.
4. Check the starch content of a leaf from each plant. The leaf that was kept in the container with the potassium hydroxide does not exhibit any signs of starch. This demonstrates the necessity of carbon dioxide in photosynthesis.

Nutrition in Animals

Nutrition in *Amoeba*: *Amoeba* is a single-celled aquatic organism. It has pseudopodia, which resemble finger-like extensions, and a food vacuole. Food is broken down and absorbed inside the food vacuole. The food that has not been digested is then expelled through the cell's exterior. This process is called phagocytosis (Fig. 3.6).

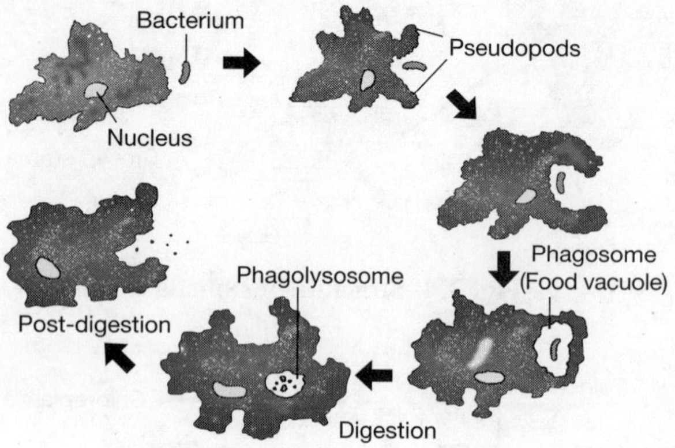

Fig. 3.6 Phagocytosis in *Amoeba*

Nutrition in humans: In humans, the digestive system processes the food. It is made up of the alimentary canal and glands that produce the enzymes required to break down food into smaller molecules.

The mouth, oesophagus, stomach, small intestine, large intestine and anus are the primary organs of the digestive system. The salivary glands, stomach glands, hepatic, pancreatic and intestinal glands are the main glands (Fig. 3.7).

- In the **mouth**, saliva from the salivary glands is combined with food that has been broken down into smaller pieces by the teeth. Salivary amylase, an enzyme that converts starch into sugar, is found in saliva. The food then passes from the mouth to the stomach through the oesophagus.
- The gastric glands in the **stomach** produce gastric juice, which contains pepsin, hydrochloric acid and mucus. Proteins are digested by pepsin. The medium becomes acidic due to hydrochloric acid, which also helps in pepsin's activity. The stomach's walls are shielded by mucus from the effects of the acid. The food then enters the small intestine.

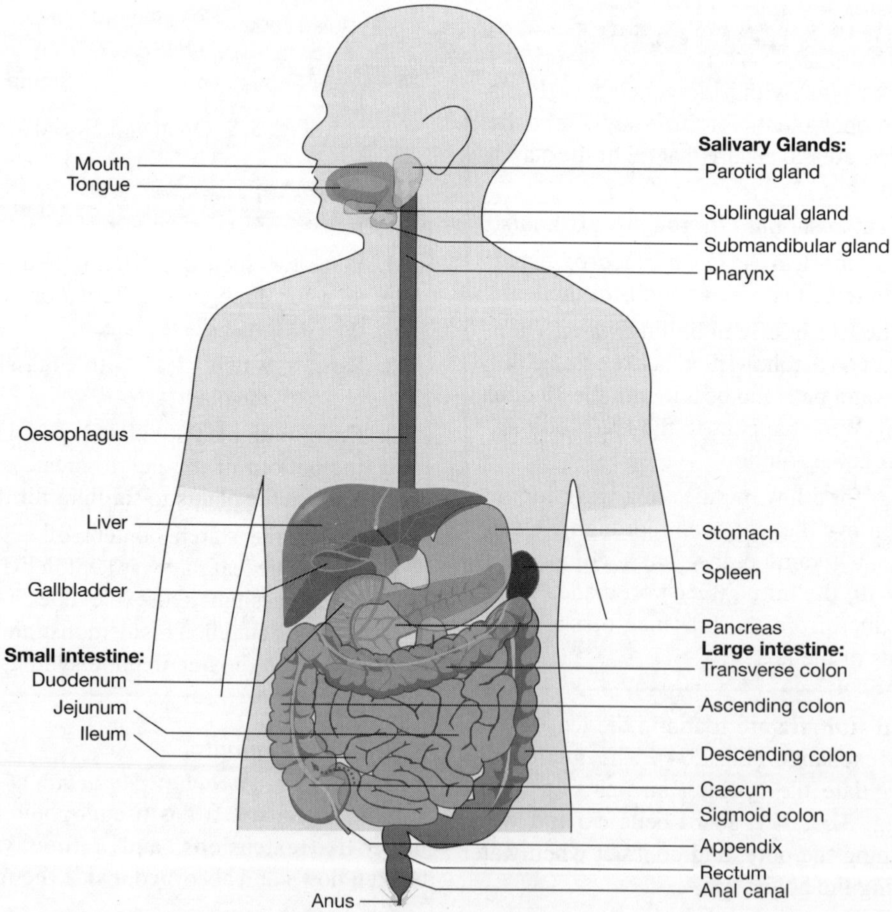

Fig. 3.7 Human digestive system

- The food is combined with pancreatic juice from the pancreas and bile from the liver in the **duodenum**, which is the upper portion of the small intestine. Bile breaks down fats into smaller globules. Trypsin and lipase are two of the enzymes found in pancreatic juice. While lipase breaks down fats, trypsin breaks down proteins.
- The **small intestine's** walls produce intestinal juice. The enzymes contained in the intestinal juice convert proteins into amino acids, lipids into fatty acids and glycerol, and carbohydrates into glucose. The small intestine's walls have numerous finger-like protrusions called villi that contain blood vessels. It aids in expanding the surface area for food to be absorbed. The digested food is absorbed by blood and transported to all the cells in the body. The undigested food goes into the large intestine.
- Water is absorbed in the **large intestine**, while waste is expelled through the anus.

Respiration

During respiration, food is burned in the body cells with the help of oxygen to release energy. This takes place in the mitochondria. ADP (adenosine diphosphate) and inorganic phosphate are used to create ATP (adenosine triphosphate) that is the energy released during respiration.

$$\text{ADP + Phosphate} \xrightarrow[\text{Energy from respiration}]{} \text{ATP}$$

ATP molecules serve as energy storage units in the cells. In the presence of water, ATP is broken down to generate ADP, which releases energy when the cell needs it.

$$\text{ATP} \xrightarrow[\text{Water}]{} \text{ADP + Energy}$$

The two primary categories of respiration (Fig. 3.8) are as follows:

- **Aerobic respiration:** This occurs when oxygen is present. It generates more power. Water, energy and carbon dioxide are the by-products. It takes place in most organisms. In aerobic respiration, oxygen-containing glucose is transformed into pyruvate in the cytoplasm; it is then converted to carbon dioxide, water and energy in the mitochondria.

$$\text{Glucose} \xrightarrow[\text{in cytoplasm}]{\text{Presence of oxygen}} \text{Pyruvate} \xrightarrow[\text{in mitochondria}]{\text{Presence of oxygen}} CO_2 + H_2O + \text{Energy}$$

- **Anaerobic respiration:** This occurs when no oxygen is present. Less energy is produced. Carbon dioxide, energy and lactic acid or ethanol are the by-products. It takes place in the muscle cells and in yeast. In of anaerobic respiration, glucose is converted into pyruvate in the muscle cells, and when oxygen is not present, pyruvate is transformed into lactic acid and energy.

$$\text{Glucose} \xrightarrow[\text{in cytoplasm}]{\text{Presence of oxygen}} \text{Pyruvate} \xrightarrow[\text{in muscle cell}]{\text{Absence of oxygen}} \text{Lactic acid + Energy}$$

In yeast, glucose is converted into pyruvate, and in the absence of oxygen, pyruvate is transformed into

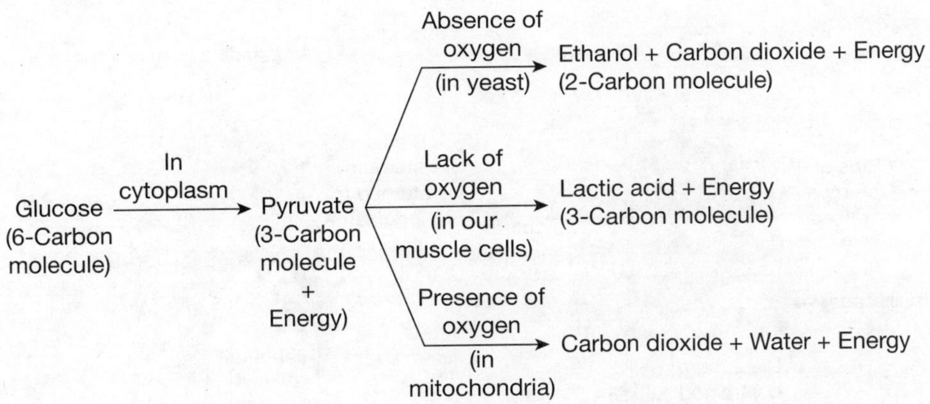

Fig. 3.8 Breakdown of glucose by various pathways

ethanol, carbon dioxide and energy. This process is called **fermentation**.

Glucose —(Presence of oxygen / in cytoplasm)→ Pyruvate —(Absence of oxygen / in yeast)→ Ethanol + CO_2 + Energy

Respiration in Humans

The main organs of the respiratory system are the nostrils, nasal cavity, pharynx, larynx, trachea, bronchi, bronchioles, lungs and diaphragm (Fig. 3.9).

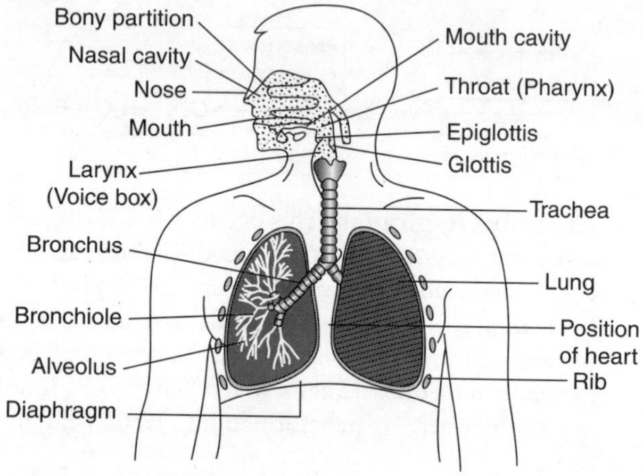

Fig. 3.9 Human respiratory system

Air enters the body through the nostrils. The dust particles are captured by hairs and mucus. Air then enters the lungs after passing through the pharynx, larynx, trachea and bronchi. When there is no air in the trachea, the cartilage rings around it keep it from collapsing. The bronchi divide into smaller tubes known as bronchioles, and these tubes end in tiny air sacs known as alveoli. Blood arteries that supply the alveoli allow for the exchange of gases (Fig. 3.10). The alveoli provide a larger surface area for gas exchange.

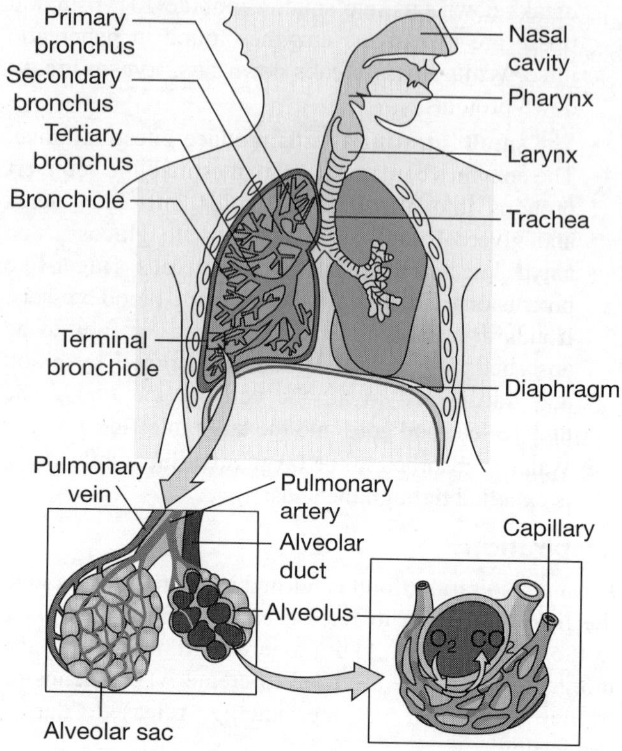

Fig. 3.10 Exchange of gases

Breathing mechanism: When we breathe in air, the diaphragm muscles contract and move downwards, the chest cavity expands and air is drawn into the lungs. When we exhale, the diaphragm's muscles relax and move upwards, the chest cavity closes and air exits the lungs (Fig. 3.11).

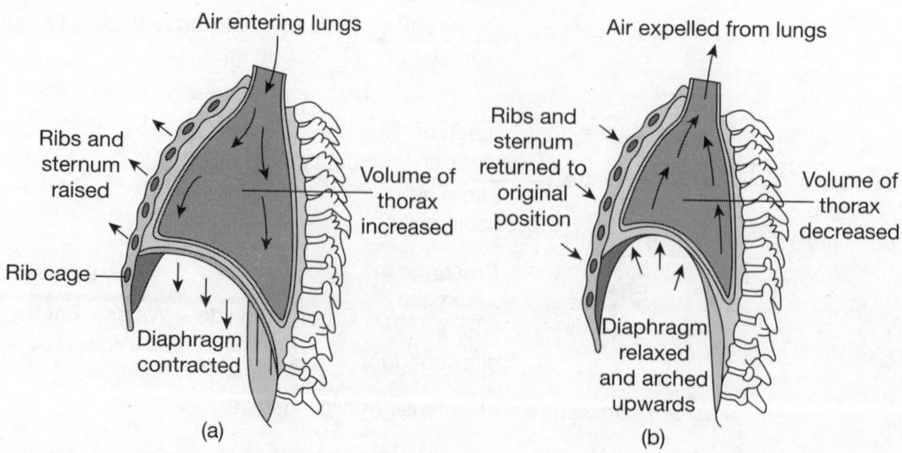

Fig. 3.11 Mechanism of breathing: (a) inhalation, (b) exhalation

Transportation

Transportation in Humans

The circulatory system serves as the primary means of transportation in humans. It is made up of the heart, blood, arteries, veins and capillaries.

- Blood carries nutrients, oxygen and waste materials. Red blood cells (RBCs), white blood cells (WBCs), platelets and plasma make up this substance.
- Food, water, carbon dioxide and nitrogenous waste are all transported by plasma.
- Oxygen is transported by RBCs. WBCs defend the body by eradicating harmful bacteria. Platelets help in clotting of blood and prevent loss of blood during injury.
- **Arteries** are vessels that distribute oxygenated blood throughout the body from the heart. They lack valves and have robust walls.
- **Veins** transport deoxygenated blood to the heart from all the parts of the body. They have valves and thin walls.
- **Capillaries** are tiny blood vessels that connect the veins and arteries. They provide for the exchange of nutrients, liquids, oxygen, carbon dioxide and other substances between the blood and cells.
- The muscular organ that pumps blood to every region of the body is the **heart**. It is made up of four chambers (Fig. 3.12). The ventricles are the lower chambers, and the atria are the upper chambers. The walls of the ventricles are thicker than those of the atria because they pump blood to various organs. The septum divides the right and left chambers. Blood with and without oxygen cannot mix as a result. There are valves between the atria and ventricles that stop blood from flowing backwards.
- **Working of the heart (circulation):** Blood that is rich in carbon dioxide is brought into the right atrium and subsequently transferred into the right ventricle. This blood is pumped from the right ventricle to the lungs, where carbon dioxide is exhaled and oxygen is taken in. Oxygen-rich blood is received in the left atrium. This blood is circulated throughout the body by the left ventricle (Fig. 3.13).

  Right atrial contraction causes the right ventricle to enlarge and receive blood. The right ventricle then contracts, forcing blood into the pulmonary artery and into the lungs. The procedure is repeated as the blood reaches the left atrium after being oxygenated and after carbon dioxide has been removed from it in the lungs. **Double circulation** refers to the fact that blood passes twice through the heart in a single cycle.
- A colourless fluid called **lymph** is found in the gaps between the cells. It is formed of the plasma that escapes from the capillaries. Lymph drains into the lymphatic capillaries, where it conjoins with other lymph fluid to form lymph channels. Digested fats

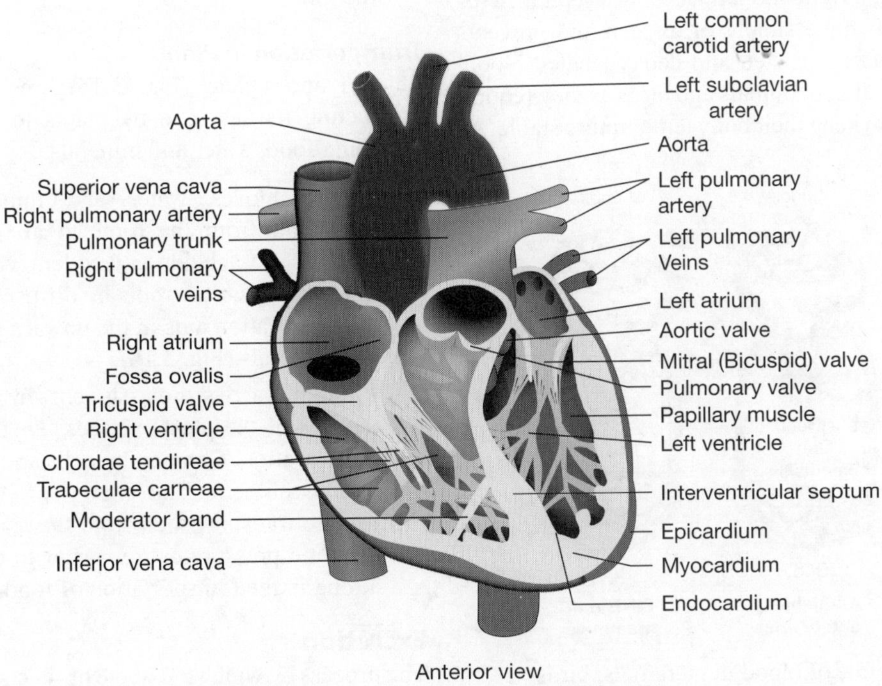

Fig. 3.12 Internal structure of human heart

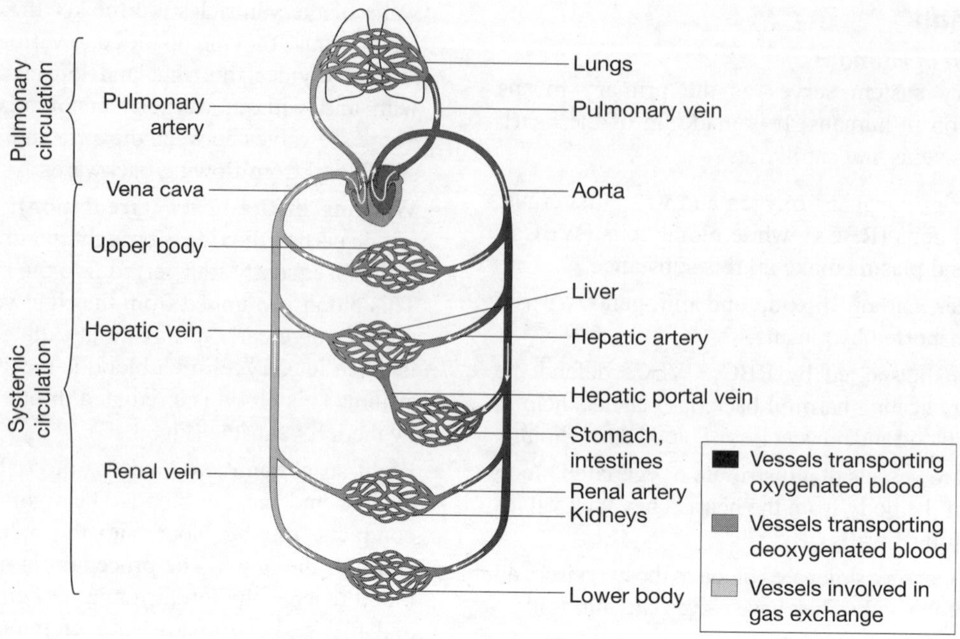

Fig. 3.13 Working of the heart

are transported by lymph, which also returns extra fluid from intercellular spaces to the circulation. It contains lymphocytes, which fight bacteria and defend the body.

- **Hearts of other animals** (Fig. 3.14): Mammals and birds have four chambers in their heart, which are divided into the right and left sides by a septum. This ensures an effective supply of oxygen and prevents the mixing of oxygenated and deoxygenated blood. This is essential in mammals and birds as they require more energy to keep their body temperature stable. As amphibians and reptiles do not use energy to maintain their body temperature, their three-chambered hearts allow some oxygenated and deoxygenated blood to mix. Their internal temperature is the same as the ambient temperature. Fish only have two chambers in their heart, and their blood is oxygenated in the gills.

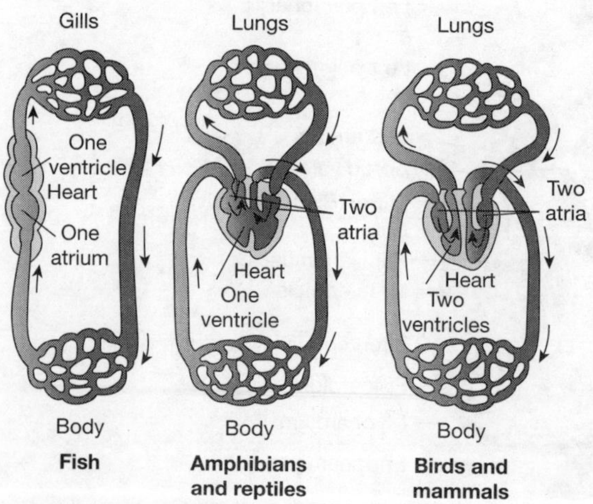

Fig. 3.14 Flow of blood in mammals, birds, amphibians, reptiles and fish

Transportation in Plants

Phloem and xylem (Fig. 3.15) are the two types of conducting tissues used by plants to transfer elements including food, water and minerals.

- **Xylem** moves water and minerals throughout the plant, from the roots to all other sections. It comprises tracheids and xylem vessels. Water and minerals reach the roots by diffusion. Suction owing to transpiration aids in the upward transport of water and minerals (Fig. 3.16).
- **Phloem** carries nourishment from the leaves to the rest of the plant's tissues. This process is called translocation. Companion cells and sieve tubes make up the phloem. The energy of ATP molecules is used to transport food from the leaves to the xylem. Osmotic pressure causes water to enter the phloem, aiding in the transportation of food.

Excretion

The process by which the waste products generated during metabolic processes are expelled from the body is known

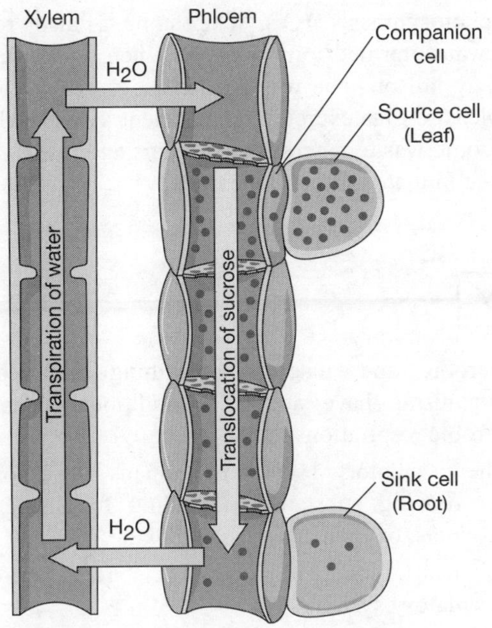

Fig. 3.15 Structure of xylem and phloem

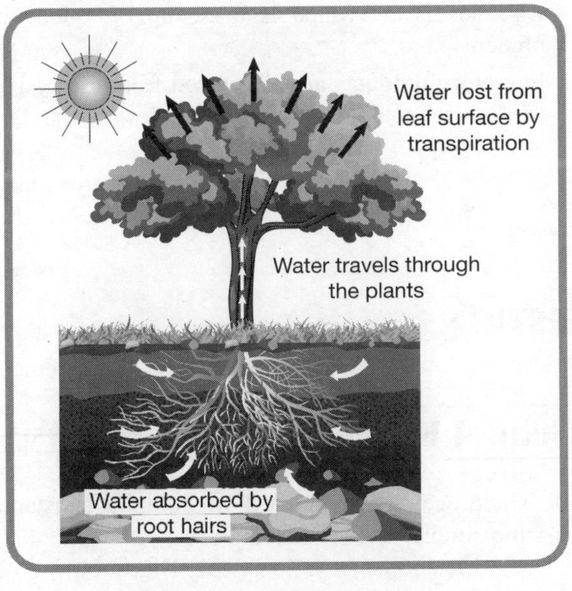

Fig. 3.16 Transport in plants

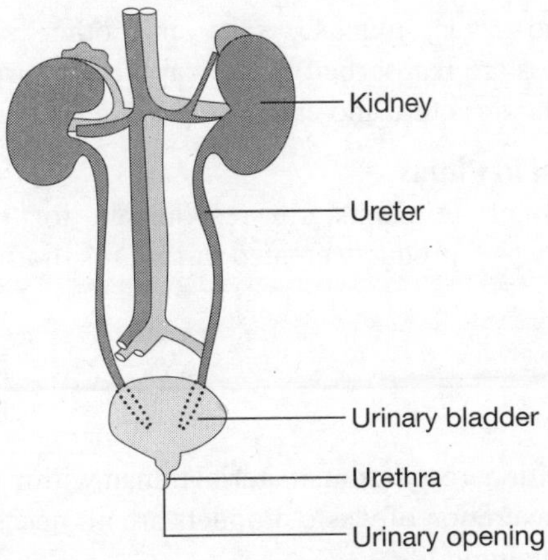

Fig. 3.17 Human excretory system

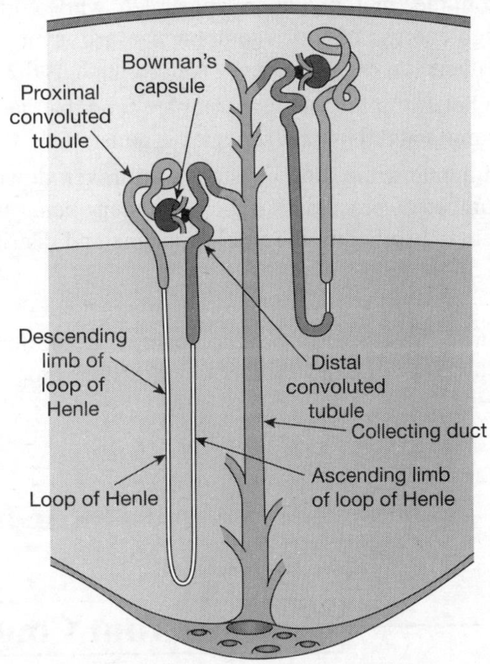

Fig. 3.18 Structure of a nephron

as excretion. In a single-celled organism, waste products are diffused out of the cells and into the environment. Waste products are eliminated from multicellular organisms by specialised organs.

Excretion in Humans

The urinary bladder, urethra, kidneys and ureters make up the excretory system (Fig. 3.17). Nephrons, or excretory organs, are found in each kidney (Fig. 3.18).

The mammalian nephron, which ranges about 35–55 mm in length, is long and tube-like. The tube is closed, folded and inflated at one end to form the Bowman's capsule, also known as the renal corpuscular capsule. This encloses the glomerulus, which is a collection of tiny blood arteries. The capsule and glomerulus together constitute the renal corpuscle. The renal artery transports extra water, salts and nitrogenous waste such as ammonia, urea and uric acid (urine) into the nephron. It filters the salts, water and nitrogenous waste that pass through the tubular structure and into the collecting duct. The waste is subsequently transported by the ureters into the bladder, where it is then released as urine through the urethra.

Amino acids, glucose, salts and other beneficial substances are reabsorbed by the capillaries surrounding the tubular structure and enter the proximal tubule.

Excretion in Plants

The stomata in plants are responsible for removing gaseous waste products created during respiration (CO_2) and photosynthesis (O_2). The stomata also remove extra water, termed transpiration. When leaves become dry, they fall off. The waste materials stored in the leaf vacuoles are thus excreted. The ancient xylem cells also store some waste materials such gums and resins. Waste is also eliminated through the roots.

SUMMARY

- Nutrition, respiration, material transit within the body and excretion of waste products are all necessary for maintaining life.
- In autotrophic feeding, simple inorganic materials from the environment are consumed, while complex, high-energy organic components are synthesised utilising an external energy source, such as the sun.
- In heterotrophic feeding, complex food that has been manufactured by other species is consumed.
- In humans, the food consumed is broken down in a number of processes in the alimentary canal before being absorbed in the small intestine and distributed to all the cells.
- Organic substances like glucose are broken down during breathing to produce energy in the form of ATP. ATP is utilised to fuel more cellular processes.
- Aerobic and anaerobic breathing are possible. Organisms have access to additional energy in aerobic respiration.
- The circulatory system in humans is responsible for moving substances including food, excretory products, oxygen and carbon dioxide.
- The heart, blood and blood vessels are all part of the circulatory system.
- Transport of water, minerals, food and other elements occurs in highly differentiated plants thanks to the vascular tissue, which is made up of xylem and phloem.
- Excretory products are eliminated from the body in the form of soluble nitrogen molecules.

PRACTICE QUESTIONS

Analyse Your Concepts (School Exam Based)

Multiple Choice Questions

1. Most of the digestion and absorption of food takes place in the:
   (a) Small intestine (b) Liver
   (c) Stomach (d) Large intestine
2. Anaerobic processes:
   (a) Take place during fermentation in yeast
   (b) Take place in the presence of oxygen
   (c) Produce only energy in the muscles of humans
   (d) Produces ethanol, oxygen and energy
3. The exchange of gases in plants occurs through:
   (a) Stomata (b) Lenticels
   (c) Roots (d) Flowers
4. The process of breaking down complex substances into simpler ones is called:
   (a) Metabolism (b) Digestion
   (c) Respiration (d) Assimilation
5. The hormone responsible for the regulation of blood sugar level in humans is:
   (a) Insulin (b) Glucagon
   (c) Adrenaline (d) Estrogen

Match the Following

Each question contains statements given in two columns which have to be matched. Statements in column I have to be matched with statements in column II.

1. Match the following with the correct response:

Column I	Column II
(1) The red colour of human blood	(A) Chlorophyll
(2) The pigment that absorbs solar energy	(B) Heart
(3) The largest gland of the human body	(C) Liver
(4) The pumping organ of the body	(D) Haemoglobin

(a) 1-A, 2-C, 3-B, 4-D (b) 1-B, 2-D, 3-A, 4-C
(c) 1-C, 2-B, 3-D, 4-A (d) 1-D, 2-A, 3-C, 4-B

2. Match the following with the correct response:

Column I	Column II
(1) Flatworms	(A) Nephridia
(2) Arthropods	(B) Flame cells
(3) Earthworm	(C) Kidney
(4) Mammals	(D) Malpighian tubules

(a) 1-C, 2-B, 3-D, 4-A (b) 1-A, 2-C, 3-B, 4-D
(c) 1-D, 2-A, 3-C, 4-B (d) 1-B, 2-D, 3-A, 4-C

3. Match the following with the correct response:

Column I	Column II
(1) Tissue fluid	(A) Lymph
(2) Atria	(B) A surface for gaseous exchange
(3) Alveoli	(C) Chamber of the heart
(4) Functional unit of kidney	(D) Nephron

(a) 1-A, 2-C, 3-B, 4-D (b) 1-C, 2-B, 3-D, 4-A
(c) 1-D, 2-A, 3-C, 4-B (d) 1-B, 2-D, 3-A, 4-C

5. Match the following with the correct response:

Column I	Column II
(1) Pulmonary artery	(A) Oxygenated blood
(2) Pulmonary vein	(B) Exchange of substances
(3) Aorta	(C) Supplies blood to body tissues/parts
(4) Capillary	(D) Deoxygenated blood

(a) 1-C, 2-B, 3-D, 4-A (b) 1-D, 2-A, 3-C, 4-B
(c) 1-B, 2-D, 3-A, 4-C (d) 1-A, 2-C, 3-B, 4-D

6. Match the following with the correct response:

Column I	Column II
(1) Insects	(A) Gills
(2) Earthworm	(B) Trachea
(3) Fish	(C) Lungs
(4) Mammals	(D) Skin

(a) 1-B, 2-D, 3-A, 4-C (b) 1-A, 2-C, 3-B, 4-D
(c) 1-D, 2-A, 3-C, 4-B (d) 1-C, 2-B, 3-D, 4-A

7. Match the following with the correct response:

Column I	Column II
(1) Saliva	(A) Sucrose
(2) Gastric juice	(B) Pepsin
(3) Pancreatic juice	(C) Amylase
(4) Succus entericus	(D) Trypsin

(a) 1-B, 2-D, 3-A, 4-C (b) 1-C, 2-B, 3-D, 4-A
(c) 1-D, 2-A, 3-C, 4-B (d) 1-A, 2-C, 3-B, 4-D

8. Match the following with the correct response:

Column I	Column II
(1) Anaerobic oxidation of glucose	(A) 2 ATP
(2) Aerobic oxidation of glucose	(B) Ethyl alcohol
(3) End product of fermentation	(C) 38 ATP
(4) Anaerobic muscular respiration	(D) Lactic acid

(a) 1-C, 2-B, 3-D, 4-A (b) 1-A, 2-C, 3-B, 4-D
(c) 1-D, 2-A, 3-C, 4-B (d) 1-B, 2-D, 3-A, 4-C

True or False

Read the following statements and write your answer as true or false.

1. Autotrophs include some bacteria as well.
2. Stomata close at very high temperature.
3. Carbohydrate is converted into starch before it can be used by plants.
4. Variegated leaves show many colours.

5. After splitting of the water molecule, oxygen is utilised for the synthesis of carbohydrates.
6. Transpiration helps in supplying water to the leaves.
7. Vultures feed on dead animals, hence they are saprotrophs.
8. Mushroom is an example of a saprotroph.
9. Salivary glands are found inside the tongue.
10. Partial digestion of protein takes place in the stomach.
11. Bile contains the digestive enzyme for fat.
12. Food needs to be broken into smaller pieces before it can be swallowed.
13. Digestion takes place in the small intestine and absorption occurs in the large intestine.
14. Tooth enamel is the hardest substance in the human body.
15. Parasites do not kill their prey in order to obtain food.

Fill in the Blanks

Complete the following statements with an appropriate word/term to be filled in the blank spaces.

1. The process by which an organism takes in food and breaks it down into simpler substances is called ________.
2. The movement of water from an area of high concentration to an area of low concentration through a semi-permeable membrane is called ________.
3. The process by which green plants make their own food is called ________.
4. The process by which cells obtain energy from glucose in the presence of oxygen is called ________.
5. The process by which cells obtain energy from glucose in the absence of oxygen is called ________.

Assertion–Reason Questions

The following questions consists of two statements: Assertion (A) and Reason (R). Answer these questions selecting the appropriate option given below:

(a) Both A and R are true and R is the correct explanation of A
(b) Both A and R are true but R is not the correct explanation of A
(c) A is true but R is false
(d) A is false but R is true

1. Assertion (A): Energy is required to carry out various life processes.
   Reason (R): Energy is found in the form of ATP in mitochondria.
2. Assertion (A): Anaerobic respiration utilises more energy than aerobic respiration.
   Reason: Mitochondria is the powerhouse of the cell.
3. Assertion (A): During the day, photosynthesis consumes the CO_2 produced during respiration.
   Reason: CO_2 is not released during the day
4. Assertion (A): The process of respiration consumes energy.
   Reason (R): The process of respiration uses energy in the form of ATP.
5. Assertion (A): Transpiration is a necessary process.
   Reason (R): Water is lost, but it also aids in the uptake and transfer of minerals and water.

One-word Answers

1. What is the process of conversion of food into soluble substances?
2. What is the pigment present in green plants that helps in photosynthesis?
3. What is the process by which glucose is converted into energy in the presence of oxygen?
4. What is the process by which plants release excess water through stomata?
5. What is the waste product of cellular respiration?

Define the Term

1. Photosynthesis.
2. Aerobic respiration
3. Excretion
4. Transpiration
5. Diffusion

Groups of Terms

Given below are certain groups of terms. In each group the first pair indicates a relationship between the two terms. Rewrite and complete the second pair on a similar basis.

1. Autotrophic : Plants :: Heterotrophic : ________
2. Mitosis : Two identical cells :: Meiosis : ________
3. Photosynthesis : Oxygen :: Respiration : ________

4. Stomata : Gaseous exchange :: Lenticels : ________

5. Chlorophyll : Green pigment :: Haemoglobin : ________

Identify the Odd One Out

1. Liver, stomach, kidneys, heart
2. Adrenaline, insulin, glucagon, urea
3. Glucose, sucrose, starch, enzyme
4. Red blood cells, white blood cells, platelets, blood plasma
5. Diffusion, osmosis, respiration, transpiration.

Image-Based Questions

1. Explain the process depicted in the diagram.

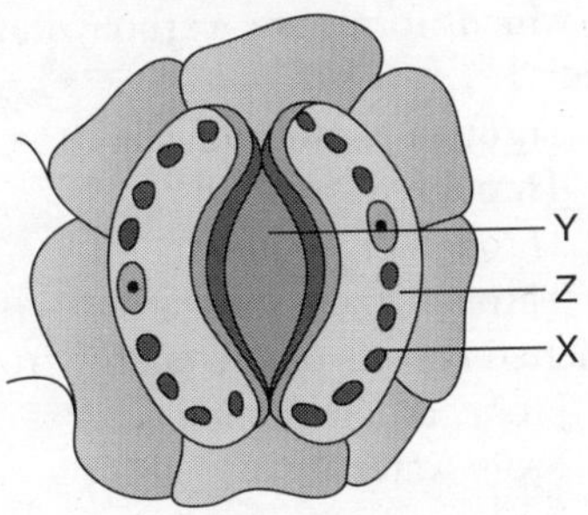

2. In the given transverse section of the leaf, identify the layer of cells where maximum photosynthesis occurs:

(a) I, II (b) II, III
(c) III, IV (d) I, IV

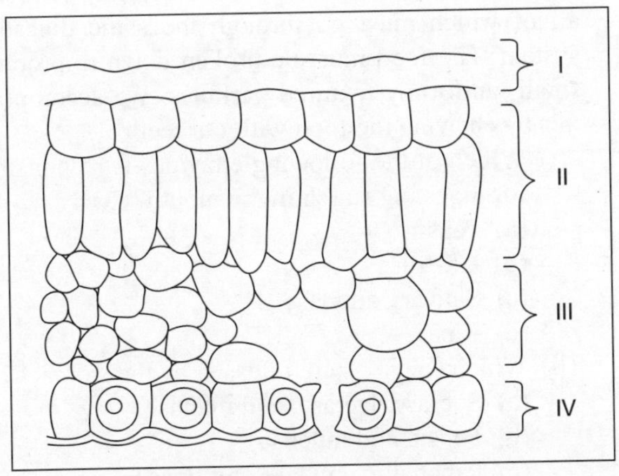

3. Identify the option that indicates the correct enzyme that is secreted in locations i, ii and iii from the given figure:

(a) i. Lipase, ii. Trypsin, iii. Pepsin
(b) i. Amylase, ii. Pepsin, iii. Trypsin
(c) i. Trypsin, ii. Amylase, iii. Carboxylase
(d) i. Permease, ii. Carboxylase, iii. Oxidase

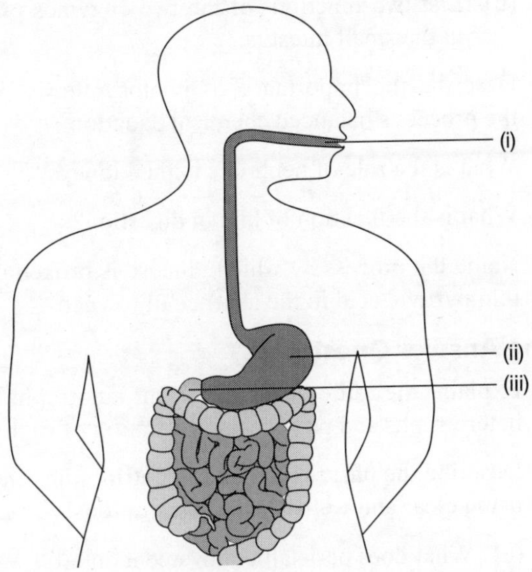

4. Explain the points in the diagram of the human digestive system.

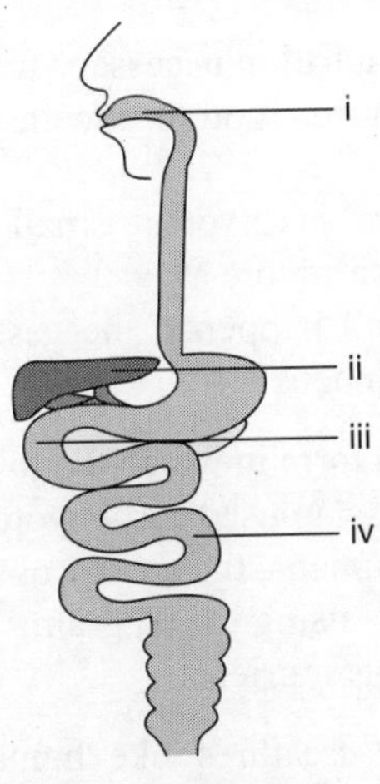

Very Short Answer Questions

1. List the essential components for photosynthesis.
2. Describe the location and purpose of the gastric glands.
3. What are the glands that discharge fluids into the stomach wall to aid in food digestion? List the three secretions that these glands produce.
4. What is the role of enzymes in digestion?
5. What is the function of haemoglobin in blood?

Short Answer Questions

1. (a) Describe how the following contribute to the digestion process:
   (i) Enzyme trypsin (ii) Enzyme lipase
   (b) List two functions of the two enzymes present in the small intestine.
2. Describe the importance of photosynthesis. Write the process's balanced chemical equation.
3. What is the role of nephrons in the kidneys?
4. What is the function of bile in digestion?
5. Name the process by which glucose is broken down into pyruvic acid in the absence of oxygen.

Long Answer Questions

1. Explain the differences between autotrophs and heterotrophs and provide an illustration of each.
2. Describe the phases involved in nutrition in amoeba using clear and well-labelled illustrations.
3. (a) What does peristaltic movement entail?
   (b) "Stomata in desert plants remain closed during the day". How do such plants carry out photosynthesis?
4. (a) Why is nutrition necessary for the human body?
   (b) What causes food to flow inside the alimentary canal?
   (c) Why are herbivores' small intestines longer than in carnivores?
   (d) What will happen if the gastric glands do not secrete mucus?
5. (a) State the form in which the following are stored:
   (i) Plants' unused carbohydrates
   (ii) In humans, the energy obtained from eating
   (b) Explain, using a diagram, how an amoeba obtains its nutrients.
6. Multicellular creatures like humans require more oxygen than diffusion can provide. List the reasons.
7. Write down two ways that the human body oxidises glucose to produce energy.
8. (a) Describe the role of alveoli in the respiratory process.
   (b) Complete the following pathway showing the break-down of glucose.

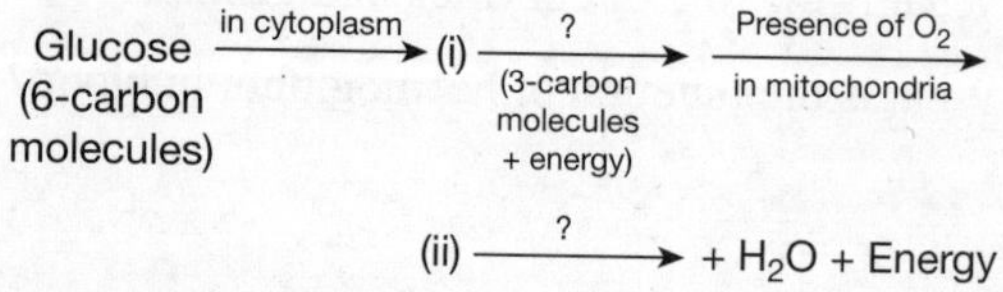

9. Draw a flow chart to show the breakdown of glucose by various pathways.
10. Write the three types of blood vessels. Give an important feature of each.
11. What do the following transport?
    (a) Xylem (b) Phloem
    (c) Pulmonary vein (d) Vena cava

Case Study-Based Questions

1. The autotrophic organism's needs for carbon and energy are met through photosynthesis. It is the method by which autotrophs consume materials from the environment and transform them into forms of energy that may be stored. This substance is consumed as carbon dioxide and water, which, when exposed to sunshine and chlorophyll, are transformed into carbohydrates. Plants obtain energy from carbohydrates.
   (i) In which form are carbohydrates stored in plants?
   (a) Through photosynthesis
   (b) Through respiration
   (c) Through digestion
   (d) Through excretion
   (ii) Where do plants obtain energy from?
   (a) From carbon dioxide
   (b) From water
   (c) From sunlight
   (d) From oxygen
2. The alimentary canal is a long tube that runs from the mouth to the anus. Various components make up the tube. Different locations are specialised to carry out particular tasks. We consume a variety of foods, all of which must go through the same digestive system. The food must be broken down to produce small, uniformly textured particles. We accomplish this by chewing the food with our teeth.
   (i) Which of the following enzymes is responsible for digesting starch in the mouth?
   (a) Pepsin
   (b) Gastrin
   (c) Salivary amylase
   (d) Lipase
   (ii) Which enzymes are released by the stomach?
   (a) Sucrase, lactase, and maltase
   (b) Pepsin and mucus
   (c) Pancreatic amylase and lipase
   (d) Trypsin and chymotrypsin
   (iii) What are villi?
   (a) Small hair-like structures found in the lungs
   (b) Finger-like projections in the inner membrane of the small intestine that increase surface area for absorption

(c) Specialised cells found in the liver that produce bile
(d) Large muscles that contract to move food through the large intestine

3. Humans have two kidneys, two ureters, a urinary bladder and a urethra as part of their excretory system. One kidney is situated on either side of the backbone in the abdomen. The urine that the kidneys make travels down the ureters and into the bladder, where it is held until it is expelled by the urethra.
   (i) What is the significance of urine?
      (a) It helps in digestion
      (b) It provides energy to the body
      (c) It removes out impurities from the blood. For example, urea.
      (d) It helps in the production of hormones.
   (ii) What do you understand by Bowman's capsule?
      (a) A muscular organ in the human body
      (b) A capillary-connected cup-shaped end of a coiled tube that collects filtrate.
      (c) A gland responsible for regulating body temperature
      (d) A type of cell in the nervous system

4. The heart is a muscular organ the size of our fist. It has separate chambers to prevent oxygen-rich blood from mixing with the blood containing carbon dioxide, since both oxygen and carbon dioxide must be delivered by blood. For carbon dioxide to be expelled from the body, the blood must first reach the lung and then oxygen-rich blood from the lungs must be transported back to the heart. The rest of the body is then supplied with this oxygen-rich blood.
   (i) How many chambers does a monkey heart have?
      (a) One (b) Four (c) Three (d) Two
   (ii) How many chambers does a reptile heart have?
      (a) One (b) Two (c) Three (d) Four
   (iii) What structure(s) carry deoxygenated blood to the heart?
      (a) Superior vena cava and inferior vena cava
      (b) Pulmonary veins
      (c) Aorta
      (d) Coronary arteries
   (iv) What device(s) is/are used to measure blood pressure?
      (a) Sphygmomanometer
      (b) Electrocardiogram
      (c) Magnetic resonance imaging
      (d) X-ray

5. The amount of blood corpuscles is changed in dengue fever. One of the common symptoms observed in people infected with dengue fever is:
   (a) Significant decrease in RBC count
   (b) Significant decrease in WBC count
   (c) Significant decrease in platelet count
   (d) Significant increase in platelet count

NCERT Corner

1. Define nutrition. What are the different modes of nutrition?

2. What is the role of saliva in the digestion of food? What happens to food in the stomach?

3. How is oxygen and carbon dioxide exchanged in the human body? Explain the process of respiration in humans.

4. Define excretion. What are the different modes of excretion in organisms?

5. What is transpiration? Explain the process of transportation of water in plants.

COMPETITION WINDOW (OBJECTIVE TYPE)

1. Where is food digested in an amoeba?
   (a) Food vacuole (b) Mitochondria
   (c) Pseudopodia (d) Chloroplast

2. Which of the following types of organisms breaks down and absorbs food ingredients outside the body?
   (a) Mushroom, green plants, amoeba
   (b) Yeast, mushroom, bread mould
   (c) *Paramecium*, *Amoeba*, *Cuscuta*
   (d) *Cuscuta*, lice, tapeworm

3. What outputs does anaerobic respiration in plants produce?

(a) Lactic acid + Energy
(b) Carbon dioxide + Water + Energy
(c) Ethanol + Carbon dioxide + Energy
(d) Pyruvate

4. What does change of rice water to blue upon addition of iodine imply?
(a) Fats (b) Complex proteins
(c) Starch (d) Simple proteins

5. What assists respiration in humans?
(a) Carotene (b) Chlorophyll
(c) Haemoglobin (d) Mitochondria

6. Which of the following describes emphysema's key feature?
(a) Damage to the lungs' air sacs and destruction of the alveolar wall
(b) Acceleration of lung tissue growth
(c) Bronchitis or inflammation of the wall of the bronchi
(d) Thickening of the walls of the pulmonary artery

7. How does fungus get its nourishment?
(a) By consuming the bread it grows on
(b) By utilising the bread's nutrients to make its own food
(c) By digesting and absorbing the nutrients in bread
(d) By letting other organisms develop on the bread before eating them

8. Stomatal pore opening and shutting are influenced by:
(a) Oxygen
(b) Temperature
(c) Water in the guard cells
(d) Concentration of CO_2

9. How do plants get rid of solid waste products?
(a) Shortening of stem
(b) Shedding of fruits
(c) Shedding of yellow leaves
(d) Extending roots into the soil

10. Where is digestion of food initiated?
(a) Mouth
(b) Oesophagus
(c) Pancreas
(d) Stomach

Answer Keys

Multiple Choice Questions

1. (a) 2. (a) 3. (a) 4. (d) 5. (a)

Match the Following

1. (d) 2. (d) 3. (a) 4. (a) 5. (b)
6. (a) 7. (b) 8. (b)

True or False

1. T 2. T 3. F 4. T 5. F
6. T 7. F 8. T 9. F 10. T
11. F 12. T 13. F 14. T 15. T

Fill in the Blanks

1. Digestion
2. Osmosis
3. Photosynthesis
4. Aerobic respiration
5. Anaerobic respiration

Assertion-Reason Questions

1. (a) 2. (d) 3. (a) 4. (d) 5. (a)

One-word Answers

1. Digestion
2. Chlorophyll
3. Respiration
4. Transpiration
5. Carbon dioxide

Define the Terms

1. Photosynthesis is the process by which green plants use sunlight, carbon dioxide and water to synthesize their own food in the form of glucose and release oxygen as a by-product.
2. Aerobic respiration is the process by which cells obtain energy from glucose in the presence of oxygen. This process involves the breakdown of glucose into carbon dioxide and water, and the release of energy in the form of ATP.
3. Excretion is the process by which metabolic waste products and excess substances are eliminated from the body of an organism. This process is essential to maintain the internal environment of the body.
4. Transpiration is the process by which excess water is released from the aerial parts of the plant, mainly through the stomata present on the leaves. This process helps in the absorption of water and minerals by the roots.
5. Diffusion is the process by which molecules or ions move from a region of higher concentration to a region of lower concentration, until they are evenly distributed. This process occurs due to the random motion of particles and does not require any external energy.

Group of Terms

1. Animals
2. Four haploid cells
3. Carbon dioxide
4. Gaseous exchange
5. Red pigment

Identify the Odd One Out

1. Heart
2. Urea
3. Enzyme
4. Blood plasma
5. Respiration

Image-Based Questions

1. X-Chloroplast, Y-Stomatal pore, Z-Guard cell
2. (b)
3. i. b. Initial phase of starch digestion.
   ii. c. Increases the efficiency of lipase enzyme action.
   iii. d. This is the site of the complete digestion of carbohydrates, proteins and fats.
   iv. The length of this depends on the food the organism eats.
4. i - Mouth, ii - Liver , iii - Duodenum, iv - Large intestine

Very Short Answer Questions

1. Carbon dioxide (CO_2), water, light and chloroplasts are the raw components needed for photosynthesis.
2. These are digestive glands found in the wall of the stomach. They release gastric fluids that contain hydrochloric acid, pepsin, rennin and mucus, which break down proteins (HCl).
3. Gastric glands are located in the muscular wall of the stomach. These glands produce gastric secretions that include mucus, the two protein-digesting enzymes rennin and pepsin, as well as diluted hydrochloric acid.
4. The role of enzymes in digestion is to break down complex food molecules into simpler ones that can be absorbed by the body.

5. The function of haemoglobin in the blood is to transport oxygen from the lungs to the body tissues.

Short Answer Questions

1. (a) (i) The inactive version of trypsin, known as trypsinogen, is produced by the pancreas. The remaining proteins are broken down by trypsin into peptones, which are further broken down into peptides and amino acids.
   (ii) The enzyme lipase is produced in the pancreas and small intestine. Fats are converted by lipase into fatty acids and glycerol.
   (b) The small intestine's inner wall is equipped with long projections that resemble fingers; these are referred to as villi. The villi significantly increase the absorptive surface area of the small intestine's inner lining. The small intestine's vast surface area aids in the quick absorption of food that has been digested.
2. Photosynthesis is significant for the following reasons:
   (i) Food: Green plants use photosynthesis to create food from basic raw ingredients such as CO_2 and H_2O. Thus, it sustains life on Earth.
   (ii) Oxygen: All living being need the oxygen produced during photosynthesis in order to breathe. Oxygen aids in the burning of fuels as well. Fossil fuels like coal, oil and natural gas were created millions of years ago through photosynthesis and are sources of stored solar energy.

   The balanced chemical equation for photosynthesis is as follows:

$$\underset{\text{Carbon dioxide}}{6CO_2} + \underset{\text{Water}}{12H_2O} \xrightarrow[\text{Sunlight, enzymes}]{\text{Chlorophyll}} \underset{\text{Glucose}}{C_6H_{12}O_6} + 6H_2O + 6O_2\uparrow$$

3. Nephrons are the functional units of the kidneys, responsible for the filtration of waste products from the blood, and the regulation of water and electrolyte balance in the body. Nephrons are made up of two main parts: the renal corpuscle and the renal tubule. The nephrons also play an important role in maintaining acid-base balance in the body by secreting hydrogen ions and reabsorbing bicarbonate ions. This helps to regulate the pH of the blood.
4. Bile is a digestive juice that is produced by the liver and stored in the gallbladder. It plays an important role in digestion by emulsifying fats into smaller droplets, which can be easily digested by lipase enzyme.When food containing fat enters the small intestine, bile is secreted into the intestine through the common bile duct. Bile contains bile salts, which help to emulsify the fats present in the food. Emulsification breaks the fats into smaller droplets, increasing their surface area, which allows the enzyme lipase to act on them more effectively, breaking them down into fatty acids and glycerol.
5. The process by which glucose is broken down into pyruvic acid in the absence of oxygen is called glycolysis. It is an anaerobic process that occurs in the cytoplasm of cells. During glycolysis, glucose is broken down into two molecules of pyruvic acid, along with the production of a small amount of ATP. The process involves a series of reactions, which are catalyzed by specific enzymes.

Long Answer Questions

1. Hint: Autotrophs are organisms that can make their own food using light energy or chemical energy, such as plants, algae and some bacteria. Heterotrophs are organisms that cannot make their own food and depend on other organisms for food, such as animals, fungi and some bacteria. Elaborate on this.
2. Hint: Nutrition in amoeba involves:
   Ingestion: taking in food by pseudopodia; Digestion: breaking down food by enzymes in food vacuoles; Absorption: passing digested food to the cytoplasm; Assimilation: using absorbed food for growth, repair and energy; Egestion: removing undigested food by cell membrane rupture.

   Elaborate on these points.
3. Hints: (a) Peristaltic movement entails the rhythmic contraction and relaxation of muscles in the walls of the alimentary canal that pushes the food forward.

   (b) Stomata in desert plants remain closed during the day to prevent water loss due to transpiration. Such plants carry out photosynthesis by storing carbon dioxide at night in organic acids and using it during the day when light is available.

   Elaborate on these points.
4. Hints: (a) Nutrition is necessary for the human body to provide energy for various activities, to supply raw materials for growth and repair of tissues, and to regulate various metabolic processes.

   (b) Food flows inside the alimentary canal due to peristaltic movement of muscles in its walls.

   (c) Herbivores' small intestines are longer than in carnivores because they need more time and surface area to digest plant matter which is rich in cellulose and complex carbohydrates.

   (d) If the gastric glands do not secrete mucus, the

gastric juice containing hydrochloric acid and digestive enzymes may damage the inner lining of stomach and cause ulcers.

Elaborate on these points.

5. Hints: (a) The form in which the following are stored are:

(i) Plants' unused carbohydrates are stored as starch in roots, stems and leaves. (ii) In humans, the energy obtained from eating is stored as glycogen in liver and muscles or as fats in adipose tissues.

(b) An amoeba obtains its nutrients by ingesting food particles through pseudopodia and forming a food vacuole around them. The food vacuole fuses with a lysosome that contains digestive enzymes. The enzymes break down the food into simpler substances that are absorbed into the cytoplasm. The undigested food is expelled out of the cell by egestion.

Elaborate on these points.

6. Hint: Multicellular organisms like humans require more oxygen than diffusion can provide because:

They have larger body size and higher metabolic rate than unicellular organisms.

They have specialised cells and tissues that perform different functions and need a constant supply of oxygen.

They have low surface area to volume ratio that limits the rate of diffusion across their body surface.

Elaborate on these points.

7. Hint: Two ways that the human body oxidises glucose to produce energy are by:

Aerobic respiration: It occurs in the presence of oxygen and involves glycolysis, Krebs cycle and electron transport chain. It produces 36 ATP molecules per glucose molecule along with carbon dioxide and water as byproducts.

Anaerobic respiration: It occurs in the absence of oxygen and involves glycolysis followed by fermentation. It produces 2 ATP molecules per glucose molecule along with lactic acid or ethanol and carbon dioxide as byproducts.

Elaborate on these points.

8. Hints: (a) The role of alveoli in the respiratory process is to provide a large surface area for gas exchange between air and blood. The alveoli are thin-walled sacs surrounded by capillaries that carry deoxygenated blood from the heart. Oxygen diffuses from alveoli into blood while carbon dioxide diffuses from blood into alveoli.

(b) The following pathway shows the break-down of glucose:

Glucose (6-carbon molecules) → Pyruvate (3-carbon molecules energy) (presence of O_2) → Acetyl-CoA (2-carbon molecules) → Krebs cycle (CO_2 + energy) → Electron transport chain (H_2O + energy)

Elaborate on these points.

9. Hint: Glucose → Glycolysis → Pyruvate → Aerobic respiration (presence of O_2)→ Krebs cycle and Electron transport chain → CO_2 (in plants and microbes) + Energy Elaborate on this pathway.

10. Hint: The three types of blood vessels are arteries, veins and capillaries. Arteries have thick, elastic and muscular walls that can withstand high pressure of blood.

Veins have thin, less elastic and less muscular walls that can collapse under low pressure of blood. They also have valves that prevent backflow of blood. Capillaries have one-cell-thick walls that facilitate diffusion of gases, nutrients and waste between blood and tissues. They connect arteries and veins. Elaborate on these points.

11. Hints: (a) Xylem transports water and mineral salts from the roots up to other parts of the plant. It consists of dead, elongated and hollow cells with thickened walls and pits.

(b) Phloem transports sucrose and amino acids from the leaves and other parts of the plant. It consists of living cells with sieve plates and companion cells.

(c) Pulmonary vein transports oxygenated blood from the lungs to the left atrium of the heart. It is one of the few veins that carry oxygen-rich blood instead of oxygen-poor blood.

(d) Vena cava transports deoxygenated blood from different parts of the body to the right atrium of the heart. It is divided into superior vena cava and inferior vena cava that drain blood from the upper and lower body, respectively.

Elaborate on these points.

Case Study-Based Questions

1. (i) (a), (ii) (c)

2. (i) (c), (ii) (b), (iii) (b)

3. (i) (c), (ii) (b)

4. (i) (b), (ii) (c), (iii) (a), (iv) (a)

5. (c)

NCERT Corner

1. Nutrition is the process by which organisms obtain and utilise nutrients for growth, maintenance, and repair. The different modes of nutrition are autotrophic, heterotrophic, and saprophytic.
2. Saliva helps in the lubrication of food, making it easier to swallow. It also contains an enzyme called salivary amylase, which breaks down starch into simpler sugars. In the stomach, food is mixed with gastric juices, which contain hydrochloric acid and enzymes such as pepsin. The acidic environment of the stomach helps in the digestion of proteins.
3. Oxygen and carbon dioxide are exchanged in the human body through the process of respiration. During inhalation, oxygen-rich air enters the lungs and diffuses into the bloodstream. This oxygen is transported by the red blood cells to the body tissues, where it is utilised in cellular respiration to produce energy. Carbon dioxide, which is a waste product of cellular respiration, diffuses out of the body tissues into the bloodstream, and is transported back to the lungs, where it is eliminated during exhalation.
4. Excretion is the process by which metabolic wastes and other harmful substances are eliminated from the body. The different modes of excretion in organisms are renal excretion (involving the kidneys), pulmonary excretion (involving the lungs), cutaneous excretion (involving the skin) and digestive excretion (involving the digestive system)
5. Transpiration is the loss of water vapour from the aerial parts of plants, mainly through the stomata on the leaves. This process helps in the absorption and transportation of water in plants. Water is absorbed by the roots and transported to the leaves through the xylem vessels. In the leaves, water evaporates through the stomata, creating a negative pressure that pulls more water up from the roots. This process is known as transpiration pull or cohesion-tension theory.

Competition Window

1. (a)	**2.** (b)	**3.** (c)	**4.** (c)	**5.** (c)
6. (a)	**7.** (c)	**8.** (c)	**9.** (c)	**10.** (a)

CHAPTER 4

Absorption by the Roots: The Processes Involved

PREREQUISITES

You must have knowledge of:

- Structure and functions of plant cells and tissues, including the root system
- Plant nutrition
- Different types of soil, their composition and the role of soil in plant growth
- Role of roots in absorption
- Transpiration
- Osmosis

LEARNING OBJECTIVES

At the end of this chapter, you will be able to:

- Explain the mechanisms involved in water absorption by the roots
- Describe the role of root hairs
- Explain the process of nutrient absorption by the roots
- List the factors affecting absorption
- Explain the importance of absorption in plant growth and development
- List the applications of absorption

INTRODUCTION

The area of biology known as **plant physiology** is concerned with how plants function in their daily lives. It includes the functioning of the organism as a whole, as well as its cells, tissues, organs and organ systems. Osmosis, one of the most fundamental processes covered in this chapter, is extremely important to all living things.

A plant's roots anchor it in the ground and provide support, but their most crucial and vital function is to take up water and mineral nutrients from the soil and conduct them up the stem so that the leaves, flowers, fruits and other parts of the plant can receive them.

In addition to being a component of the protoplasm, water is required inside the plant body for four processes:

- **Photosynthesis:** During photosynthesis, water is used by the green leaves to make glucose.
- **Transpiration:** During transpiration, a substantial amount of water is expelled as water vapour for cooling the plant in hot weather, producing a suction force, and so on.
- **Transportation:** The movement of substances dissolved in water from the roots into the shoots and up to the leaves or from the leaves to other parts is known as transportation.
- **Turgidity:** This is a fully distended condition, which is required for stiffness of plant tissues. It is enabled by water.

Only the roots can pull the mineral nutrients from the soil that the plant needs. Some of these nutrients (such as nitrates, phosphates and sulphates) are absorbed as salts, while others are absorbed as ions (potassium, calcium, magnesium and chlorine). These substances are necessary for the synthesis of numerous compounds or enzymes inside the cell, as well as for the construction of cells and cell organelles.

CHARACTERISTICS OF ROOTS FOR ABSORBING WATER

Three factors determine a root's capacity to draw water from the soil: a large surface area provided by rootlets and root hairs, the solution (cell sap) being present in the root hairs at a concentration greater than that of the surrounding soil water and the root hairs having thin walls.

- The surface area of the roots is vast. Even a small garden plant, like balsam, reveals a thick bundle of rootlets when pulled gently from the earth (branch roots). On close inspection, each rootlet displays hundreds of root hairs. If the entire root system of this plant were laid end to end, it would stretch for many kilometres. A four-month-old rye plant's aggregate roots were estimated by botanist H.J. Dittmer to be roughly 600 km long in 1937. There were more than 14 billion root hairs present, with a possible total length of more than 10,000 kilometres. Therefore, any plant's roots offer a huge surface area to aid in water absorption.
- The root hairs contain more cell sap than the surrounding water does. Root hairs are extensions of the root's exterior (epidermal) cells. They also feature large vacuoles that contain cell sap. The concentration of salts in the cell sap is often higher than that of the water surrounding it because some salts are dissolved in it. This feature is a critical requirement to draw in outside water or for the occurrence of osmosis, which is explained later in this chapter.
- The walls of the root hairs are quite thin. Like all plant cells, root hairs contain two outer layers consisting of a cell wall and a cell membrane. The cell wall is very thin and permeable. It allows unfettered passage of water molecules and dissolved chemicals into and out of the cell. The very thin and somewhat permeable cell membrane allows water molecules to flow through but not the bigger salt molecules. The ability of roots to absorb water from the soil is mostly due to this characteristic.

ABSORPTION AND CONDUCTION OF WATER AND MINERALS

Five primary phenomena combine to form the overall process by which water and minerals are taken up by the roots, moved through the thickness of the root, and then conducted upwards through the stem: imbibition, diffusion, osmosis, active transport, turgidity and flaccidity (plasmolysis). Let us study each of these in detail.

Imbibition

Water is drawn to the surface of living or dead plant cells in a process known as imbibition, which causes the cells to absorb the water. Proteins and materials consisting of cellulose are examples of hydrophilic materials (strong affinity for water). When exposed to damp air or when they consume water or moisture, dry seeds, wooden doors and other items swell. The seed coat rips during germination (Fig. 4.1) due to imbibitional pressure. It is important for the sap's development as well.

Fig. 4.1 Germinating seed
(*Carduus nutans* (seedling)…; Krzysztof Ziarnek, Kenraiz; 2020; https://commons.wikimedia.org/wiki/File:Carduus_nutans_kz04.jpg. Licensed under: CC-BY-SA-4.0)

Diffusion

When two substances are in direct contact, there is free movement of molecules of a substance (solute or solvent, gas or liquid) from the region of their higher concentration to the region of their lower concentration. This is called diffusion. An easy activity can be used to demonstrate diffusion.

In a water-filled beaker, add a sugar cube, a small tablet of a soluble dye or a crystal of potassium permanganate (Fig. 4.2).

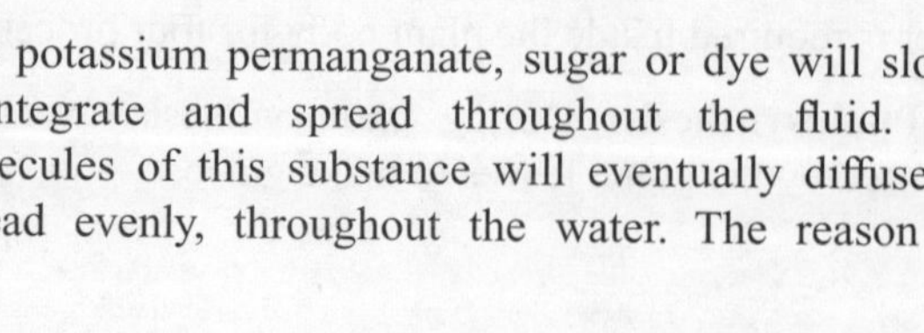

Fig. 4.2 Process of diffusion

The potassium permanganate, sugar or dye will slowly disintegrate and spread throughout the fluid. The molecules of this substance will eventually diffuse, or spread evenly, throughout the water. The reason for this is as follows: In and around the dye tablet, the dye molecules are more closely packed (more concentrated). In areas where they are scarcer or absent, these molecules disperse farther and farther (less concentrated). The molecules are moved in this manner until they are evenly distributed. The uniform diffusion process is accelerated by stirring the fluid with a spoon or glass rod, which also makes it easier to obtain a homogeneous solution.

Osmosis

Osmosis is the process by which water molecules pass through a semi-permeable membrane from their region of higher concentration (dilute solution) to their region of lower concentration (concentrated solution). The most crucial point is that here, only water molecules shift from their higher concentration to their lower concentration (whether in pure water or a diluted solution).

According to the **degree of concentration** of the solution around a cell, osmosis can occur in the inward or outward directions:

- When the surrounding solution is less concentrated, there is the inward diffusion of water through a semi-permeable membrane. This is called **endosmosis**. The cell enlarges as a result of this.
- When the surrounding solution is more concentrated, there is outward movement of water through a semi-permeable membrane. This is called **exosmosis**. The cell typically shrinks as a result of this.

To demonstrate this, take a thistle funnel and a strong sugar solution. A cellophane paper, egg membrane or animal bladder can be used to cover the thistle funnel's mouth, and it should be attached tightly. As shown in Fig. 4.3, the thistle funnel needs to be suspended upside down in a water-filled beaker. Take note of how much sugar solution is spread around the thistle funnel's stem. The experiment is set up in this manner.

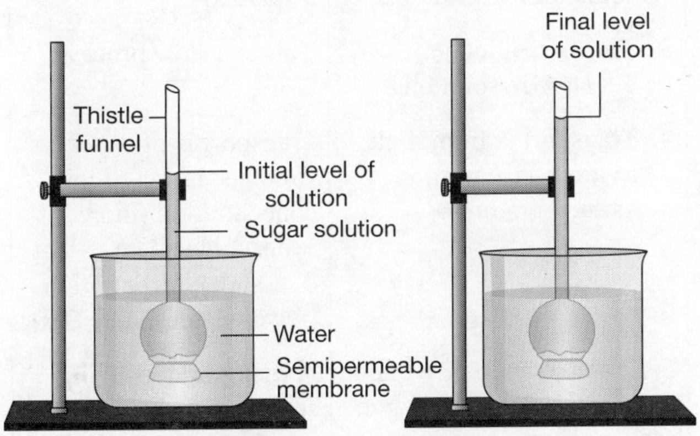

Fig. 4.3 Process of osmosis

As a control, use a different thistle funnel, add plain water to it, and place it in another water-filled beaker (for comparison). Mark the level again on its stem.

The sugar solution level in the thistle funnel in the test set-up will increase after a few hours.

Three key findings may be drawn from this study:

- In the experimental set-up, water from the beaker was funnelled through the cellophane paper and into the thistle funnel containing the sugar solution.
- No sugar from the thistle funnel flowed into the beaker.
- The cellophane paper serves as a selectively or variably permeable membrane. Water molecules can flow through but not sugar molecules.

You can modify the experimental set-up by using a more concentrated sugar solution in the thistle funnel and a very dilute or less concentrated sugar solution in the beaker. Even then, the result would be the same, that is, the level of the fluid would rise. Another explanation for this is that some water from the less concentrated sugar solution in the beaker spilled through the cellophane paper into the more concentrated sugar solution in the thistle funnel.

What would have happened if you had used a rubber sheet instead of cellophane and a similar arrangement with a muslin cloth? In the first case, there would obviously be no change in the level of the sugar solution because the rubber sheet is impermeable and would prevent water molecules from the beaker from passing through to the other side. In the second scenario, gravity would drive the entire sugar solution to sink to a common level since the meshes or pores of the muslin fabric are so large that even the sugar molecules cannot be held back. As a result, the muslin fabric would allow the sugar solution to pass through easily.

Two key points from the above experiments:

- Two liquids exist, each with a distinct concentration. While there may be fewer or no sugar molecules in the beaker, the sugar solution contains a larger concentration of sugar molecules. It is crucial to note that the concentration may also be seen from the side of the solvent; in Fig. 4.3, the beaker has more water molecules per unit volume than the thistle funnel contains of sugar solution. As a result, there exist two areas of varying concentrations (tonicity), which you can represent as follows:
  - Two regions of different concentrations of water molecules, or
  - Two regions of different concentrations of sugar molecules

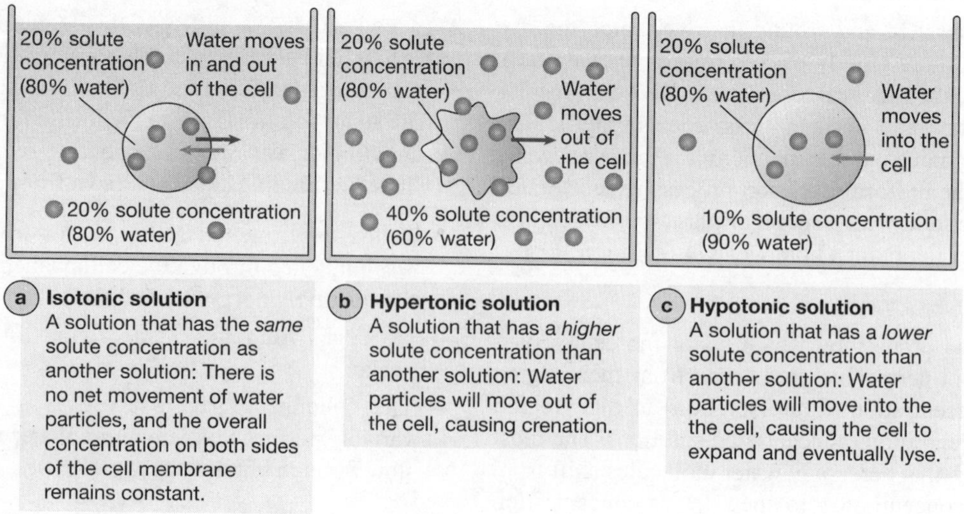

Fig. 4.4 Tonicity

- The cellophane paper that acts as a semi-permeable membrane separates the two liquids. Selective molecule transit is possible through a semi-permeable membrane. The solute (molecules of sugar or salt in the solution) cannot pass through it freely, but a solvent (such as water molecules) may do so.

Osmotic Pressure

If an airtight piston bearing some weight is added to the thistle funnel from the beginning of the experiment, the level of the solution would not rise at all, demonstrating that no water had entered. Thus, when two fluids are separated by a semi-permeable barrier, the osmotic pressure is the least amount of pressure required to stop the pure solvent from permeating the solution. In other words, the osmotic pressure of a solution is a measure of its tendency to take in water by osmosis. It is equal to the weight or pressure required to nullify osmosis.

Tonicity

Tonicity is a measure of the concentration of solutes in a solution that determines the direction and extent of osmosis. It refers to the ability of a solution to cause water to move into or out of a cell by osmosis. Tonicity is classified as follows (Fig. 4.4):

- **Isotonic:** On either side of the cell membrane, the relative concentration of water molecules and the solute is the same. There is no net flow of water molecules across the cell membrane in such a solution.
- **Hypotonic:** In this situation, the solute concentration of the solution outside the cell is lower than that of the fluids inside the cell. As a result, water molecules from outside will enter the cell (endosmosis).
- **Hypertonic:** In this situation, the solute concentration of the solution outside the cell is higher than that of the fluids inside the cell. As a result, the water molecules inside the cell will flow outside (exosmosis).

In order to emphasise the numerical basis of the total moles of solute per kilogramme of water, the three terms isotonic, hypotonic and hypertonic are alternatively expressed as isoosmotic, hypoosmotic and hyperosmotic.

Table 4.1 lists the differences between diffusion and osmosis.

Table 4.1 Comparison of diffusion and osmosis

Diffusion	Osmosis
1. Liquids and gases can diffuse over considerable distances.	Water can be transported only over a short distance.
2. There is movement of the molecules of the solute or solvent.	There is movement of water molecules only as a solvent.
3. This is rapid in gases, but slow in solutions.	This is a slow process.
4. Transport is from high to low concentration along a gradient.	Transport is of water from a solution of low concentration (more water molecules) to that of a high concentration (fewer water molecules).
5. It occurs with or without a non-living permeable membrane.	A living or non-living semi-permeable membrane is required.

Active Transport

Unlike diffusion, active transport is the movement of a material (such as a salt or an ion) through a live cell membrane. This requires energy from the cell. Active transport moves in the opposite direction to diffusion.

The cell membrane of the root cells makes it difficult for several nutrients, such as potassium, zinc, manganese, ions of nitrates and sulphates, to pass through. This is due to the fact that their concentration is higher within the root cells and that is maintained as such in order to provide osmotic pressure for water absorption. The concentration gradient of these ions is in opposition to diffusion. In other words, in order to obtain them, these ions must be 'forcefully' transported from the area of their higher concentration inside to the region of their lower concentration outside. For this, ATP, which is the energy generated by the cell, is needed.

Diffusion and passive transport are identical in that they do not require any energy input. Molecules can travel freely from areas of higher concentration to areas of lower concentration.

Turgidity and Flaccidity (Plasmolysis)

Every living plant cell comes in contact with liquids, either directly or indirectly. In particular, the root hairs are surrounded by soil and water. As long as the cell sap is more concentrated than the surrounding fluids, water will continue to enter all of these cells through osmosis.

A cell is said to be turgid and the condition is referred to as **turgidity** when it reaches the point where it can no longer hold any more water, that is, when it is fully distended. When a cell is turgid, its wall is stretched by internal pressure, which pushes its contents towards the centre of the cell.

Wall pressure refers to the force the cell wall exerts on the cell content, whereas **turgor pressure** refers to the force the cell contents exert on the cell wall. The cell wall bursts and the contents spill out if it is unable to withstand the turgor pressure.

When the cell is turgid, it is in a somewhat balanced state since no additional water can enter or leave it. Even though the concentration of solutes inside the cell may be higher than that outside the cell, its turgor pressure counteracts the wall pressure, preventing further water absorption.

When a living cell is submerged in freshwater, like the cells of an aquatic plant's leaf, it maintains its completely distended state. The plasma membrane presses on the cell wall and stays in close touch with it (Fig. 4.5) much like a football's rubber bladder presses against its leather skin. The cytoplasm will contract, the plasma membrane will

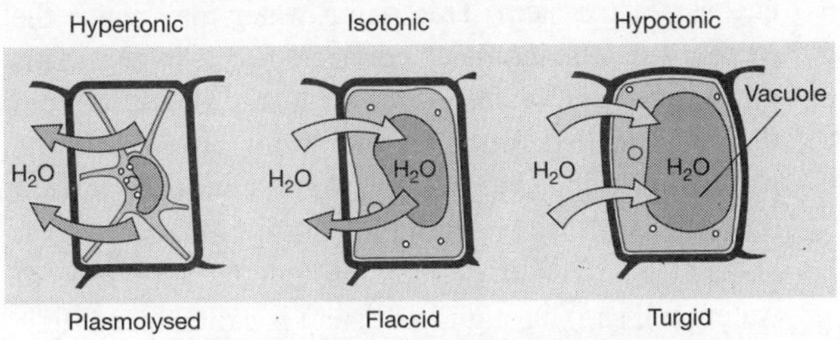

Fig. 4.5 Turgidity and flaccidity

pull away from the cell wall, and the plant cell will lose its inflated appearance if it is held in a solution of 5% salt for a few minutes.

Plasmolysis, the term for this contraction of the cell wall, causes the cells to become limp or flaccid (the condition is called flaccidity). **Flaccidity** is the reverse of turgidity. If, however, a plasmolysed (flaccid) cell is returned to water before it is dead, its protoplasm will again swell up and press against the cell wall. The recovery or reversal of plasmolysis is called **deplasmolysis**. Thus, plasmolysis is the result of outflow of water from the cell and deplasmolysis is the result of its re-entry.

Uses of Turgidity

- Turgidity provides stiffness to soft tissues such as the leaves. Lack of water causes a leaf to wilt, which causes the petiole and lamina to become loose and the leaf to droop. When a plant is exposed to the hot afternoon sun, the amount of water lost through transpiration is typically greater than the amount of water absorbed through the roots, causing the leaves to wilt. However, in the evening, when transpiration is reduced and the amount of water absorbed is greater than the amount of water lost through transpiration, the turgidity of the leaf cells is restored and the leaves once again stand out.
- Weeds can be killed in a playground by sprinkling excessive salts around their base. Excessive application of fertilisers in the agricultural fields may similarly damage the roots and diminish the yield.
- Turgor pressure, which is present in both mushrooms and seedlings, helps them break through hard soils. Certain tree roots have occasionally been observed to cause cracks in a building's concrete floor or wall. Turgor pressure is again responsible for this.
- Root pressure is increased by turgor in the root cell walls. Cutting a well-watered houseplant (like balsam) a few centimetres above the soil and attaching a glass tube to it immediately with a rubber connection will cause water to begin rising in the glass tube from the

cut end of the stem. This rising water may cause the mercury in a manometer connected to it to rise. This upward flow of water is due to heavy pressure from the roots, called root pressure. Root pressure is the pressure developed in the roots due to continued inward movement of water through cell-to-cell osmosis. It helps in the ascent of cell sap through the stem. Loss of water (cell sap) through a cut stem is called bleeding.

- The opening and closing of the stomata in a plant are influenced by its guard cells. Every guard cell has a wall that is thicker on the side facing the stoma and thinner on the other. The guard cells' contents experience an increase in osmotic pressure as a result of the production of glucose during photosynthesis and other chemical changes. As a result, they take in more water from their surrounding cells and become turgid. Turgor causes the guard cells to arch outward further and the gap between them to grow, which opens the stoma. At night, or when there is shortage of water in the leaf, the guard cells turn flaccid and their inner rigid walls become straight, thus closing the stomatal aperture.
- **Turgor movement:** The rapid drooping of the leaves of sensitive plants (such as *Mimosa pudica*) is an outstanding example of turgor movement. If one of the leaves is touched, even lightly, the leaflets fold up and within 2 to 3 seconds, the entire leaf droops. If the leaf is touched somewhat strongly, the wave of folding and drooping spreads from the stimulated leaf to all the neighbouring leaves. Slowly, the leaves recover and again stand erect. In this plant, the stimulus of touch leads to loss of turgor at the base of the leaflets and at the base of the petioles called pulvinus. Somewhat similar turgor movements are found in insectivorous plants whose leaves close up to entrap a living prey. The bending movement of certain flowers towards the sun and the sleep movements of the leaves of certain plants at night are also due to turgor movements.

ROOT PRESSURE

The pressure that builds up in the roots as a result of water infiltration is known as root pressure. This pressure is caused by the cortex and root hair cells' alternating between turgidity and flaccidity, which aids in forcing the plant sap upwards.

Due to cell-to-cell osmosis in the root tissue, root pressure builds up (Fig. 4.6). Water flow rises inward as one turgid cell presses the next cell. Water enters the pores of the thick walls of the xylem vessels, which are vertical tubes that are positioned in the centre of the plant. Thus, one of the factors that raises water up through the stem and into the leaves is root pressure. However, this force alone cannot push water all the way to the top of a tall tree.

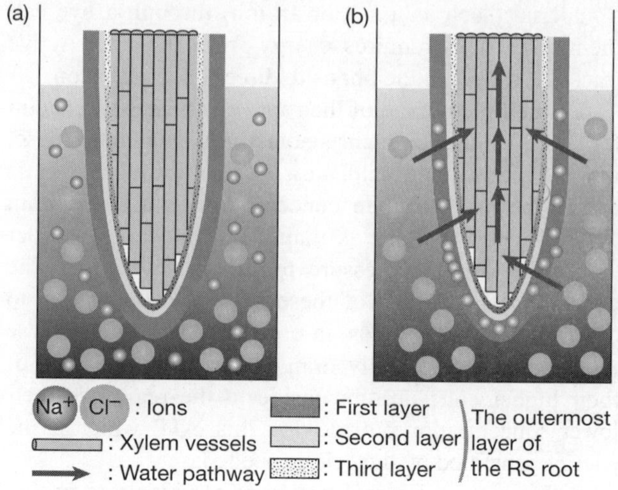

Fig. 4.6 Root pressure ((a) represents the turgor movement of ions in the root cell wall and (b) represents the upward flow of water due to heavy pressure from the roots)

Imbibition is the process by which compounds like cellulose (found in cell walls) and starch passively absorb water. Turgor is the pressure that builds up inside plant cells as a result of the hydrostatic pressure that incoming water exerts on the cell wall as a result of endosmosis.

When soaked in water, seeds and grains expand as a result of imbibition and endosmosis. The force produced by the water that has been absorbed in this way is sufficient to cause the seed coverings to rupture.

Here are some additional instances that have been noted:

- Soaked seeds exert a lot of pressure when kept in a packed, closed container. Basement godowns that were loaded to the brim with food grain bags had cracks in their walls where rainwater had entered it.
- In certain plants, like tomato, grass, banana or ferns, the root pressure is high enough to force the water all the way through the stem and out through the ends of leaf veins. This water appears as tiny drops along the margins or the tips of the leaves, especially in the early morning. This loss of excessive water is called **guttation**.

IMPORTANCE OF ROOT HAIRS AND THE UPWARD MOVEMENT OF ABSORBED WATER AND MINERALS

Root hairs are used by the root to absorb water. The cell sap found in the root hairs has a higher salt content than the soil water found outside the root. Osmosis is triggered by this differential, and outside water diffuses into the root hair. Water continues to flow from the cell bearing root hairs to the neighbouring cells, one after the other, until it ultimately enters the xylem vessels. Water is pushed upward through the xylem vessels because of the turgidity the cells acquire during the process.

Cells actively carry mineral elements from the earth into their bodies. Instead of as salts, minerals can also be absorbed as ions. The dilute solution of water and mineral salts, absorbed from the soil by the roots, can be used for food manufacture in the leaves, only if it can travel up to the highest points of the plant. This upward flow occurs through the xylem.

Forces Contributing to Ascent of Sap

There are four main forces which contribute to the upward movement (ascent) of sap:

- For herbaceous plants, **root pressure** may be sufficient to push the sap in the xylem vessels up to a particular height.
- Water from a lower level rises due to the capillarity (small diameter) of the xylem vessels to cover the space left by the water lost during transpiration from the leaves. More water will rise higher in a tube with a smaller diameter, creating a force known as **capillary force**.
- Due to the tendency of water molecules to stick together (**cohesion**), more water molecules are drawn up when water is lost from the leaf surface through transpiration, creating a continuous column of water through the stem.
- This pulling force (or **suction force**) provided by the leaves is especially important in tall trees, such as pines, which do not have enough root pressure.

Downward movement of sap is relatively simple. The food manufactured in the leaves is dissolved in water and it flows down mainly due to gravity.

Cohesive forces are the forces of attraction between water molecules, which help to hold them together in a continuous column within the plant's vascular system. Adhesive forces are the forces of attraction between water molecules and the walls of the xylem vessels, which help to hold the water column in place. Transpirational pull refers to the force created by the evaporation of water from the leaves, which pulls water up through the plant from the roots.

SUMMARY

- Plants require structural support to remain stable in the ground.
- Absorption of water and minerals from the surrounding soil is crucial for plant growth and development.
- Imbibition is a surface attraction-driven process involving water-loving compounds such as proteins and cellulose.
- Diffusion is the free movement of any gas, liquid or solvent from an area of high concentration to an area of low concentration.
- Osmosis is a specific type of diffusion that involves the movement of solvent or water molecules through a semi-permeable membrane from an area of higher concentration to an area of lower concentration.
- Osmotic pressure is the minimum pressure required to stop pure solvent from entering a solution, and it plays a critical role in the movement of water and minerals into and out of plant cells.
- Plant roots have specialised structures called root hairs, which greatly increase their surface area. These root hairs are in close proximity to soil particles and create an interface for the exchange of water and nutrients.
- The absorption of water occurs through a process called osmosis. When the concentration of solutes is higher in the soil than in the root cells, water moves into the roots through osmosis, facilitated by the presence of aquaporins, which are specialised water channels in the cell membrane.
- Along with water, plants also absorb essential mineral nutrients from the soil. This process involves active transport mechanisms. Active transport is the movement of substances against the concentration gradient, requiring energy. Plants use ATP (adenosine triphosphate) to power these transport processes.
- Different ions, such as potassium (K^+), nitrate (NO_3^-) and phosphate (PO_4^{3-}), are taken up by specific transport proteins located on the plasma membrane of root cells. These transport proteins help in the selective uptake of nutrients.
- Once inside the root cells, the absorbed water and nutrients are transported upward through the plant through the xylem, a specialised tissue responsible for water and mineral transport.
- The movement of water and dissolved nutrients in the xylem is driven by transpiration, the loss of water vapour through tiny openings called stomata on the leaves.

PRACTICE QUESTIONS

Analyse Your Concepts (School Exam Based)

Multiple Choice Questions

1. Absorption of water by the plant cells by surface attraction is called:
(a) Diffusion (b) Osmosis
(c) Imbibition (d) Endosmosis

2. A plant cell placed in a certain solution is plasmolysed. What kind of solution was it?
(a) Isotonic sugar solution
(b) Hypotonic salt solution
(c) Hypertonic salt solution
(d) Isotonic salt solution

3. The state of a cell in which the cell wall is rigid and stretched by increase in volume due to absorption of water is called:
(a) Flaccidity (b) Turgidity
(c) Capillarity (d) Tonicity

4. Which one of the following is a characteristic not related with the suitability of the roots to absorb water?
(a) Tremendous surface area
(b) Contain cell sap at a higher concentration than the surrounding soil water
(c) Root hairs have thin cell walls
(d) Grow downward into the soil

5. Movement of molecules of a substance from a region of higher concentration to a region of lower concentration without the involvement of a separating membrane is called:
(a) Osmosis (b) Diffusion
(c) Active transport (d) Capillarity

6. Osmosis and diffusion are the same, except that in osmosis, there is:
(a) A freely permeable membrane
(b) A cell wall in between
(c) A selectively permeable membrane in between
(d) An endless inflow of water into a cell

7. The highest water potential (capacity to move out to a higher concentrated solution) is that of:
(a) Pure water (b) 10% salt solution
(c) Honey (d) 50% sugar solution

8. The space between the cell wall and the plasma membrane in a plasmolysed cell is filled with:
(a) Isotonic solution
(b) Hypotonic solution
(c) Hypertonic solution
(d) Water

9. What is responsible for guttation?
(a) Osmotic pressure (b) Root pressure
(c) Suction pressure (d) Capillarity

10. The most appropriate characteristic of a semi-permeable membrane is that:
(a) It has minute pores
(b) It has no pores
(c) It allows the solute to pass through but not the solvent
(d) It allows a solvent to pass through freely but prevents passage of the solute

Match the Following

Each question contains statements given in two columns which have to be matched. Statements in column I have to be matched with statements in column II.

Column I	Column II
(1) Semi-permeable	(a) Xylem
(2) Permeable	(b) Phloem
(3) Downward flow of sap	(c) Cell membrane
(4) Upward flow of water	(d) Root pressure
(5) Guttation	(e) Cell wall

True or False

Read the following statements and write your answer as true or false.

1. A plant cell placed in hypotonic solution becomes plasmolysed.

2. The addition of salt to pickles prevents the growth of bacteria because they turn turgid.

3. Cells that have lost their water content are said to be deplasmolysed.
4. Xylem is the water conducting tissue in plants.
5. The cell swells and may burst due to an influx of water in a hypotonic solution.
6. The cell wall of the root cell is a differentially permeable membrane.
7. Plasmolysis is the reverse of deplasmolysis.
8. Guttation is another name for bleeding in plants.
9. Soaked seeds burst their seed coats.
10. If the phloem of a twig is removed keeping the xylem intact, the leaves of the twig will wilt.
11. Guttation in plants is maximum at mid-day.
12. Dry seeds when submerged in water swell up due to endosmosis.

Fill in the Blanks

Complete the following statements with an appropriate word/term to be filled in the blank spaces.

1. When placed in a more concentrated solution, the cell contents will ________.
2. The pressure by which the ________ molecules tend to cross the semi-permeable membrane is called osmotic pressure.
3. Active transport is in a direction ________ to that of diffusion.
4. The absorption of water by plant roots occurs through the process of ____________.
5. Root hairs increase the surface area of roots, facilitating ____________ with soil particles.
6. The movement of water and dissolved nutrients in the xylem is driven by ____________, which is the loss of water vapour through the stomata.

Assertion–Reason Questions

The following questions consist of two statements: Assertion (A) and Reason (R). Answer these questions selecting the appropriate option given below:

(a) Both A and R are true and R is the correct explanation of A
(b) Both A and R are true but R is not the correct explanation of A
(c) A is true but R is false
(d) A is false but R is true

1. Assertion (A): Water absorption by the roots from the soil is often an active process.
Reason (R): Because of concentration gradients, the majority of the water is absorbed by the plants in osmotic mode, which uses no metabolic energy.
2. Assertion (A): The roots of the plant absorb water and minerals, which are then carried upward to different areas of the plant.
Reason (R): Active transport is the method by which a substance received from a plant's roots is transferred to the top of the plant.
3. Assertions (A)
(a) Plants produce their own food, and their roots allow them to take up water and minerals.
(b) They must carry food and water to all plant parts in order to survive.
(c) The movement of substances absorbed in one area of the plant to another area of the plant is a crucial step for survival.
(d) Xylem tubes travel from the roots to the top of the plant carrying water and absorbed minerals.
Reasons (R)
(a) Plants may take up water and other minerals from the soil by osmosis, diffusion and active transport.
(b) Capillary action and the pull produced by water evaporation from the roots transports water to all other sections of the plant. This is passive movement of the water as it does not use ATP for transport.
(c) In active transport, ATP is used in the process.
(d) Hence, the process by which a substance absorbed from the roots of the plant is moved to the upper part of the plant is not active transport.
4. Assertion (A): Root hairs increase the surface area of roots, facilitating absorption of water and nutrients.
Reason (R): The increased surface area provided by root hairs allows for more contact with the soil particles, enhancing the uptake of water and nutrients.
5. Assertion (A): Active transport is involved in the absorption of mineral nutrients by plant roots.
Reason (R): Active transport allows plants to take up mineral nutrients against the concentration gradient, ensuring selective uptake of essential ions necessary for plant growth.
6. Assertion (A): Transpiration plays a crucial role in the upward movement of water and dissolved nutrients in the xylem.

Reason (R): Transpiration creates a negative pressure gradient within the plant, causing water to be pulled up from the roots to the leaves, facilitating the transport of nutrients through the xylem.

One-word Answers

1. The condition of a cell placed in a hypotonic solution.
2. The process by which intact plants lose water in the form of droplets from the leaf margins.
3. The process by which water enters the root hairs.
4. The tissue concerned with the upward conduction of water in plants.
5. The term for the inward movement of solvent molecules through the plasma membrane of a cell.
6. The process by which molecules distribute themselves evenly within the space they occupy.
7. The pressure which is responsible for the movement of water molecules across the cortical cells of the root.
8. The pressure exerted by the cell contents on the cell wall.
9. The condition in which the cell contents are shrunken.
10. Loss of water through a cut stem.

Define the Term

1. Diffusion
2. Osmosis
3. Transpiration
4. Active transport
5. Root hairs

Groups of Terms

Given below are certain groups of terms. In each group the first pair indicates a relationship between the two terms. Rewrite and complete the second pair on a similar basis.

1. Photosynthesis : Chloroplasts :: Respiration : ________
2. Nucleotide : DNA :: Amino Acid : ________
3. Stamen : Male Reproductive Organ :: Pistil : ________
4. Prokaryotic Cell : No Membrane-bound Nucleus :: Eukaryotic Cell : ________

Identify the Odd One Out

1. Which one of the following is the odd one out?
   (a) Diffusion (b) Osmosis
   (c) Transpiration (d) Active transport
2. Which one of the following is the odd one out?
   (a) Root hairs (b) Epidermis
   (c) Cortex (d) Stomata
3. Which one of the following is the odd one out?
   (a) Water potential (b) Turgor pressure
   (c) Osmotic pressure (d) Chemical potential
4. Which one of the following is the odd one out?
   (a) Hypertonic solution (b) Hypotonic solution
   (c) Isotonic solution (d) Osmotic solution
5. Which one of the following is the odd one out?
   (a) Endodermis (b) Xylem
   (c) Phloem (d) Pericycle

Image-Based Questions

1. The following diagram represents a plant cell after being placed in a strong sugar solution.

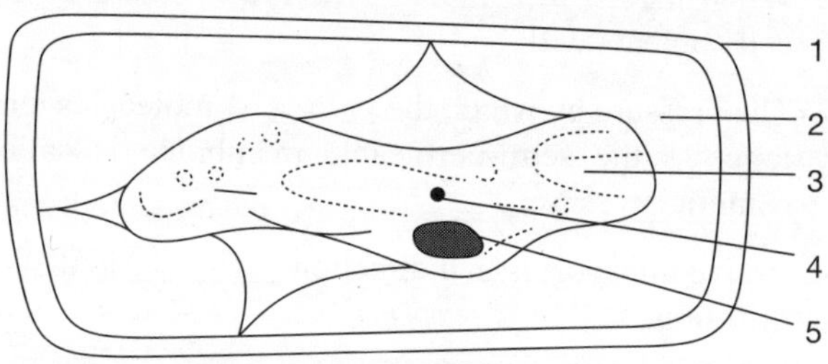

Labels 1 to 5 indicate the following:
(1) Cell wall
(2) Strong sugar solution
(3) Protoplasm
(4) Large vacuole
(5) Nucleus
(a) What is the state of the cell shown in the diagram?
(b) Name the structure which acts as a selectively permeable membrane.
(c) If the cell had been placed in distilled water instead of strong sugar solution, which feature would not have been seen?
(d) Name any one feature of this plant cell which is not present in an animal cell.

2. The diagram given below represents an experimental set-up to demonstrate a certain process. Study the same and answer the questions that follow:

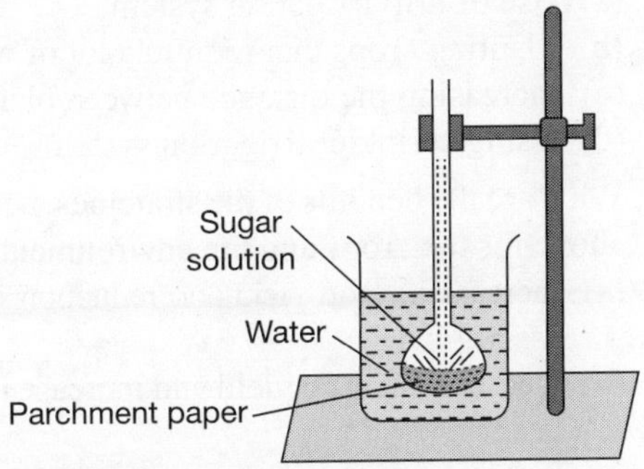

(a) Name the process.
(b) Define the above-named process.
(c) What would you observe in the experimental set-up after an hour or so?
(d) What control experiment can be set up for comparison?
(e) Keeping in mind the root hair, cell and its surroundings, name the parts that correspond to
  (i) concentrated sugar solution
  (ii) parchment paper
  (iii) water in the beaker
(f) Name any other substance that can be used instead of parchment paper in the above experiment.
(g) Mention two advantages of the process to plants.

3. The diagram below represents a layer of epidermal cells showing a fully grown root hair. Study the diagram and answer the questions that follow:

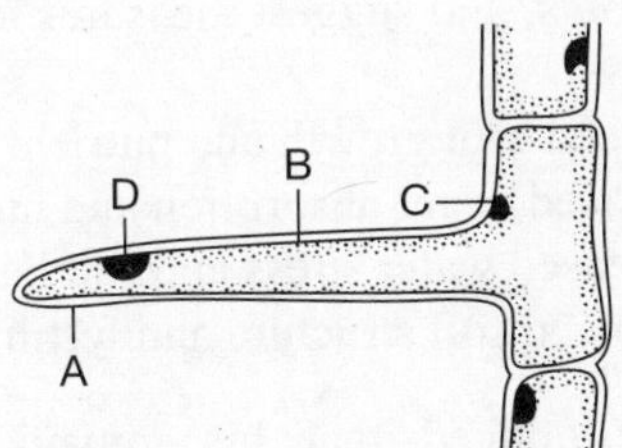

(a) Name the parts labelled A, B, C and D.
(b) The root hair cell is in a turgid state. Name and explain the process that caused this state.
(c) Mention one distinct difference between the parts labelled A and B.

4. Study the diagram given below and answer the questions that follow:

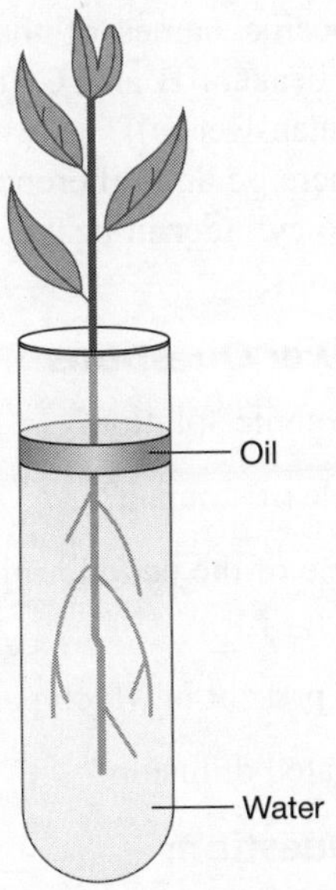

(a) Name the process being studied in the above experiment.
(b) Explain the process mentioned in (a).
(c) Why is oil placed over water?
(d) What do we observe with regard to the level of water when this set up is placed in (1) bright sunlight (2) humid conditions (3) windy day?
(e) Mention any three adaptations found in the roots of the plant which enable them to carry out the process mentioned in (a).

5. Three cylinders of potato were carefully dried on a blotting paper and weighed. Each piece weighed 3 grams. Each one was placed in the beaker as shown below:

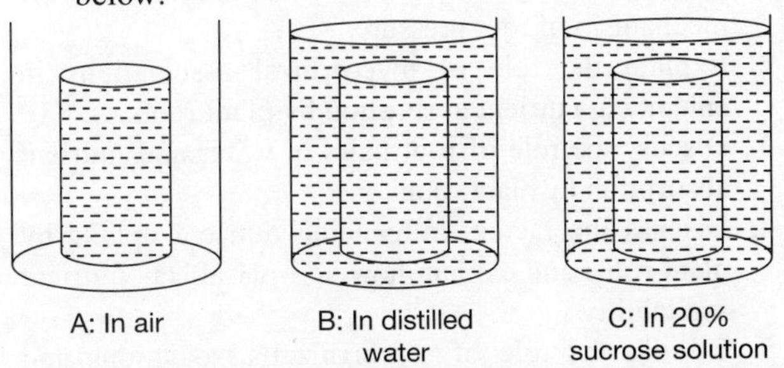

(a) After 48 hours, which potato cylinder would be the heaviest?
(b) The movement of which substance is mainly responsible for the weight change in the potato cylinders?
(c) Name and define the process which is responsible for the movement of substance mentioned in answer (b).

(d) Write specific names of the processes which occur in beakers B and C (types of processes defined in answer (c)).

(e) Would there be any difference in the weight of the potato cylinder in beaker A after 48 hours? Give reasons.

Very Short Answer Questions

1. What is a hypotonic solution?
2. What is the role of stomata?
3. What is the role of the endodermis in absorption by roots?
4. Root hairs are present in which part of the root?
5. What is facilitated diffusion?

Short Answer Questions

1. Explain root pressure.
2. List two instances of plant turgor motions.
3. The root hairs have a larger concentration of mineral nutritional components than the surrounding soil. How are these components ingested from the soil by the roots?
4. How does transpiration pull work?
5. Define absorption by roots and list the different processes involved.

Long Answer Questions

1. Describe the process of water absorption in plant roots, including the role of the endodermis and the mechanism of root pressure.
2. Explain the role of mycorrhizal associations in enhancing nutrient absorption by plant roots.
3. Discuss the role of root hairs in water and nutrient absorption by plant roots.
4. Discuss the factors influencing nutrient uptake by plant roots and explain how soil pH affects nutrient availability.
5. Discuss the role of mycorrhizal associations and *Rhizobium* bacteria in enhancing nutrient uptake by plant roots, highlighting their specific mechanisms and benefits to plants.

Case Study-Based Questions

1. We want to grow crops in an arid region with limited water resources and improve the efficiency of water uptake by the crops to maximise their yield while minimising water usage.

   (i) Which of the following are strategies that you can use to improve the water uptake efficiency of the crops?
   (a) Use of drip irrigation system
   (b) Planting crops that require a lot of water
   (c) Increasing the distance between plants
   (d) Using sprinkler irrigation system

   (ii) What are the benefits of the strategies mentioned above for the crops and the environment?
   (a) Increase in crop yield and reduction in water usage
   (b) Decrease in crop yield and increase in water usage
   (c) Increase in crop yield and increase in water usage
   (d) Decrease in crop yield and reduction in water usage

   (iii) Which of the following are potential drawbacks or limitations to the strategies mentioned above, and how can they be mitigated?
   (a) Increased cost of irrigation systems, use of chemicals for fertigation
   (b) Decrease in crop yield due to lack of water, soil erosion due to over-watering
   (c) Increased competition for water resources, waterlogging and root diseases
   (d) None of the above

2. A farmer noticed that his crops were not growing well despite regular watering and fertilisation. On investigation, he found that the soil in his field was compacted and had poor drainage. Explain how this could affect the absorption of water and nutrients by plant roots, and suggest measures to improve the situation.
   (a) Reduced root growth and nutrient uptake
   (b) Increased water absorption and nutrient uptake
   (c) Decreased water stress and salt build-up
   (d) Enhanced soil structure and fertility

3. A farmer noticed that his tomato plants were wilting and showing signs of nutrient deficiency, despite regular watering and fertilisation. Upon investigation, he found that the pH of the soil was very low. What is the likely cause of the nutrient deficiency in the tomato plants?
   (a) Overwatering
   (b) Under-fertilisation
   (c) Low pH of the soil
   (d) High pH of the soil

NCERT Corner

1. What are the factors affecting the rate of diffusion?
2. What are porins? What role do they play in diffusion?
3. Describe the role played by protein pumps during active transport in plants.
4. Explain why pure water has the maximum water potential.
5. Explain why xylem transport is unidirectional and phloem transport bi-directional.
6. Explain the pressure flow hypothesis of translocation of sugars in plants.
7. How is mycorrhizal association helpful in the absorption of water and minerals in plants?

COMPETITION WINDOW (OBJECTIVE TYPE)

1. Water-soaked dry seeds swell as a result of:
   (a) Absorption (b) Diffusion
   (c) Imbibition (d) None of these
2. Choose the correct options. In active transportation, pumps are:
   A. Lipids that bind molecules to pass them through the membrane
   B. Transport molecules uphill
   C. Specific transporters
   D. Sensitive to inhibitors
   (a) A and C (b) A, B and D
   (c) B, C and D (d) C and D
3. Among which of the following is the unit of water potential ψ?
   (a) Pascal (b) Joule
   (c) Newton (d) Electron volt
4. Two cells A and B are contiguous. Cell A has osmotic pressure 10 atm, turgor pressure 7 atm and diffusion pressure deficit 3 atm. Cell B has osmotic pressure 8 atm, turgor pressure 3 atm and diffusion pressure deficit 5 atm. The result will be:
   (a) No movement of water
   (b) Movement of water from cell B to A
   (c) Movement of water from cell A to B
   (d) Equilibrium between the two
5. Which of the following is decreased during increase in humidity in the atmosphere?
   (a) Transpiration (b) Photosynthesis
   (c) Respiration (d) Glycolysis
6. Solute potential Ψs is always:
   (a) Positive (b) Zero
   (c) Negative (d) One
7. Two types of molecules cross the membrane in the same direction. This is called:
   (a) Uniport (b) Symport
   (c) Antiport (d) Multiport
8. Diffusion rate is influenced by:
   (a) Temperature
   (b) Membrane permeability
   (c) Gradient of concentration
   (d) All of these
9. Cells shrink in ________ solution:
   (a) Isotonic
   (b) Hypotonic
   (c) Hypertonic
   (d) Hypotonic and hypertonic
10. Which of the following statements regarding diffusion is false?
    (a) There is no expenditure of energy.
    (b) Molecules move in Brownian motion.
    (c) Substances moves from the region of higher concentration to the region of lower concentration.
    (d) Diffusion depends on the type of living system.
11. Which statement about osmosis is precise?
    (a) There is expenditure of energy.
    (b) Molecules cannot move at random during osmosis.
    (c) Substances moves from the region of higher concentration to the region of lower concentration.
    (d) Osmosis depends on the type of membrane.
12. In some plants, liquid water is lost through hydathodes. Such hydathodes:

(a) Do not show any specificity in opening and closing
(b) Remain closed during day
(c) Remain closed at night
(d) Remain open always

13. If pressure greater than atmospheric pressure is applied to pure water, the water potential:
(a) Increases
(b) Decreases
(c) Remains the same
(d) Atmospheric pressure does not affect water potential

14. Choose the correct sequence of water movement in plants:
(a) Cortex, protoxylem, endodermis, root hair, pericycle, metaxylem
(b) Root hair, cortex, endodermis, pericycle, protoxylem, metaxylem
(c) Root hair, protoxylem, metaxylem, pericycle, cortex, endodermis
(d) Root hair, pericycle, endodermis, cortex, protoxylem, metaxylem

15. Wilting of a plant results from excessive:
(a) Respiration (b) Photosynthesis
(c) Absorption (d) Transpiration

16. The cohesion theory of plant water transport was proposed by:
(a) Henry Dixon (b) FF Blackman
(c) Melvin cycle (d) Hans A Krebs

17. Osmotic pressure is affected by:
(a) Temperature
(b) Membrane permeability
(c) Gradient of concentration
(d) All of these

18. When a plant is girdled (ringed):
(a) The root and shoot die at the same time
(b) The shoot dies first
(c) The root dies first
(d) Neither the root nor the shoot will die

19. Choose the correct options. Active transport:
A. Has molecule-binding lipids that allow molecules to flow through membranes
B. Moves molecules upwards
C. Has specific transporters
D. Is inhibitor-sensitive
(a) A and C (b) A, B and D
(c) B, C and D (d) C and D

20. Guttation takes place:
(a) During the day (b) At night
(c) Early morning (d) All the time

Answer Keys

Multiple Choice Questions

1. (c) 2. (c) 3. (b) 4. (b) 5. (b)
6. (c) 7. (a) 8. (c) 9. (b) 10. (d)

Match the Following

1. (c) 2. (e) 3. (b) 4. (a) 5. (d)

True or False

1. T 2. F 3. F 4. T 5. T
6. T 7. F 8. T 9. T 10. T
11. F 12. T

Fill in the Blanks

1. Shrink or undergo plasmolysis
2. Solute
3. Opposite
4. Osmosis
5. Exchange
6. Transpiration

Assertion–Reason Questions

1. (d) 2. (a) 3. (c) 4. (a) 5. (a)
6. (a)

One-word Answers

1. Turgidity
2. Guttation
3. Osmosis
4. Xylem
5. Endocytosis
6. Diffusion
7. Root pressure
8. Turgor pressure
9. Plasmolysis
10. Transpiration

Define the Term

1. The movement of particles from an area of high concentration to an area of low concentration. In plant roots, diffusion is one of the mechanisms by which water and nutrients move from the soil into the roots.
2. The movement of water molecules across a selectively permeable membrane from an area of high water concentration to an area of low water concentration. In plant roots, osmosis is the process by which water moves from the soil into the root cells.
3. The loss of water vapour from plant leaves through small pores called stomata. This process creates a negative pressure in the plant's water transport system, which helps pull water and nutrients from the soil into the roots.
4. The movement of particles across a cell membrane against a concentration gradient, requiring energy in the form of ATP. In plant roots, active transport is used to move ions and other nutrients from the soil into the root cells.
5. Tiny, hair-like structures that extend from the surface of plant roots and increase the root's surface area for water and nutrient absorption. Root hairs play an important role in the uptake of water and nutrients from the soil.

Groups of Terms

1 Mitochondria
2. Protein
3. Female Reproductive Organ
4. Membrane-bound Nucleus

Identify the Odd One Out

1. (c) 2. (d) 3. (d) 4. (d) 5. (c)

Image-Based Questions

1. (a) Based on the given structures, the cell shown in the diagram is a typical plant cell.
   (b) The cell wall acts as a selectively permeable membrane, controlling what enters and leaves the cell.
   (c) If the cell had been placed in distilled water instead of a strong sugar solution, the large vacuole would not have been seen. In a hypotonic solution like distilled water, water moves into the cell by osmosis, causing the vacuole to expand and push against the cell wall. In a hypertonic solution like a strong sugar solution, water moves out of the cell by osmosis, causing the vacuole to shrink.
   (d) One feature of this plant cell that is not present in an animal cell is the cell wall. Plant cells have a rigid cell wall made of cellulose that provides structural support and protection, while animal cells have a flexible cell membrane made of

lipids and proteins.

2. (a) Osmosis refers to the movement of water molecules from a region of higher concentration to a region of lower concentration through a semi-permeable membrane.

(b) The process of osmosis involves the movement of water molecules from a less concentrated solution (with a lower solute concentration) to a more concentrated solution (with a higher solute concentration) through a semi-permeable membrane.

(c) The sugar solution level in the thistle funnel will increase, and the water level in the beaker will decrease slightly after an hour.

(d) To have a control, another thistle funnel filled with plain water can be suspended in another beaker containing water, and the level on its stem can also be marked.

(e) The corresponding parts are as follows: The concentrated sugar solution represents the cell sap with a higher concentration than the surrounding water in the root hair. The parchment paper symbolises the cell membrane of the root hair. The water in the beaker is analogous to the water in the soil.

(f) An alternative substance that can be used in place of parchment paper in the above experiment is cellophane paper or an animal bladder.

(g) Osmosis offers several advantages, such as regulating the absorption of water by root hairs from the soil and controlling the opening and closing of stomata during transpiration through the regulation of guard cell turgidity.

3. (a) The labelled parts are: A as the cell wall, B as the cell membrane, C as the cytoplasm, and D as the nucleus.

(b) When water is absorbed from the surrounding, the root hair becomes turgid. This process is achieved through osmosis, where the root hair's higher concentration of salts in its cell sap compared to the outside soil water causes water from the surrounding to move in through endosmosis, thus making the root hair turgid.

(c) The root hair's cell wall is freely permeable, allowing both salt and water to pass through. However, the cell membrane of a root hair is semi-permeable, and does not permit large dissolved salt molecules to pass through.

4. (a) The process of absorption of water by plant roots through osmosis is being studied here.

(b) Absorption of water by the root is by means of root hairs. A root hair contains cell sap which has a higher concentration of salts as compared to the outside soil water. This difference sets off osmosis and the outside soil water diffuses into the root hair. From the cell bearing root hair, water continues to pass to adjoining cells one after another to finally enter the xylem vessels. The turgidity acquired by the cells in the process also helps to push the water upwards through the xylem vessels.

(c) The surface of the water is covered with oil to prevent any loss of water by evaporation.

(d) Roots absorb water and hence, the level of water in set up A falls. Since the surface of water was covered with oil, there will be no effect of factors such as bright sunlight, humid conditions and windy day on the given set up.

Below are the observations with regard to the level of water for the different situations:

Bright sunlight: When this set-up is placed in bright sunlight, the water level in the test tube is seen to be less when compared to its initial marking as the rate of transpiration is very high.

Humid conditions: When this set-up is placed in humid conditions, the water level in the test tube decreases from its initial mark, but at a very slow rate as the rate of transpiration is reduced.

Windy day: When this set-up is placed in windy day conditions, the rate of transpiration increases rapidly, and thus the level of the water in the test tube is seen to decrease fast from its initial marking.

(e) Adaptations in plants to foster the process of absorption of water by plant roots:

- Large surface area provided by rootlets and root hairs.
- Root hairs contain cell sap at a higher concentration than that of the surrounding water

Root hairs with thin walls.

5. (a) After 48 hours, cylinder B kept in distilled water will be the heaviest as distilled water will act as a hypotonic solution for the potato cylinder. The potato cylinder will absorb distilled water from the beaker through the process of endosmosis and its weight will increase.

(b) Movement of water between the potato cylinder and the beaker is mainly responsible for the weight change in the potato cylinder.

(c) The physical process which is responsible for the movement of the substance mentioned in answer (b) is osmosis. Osmosis is the movement of water molecules from their region of higher concentration to their region of lower concentration through a semi-permeable membrane.

(d) The process that takes place in beaker B is osmosis, which is diffusion of water across the membrane of the potato slice cells. Plasmolysis occurs in beaker C as the potato shrinks due to a hypotonic solution of 20% sucrose solution.

Beaker B → Endosmosis

Beaker C → Exosmosis

(e) There will be no change in the weight of the potato cylinder in beaker A because it is kept in air. As there is no movement of water between the potato cylinder and the beaker, no significant change in the weight of the potato cylinder is observed in beaker A.

Very Short Answer Questions

1. A hypotonic solution is one in which the solution outside the cell has a lower solute concentration than the one inside the cell. In other words, the concentration of solutes (such as salt or sugar) in the solution outside the cell is lower than the concentration of solutes inside the cell.
2. Stomata are small pores found on the surface of leaves, stems and other plant organs that allow gases, such as carbon dioxide (CO_2) and oxygen (O_2), to enter and exit the plant. Stomata are surrounded by two specialised cells called guard cells, which can open and close the pore to regulate gas exchange and water loss.
3. The endodermis acts as a selective barrier, controlling the movement of water and solutes into the xylem, forcing them to pass through the cell membranes and cytoplasm of the endodermal cells, allowing for selective uptake of nutrients and water.
4. Root hairs are present in the zone of maturation, just behind the root tip, where they increase the surface area of the root for absorption of water and nutrients.
5. Facilitated diffusion is a passive process in which molecules or ions move across a cell membrane from high to low concentration with the help of specific membrane proteins called transporter or carrier proteins, which allow for the selective uptake of large or polar molecules.

Short Answer Questions

1. The pressure which is responsible for the movement of water molecules across the cortical cells of the root is called root pressure. Root pressure is the hydrostatic pressure that builds up in the roots of plants as a result of the active transport of mineral ions into the root cells.
2. Plant turgor motions are movements caused by changes in turgor pressure in plant cells. Examples are closing of the Venus flytrap and opening and closing of stomata.
3. Root hairs transport minerals against the concentration gradient by using ATP-powered pumps and creating a negative electrical potential.
4. Transpiration pull is a key mechanism by which plants move water from their roots up to their leaves against the force of gravity. This process relies on a combination of physical and biological factors, including the cohesion and adhesion of water molecules, the properties of plant cells, and the movement of water through the soil and plant tissues.
5. Absorption by roots refers to the process of uptake of water and essential minerals by the roots of a plant from the soil. The different processes involved in absorption by roots are:

   (i) Diffusion (ii) Osmosis
   (iii) Active transport (iv) Mass flow
   (v) Root pressure

Long Answer Questions

1. Hints: (a) Plasmolysis: Cytoplasm shrinks and detaches from the cell wall due to water loss.

   Deplasmolysis: The process of reversing plasmolysis by reabsorption of water, leading to the cytoplasm returning to its normal state.

   (b) Turgor pressure: The pressure exerted by the fluid (water) inside the plant cell against the cell wall.

   Wall pressure: The force exerted by the cell wall against the protoplast.

   (c) Guttation: The exudation of liquid droplets from the tips or edges of leaves, usually due to root pressure.

   Bleeding: The flow of sap from a cut or injured plant tissue, typically from the phloem or xylem.

   Elaborate on these points.
2. Hint: 'Flaccid' refers to a state where plant cells lack turgidity and appear limp or wilted. 'Turgid' describes a state of swollen or firm plant cells due to the presence of water or high internal pressure.

   Example of flaccid conditions: When a plant does not receive enough water or experiences drought, its cells lose water and become flaccid, causing the plant to wilt.

   Elaborate on these points.

3. Hints: (a) Common salt (sodium chloride) increases the osmotic pressure in the plant cells, leading to water loss through osmosis, ultimately killing the grass.

   (b) Uprooting a plant disrupts the water supply to the leaves, resulting in lack of water uptake and subsequent wilting.

   (c) Transplanting seedlings in the evening allows them to establish roots and absorb water overnight before being exposed to the heat and stress of the daytime.

   Elaborate on these points.

4. Hints: (a) A hypertonic salt solution has a higher solute concentration than the plant cell's cytoplasm. Water moves out of the cell through osmosis, causing the cell to lose turgidity and become flaccid.

   (b) Potato cubes have a lower solute concentration than water. As a result, water moves into the potato cells through osmosis, leading to an increase in cell volume and firmness.

   Elaborate on these points.

5. Hint: Root absorption in plants occurs through the following factors:

   Active transport: Plants use energy to transport ions against their concentration gradient.

   Osmosis: Water moves from an area of low solute concentration to an area of high solute concentration.

   Root hairs: Increase the surface area for absorption.

   Mycorrhizal associations: Symbiotic fungi enhance nutrient uptake.

   Factors affecting absorption: Temperature, water availability, nutrient concentration, pH, and presence of root diseases.

   Elaborate on these points.

Case Study-Based Questions

1. (d) 2. (a) 3. (c)

NCERT Corner

1. The factors affecting the rate of diffusion include concentration gradient, temperature, pressure, surface area, and the nature of the molecules or ions involved.
2. Porins are proteins that form channels in the outer membrane of some bacteria, mitochondria, and chloroplasts. They allow the passage of small molecules and ions across the membrane, thereby facilitating diffusion.
3. Protein pumps are ATP-powered pumps that transport molecules or ions across the cell membrane against their concentration gradient. In plants, protein pumps are involved in actively transporting minerals from the soil into the root hairs.
4. Pure water has the maximum water potential because it has no solutes dissolved in it. Water always flows from a region of higher water potential to a region of lower water potential.
5. Xylem transport is unidirectional because it is driven by transpiration, which creates a negative pressure gradient that pulls water and minerals up from the roots to the leaves. Phloem transport is bi-directional because it involves the movement of sugars and other organic molecules both from the source, where they are produced, to the sink, where they are utilised, and vice versa.
6. The pressure flow hypothesis explains the translocation of sugars in plants. According to this hypothesis, sugars are loaded into the phloem at the source, creating a high concentration gradient. This causes water to flow into the phloem, increasing the pressure. The pressure gradient then drives the movement of sugars from the source to the sink, where they are unloaded and utilised.
7. Mycorrhizal association is a symbiotic relationship between fungi and plant roots. The fungi help in the absorption of water and minerals by increasing the surface area of the root system and producing enzymes that break down organic matter in the soil. In return, the plant provides the fungi with carbohydrates produced by photosynthesis.

Competition Window

1. (c)	2. (c)	3. (b)	4. (b)	5. (a)
6. (a)	7. (b)	8. (d)	9. (c)	10. (d)
11. (d)	12. (a)	13. (d)	14. (b)	15. (d)
16. (a)	17. (d)	18. (b)	19. (c)	20. (c)

CHAPTER

5 Transpiration

PREREQUISITES

You must have knowledge of:

- Structure and functions of plant cells, including the cell wall, cell membrane and cytoplasm
- Process of photosynthesis and the role of water
- Structure and functions of leaves, including the epidermis, stomata and mesophyll cells
- Plant physiology, including the róles of xylem and phloem in transporting water and nutrients within the plant
- Environmental factors that affect plant growth and survival, including temperature, light and water availability

LEARNING OBJECTIVES

At the end of this chapter, you will be able to:

- Explain the process of transpiration and how it occurs in plants
- Explain the movement of water through the plant, the role of stomata and the process of water loss from the leaves
- Describe the significance of transpiration in plant growth and development
- Explain the importance of water conservation in agriculture and the environment

INTRODUCTION

Transpiration is the process by which water from a plant's aerial portions (leaves and stem) is lost by evaporation. In other words, it is the term used to describe the loss of water from a plant's leaves and aerial parts in the form of water vapour. It is a very important process for plants. It serves two important functions: first, it creates a suction force in the stem that enables the roots to absorb water and mineral nutrients, and second, it serves to cool the plant in hot weather. Several experiments have been used to illustrate and study the process.

All plants continuously absorb water through their roots. The aerial parts, including the leaves, receive this water, which is transferred upward through the stem. Approximately 2% of this water is utilised by the plant for photosynthesis and other processes. The remainder is almost completely lost through transpiration as water vapour to the environment.

Experiment 1: Take a medium-sized potted plant that is well-watered. Wrap the plant's mouth around the stem's base and enclose it in a clear plastic bag (Fig. 5.1). Place the plant for an hour or two in the sun. Due to saturation of water vapour released by the leaves, drops of water will soon emerge in the interior of the bag (the water vapour will condense only if the outside temperature is cool enough). No water drops will be visible in a similar empty polythene bag with its mouth closed and placed in direct sunshine. This is a control to demonstrate that plants exhale water as vapour. The drops will only be identified as water when tested on dry cobalt chloride paper (the blue paper must turn pink).

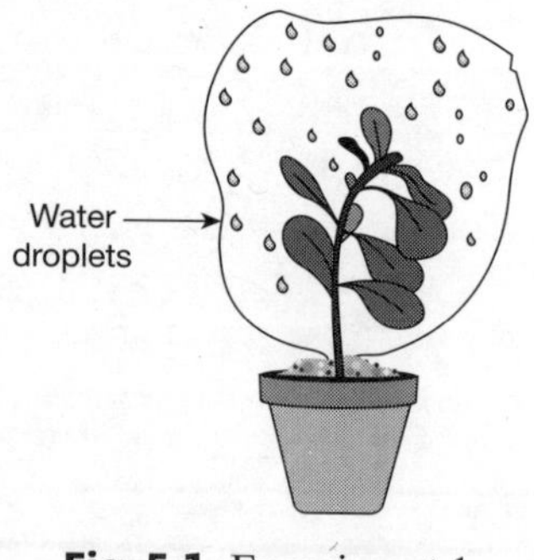

Fig. 5.1 Experiment 1

Experiment 2: Arrange three set-ups A, B and C as follows:

- **Set-up A:** Select a modest, well-watered houseplant with broad leaves. Place the pot entirely inside a polythene bag, and then tightly tie the bag's mouth to the pot's base. This would stop water vapour from escaping from the pot. Place a bell jar over the entire plant.
- **Set-up B:** Arrange a second plant, like in set-up A, and cover it with a bell jar in the same way, but this time, keep a piece of dry cobalt chloride paper next to the plant. The paper can be fastened to a cork sheet or a wooden stick.
- **Set-up C:** Select up a third bell jar that is devoid of the plant but still contains a piece of cobalt chloride paper.

Keep the three bell jars together in the sun. The first bell jar (A) would display water vapour condensing on its inner walls after about 30 minutes. Similar condensation would appear in the second bell jar (B), and the paper inside would change colour from blue to pink at the same time. In the third bell jar (C), neither the cobalt chloride paper's blue hue nor the presence of water drops on the jar's inner walls will change in any way. In this experiment, the third bell jar serves as a control, demonstrating that no transpiration took place and, therefore, there was no moisture in the air.

The condensation of water vapour into droplets and the change in colour of the cobalt chloride paper provide a double visual demonstration of transpiration. You can simplify this experiment by using only set-ups B and C or by using set-up A and another empty bell jar that does not hold a plant or a piece of paper.

MEASUREMENT OF TRANSPIRATION

A number of methods are used for measuring transpiration. Some of these are as follows:

- **Weighing method:** Here, a small, lightweight potted plant is weighed before and after a predetermined time period. To stop evaporation from surfaces other than the plant, the soil surface and the pot should be completely covered. The plant's weight loss during that time is caused by the transpiration of water. A glass bottle with a graduated side tube filled with water that has a tube placed in it can be used to improve the weighing technique. The side tube's water level drops to show how much water is lost due to transpiration from the leaves.

  Using a weighing scale or translating cubic centimetre into grams would show the amount of water loss that may be compared with weight loss (1 cc of water weighs 1 g). Another weighing experiment can be performed by placing a leafy shoot (without roots) in a test tube filled with water, covering the surface with oil to prevent evaporation of water, and weighing the tube. The test tube should be placed in a small beaker and both should be weighed together. The test tube should be removed and left in the test tube stand for a few hours. Maintaining it in the same beaker, it should be weighed once more. Any change in weight will show that the shoot has lost water (due to transpiration).
- **Potometer method:** A potometer is a tool that measures the rate at which a plant consumes water (from the Greek '*poton*' meaning 'to drink' and '*metre*' meaning 'to measure'), and this water intake is nearly equal to the water lost through transpiration.

  Farmer and Ganong are two examples of potometers created by scientists to investigate and measure the rate of transpiration, which serves to estimate the rate of water intake by a plant. Darwin's potometer and the Garreau potometer show the suction force produced by transpiration and uneven transpiration from the two surfaces of a dorsiventral leaf, respectively. The Ganong potometer is the most popular.

  - **Ganong potometer:** A sharply cut branch of an appropriate plant, like *Coleus*, is fixed inside the device (Fig. 5.2). To ensure that there are no air pockets, the entire device is filled with water. An air bubble is introduced into a horizontal graded capillary tube dipped into a water-filled beaker. To achieve this, lift the bent capillary tube so that air may be drawn in by suction pull. The tube is then lowered back into the water. A suction force is created as a result of transpiration, or the loss of water from the twig, and it draws water from the beaker while moving the bubble in the capillary tube. The capillary tube's measurements would indicate how much water was lost over a specified period of time. By opening the stopcock and allowing more water from the reservoir to enter the capillary tube, the air bubble can be returned to its previous position.

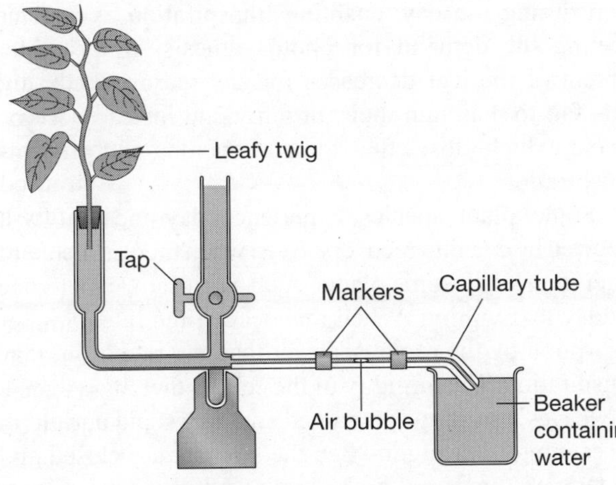

Fig. 5.2 Ganong potometer

Potometers measure the amount of water taken in by the cut shoot, and not the amount of water lost through transpiration. The cells use a portion of the water for various purposes, such as the production of food (photosynthesis). To prevent suction of the air bubble into the potometer, which would prevent absorption of water into the xylem, the twig should be cut obliquely (to provide for a greater surface area for water intake).

Limitations in the use of potometers:

- It is not easy to introduce the air bubble.
- The twig may remain fully alive for the time required.
- Any variations in the temperature of the ambient air may have an impact on where the air bubble is located within the capillary tube.

TYPES OF TRANSPIRATION

Three separate areas of a plant's aerial components are responsible for transpiration:

- Through the stomata in the leaves (stomatal transpiration)
- Directly from the surfaces of the leaves and stems' (cuticular transpiration)
- From the tiny apertures called lenticels that are found on the surface of ancient stems (lenticular transpiration)

Stomata are responsible for the majority of transpiration, with only a small amount taking place through the other two types.

Stomatal Transpiration

In the epidermal layer of leaves, there are tiny holes called stomata. Two guard cells in the shape of beans surround a stoma. The primary purpose of the open stomata during the day is to allow CO_2 to enter for photosynthesis. Stomata can have a density of 1,000–10,000 per cm^2. The water vapour in stomatal transpiration leaves the leaf through the stomata (Fig. 5.3). Water rises through the stem and flows through the veins to the tissues of the leaves, after being absorbed by the roots from the soil.

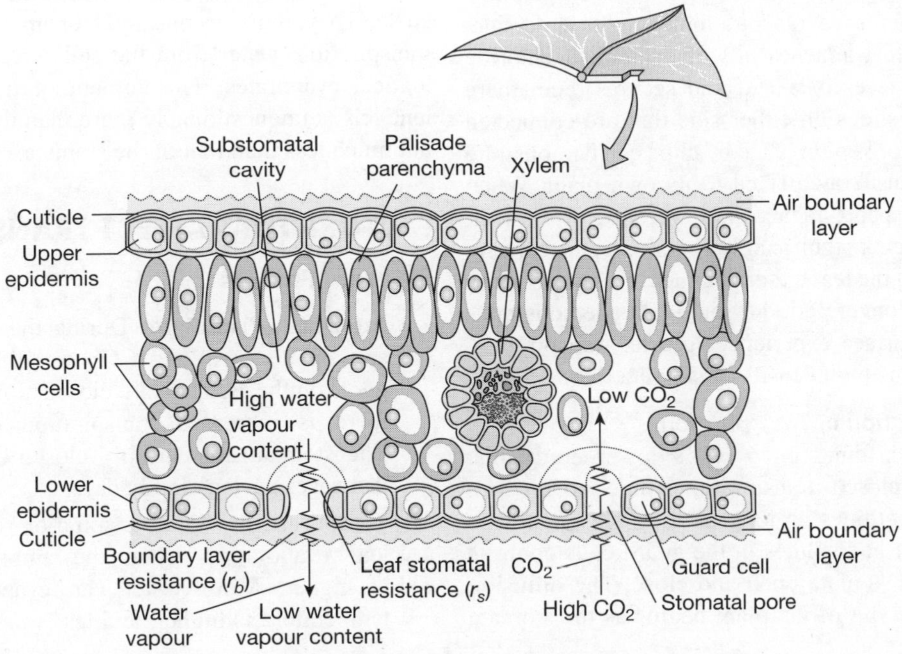

Fig. 5.3 Stomatal transpiration

The intercellular spaces are exposed to the surfaces of a significant proportion of spongy mesophyll cells in the leaves. Some of the water is released as a thin coating from these cell surfaces. As this layer dries out, water evaporates from it, and water vapour is created that saturates the air of the intercellular gaps. After diffusing through additional intercellular spaces that connect to it, the vapour eventually reaches the sub-stomatal space, where it leaves the cell through the stomatal opening. Diffusion is responsible for the water vapour's migration from the cell's surface into the atmosphere outside. Like those of any other gas, the molecules of water vapour travel from an area of higher concentration to a region of lower concentration.

When water flows through a leaf, turgor pressure is applied to the cell wall by the cell sap in every cell. The air between the cells is forced to receive some water that was forced through the cell wall by this pressure. Here, water vapour diffuses through the air spaces between the mesophyll cells into the sub-stomatal region, where it eventually exits through diffusion through the stomata. By pulling extra water from the nearby vein, the cells that lose water in this way replace it. The majority of this water enters the cell through osmosis, while very little enters through imbibition along the cell walls.

This causes more water to be drawn from below through the xylem vessels as thousands of leaf cells evaporate water in this manner. The resulting transpiration pull can raise water in tall trees to a height of at least 50 metres.

The underside of a leaf produces more transpiration. A dicot leaf's underside has more stomatal openings, which leads to increased transpiration from the underside. An experiment that attaches bits of dry cobalt chloride paper to the surfaces of a leaf and secures them there using two glass slides on either side that are connected by elastic bands or kept in place by clips can demonstrate this. The leaf must stay affixed to its own plant. When compared to the paper facing the lower side of the leaf, which becomes pink significantly faster, the paper facing the top surface of the leaf either does not turn pink or does so after a much longer period of time. This demonstrates that the lower surface experiences greater transpiration, which is due to the high density of stomata there.

Stomatal Regulation of Transpiration

The stomata are found on a leaf's lower epidermis. When they are closed, transpiration comes to an end. Transpiration continues as long as they are open. The quantity of water and solutes in the guard cells controls how quickly the stomata open and close. The diffusion of gases into and out of the plant begins as the stomata open during the day, enabling transpiration as well as meeting the demand for photosynthesis. If the water content of the leaf decreases for any reason, the guard cells fail to maintain their turgidity and instead become flaccid, which closes the stomatal opening and prevents transpiration.

Some plant species experience daytime leaf wilt followed by evening recovery. Even when there is plenty of water in the soil, some plants, such as balsam, experience midday leaf wilting. When this occurs, the rate of water absorption by the roots is slower than the rate of midday transpiration. The turgidity of the cells is thereby reduced. The leaves' turgidity is regained and they stand upright in the evening or at night when the stomata are closed and the temperature is not high. As a result, there is no water loss through transpiration. By altering the stoma's size, a plant can regulate stomatal transpiration.

Cuticular Transpiration

On the leaf surface, the epidermis secretes a waxy layer known as the cuticle. The thickness of this cuticle varies from plant to plant. The cuticle's primary purpose is to stop evaporation of water from the leaf surfaces. However, some evaporation does take place. The rate of evaporation decreases with increasing cuticle thickness. The cuticles of desert plants are often thicker to reduce transpiration.

Lenticular Transpiration

Special holes called lenticels form on elder stems in place of stomata. These permit the passage of gases for both photosynthesis and respiration. Lenticels never close. They remain constantly open. In order to effect transpiration, water from the cell surfaces that face the lenticel evaporates. The amount of transpiration from lenticels is unquestionably more than that of the cuticle, but much less than that of the stomata.

FACTORS THAT AFFECT TRANSPIRATION

External Factors

- **Intensity of sunlight:** During the day, the stomata are open to allow CO_2 to diffuse inside for photosynthesis. They are closed at night. As a result, there is increased transpiration during the day. The stomata are partially closed and transpiration decreases during cloudy days.
- **Temperature:** More transpiration results from more leaf evaporation when the ambient temperature is higher. More water can evaporate when the temperature is higher, and less water may evaporate

when the temperature is lower. Compared to cold air, warm air can store more water.

- **Velocity of wind:** Transpiration increases with increasing wind speed. The area outside the leaf does not become saturated with water vapour when the wind blows strongly because it removes the water vapour produced during transpiration more quickly.
- **Humidity:** When the air outside is humid, less sweat is produced. High air humidity slows down the rate of internal water vapour diffusion across stomata, which lowers the rate of transpiration.
- **Carbon dioxide:** Stomatal closure and decrease in transpiration are caused by a CO_2 level in the ambient air that is greater than the typical threshold of 0.03%.
- **Atmospheric pressure:** As air pressure drops, the rate of transpiration rises. As a result, it improves water vapour diffusion.

Internal Factor

If the leaves' water content drops because the roots are not absorbing enough water, the leaves will wilt and their ability to transpire will be diminished. This decrease in transpiration is a result of the stomata closing, which is a normal way for the plant to conserve water.

ADAPTATIONS TO REDUCE EXCESSIVE TRANSPIRATION

Numerous plants, especially those that thrive in arid environments, have developed a variety of long-lasting adaptations to reduce transpiration. Here are a few of these adaptations:

- **Sunken stomata:** The stomata may be sunken or covered by hairs (example, *Nerium*).
- **Fewer stomata:** The number of stomata may be reduced.
- **Narrow leaves:** The leaves may become narrower to reduce the surface area (example, *Nerium*).
- **Reduced exposed surfaces:** In some cases, the leaves may become wavy, rolled or folded to reduce the exposed surface.
- **Loss of leaves:** In some cases, leaves may be dropped, be absent or changed into spines, as in most cacti.
- **Thick cuticle:** The leaves may be covered by a thick cuticle (example, banyan and most evergreen trees).

SIGNIFICANCE OF TRANSPIRATION

Transpiration has great significance for plants. Its main advantages are:

- **Cooling effect:** The temperature of the leaf surface is lowered by evaporation. Therefore, during hot, sunny days, transpiration is beneficial to plants (enzymes are killed by high heat).
- **Suction force:** Transpiration creates a suction force that acts from the top of a plant and aids in the ascent of sap. Cellular sap is concentrated and its osmotic pressure is raised by evaporation from the leaves. This sequentially removes water from lower level cells (ascent of sap) and ultimately encourages root absorption of water from the soil.
- **Distribution of water and mineral salts:** All branches and twigs have leaves at the tips, which causes transpiration from their surfaces to draw water towards them. This aids in the distribution of water throughout the plant body. The rate at which water and other solutes are absorbed from the soil increases with transpiration.

It is incorrect to equate transpiration with excretion or the removal of "excess water". Wherever it takes place, excretion is an intentional, active activity that organisms use to get rid of undesired and metabolic waste products.

The differences between evaporation and transpiration are given in Table 5.1.

Table 5.1 Comparison of evaporation and transpiration

Evaporation	Transpiration
Loss of water from the surface of water bodies in the form of vapour	Loss of water in the form of vapour from the aerial parts of the plant
A physical change controlled by the temperature and humidity of the atmosphere	A vital and partly a physical process controlled by both internal and external factors
A fast process	A slow process

DIRECT LOSS OF WATER BY PLANTS

Some plants may lose water directly, that is, in liquid form and not as water vapour. This occurs in two ways:

- **Guttation:** In the morning, the edges of some plants' leaves display water droplets. This occurs particularly in warm, humid environments. When the air is humid, transpiration is hampered while the roots continue to take in water from the ground. This causes a significant increase in hydrostatic pressure within the plant, which 'forces out' the extra water from the vein tips in the leaf (Fig. 5.4).

- **Bleeding:** This occurs when a plant becomes injured. The punctured or damaged surfaces of a plant allow the plant sap to escape. The plant's root pressure contributes to bleeding.

Fig. 5.4 Guttation
(Courtesy: Schwester M Jutta, https://pixabay.com/photos/leaf-plant-nature-medicine-3357681/)

SUMMARY

- Water is lost from the aerial regions of plants in the form of water vapour through a biological process called transpiration.
- Water is absorbed by the roots of a plant and transported up to the leaves through the xylem tissue. It then evaporates from the surface of the leaves through small openings called stomata.
- Transpiration is driven by the process of photosynthesis, which requires water as one of its reactants.
- Factors affecting transpiration include temperature, humidity, wind, light intensity and soil moisture. Stomata play a crucial role in regulating transpiration, opening and closing in response to environmental conditions to balance the plant's water needs with its need to conserve water.
- Transpiration has several important functions in plants, including cooling the plant, supplying water and minerals to the leaves for photosynthesis, and maintaining the plant's water balance. However, excessive transpiration can also lead to water stress and wilting in plants.
- Transpiration can be measured through techniques such as potometer, gravimetric method or the cobalt chloride paper test.
- Understanding transpiration is important for farmers and gardeners to manage their water resources effectively and to promote healthy plant growth.
- Cuticular transpiration is the process of water loss through the cuticle, which is a waxy layer on the surface of leaves and stems that helps to reduce water loss from the plant
- Lenticels are little pores in the bark of twigs and branches. All plants do not contain lenticels. Lenticels cause negligible loss of water.

PRACTICE QUESTIONS

Analyse Your Concepts (School Exam Based)

Multiple Choice Questions

1. The rate of transpiration is more when:
   (a) Atmosphere is dry
   (b) Temperature is high
   (c) Humidity is high
   (d) Atmosphere is dry and temperature is high

2. Transpiration pull will be maximum under which set of conditions?
   (a) Open stomata, dry atmosphere and moist soil
   (b) Open stomata, high humid atmosphere and well irrigated soil
   (c) Open stomata, high humid atmosphere and dry soil
   (d) Closed stomata, dry atmosphere and dry soil

3. With decrease in atmospheric pressure, the rate of transpiration will:
   (a) Increase (b) Decrease rapidly
   (c) Decrease slowly (d) Remain the same

4. One of the internal factors which affect the rate of transpiration is:
   (a) Large size of the leaf (b) Colour of the leaf
   (c) Sunken stomata (d) Sunny day

5. Guttation takes place through:
   (a) Stomata
   (b) Lenticels
   (c) Lower epidermis of leaves
   (d) Hydathodes

6. The loss of water as water vapour from the aerial parts of a plant is known as:
   (a) Evaporation (b) Perspiration
   (c) Guttation (d) Transpiration

7. Transpiration will be fastest when the day is:
   (a) Cool, humid and windy
   (b) Hot, humid and still
   (c) Hot, humid and windy
   (d) Hot, dry and windy

8. Transpiration is best defined as:
   (a) Loss of water by the plant
   (b) Evaporation of water from the aerial surfaces of a plant
   (c) Loss of water, as water vapour, by a plant
   (d) Release of water by a plant into the atmosphere

9. Most of the transpiration in tall trees occurs through:
   (a) Stomata (b) Lenticels
   (c) Cuticle (d) Bark

Match the Following

Each question contains statements given in two columns which have to be matched. Statements in column I have to be matched with statements in column II.

Column I	Column II
1. Cohesion theory	A. Transpiration
2. Stomata	B. Xylem
3. Movement of water	C. Translocation
4. Phloem	D. Transpiration pull
5. Water absorption	E. Transpiration stream

True or False

Read the following statements and write your answer as true or false.

1. Voltmeter is an instrument used for measuring the rate of transpiration in green plants.
2. Wind velocity has an effect on transpiration.
3. Transpiration creates a pull for upward movement of sap.
4. The maximum transpiration occurs at midnight.
5. Facilitated diffusion is an active process that requires energy expenditure by the cell.

Fill in the Blanks

Complete the following statements with an appropriate word/term to be filled in the blank spaces.

1. Transpiration is the loss of water as water ________ from the parts of the plant.
2. Closing of ________ and shedding of leaves reduces ________.
3. Transpiration helps in creating ________ force and in eliminating excess ________.

4. The active uptake of minerals by the root hairs is an example of _________ transport.

5. The process by which plants lose water in the form of droplets is called _________.

Assertion–Reason Questions

The following questions consist of two statements: Assertion (A) and Reason (R). Answer these questions selecting the appropriate option given below:

(a) Both A and R are true and R is the correct explanation of A
(b) Both A and R are true but R is not the correct explanation of A
(c) A is true but R is false
(d) A is false but R is true

1. Assertion (A): Transpiration is a biological activity.
   Reason (R): Transpiration occurs through the living tissues of the aerial parts of plants.

2. Assertion (A): Transpiration is an important process in plants as it helps in the absorption of minerals and water from the soil.
   Reason (R): The loss of water through transpiration creates a negative pressure in the leaves, which helps to pull water and minerals up from the soil. The process also helps to regulate the temperature of the plant and aids in the transportation of organic compounds throughout the plant.

3. Assertion (A): The rate of transpiration is affected by the temperature of the surrounding air.
   Reason (R): Increase in temperature causes an increase in the kinetic energy of water molecules, leading to an increase in the rate of evaporation from the leaf surface.

4. Assertion (A): Wilting of plants occurs when the rate of transpiration exceeds the rate of absorption of water by the roots.
   Reason (R): Transpiration results in the loss of water vapour from the plant, which leads to a decrease in the water potential of the plant tissues.

5. Assertion (A): Plants adapted to arid conditions have thick cuticles and sunken stomata.
   Reason (R): The thick cuticle reduces water loss through the epidermis, while sunken stomata trap moisture around the stomatal opening, reducing water loss.

6. Assertion (A): Guttation is a process that occurs when water is lost from the leaves of plants.
   Reason (R): Guttation is the exudation of water droplets from the tips or margins of leaves of some vascular plants, caused by root pressure.

One-word Answers

1. The stem's openings where transpiration takes place.
2. The process through which water is lost from intact plant in droplets.
3. A device for measuring transpiration rates.
4. A plant with sunken stomata.
5. Any two leaf portions that permit transpiration.
6. The component of a leaf that permits guttation.

Define the Term

1. Xylem
2. Hydathodes
3. Stomata
4. Lenticels
5. Leaf spines

Groups of Terms

Given below are certain groups of terms. In each group the first pair indicates a relationship between the two terms. Rewrite and complete the second pair on a similar basis.

1. Cuticle: Water conservation :: Guard cells: _______
2. Transpiration: Cooling effect :: Transpiration: _______
3. Xylem: Cohesion-tension theory :: Stomata: _______
4. Transpiration: Water transport :: Transpiration: _______
5. Transpiration: Wilting prevention :: Transpiration: _______

Identify the Odd One Out

1. Stomata, Lenticels, Chloroplast, Hydathodes
2. Active transport, Passive transport, Endocytosis, Facilitated diffusion
3. Water potential, Osmotic pressure, Atmospheric pressure, Root pressure
4. RNA, DNA, ATP, NADP+
5. Plasmolysis, Imbibition, Diffusion, Transpiration

Image-Based Questions

1. The illustration below shows an experimental set-up with a weighing device to illustrate a specific plant process. The test set-up was positioned in direct sunlight. Examine the diagram and respond to the questions.

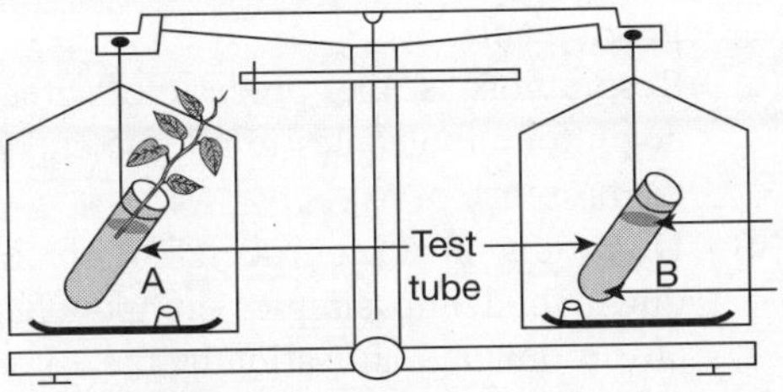

(a) Which process is being studied in the experimental set-up?
1. Photosynthesis
2. Respiration
3. Transpiration
4. Mitosis

(b) What is the definition of the process being studied in the experimental set-up?
1. The process of converting light energy into chemical energy
2. The process of breaking down food molecules to release energy
3. The process of loss of water vapour from plants
4. The process of cell division

(c) What change is observed in the weight of test tubes A and B before and after the experiment?
1. Increase in weight
2. No change in weight
3. Decrease in weight of test tube A and increase in weight of test tube B
4. Decrease in weight of both test tubes A and B

(d) What is the purpose of keeping test tube B in the experimental set-up?
1. To act as a control
2. To provide oxygen for the process
3. To absorb the carbon dioxide produced during the process
4. To measure the rate of the process

2. Given is the diagram of an experimental set-up to study the process of transpiration in plants. Study the same and then answer the questions that follow:

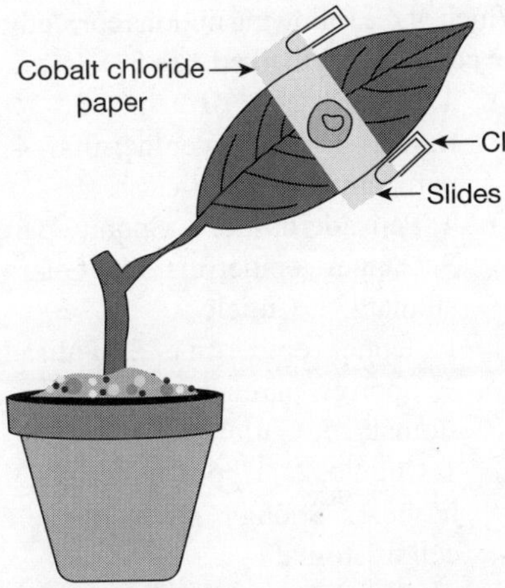

(a) What is the color of dry cobalt chloride paper?
A. Blue B. Green
C. Red D. Yellow

(b) Why are glass slides placed over the dry cobalt chloride papers?
A. To hold the paper in place
B. To prevent exposure to light
C. To prevent air circulation
D. To prevent water vapour from being formed

(c) After about half an hour, what change would you expect to find in the cobalt chloride paper placed on the dorsal and ventral sides of the leaf?
A. No change in colour
B. Both turn blue
C. Dorsal side show less colour, ventral side turns pink
D. Dorsal side turns pink, ventral side shows mixed colour.

3. The given figure represents a leaf section:

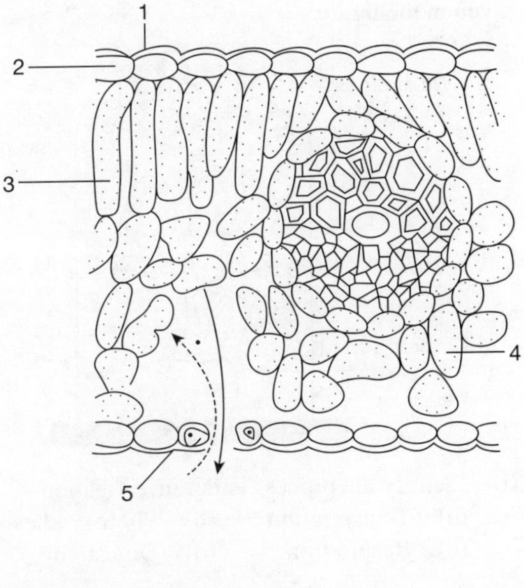

(i) Which of the following options correctly identifies the components marked 1 to 5?
(a) 1. Guard cell of stomata, 2. Palisade tissue, 3. Upper epidermis, 4. Spongy parenchyma, 5. Cuticle
(b) 1. Palisade tissue, 2. Spongy parenchyma, 3. Upper epidermis, 4. Guard cell of stomata, 5. Cuticle
(c) 1. Upper epidermis, 2. Palisade tissue, 3. Spongy parenchyma, 4. Guard cell of stomata, 5. Cuticle
(d) 1. Cuticle, 2. . Upper epidermis,3. Palisade tissue,4. Spongy parenchyma 5. Guard cell of stomata

(ii) What do the dotted and solid arrows represent during the day and night in a leaf?
(a) Dotted arrow shows the path of CO_2 and solid arrow shows the path of oxygen during the day, while dotted arrow shows the path of oxygen and solid arrow shows the path of CO_2 during night.
(b) Dotted arrow shows the path of oxygen and solid arrow shows the path of CO_2 during the day, while dotted arrow shows the path of CO_2 and solid arrow shows the path of oxygen during night.
(c) Dotted arrow shows the path of CO_2 and solid arrow shows the path of CO during the day, while dotted arrow shows the path of oxygen and solid arrow shows the path of oxygen during night.
(d) Dotted arrow shows the path of oxygen and solid arrow shows the path of CO_2 during day and night.

4. Answer the questions by analysing the experiment given in the figure.

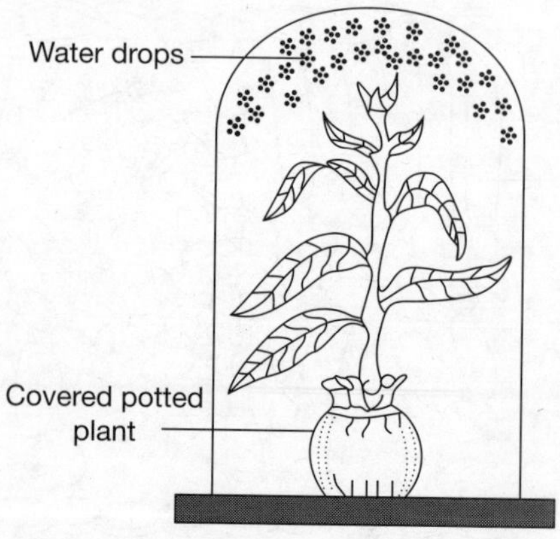

(i) Identify the process under investigation.
(a) Transpiration (b) Photosynthesis
(c) Respiration (d) Digestion

(ii) Elucidate the process mentioned in (i).
(a) Transpiration is the process of loss of water vapour from the aerial parts of the plant, mainly through the stomata present in the leaves.
(b) Photosynthesis is the process of conversion of light energy into chemical energy by green plants.
(c) Respiration is the process of breaking down food molecules to release energy in the presence of oxygen.
(d) Digestion is the process of breaking down food into simpler components for absorption and utilisation by the body.

(iii) What is the reason for covering the pot with a plastic sheet?
(a) To prevent loss of water from the soil
(b) To increase the temperature inside the pot
(c) To provide shade to the plant
(d) To protect the plant from insects

(iv) State one advantage of this process for the plant.
(a) It helps in the absorption of nutrients from the soil
(b) It helps in the production of food by the plant
(c) It facilitates the upward movement of sap in the plant
(d) It helps in the exchange of gases

5. The diagram illustrates Garreau's potometer, which is used to exhibit unequal transpiration in a dorsiventral leaf. Two small vials of anhydrous calcium chloride ($CaCl_2$) were weighed and put in the cups before placing the leaf in between them. Mercury manometers were attached to both cups through corks to seal their ends. After a few hours, the $CaCl_2$ vials were removed and weighed again.

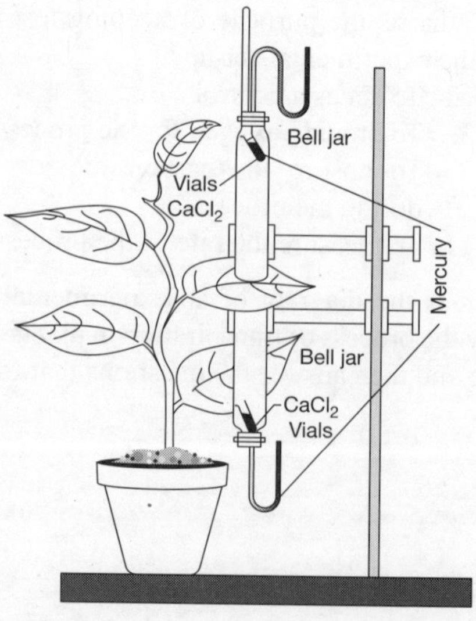

(i) Why are $CaCl_2$ vials placed in the cups of Garreau's potometer during the experiment?
   (a) To absorb moisture
   (b) To increase the rate of transpiration
   (c) To decrease the rate of transpiration
   (d) To measure the rate of transpiration

(ii) When the $CaCl_2$ vials are weighed again after a few hours, do you expect any change in weight? If yes, what could be the reason for it?
   (a) No, there will be no change in weight
   (b) Yes, weight will increase due to absorption of moisture
   (c) Yes, weight will decrease due to loss of moisture
   (d) Yes, weight will remain the same

(iii) What is the purpose of using a manometer in Garreau's potometer?
   (a) To measure the rate of transpiration
   (b) To measure the water potential of the plant
   (c) To determine the water uptake by the roots
   (d) To measure the atmospheric pressure

Very Short Answer Questions

1. Most transpiration occurs during midday. Why?
2. Describe the potometer.
3. It is a common misconception that forests tend to provide more rainy days. Can you provide a scientific explanation?
4. Transplanting seedlings to flowerbeds in the evening is better than doing so in the morning. Why?
5. Land plants die if their roots are water-logged. Why?
6. Why do plants absorb more water than required?

Short Answer Questions

1. On windy days compared to calm days, a higher rate of transpiration is seen. What is the relationship between wind and transpiration rate?
2. Does a dorsiventral leaf's lower surface contribute more to transpiration than the upper surface?
3. Tree bark and cork aid in limiting water loss. How?
4. What is the role of sweating and transpiration in regulating an organism's body temperature?
5. Some desert plants have leaves with sunken stomata. Why?

Long Answer Questions

1. How do forests contribute to the production of rain?
2. In what ways are hydathodes and stomata comparable in plant physiology?
3. What causes wilting? Even when the soil is well irrigated, some plants exhibit leaf wilting at noon. Why is that?
4. Why do most plants have more stomata on the lower surface of the leaf as opposed to the upper surface?
5. Imagine that you have a miniature rose plant flourishing in a pot. In what way would you illustrate transpiration?
6. How does lenticular transpiration work? Name one key distinction between lenticular and stomatal transpiration.
7. Mention any three significant variables that help to speed up transpiration.
8. Summarise the four benefits that transpiration brings to plants.
9. List any three ways by which plants typically limit transpiration.
10. Briefly explain how the rate of transpiration is affected by
    (i) Intensity of light
    (ii) Humidity of the atmosphere

Case Study-Based Questions

1. A farmer noticed that his plants were wilting even though he had watered them thoroughly the previous day. He was worried and decided to consult a botanist. What could be the reason for the plants wilting even though they were watered thoroughly?
   (a) The plants might have been infected with a disease.
   (b) The farmer might have overwatered the plants.
   (c) The plants might be experiencing excessive transpiration.
   (d) The soil might be too dry and unable to retain water.
2. A gardener notices that his plants are wilting, despite the fact that the soil is moist. He suspects that there may be a problem with the transpiration process in the plants.
   (i) Which of the following is the most likely cause of the wilting?
      (a) Overwatering
      (b) Underwatering
      (c) Lack of sunlight
      (d) Too much sunlight

(ii) What is the function of the stomata in the leaves of plants?
(a) To absorb carbon dioxide
(b) To release oxygen
(c) To release water vapour
(d) All of the above

(iii) What is the name of the process by which water is lost through the stomata in the leaves of plants?
(a) Respiration (b) Photosynthesis
(c) Transpiration (d) None of the above

(iv) How does transpiration help in the absorption of water and minerals from the soil?
(a) It creates a positive pressure in the roots, which pushes water and minerals up into the leaves.
(b) It creates a negative pressure in the leaves, which pulls water and minerals up from the roots.
(c) It helps to regulate the temperature of the plant, which in turn aids in the absorption of water and minerals.
(d) None of the above.

(v) What factors can affect the rate of transpiration in plants?
(a) Temperature, humidity, and wind
(b) Soil pH and nutrient levels
(c) The age of the plant and the type of plant
(d) All of the above

3. A student performed an experiment to investigate the rate of transpiration in a plant. She used a potometer and observed the movement of air bubbles in the capillary tube. However, after a few minutes, the air bubbles stopped moving. What could be the reason for this?
(a) The plant ran out of water and transpiration stopped
(b) The potometer was not properly sealed
(c) The temperature was too low to facilitate transpiration
(d) The air bubbles got stuck in the capillary tube due to an air leak

4. A farmer was worried about his crops in a dry season. He decided to water his crops at night to reduce the rate of transpiration. Will his strategy be successful? Why or why not?
(a) Yes, it will be successful because transpiration does not occur at night
(b) No, it will not be successful because water uptake by the roots is also reduced at night
(c) Yes, it will be successful because the temperature is lower at night, reducing the rate of transpiration
(d) No, it will not be successful because plants can still lose water through stomatal transpiration at night

5. A student observed that the leaves of a plant wilted during the day but regained their turgidity at night. She hypothesized that the plant was losing too much water through transpiration during the day. However, when she measured the rate of transpiration, she found that it was not abnormally high. What other factor could be responsible for the wilting of the leaves during the day?
(a) Lack of sunlight
(b) Excessive heat
(c) Low atmospheric pressure
(d) Salt accumulation in the soil

NCERT Corner

1. In what way do plants transport water and minerals?

2. How does the structure of the leaf help in the process of transpiration?

3. How does temperature affect the rate of transpiration in plants?

4. Describe an experiment to show the effect of light on the rate of transpiration.

5. What is wilting? How does excessive transpiration cause wilting in plants?

COMPETITION WINDOW (OBJECTIVE TYPE)

1. The tissue that conducts water is:
(a) Phloem (b) Xylem
(c) Companion cells (d) Transfusion tissue

2. Exudation of sap from injured parts is called:
(a) Guttation (b) Secretion
(c) Bleeding (d) Active transport

3. A plant with sunken stomata is:
(a) Mango (b) *Nerium*
(c) Neem (d) Shrubs

4. The leaf modification in cacti to check transpiration is:
(a) Thorns (b) Phylloclade
(c) Prickles (d) Spines

5. The factor that does not affect the rate of transpiration is:
(a) Intensity of light (b) Velocity of wind
(c) Carbon dioxide (d) Oxygen

6. Plastic is draped over a piece of blue cobalt chloride paper that has been clipped to the underside of a new leaf. Part of the paper begins to turn pink after a few minutes, indicating the presence of water. Which process does the foliage engage in to colour the paper pink?
(a) Absorption (b) Photosynthesis
(c) Respiration (d) Transpiration

7. Which of the following is the enzyme responsible for catalyzing the process of transcription in cells?
(a) DNA polymerase (b) RNA polymerase
(c) Ribosome (d) Helicase

8. With decrease in atmospheric pressure, the rate of transpiration will:
(a) Increase (b) Decrease rapidly
(c) Decrease slowly (d) Remain the same

9. How does water vapour flow through a leaf's stomata on a dry, sunny day?
(a) Into the leaf by diffusion
(b) Into the leaf by respiration
(c) Out of the leaf by diffusion
(d) Out of the leaf by respiration

10. Transpiration is highest during:
(a) Rainy season (b) Winter
(c) Summer (d) Autumn

11. Which one of the following does not affect the rate of transpiration?
(a) Light (b) Humidity
(c) Wind (d) Age of the plant

12. The rate of transpiration increases with:
(a) Increase in humidity
(b) Increase in wind velocity
(c) Reduced light intensity
(d) Increase in the CO_2 level

13. Which of the following is not true for transpiration?
(a) It cools the plants
(b) Water lost contains minerals
(c) Water is lost as vapour
(d) Maximum loss through stomata

14. Where on a leaf does the majority of the water evaporate during transpiration occur?
(a) The cuticle
(b) The guard cells
(c) The spongy mesophyll cells
(d) The xylem vessels

15. A decrease in which factor normally causes the transpiration rate to increase?
(a) Humidity (b) Light intensity
(c) Stomatal opening (d) Temperature

Answer Keys

Multiple Choice Questions

1. (b) **2.** (a) **3.** (a) **4.** (c) **5.** (d)
6. (d) **7.** (d) **8.** (b) **9.** (b)

Match the Following

1. D **2.** A **3.** E **4.** C **5.** B

True or False

1. F **2.** T **3.** T **4.** F **5.** F

Fill in the Blanks

1. Vapour
2. Stomata and transpiration
3. Transpirational and water
4. Active
5. Guttation

Assertion–Reason Questions

1. (a) **2.** (a) **3.** (a) **4.** (a) **5.** (a)
6. (a)

One-word Answers

1. Stomata
2. Guttation
3. Potometer
4. Xerophyte
5. Stomata and cuticle
6. Hydathode

Define the Term

1. Xylem is a type of plant tissue that is responsible for transporting water and dissolved minerals from the roots to the rest of the plant.
2. Hydathodes are specialised pores or glands on the leaves of some plants that secrete water droplets. They are also called water glands or water stomata.
3. Stomata are microscopic openings or pores found mainly on the leaves and stems of plants. They regulate the exchange of gases and water vapour between the plant and the environment.
4. Lenticels are small, raised pores found on the bark of woody plants that allow for gas exchange between the plant and the external environment.
5. Leaf spines are modified leaves that have evolved to protect the plant from herbivores and other threats. They are usually stiff and pointed, and can be found on a variety of plants, including cacti, succulents and some shrubs.

Groups of Terms

1. Stomatal regulation
2. Nutrient uptake
3. Transpiration pull
4. Gas exchange
5. Mineral absorption

Identify the Odd One Out

1. Chloroplast
2. Endocytosis
3. Atmospheric pressure
4. ATP
5. Transpiration

Image-Based Questions

1. (a) 1. Transpiration
(b) 3. The process of loss of water vapour from plants
(c) No change in weight
(d) To act as a control

2. (a) D. Yellow
(b) D. To prevent water vapour from being formed
(c) C. Dorsal side show less colour, ventral side turns pink

3. (i) (d) 1. Cuticle, 2. Upper epidermis, 3. Palisade tissue, 4. Spongy parenchyma, 5. Guard cell of stomata
(ii) (a) Dotted arrow shows the path of CO_2 and solid arrow shows the path of oxygen during the day, while dotted arrow shows the path of oxygen and solid arrow shows the path of CO_2 during night.

4. (i) (a) Transpiration
(ii) (a) Transpiration is the process of loss of water vapour from the aerial parts of the plant, mainly through the stomata present in the leaves.
(iii) (a) To prevent loss of water from the soil
(iv) (c) It facilitates the upward movement of sap in the plant.

5. (i) (c) To decrease the rate of transpiration
   (ii) (b) Yes, weight will increase due to absorption of moisture
   (iii) (a) To measure the rate of transpiration

Very Short Answer Questions

1. During midday, the intensity of sunlight is at its peak, leading to high temperatures and an increase in the rate of transpiration.
2. A potometer is an instrument used to measure the rate of water uptake by a leafy shoot, which is an indirect way of measuring transpiration.
3. Forests release water vapour into the atmosphere through transpiration, which can lead to the formation of clouds and, in some cases, precipitation. However, forests can also reduce rainfall in some circumstances.
4. The rate of transpiration is particularly high during the day. Excessive transpiration will cause the seedling to wilt and eventually die. The seedlings will, however, continue to be transplanted in the evening because transpiration rates are very low and they may retain the water taken from the soil.
5. The root hairs are absent from older sections of the root. They are protected by a layer of dead calls with minuscule openings known as lenticels. Gaseous exchange takes place between the earth and the interior living cells through these lenticels. Land plants will perish if their roots are left soggy with water since their lenticels will not be able to exchange gases.
6. The plants absorb much more water than their requirement because:
   (a) They require more mineral salts for their growth and minerals are present in very small quantity in water
   (b) Most of the water is lost through transpiration. Therefore, to prevent wilting they require excess of water

Short Answer Questions

1. Windy conditions increase the movement of air around the leaves, which results in increased evaporation of water from the leaves. This leads to a higher rate of transpiration compared to calm days when there is less air movement around the leaves.
2. Dorsiventral leaves, which have distinct upper and lower surfaces, tend to have more stomata on the lower surface, which results in greater transpiration from that side. The lower surface is also cooler and more humid, which can further increase transpiration rates.
3. Tree bark and cork cells are both composed of dead cells and serve as protective layers to prevent water loss and protect the plant from external damage. Bark and cork cells are also impervious to gases, which further helps to reduce water loss through transpiration.
4. Sweating is the process of secreting water and salt through sweat glands on the skin's surface, which cools the body as the sweat evaporates. Transpiration is the process by which plants lose water vapour through their leaves, which helps to cool the plant and prevent overheating. Both processes involve the evaporation of water to remove excess heat from the organism.
5. Sunken stomata are situated in pits or grooves on the surface of the leaf, which helps to create a microclimate that is more humid and less prone to evaporation than the surrounding air. This reduces the rate of water loss from the plant and helps to conserve water.

Long Answer Questions

1. Hint: Forests contribute to rain production through transpiration and the release of water vapour into the atmosphere, which can form clouds and eventually lead to rainfall. Elaborate on this point.
2. Hint: Hydathodes and stomata are involved in the regulation of water loss in plants. Elaborate on this point.
3. Hint: Wilting in plants can be caused by excessive transpiration, which can occur even when the soil is well-irrigated. This can be due to factors such as high temperature, low humidity, or inadequate water uptake by the roots. Elaborate on these points.
4. Hint: Most plants have more stomata on the lower leaf surface to reduce water loss by minimizing direct exposure to sunlight and wind. Elaborate on this point.
5. Hint: Transpiration in a miniature rose plant can be illustrated by the water vapour escaping from the tiny pores (stomata) on the leaves. Elaborate on this point.
6. Hint: Lenticular transpiration occurs through the surface of stems or other specialised structures, and a key distinction from stomatal transpiration is that it does not involve the opening and closing of stomata. Elaborate on these points.
7. Hint: Three variables that can speed up transpiration are high temperature, low humidity, and increased air movement. Elaborate on these points.
8. Hint: The four benefits of transpiration for plants include cooling, nutrient uptake, movement of water

and dissolved minerals, and maintenance of cell turgidity. Elaborate on these points.

9. Hint: Plants typically limit transpiration through adaptations such as having a waxy cuticle on leaves, closing stomata during hot and dry conditions, and developing specialised structures like spines or hairs to reduce airflow around the leaves. Elaborate on these points.
10. Hints: (i) The rate of transpiration increases with higher light intensity.

    (ii) The rate of transpiration decreases with higher humidity levels.

    Elaborate on these points.

Case Study-Based Questions

1. (c)
2. (i) (b), (ii) (d), (iii) (c), (iv) (b), (v) (a)
3. (b)
4. (d)
5. (b)

NCERT Corner

1. Plants transport water and minerals through their roots, which absorb the water and minerals from the soil. The water and minerals then travel up the stem and into the leaves through the xylem vessels.
2. The structure of the leaf helps in the process of transpiration by providing a large surface area for water evaporation through the stomata, which are tiny pores on the underside of the leaf. The presence of air spaces between the spongy mesophyll cells also aids in the movement of water vapour out of the leaf.
3. Temperature affects the rate of transpiration in plants because as the temperature increases, the rate of evaporation of water from the leaf surface also increases. This leads to an increase in the rate of transpiration.
4. To show the effect of light on the rate of transpiration, one can set up an experiment by taking a potted plant with a plastic bag covering its leaves. Make a small hole in the bag, insert a capillary tube and fill it with water. Then expose the plant to different levels of light, record the water loss from the capillary tube and analyse the data to determine the effect of light on the rate of transpiration.
5. Wilting is the drooping of plant leaves due to a lack of water uptake, which causes a decrease in turgor pressure within the plant cells. Excessive transpiration causes wilting in plants as it leads to a water deficit within the plant, which in turn causes a decrease in turgor pressure and the wilting of leaves.

Competition Window

1. (b)	2. (c)	3. (b)	4. (d)	5. (d)
6. (d)	7. (a)	8. (a)	9. (c)	10. (c)
11. (d)	12. (b)	13. (b)	14. (c)	15. (a)

CHAPTER 6

Photosynthesis: Provider of Food for All

PREREQUISITES

You must have knowledge of:

- Structure of plants, including the functions of different plant organs such as leaves, stems and roots
- Basic properties of light, including the electromagnetic spectrum, wavelengths and energy
- Environmental factors such as temperature, light intensity and carbon dioxide levels

LEARNING OBJECTIVES

At the end of this chapter, you will be able to:

- Describe the process of photosynthesis
- Explain the relevance of photosynthesis to other living things
- Identify the reactants and products of photosynthesis
- Describe the main structures involved in photosynthesis
- Identify the factors that can affect the rate of photosynthesis, such as light intensity, temperature and the availability of water and nutrients
- Explain the relationship between photosynthesis and cellular respiration

INTRODUCTION

All living organisms require food to survive. Animals consume plants as food. Plants manufacture food through the process of photosynthesis. Photosynthesis is the process by which plants, algae and some bacteria convert light energy from the sun into chemical energy in the form of glucose. It is a critical process that sustains life on earth, as it is the primary means by which energy from the sun is converted into a usable form for living organisms.

During photosynthesis, chlorophyll pigments located in the chloroplasts of plant cells absorb light energy from the sun. This energy is then used to power a series of chemical reactions that convert carbon dioxide and water into glucose and oxygen. The glucose serves as a source of energy for the plant, while the oxygen is released into the atmosphere.

Photosynthesis can be divided into two stages: the light-dependent reactions and the light-independent reactions. The light-dependent reactions occur in the thylakoid membrane of the chloroplasts and involve the absorption of light by the chlorophyll pigments to produce ATP and NADPH. The light-independent reactions occur in the stroma of the chloroplasts and involve the use of these energy carriers to convert carbon dioxide into glucose.

PHOTOSYNTHESIS

All green plants carry out photosynthesis. It is the process by which food is created from carbon dioxide and water in the presence of chlorophyll and light energy. All green plants go through the same fundamental chemical steps in this process. During photosynthesis, plants exhale oxygen as a waste product.

Importance of Photosynthesis

- **Access to food.** Photosynthesis ultimately provides all living things with energy and food, directly for plants and indirectly for animals and humans who consume plants or other animals that consume plants.
- **Breathing in oxygen.** The only biological process that releases oxygen into the atmosphere is photosynthesis. All life on earth needs oxygen to live.

Chlorophyll: The Vital Plant Pigment

Chlorophyll ('*chloro*' meaning 'green', '*phyll*' meaning 'leaf') is the green-coloured matter found in plants. It is found in the microscopic cell organelles called chloroplasts.

The interior of the chloroplast, which is a tiny oval body with a double membrane, is made up of densely clustered flattened sacs (thylakoids) arranged in piles (grana) on top of a stromal substance that is colourless (Fig. 6.1). A cell typically contains 40–50 chloroplasts. The walls of the thylakoids contain the pigment chlorophyll. Carbon, hydrogen, oxygen, nitrogen and magnesium make up this extremely complex substance.

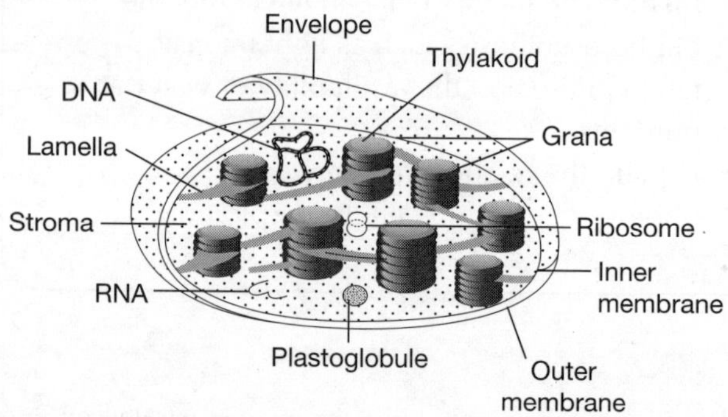

Fig. 6.1 Structure of the chloroplast

Palisade cells and spongy cells, which are mesophyll cells situated between the upper and lower epidermis of the leaves, are the main cells that contain chloroplasts. These are present in the outer layers of young green stems as well as in the guard cells of stomata. There could be over 500,000 chloroplasts per sq. mm. The fact that there are nine different types of chlorophyll may surprise you. The most well-known and prevalent of these are chlorophylls a and b.

Light from both ends of the visible spectrum are absorbed by chlorophyll. Green light is reflected away by blue and red light. Chlorophyll appears green because of this. For photosynthesis, the absorbed blue and red light is most useful.

Due to its high sensitivity to light, chlorophyll can be destroyed by excessive light. However, exposure to light is necessary for the plant to produce chlorophyll. For example, grass growing in the shade beneath a stone turns yellow due to lack of new chlorophyll and breakdown of the older one in the absence of light.

Regulation of Stomatal Openings

The primary purpose of stomata is to allow CO_2 from the atmosphere into the plant for photosynthesis. When they are not used for photosynthesis, they tend to close their openings at night to reduce water loss from the leaves through transpiration. They reopen to let CO_2 diffuse in when there is light, such as just after sunrise. The price a plant pays for photosynthesis is transpiration.

The water moving into and out of cells is what causes the stomata to close and open (Fig. 6.2). Their cytoplasm contains chloroplasts, and they have a thick inner wall that faces the opening and a thin outer wall on the other side.

Opening and Closing of Stomata

There are two hypotheses about how stomata open and close:

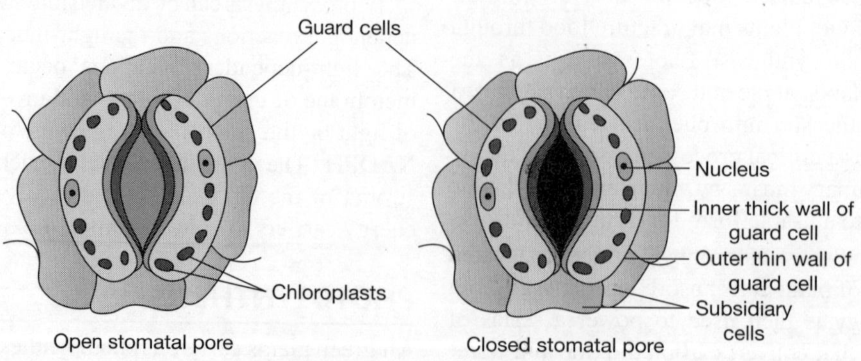

Fig. 6.2 Open and closed stomata

- **Sugar concentration hypothesis:** During the day, the guard cells carry out photosynthesis, and the sugar (glucose) produced raises the osmotic pressure. This causes water to be drawn from neighbouring cells through endosmosis. As the guard cells have a thin outer wall, they become turgid and protrude outwards, which widens the stomatal opening between them. Diffusion of gases into and out of the plant begins as soon as the stomata open, enabling transpiration and supplying the necessary gases for photosynthesis.

  If the amount of water in the leaf decreases for any reason, exosmosis draws the water out of the guard cells, causing them to become flaccid. Their inner, thick walls straighten, as a result, leading to closing of the stomata.
- **K^+ ion concentration theory:** This theory states that the production of a potassium ion (K^+) gradient is necessary for the opening and closing of stomata. The guard cells' chloroplasts carry out photosynthesis during the day, which results in the synthesis of ATP. Potassium ions from nearby cells are actively pumped into the guard cells with the help of this ATP. The guard cells become hypertonic at night due to increase in K^+ concentration. As a result, more water from neighbouring cells is drawn in, making the cells more turgid. They then move outward to open the stomatal pore. The K^+ ions leak out, thereby reducing the turgor of guard cells and the stomatal pore closes.

Process of Photosynthesis

The primary hubs of this activity are the mesophyll cells in leaves, both palisade and spongy. When sunlight strikes a leaf during the day, the upper layers of the mesophyll, particularly the palisade cells, trap the light energy. This energy is used for the chemical processes involved in the production of food, where the raw materials are carbon dioxide and water (Fig. 6.3).

Through the stomata, carbon dioxide from the atmosphere diffuses into the leaf along a concentration gradient (higher concentration outside, lower concentration inside). Water is distributed in the mesophyll tissue of the leaves after being absorbed by the roots, and sent up through the stem and by the roots to the soil.

The chemical equation to represent this process is as follows:

$$6CO_2 + 12H_2O \longrightarrow C_6H_{12}O_6 + 6H_2O + 6O_2$$

The glucose molecule $C_6H_{12}O_6$ is a simple sugar readily soluble in water.

The 6 molecules of H_2O liberated at the end of the process are those that are re-formed during a chain of reactions and not from the original ones.

Phases of Photosynthesis

The two main phases of photosynthesis are light-dependent and light-independent.

Light-Dependent Phase (Photochemical Phase): Light is incredibly important in this stage. The photochemical phase is so-called because it is characterised by a rapid succession of chemical reactions that are brought about by light. The thylakoids of the chloroplast, which contain chlorophyll, are where the light reaction takes place.

The light reaction (photochemical phase) occurs in two main steps:

1. **Activation of chlorophyll:** The chlorophyll on exposure to light energy becomes activated by

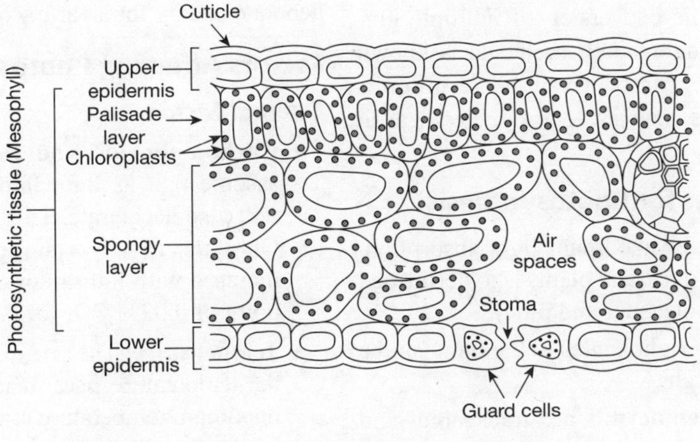

Fig. 6.3 Photosynthetic tissue of mesophyll cell

absorbing photons (photon is the smallest unit of light energy).

2. **Splitting of water:** The absorbed energy is used to split the water molecule (H_2O) into its two components (hydrogen and oxygen) and release electrons.

$$2H_2O \longrightarrow 4H^+ + 4e^- + O_2$$

The reaction is known as **photolysis**. It is defined as the splitting of H_2O molecules into hydrogen ions and oxygen in the presence of light and grana.

End result of the products of photolysis

- The hydrogen ions (H^+) are picked up by a compound NADP (nicotinamide adenine dinucleotide phosphate) to form NADPH

$$NADP^+ + e^- + H+ \longrightarrow NADPH$$

- The oxygen (O) component is given out as molecular oxygen (O_2).

$$2O \longrightarrow O_2$$

- The electrons (e^-) are used in converting ADP (adenosine diphosphate) into energy-rich compound ATP (adenosine triphosphate) by adding one phosphate group P_i (inorganic phosphate).

$$ADP + Phosphate \longrightarrow ATP$$

This process is called **phosphorylation** (addition of phosphate) and since the energy used in it comes from light (photons), the process is termed **photophosphorylation**.

Light-Independent (Dark) Phase (Biosynthesis Phase): Here, glucose is transformed into starch and other chemicals (Fig. 6.4). Most green plants immediately transform glucose into starch after it is created during photosynthesis. The conversion of multiple glucose molecules into a single starch molecule is known as **polymerisation**. Plants can convert glucose into oils or sucrose, also known as cane sugar or double sugar in chemistry.

Adaptations in Leaves for Photosynthesis

- **Large surface area** for maximum light absorption. (Narrowing of leaves in certain plants is to performed conserve water, as in oleander and pine).
- **Leaf arrangement** at right angles to the light source to obtain maximum light.
- **Cuticle and upper epidermis** are transparent and waterproof to allow light to enter freely.
- **Numerous stomata** allow for rapid exchange of gases (oxygen and carbon dioxide).
- The **thinness of leaves** reduces the distance between cells, facilitating rapid transport.
- The **chloroplasts** are concentrated in the upper layers of the leaf so as to obtain light energy quickly.

End Result of the Products of Photosynthesis

- **Glucose:** The simple sugar glucose is used in four different ways, as required by the plants:
  - Immediately consumed by the plant cells
  - Stored in the form of insoluble starch
  - Converted into sucrose
  - Used in synthesizing fats, proteins, etc.
- **Water:** The water produced in the process may be re-utilised in the continuance of photosynthesis.
- **Oxygen:** Most of the oxygen is not needed and diffuses out into the atmosphere through the stomata. However, some oxygen may be used in respiration in the leaf cells (a process known as **photorespiration**). In a way, even this oxygen is not wasted because it is necessary for all living things, including plants that need it at night.

Utilisation of artificial food and its transfer: All other plant parts require food produced in the leaf in order to function. During photosynthesis, glucose is created in the leaf very quickly, but it cannot be transported to other areas as quickly. In order to store it temporarily in the leaf, it is transformed into an insoluble starch. The starch is converted back into soluble sugar at night and is then transported in solution through the veins of the leaf and the stem's phloem. This allows it to be moved (translocated) to various areas of the plant, where it may be converted back into starch for storage (like in potatoes) in order to generate energy for a variety of tasks within the plant.

Factors Affecting Photosynthesis

External Factors

- **Carbon dioxide and brightness:** It is natural to assume that as light intensity rises, photosynthesis will also accelerate. This is accurate to some extent. Only up to a certain point does photosynthesis increase with light intensity before stabilising at the point of 0.02% CO_2 for the two factors combined.
- **Temperature:** The rate of photosynthesis increases as the temperature rises, reaching its peak at 35°C (the maximum temperature at which photosynthesis works best), after which it begins to decline and ceases above 40°C. The enzymes are destroyed at this maximum

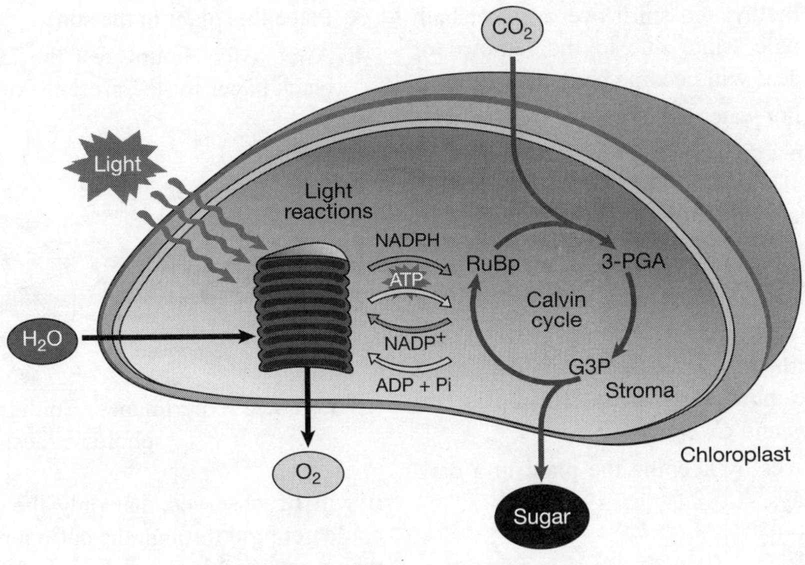

Fig. 6.4 Photosynthesis in plants

limit. The rate of photosynthesis doubles for every 10°C increase up to the ideal temperature (35°C). For instance, the rate of photosynthesis doubles when the temperature increases from 20°C to 30°C, 22°C to 32°C, or 25°C to 35°C.

- **Water content:** By lowering diffusive capacity (CO_2 intake) due to closure of stomata, a lack of water caused by decreased soil absorption or excessive transpiration decreases the rate of photosynthesis. Only 1% of the water taken in by the root is used for photosynthesis.

Internal Factors

- **Chlorophyll:** Nutritional deficiencies of minerals cause loss of chlorophyll, and hence, a drop in trapping solar energy.
- **Protoplasm:** Dehydration of the protoplasm reduces the rate of photosynthesis. Similarly, the accumulation of carbohydrates (sugar and starch) also reduces the rate of photosynthesis.
- **Structure of the leaf:** The thickness of the cuticle, distribution of stomata and size of the leaf influence the amount of light and CO_2 entering the leaf.

Dawn to dusk natural changes in environmental conditions and photosynthesis

Plants experience a predictable cycle of changes in light intensity and photosynthesis rate every 24 hours. It rises from dawn until midday and falls as dusk draws near. Like other living things, plants breathe, consuming oxygen and exhaling carbon dioxide. In the presence of light, photosynthesis masks the effects of respiratory activity, resulting in net oxygen output. In the dark, changes in the surrounding atmosphere's gaseous composition are solely brought on by respiration.

Some individuals think that peepal trees are the only plants that produce oxygen even at night. As of now, neither the release of such oxygen nor a similar viewpoint can be supported by scientific evidence.

EXPERIMENTS

The various conditions and requirements necessary for photosynthesis can be demonstrated through a variety of experiments.

A plant used in a photosynthesis experiment needs to first be de-starched (starch removed) by being left in the dark for 24–48 hours. The leaves will not display any starch during this time because all of the starch would have been removed to the storage organs.

Experiment 1: To test a leaf for starch (iodine test)

1. Dip the leaf in boiling water for a minute to kill the cells (Fig. 6.5).

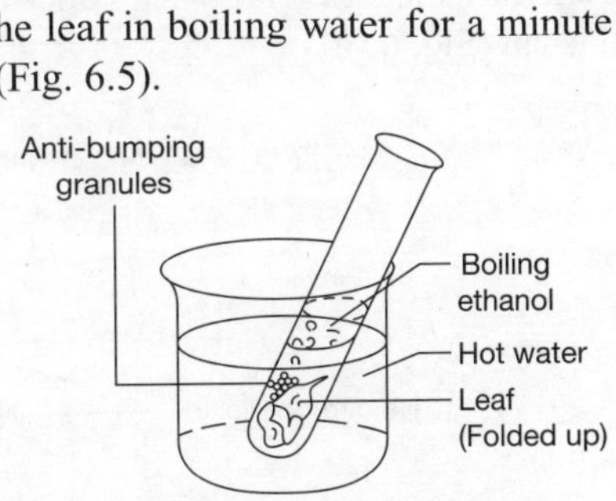

Fig. 6.5 Experiment 1: Iodine test

2. Boil the leaf in methylated spirit over a water bath till it becomes pale white due to the removal of chlorophyll. The leaf will become hard and brittle.
3. Place it again in hot water to soften it.
4. Spread the leaf in a dish and pour iodine solution on it. The presence of starch will be indicated by a blue-black colour. A leaf without starch will show brown colouration.

Experiment 2: To show that chlorophyll is necessary for photosynthesis

1. Take a plant with variegated leaves having some green and some non-green areas. Examples are *Coleus*, *Geranium* and *Croton*.
2. De-starch the leaves by keeping the plant in a dark room for a few days.
3. Place the plant in the sun.
4. Draw its outline on paper and mark the green parts of the leaf that turn bluish, showing the presence of starch (Fig. 6.6)

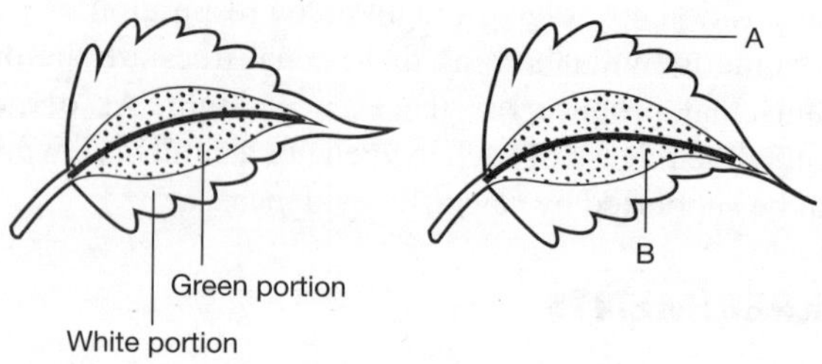

Fig. 6.6 Experiment 2: Chlorophyll is necessary for photosynthesis (After performing the starch test, part A turns blue-black with iodine solution and part B turns brown with iodine solution)

Experiment 3: To show that sunlight is necessary for photosynthesis

1. Take a plant with de-starched leaves.
2. Cover one of its leaves with black paper on which a design is cut (Fig. 6.7).
3. Place this plant in the sun.
4. After a few hours, test the leaf that is covered by black paper for the presence of starch.

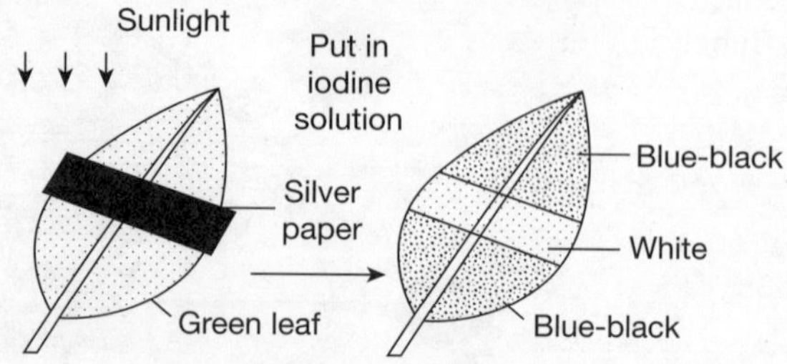

Fig. 6.7 Experiment 3: Sunlight is necessary for photosynthesis

It will be observed that only the parts of the leaf that could get light through the cut-out design as well as those that were left uncovered turn blue-black, showing the presence of starch.

Experiment 4: To show that carbon dioxide is necessary for photosynthesis

1. Take a plant with de-starched leaves.
2. Insert a part (half) of one of its leaves (through a split cork) into a conical flask containing potassium hydroxide (potassium hydroxide absorbs carbon dioxide).
3. Leave the plant in sunlight (Fig. 6.8).
4. After a few hours, test the leaf for starch.

It can be seen that photosynthesis requires carbon dioxide because the portion of the leaf that was exposed to atmospheric air turns blue-black while the portion of the leaf inside the flask containing potassium hydroxide does not. The portion of the leaf that is not inside the flask serves as the control.

Experiment 5: To show that oxygen is produced during photosynthesis

1. Place some water plants (*Elodea* or *Hydrilla*) in a beaker containing pond water and cover them with a short-stemmed funnel (Fig. 6.9).

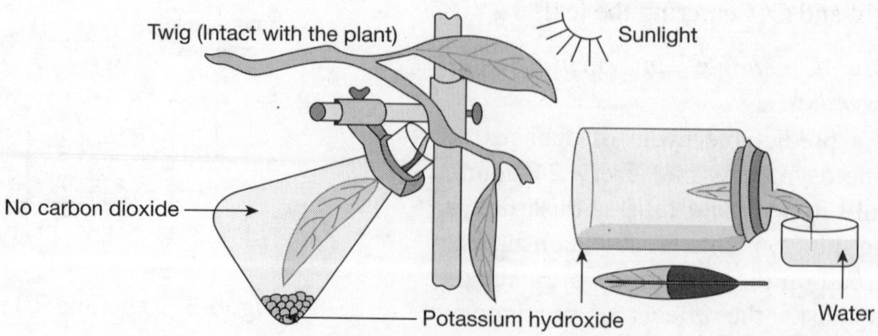

Fig. 6.8 Experiment 4: Carbon dioxide is necessary for photosynthesis

2. Invert a test tube full of water over the stem of the funnel. (Ensure that the level of water in the beaker is above the level of stem of the inverted funnel).
3. Place the apparatus in the sun for a few hours. Bubbles of the gas will collect in the test tube.

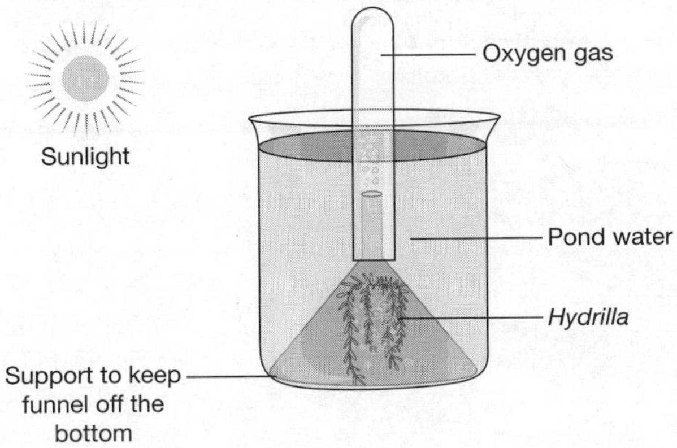

Fig. 6.9 Experiment 5: Oxygen is produced during photosynthesis

4. Test the gas in the test tube. A glowing splinter will burst into flames, which shows the presence of oxygen.

Importance of Photosynthesis

- **Provides food:** Plants are ultimately the source of nutrition for all animals, including humans. A carnivore (flesh-eater) consumes the flesh of an animal that was once a herbivore, while a herbivore (plant-eater) consumes plant parts directly. The term **food chain** (Fig, 6.10) refers to this network of reliance on food. It may include a number of intermediate stages, but a plant (a producer) that produces its own food through photosynthesis is always the chain's initial link. You can imagine several food chains. For example:

  Corn → Rat → Snake → Hawk
  Grass → Rabbit → Tiger
  Grass → Grasshopper → Lizard → Kite
  Grass → Insects → Frog → Snake → Peacock
  Wheat → Humans → Mosquito

  Fungi and bacteria, which are not green plants, feed on the environment's decomposing organic matter. This material is derived from decomposing plants and animals that were themselves dependent on photosynthesis.
- **Provides oxygen:** Photosynthesis is the only process that makes oxygen, the gas that sustains life, available in the atmosphere in a free state. Most scientists concur that the atmosphere was devoid of free oxygen about 2 billion years ago, when there was no life of any kind. The appearance of green plants paved the way for the appearance of animals. Almost all living things require oxygen to breathe.

Carbon Cycle

The carbon cycle is a series of chemical reactions in which the chemical element carbon (found in CO_2) is extracted from the air, used by living things for physiological functions, and then re-introduced in the atmosphere (Fig. 6.11).

Essential Steps:

1. **Photosynthesis:** Green plants (producers) use carbon dioxide from the air to produce carbohydrates (sugar, starch, etc.)

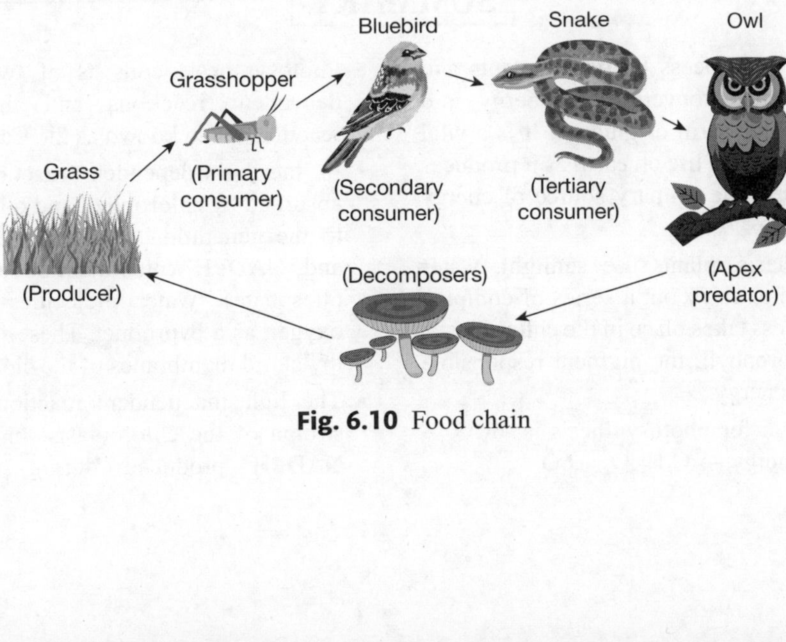

Fig. 6.10 Food chain

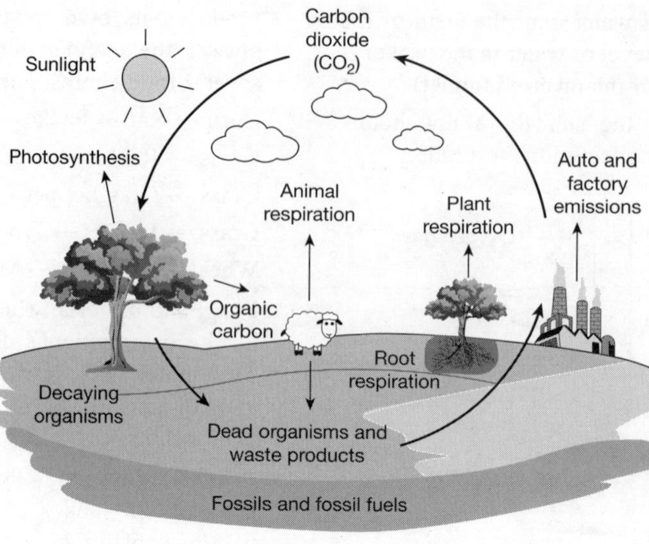

Fig. 6.11 Carbon cycle

2. **Food chains:** Through food chains, the food (containing carbon in carbohydrates and other products) is passed on from plants to animals (consumers). Carbon in an essential element of the body tissues of both plants and animals.
3. **Respiration:** All plants and animals respire by oxidising carbohydrates in their cells to produce energy and give out carbon dioxide into the atmosphere.
4. **Decay:** The dead remains of plants and animals are consumed by bacteria and fungi (decomposers) and, in the process, they break down organic matter, releasing carbon dioxide back into the atmosphere.
5. **Combustion (burning):** When a fuel such as wood or a fossil fuel like coal, petroleum or natural gas is burnt, the carbon contained in it is oxidised to carbon dioxide, which is returned to the atmosphere. All the fuels named above originally came from living organisms. A certain amount of carbon dioxide is released during heating or burning of limestone in lime kilns.

SUMMARY

- Photosynthesis is the process by which plants and some other organisms convert light energy into chemical energy in the form of glucose. It is a vital process for the survival of life on earth as it produces oxygen and serves as the primary source of energy for most ecosystems.
- During photosynthesis, plants use sunlight, water and carbon dioxide to carry out a series of complex reactions. The process takes place in the chloroplasts, which contain chlorophyll, the pigment responsible for capturing light energy.
- The overall equation for photosynthesis is: $6CO_2 + 6H_2O + \text{Sunlight energy} \rightarrow C_6H_{12}O_6 + 6O_2$
- Photosynthesis consists of two stages: the light-dependent reactions and the light-independent reactions (also known as the Calvin cycle).
- In the light-dependent reactions, light energy is absorbed by chlorophyll and other pigments, leading to the generation of ATP (adenosine triphosphate) and NADPH (nicotinamide adenine dinucleotide phosphate). Water molecules are split, releasing oxygen as a byproduct. These reactions occur in the thylakoid membranes of the chloroplasts.
- The light-independent reactions take place in the stroma of the chloroplasts. In this stage, ATP and NADPH produced during the light-dependent

reactions are used to convert carbon dioxide into glucose through a series of enzyme-catalyzed reactions known as the Calvin cycle.

- Factors such as light intensity, temperature and the availability of water and carbon dioxide influence the rate of photosynthesis. Plants have adapted various mechanisms to optimise photosynthesis in different environmental conditions, such as opening and closing stomata to regulate gas exchange and minimise water loss.
- Understanding photosynthesis is crucial for studying plant biology, ecosystem dynamics and the earth's carbon cycle. It provides the basis for understanding how plants generate energy and produce oxygen, which are fundamental for supporting life on our planet.

PRACTICE QUESTIONS

Analyse Your Concepts (School Exam Based)

Multiple Choice Questions

1. Chlorophyll is present:
(a) In the grana of chloroplasts
(b) On the surface of chloroplasts
(c) Dispersed throughout the chloroplasts
(d) In the stroma of chloroplasts

2. The specific function of light energy in the process of photosynthesis is to:
(a) Reduce carbon dioxide
(b) Synthesise glucose
(c) Activate chlorophyll
(d) Split water molecules

3. Which one of the following is not a limiting factor for photosynthesis?
(a) Oxygen (b) Light
(c) Carbon dioxide (d) Chlorophyll

4. A cell that lacks chloroplast does not:
(a) Evolve carbon dioxide
(b) Liberate oxygen
(c) Require water
(d) Utilise light energy

5. Which would do maximum harm to a tree?
(a) Loss of half of its branches
(b) Loss of half of its flowers
(c) Loss of all of its leaves
(d) Loss of a little bark

6. The production of starch, and not glucose, is often used as a measure of photosynthesis in leaves because:
(a) Starch is an immediate product of photosynthesis
(b) Glucose formed in photosynthesis is soon converted to starch
(c) Starch is soluble in water
(d) Glucose cannot be used

7. The number of water molecules required in chemical reactions to produce one molecule of glucose during photosynthesis is:
(a) 6 (b) 12
(c) 18 (d) 24

8. The rate of photosynthesis is not affected by:
(a) Light intensity (b) Humidity
(c) Temperature (d) CO_2 concentration

9. Chlorophyll in a leaf is required for:
(a) Breaking down water into hydrogen and oxygen
(b) Emitting green light
(c) Trapping light energy
(d) Storing starch in the leaves

10. If the rate of respiration becomes more than the rate of photosynthesis, plants will:
(a) Continue to live, but will not be able to store food
(b) Be killed instantly
(c) Grow more vigorously because more energy will be available
(d) Stop growing and die gradually of starvation

11. Which chemical reaction occurs during photosynthesis?
(a) Carbon dioxide is reduced and water is oxidised
(b) Water is reduced and carbon dioxide is oxidised
(c) Both carbon dioxide and water are oxidised
(d) Both carbon dioxide and water are reduced

12. A plant is kept in a dark cupboard for 48 hours before conducting any experiment on photosynthesis in order to:

(a) Remove chlorophyll from leaves
(b) Remove starch from the leaves
(c) Ensure that no photosynthesis occurred
(d) Ensure that the leaves are free from starch

13. During photosynthesis, the oxygen in glucose comes from:
(a) CO_2
(b) Water
(c) CO_2 and water
(d) Oxygen through air

Match the Following

Each question contains statements given in two columns which have to be matched. Statements in column I have to be matched with statements in column II.

Column I	Column II
(i) Grana	(a) A green pigment
(ii) Chloroplast	(b) Source of energy
(iii) CO_2 and H_2O	(c) Power
(iv) Glucose	(d) Organelles found in autotrophs
(v) Chlorophyll	(e) Light reaction

True or False

Read the following statements and write your answer as true or false.

1. The overall goal of photosynthesis's light reaction is to transform solar energy into kinetic energy that cells can use to do work.
2. Carnivores, such as lions, are dependent on photosynthesis to survive because lions eat animals that eat plants.
3. Chlorophyll is a specialised pigment that is necessary for photosynthesis to occur as it absorbs light and transforms it into chemical energy.
4. Photosynthesis occurs only during the day when the sun is shining.
5. The energy required for photosynthesis comes from the sun.

Fill in the Blanks

Complete the following statements with an appropriate word/term to be filled in the blank spaces.

1. The site of light reaction in the cells of a leaf is ________.
2. The chemical substance used to test the presence of starch in the cell of a leaf is ________.
3. Stroma is the ground substance in the ________.
4. The dark reaction of photosynthesis is known as ________.
5. In flowering plants, food is transported in the form of ________.

Assertion-Reason Questions

The following questions consist of two statements: Assertion (A) and Reason (R). Answer these questions selecting the appropriate option given below:

(a) Both A and R are true and R is the correct explanation of A
(b) Both A and R are true but R is not the correct explanation of A
(c) A is true but R is false
(d) A is false but R is true

1. Assertion (A): Chloroplasts mostly occur in mesophyll cells along their walls inside the leaves.
Reason (R): The membrane system of chloroplasts is responsible for trapping light energy and also for the synthesis of ATP and NADPH.
2. Assertion (A): *Rhoeo* leaves contain anthocyanin pigments in epidermal cells.
Reason (R): Anthocyanins are accessory photosynthetic pigments.
3. Assertion (A): The presence of four pigments – chlorophyll a, chlorophyll b, xanthophylls and carotenoids – is what gives leaves their colour.
Reason (R): The primary pigment involved in photosynthesis is chlorophyll b.
4. Assertion (A): Bacterial photosynthesis occurs by utilising wavelengths longer than 700 nm.
Reason (R): Here, the reaction centre is B-890.
5. Assertion (A): There is a decrease in photosynthesis, if the photosynthetic cells are illuminated by light with a wavelength of P680 or more.
Reason (R): In red drop phenomenon, the rate of photosynthesis decreases.

One-word Answers

1. The organisms that prepare their own food from simple raw materials.

2. The type of plastids present in the leaf's mesophyll cells.
3. The substance that the cells use to store energy.
4. The initial food substance created by photosynthesis.
5. The organisms that are referred to as 'natural air cleansers'.
6. Where do aquatic plants get their carbon dioxide from?
7. The region of the chloroplast where photosynthesis's dark reaction takes place.
8. The tissue that carries synthetic starch from the leaves to every portion of the plant.

Define the Term

1. Stroma
2. Autotroph
3. Chlorophyll
4. Photons
5. Light reaction

Groups of Terms

Given below are certain groups of terms. In each group the first pair indicates a relationship between the two terms. Rewrite and complete the second pair on a similar basis.

1. Chloroplast: Photosynthesis :: Mitochondria: ___
2. Light-dependent reactions: Thylakoid membrane :: Light-independent reactions: ___
3. Chlorophyll: Pigment :: Carotenoid: ____
4. ATP: Energy currency :: NADPH: _____
5. Calvin cycle: Carbon fixation :: Electron transport chain: ____

Identify the Odd One Out

1. Stomata, Xylem, Chloroplast, Mitochondria
2. Photosystem I, Photosystem II, ATP Synthase, Ribosomes
3. Autotrophs, Heterotrophs, Consumers, Producers
4. Absorption spectrum, Action spectrum, Photosynthesis, Respiration
5. Photosynthesis, Cellular respiration, Transpiration, Digestion

Image-Based Questions

1. How would you prove that when green plants are exposed to light, oxygen is released?

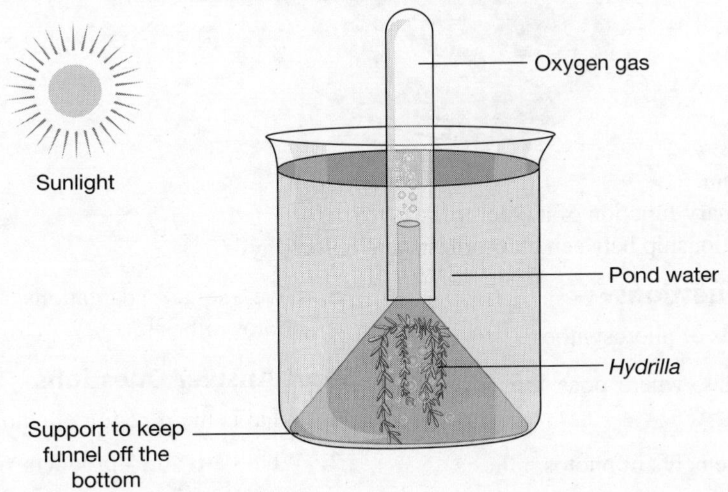

2. Answer the question about the diagram:

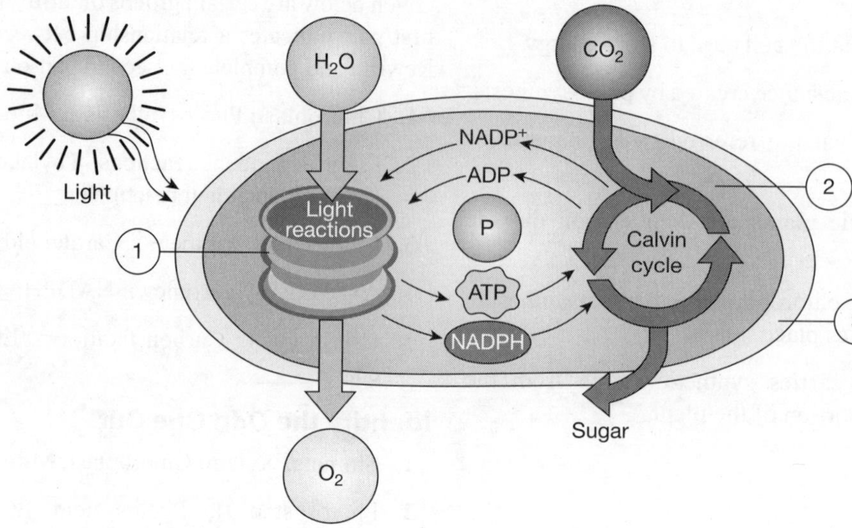

(a) Label the diagram.
(b) What are the two stages of photosynthesis called?
(c) What is the product of the light-dependent stage of photosynthesis?

3. Answer the question about the diagram:

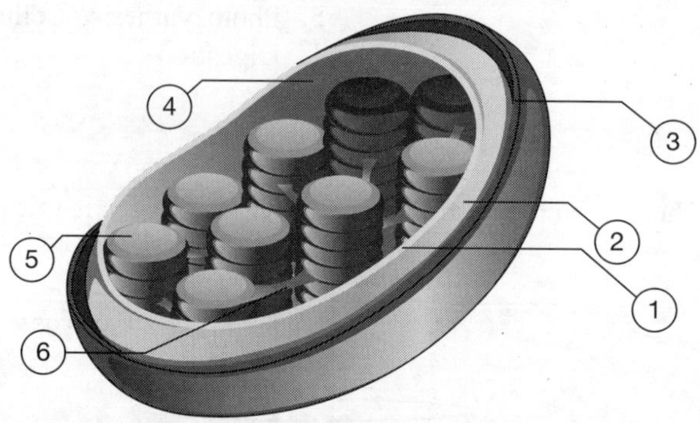

(a) Label the diagram.
(b) What is the primary function of a chloroplast?
(c) What is the relationship between chloroplasts and chlorophyll?

Very Short Answer Questions

1. What are the reactants of photosynthesis?
2. During photosynthesis, where does the oxygen in glucose come from?
3. Give the basic requirements of photosynthesis.
4. (i) Where are the chlorophyll pigments present in a cell?
   (ii) Name the membrane that connects the thylakoid of one granum with another granum.
5. Give some adaptations in a green leaf for photosynthesis.

Short Answer Questions

1. What is the function of chloroplasts?
2. What are the products of photosynthesis and respiration?
3. What is the role of sunlight in respiration?
4. What is the importance of photosynthesis in the life of the following:

(i) Green plants (ii) Non-green plants (iii) Animals

5. A leaf is a food factory. Explain.
6. How do non-green plants such as fungi and bacteria obtain their nourishment?

Long Answer Questions

1. Explain the process of photosynthesis in detail, highlighting the roles of key components such as chloroplasts, chlorophyll, and the Calvin cycle.
2. Discuss the factors that affect the rate of photosynthesis. Explain how light intensity, carbon dioxide concentration and temperature influence the process.
3. Explain the concept of chlorophyll and its role in photosynthesis. Discuss the different types of chlorophyll and their absorption spectra.
4. Describe the role of ATP and NADPH in the process of photosynthesis. Explain how these energy carriers are produced and utilised in the light-dependent and light-independent reactions.
5. Discuss the factors that affect the rate of photosynthesis. Explain how light intensity, carbon dioxide concentration and temperature influence the process.

Case Study-Based Questions

1. Given below is a representation of a certain phenomenon in nature, with four organisms.

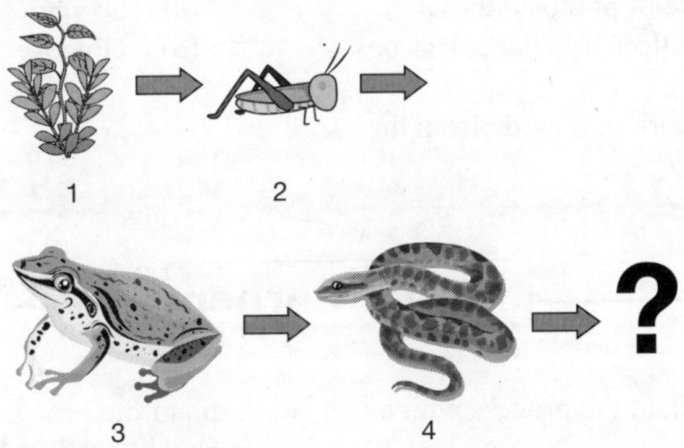

(i) Name the phenomenon represented:
 (a) Food chain (b) Decomposition
 (c) Digestion (d) None of the above

(ii) Which of the following organisms is a primary producer in a food chain?
 (a) Mosquito (b) Deer
 (c) Grass (d) Snake

(iii) Name any one organism that could be shown at position no. 5:
 (a) Deer (b) Giraffe
 (c) Eagle (d) None of the above

(iv) Which biological process is the starting point of the food chain?
 (a) Photosynthesis
 (b) Cellular respiration
 (c) Protein synthesis
 (d) DNA replication

(v) Name one natural element which organisms 2–4 and even 5 obtain from 1 for their survival:
 (a) Oxygen
 (b) Carbon
 (c) Nitrogen
 (d) Hydrogen

2. A potted plant was used to demonstrate a requirement for photosynthesis. For 24 hours, the potted plant was left in complete darkness. The centre of one of the leaflets was lined with a piece of black paper. After that, the potted plant spent a few hours in direct sunlight. Respond to the questions below.
   (i) What is the purpose of lining the centre of the leaflet with a piece of black paper?
      (a) To prevent photosynthesis from occurring in that area
      (b) To enhance photosynthesis in that area
      (c) To measure the amount of oxygen produced by that area
      (d) To make the leaflet look attractive
   (ii) Why was the potted plant left in complete darkness for 24 hours?
      (a) To reduce the rate of respiration
      (b) To speed up the rate of photosynthesis
      (c) To observe the effect of darkness on photosynthesis
      (d) To remove excess carbon dioxide from the plant
   (iii) What is the importance of direct sunlight in this experiment?
      (a) To provide energy for photosynthesis
      (b) To reduce the rate of respiration
      (c) To observe the effect of light on photosynthesis
      (d) To remove excess oxygen from the plant
   (iv) If the black paper was removed after a few hours in direct sunlight, what would happen to the rate of photosynthesis in that area?
      (a) It would increase
      (b) It would decrease
      (c) It would remain the same
      (d) It would stop completely
   (v) What is the final product of photosynthesis?
      (a) Carbon dioxide
      (b) Water
      (c) Oxygen
      (d) Glucose

NCERT Corner

1. What is photosynthesis? Explain the process with a balanced chemical equation.
2. Why is photosynthesis important for life on earth? Explain.
3. Name the pigment involved in photosynthesis. What is its role in the process?
4. Differentiate between autotrophs and heterotrophs. Give examples of each.
5. How do plants obtain the raw materials necessary for photosynthesis?
6. Explain the process of the light-dependent reaction in photosynthesis.
7. What is the role of ATP and NADPH in the light-independent reaction of photosynthesis?
8. Describe the factors affecting the rate of photosynthesis.
9. Explain the significance of the Calvin cycle in photosynthesis.
10. What happens to the glucose produced during photosynthesis?

COMPETITION WINDOW (OBJECTIVE TYPE)

1. Plants close their stomata to preserve water on a hot, dry day. How would this affect the process of photosynthesis?
   (a) The rate of photosynthesis will stay the same.
   (b) The rate of photosynthesis will speed up.
   (c) The rate of photosynthesis will slow down.
   (d) The rate of photosynthesis will stop.
2. What two products result from photosynthesis?
   (a) Water and oxygen
   (b) Water and carbon dioxide
   (c) Glucose and carbon dioxide
   (d) Glucose and oxygen
3. All of the statements about thylakoids are correct, except:

(a) Thylakoids are assembled into stacks.
(b) Thylakoids exist as a maze of folded membranes.
(c) The space surrounding thylakoids is called stroma.
(d) Thylakoids contain chlorophyll.

4. From where does a heterotroph directly obtain its energy?
(a) Eating other organisms
(b) The sun
(c) Simple chemicals in the environment
(d) The sun and eating other organisms

5. The plastids which impart green colour to leaves are:
(a) Chloroplast (b) Xanthophyll
(c) Leucoplast (d) Carotenes

6. The plastids which store starch in plants are:
(a) Chromoplasts (b) Chloroplasts
(c) Leucoplast (d) None of these

7. The part of the chloroplast where the biosynthetic phase occurs is:
(a) Thylakoid (b) Mitochondria
(c) Granum (d) Stroma

8. Splitting of water is caused by chemical reactions brought on by light, called:
(a) Photolysis
(b) Photo-oxidation
(c) Electrolysis
(d) Activation

9. Photophosphorylation is:
(a) A process in which a phosphate group is added to a molecule, such as a sugar or a protein.
(b) The process of synthesising energy-rich ATP molecules from ADP in the presence of sunlight.
(c) A biochemical process that involves the addition of phosphate to an organic compound.
(d) A biochemical process that involves the addition of phosphate to an inorganic compound.

10. Stroma is the:
(a) Site of photolysis of photosynthesis
(b) Site of photochemical phase of photosynthesis
(c) Site of light-dependent phase of photosynthesis
(d) Site of light-independent phase of photosynthesis

Answer Keys

Multiple Choice Questions

1. (a) 2. (c) 3. (a) 4. (b) 5. (c)
6. (b) 7. (b) 8. (b) 9. (c) 10. (a)
11. (a) 12. (c) 13. (d) 14. (a)

Match the Following

1. (c) 2. (d) 3. (e) 4. (b) 5. (a)

True or False

1. F 2. T 3. T 4. F 5. T

Fill in the Blanks

1. Grana 2. Iodine solution
3. Chloroplast 4. Calvin cycle
5. Sucrose

Assertion-Reason Questions

1. (b) 2. (c) 3. (c) 4. (b) 5. (b)

One-word Answers

1. Autotrophs
2. Chloroplasts
3. ATP
4. Glucose
5. Plants
6. Water
7. Stroma
8. Phloem

Define the Term

1. Stroma refers to the fluid-filled region within the chloroplast where the light-independent reactions of photosynthesis take place.
2. An autotroph is an organism capable of producing its own food using light energy or chemical energy, without needing to consume other organisms.
3. Chlorophyll is a green pigment found in chloroplasts that plays a key role in photosynthesis by absorbing light energy and converting it into chemical energy.
4. Photons are the fundamental particles of light and electromagnetic radiation, which are absorbed by chlorophyll during the process of photosynthesis.
5. Light reaction is the first stage of photosynthesis that takes place in the thylakoid membrane of the chloroplast, where light energy is absorbed by chlorophyll and converted into chemical energy in the form of ATP and NADPH.

Groups of Terms

1. Cellular respiration
2. Stroma
3. Accessory pigment
4. Reducing power
5. ATP synthesis

Identify the ODD One Out

1. Mitochondria
2. Ribosomes
3. Consumers
4. Respiration
5. Digestion

Image-Based Questions

1. The release of oxygen during photosynthesis can be demonstrated by an experiment involving aquatic plants such as *Hydrilla*. The experiment involves the following steps:
   (i) Take a beaker of water and add some sprigs of *Hydrilla* to it.
   (ii) Place the beaker in bright sunlight.
   (iii) After a few hours, bubbles will start to appear on the leaves of the *Hydrilla* and rise to the surface of the water.
   (iv) Collect some of the gas released in a test tube and add a glowing splint to it.
   (v) If the gas is oxygen, the glowing splint will relight.

   This experiment shows that in the presence of sunlight, *Hydrilla* undergoes photosynthesis and releases oxygen gas as a byproduct.
2. (a) 1 - Thylakoid, 2 - Stroma, 3 - Chloroplast
   (b) The two stages of photosynthesis are called the light-dependent reaction (or light reaction) and the light-independent reaction (or dark reaction, or Calvin cycle).
   (c) The product of the light-dependent stage of photosynthesis is ATP (adenosine triphosphate) and NADPH (nicotinamide adenine dinucleotide phosphate), which are used as energy and electron carriers in the light-independent reaction. Oxygen (O_2) is also produced as a byproduct of the light-dependent

reaction.

3. (a) 1 - Inner membrane, 2 - Intermembrane space, 3 - Outer membrane, 4 - Stroma, 5 - Thylakoid, 6 - Lamella

(b) The primary function of a chloroplast is to carry out photosynthesis, which involves converting light energy into chemical energy in the form of organic compounds.

(c) Chloroplasts contain the pigment chlorophyll, which is responsible for absorbing light energy and converting it into chemical energy during the process of photosynthesis. Chlorophyll is essential for photosynthesis to occur, and without chloroplasts, there would be no way for plants to carry out this important process.

Very Short Answer Questions

1. The reactants of photosynthesis are light energy, water, carbon dioxide and chlorophyll.
2. The splitting of water during the light-dependent reaction releases oxygen.
3. The basic requirements of photosynthesis are solar energy, green chlorophyll, carbon dioxide from the air and water.
4. (i) In the cell organelle called plastids (or chloroplasts)
   (ii) Stroma lamellae
5. (i) A lot of surface area for water absorption.
   (ii) In order to get maximum light, it is positioned at right angles to the light.
   (iii) There are many stomata to allow for quick gas exchange.
   (iv) Materials can move or diffuse more quickly due to the thinness of leaves.

Short Answer Questions

1. The main function of chloroplasts is to carry out photosynthesis, which involves converting light energy into relatively stable chemical energy. Chloroplasts are also responsible for the biosynthesis of active compounds such as amino acids, phytohormones, nucleotides, vitamins, lipids and secondary metabolites. Additionally, they are essential for ensuring the growth and survival of plants.
2. The products of photosynthesis are glucose and oxygen, while the products of cellular respiration are water and carbon dioxide. Photosynthesis involves the use of energy from sunlight, water and carbon dioxide to produce glucose and oxygen, while cellular respiration uses glucose and oxygen to produce water and carbon dioxide.
3. Sunlight is essential for photosynthesis, which converts light energy into chemical energy in the form of ATP (adenosine triphosphate). This chemical energy is then used to produce glucose and oxygen, which are necessary for cellular respiration.
4. (i) Complex, energy-rich carbohydrates can be built up by green plants and then utilised for various cell metabolic processes.
   (ii) Non-green plants like saprophytes and parasites eat the food that green plants make during photosynthesis in order to sustain themselves.
   (iii) Animals eat green plants or other animals that feed on green plants.
5. Green plants have leaves that are uniquely adapted to the task of food synthesis. The raw components can be collected most effectively using the anatomy of leaves. Through its stomatal holes, the leaf absorbs CO_2 from the surrounding atmosphere. The stem provides water for it. The real locations of light- and dark-dependent photosynthetic processes, which produce glucose, are the chloroplasts. As a result, it is referred to as the plant's food factory because of its capacity to gather raw materials and conduct chemical reactions to create food.
6. Fungi, bacteria and other non-green plants feed off the organic matter in their surroundings that has decomposed from dead animals and plants. They, therefore, rely on photosynthesis.

Long Answer Questions

1. Hints: (a) Respiration releases carbon dioxide, while photosynthesis produces oxygen.
   (b) Light reactions produce ATP and NADPH, while dark reactions produce glucose.
   (c) Chlorophyll is a pigment involved in capturing light energy, while chloroplast is the organelle where photosynthesis takes place.
   Elaborate on these points.
2. Hint: Respiration requires oxygen and occurs in the presence of sunlight, so it cannot be directly observed in a green plant kept in sunlight. Elaborate on this point.
3. Hint: Photosynthesis and respiration are interconnected processes in plants. Photosynthesis utilises carbon dioxide and sunlight to produce glucose and

oxygen, while respiration uses oxygen and glucose to produce carbon dioxide and energy (ATP). The products of one process serve as the reactants for the other, creating a cycle of energy and material exchange. Elaborate on these points.

4. Hint: Putting a plant in the dark before starting a photosynthesis experiment is important to ensure that only photosynthesis, not respiration, is occurring. In the dark, the plant stops producing oxygen through photosynthesis and relies solely on respiration, allowing researchers to measure the net production or consumption of oxygen during the experiment. Elaborate on these points.
5. Hint: Plants obtain the raw materials necessary for photosynthesis through their roots and leaves. Water is absorbed from the soil through the roots, while carbon dioxide is taken in from the atmosphere through tiny pores called stomata on the leaves. These raw materials, along with sunlight, are used by the plant to carry out photosynthesis. Elaborate on these points.

Case Study-Based Questions

1. (i) (a), (ii) (c), (iii) (c), (iv) (a), (v) b
2. (i) (a), (ii) (c), (iii) (a), (iv) (a) (v) (d)

NCERT Corner

1. Photosynthesis is the process by which plants, algae and some bacteria convert light energy into chemical energy in the form of glucose. The balanced chemical equation for photosynthesis is: 6 CO_2 + 6 H_2O + Light energy $\rightarrow C_6H_{12}O_6 + 6\ O_2$
2. Photosynthesis is important for life on earth because it is the primary source of energy for most living organisms. The process releases oxygen into the atmosphere, which is essential for respiration and the survival of many organisms. Additionally, photosynthesis removes carbon dioxide from the atmosphere, which helps regulate the earth's climate.
3. The pigment involved in photosynthesis is chlorophyll. Its role is to absorb light energy from the sun and transfer it to the reaction centre of the photosystems, where it is converted into chemical energy.
4. Autotrophs are organisms that can produce their own food through photosynthesis, while heterotrophs depend on other organisms for their food. Examples of autotrophs include plants, algae and some bacteria, while examples of heterotrophs include animals and fungi.
5. Plants obtain the raw materials necessary for photosynthesis as follows: water is absorbed by the roots from the soil, carbon dioxide enters through the stomata in the leaves, and chlorophyll is already present in the leaves. Sunlight is obtained from the sun.
6. The light-dependent reaction in photosynthesis occurs in the thylakoid membrane of chloroplasts. The process involves the absorption of light energy by chlorophyll, which is then used to generate ATP and NADPH.
7. In the light-independent reaction, ATP and NADPH are used to power the conversion of carbon dioxide into glucose through the Calvin cycle. These molecules provide the energy and reducing power needed for this process.
8. Factors affecting the rate of photosynthesis include light intensity, temperature and the availability of water and carbon dioxide.
9. The Calvin cycle is significant in photosynthesis because it converts carbon dioxide into glucose, which is used as an energy source by plants and other organisms. The cycle also regenerates the molecules needed to continue the process of carbon fixation.
10. Glucose produced during photosynthesis is either used by the plant as an energy source or stored as starch for later use. Some of it may also be used by other organisms in the ecosystem.

Competition Window

1. (c)	**2.** (d)	**3.** (b)	**4.** (a)	**5.** (a)
6. (c)	**7.** (d)	**8.** (a)	**9.** (b)	**10.** (d)

CHAPTER

7 Genetics and Evolution

PREREQUISITES

You must have knowledge of:

- Cell structure and functions
- Chromosomes and DNA
- Meiosis and mitosis
- Natural selection and adaptation
- Origin of life and its theories
- Scientific methods and laboratory techniques

LEARNING OBJECTIVES

At the end of this chapter, you will be able to:

- Define the terms heredity, variation and genetics
- Explain the pattern of Mendelian inheritance
- Describe evolution
- Explain the features of evolution

INTRODUCTION

Why does a human baby look like a human being and also resemble his/her parents, grandparents or even cousins? Why does a kitten resemble a cat? Why does a seedling develop the same leaves, stems and flowers as its parents? Why do all organisms, for that matter, have the same structure as their parents?

Heredity refers to the passing down of comparable characteristics from generation to generation. In other words, it is the transmission of characteristics/traits from one generation to the next. Genes are responsible for inheritance. Even among members of the same family, different gene combinations cause 'variations' or distinctions. **Genetics** is the branch of biology that studies heredity and variation.

Variation refers to the differences between individuals within a species or population. For example, in a rose garden, the colour of the rose blossoms on different plants will vary and the coat colour of puppies from the same mother will vary. Individual variations are referred to as variants. Variations can be attributed to genes or the environment.

GENES AND ALLELES

Every gene often has two distinct characters that have different consequences. The term 'allele' refers to these alternative forms. For instance, there are two outcomes for the behaviour known as tongue rolling, which involves extending the tongue in the shape of a tube from the mouth. As a result, this gene has two alleles, one for rolling and the other for not rolling (Fig. 7.1).

Examine your ear lobes, as well as those of your acquaintances and family members. As shown in (Fig. 7.2), the lower end of the ear lobe can be attached or free. This characteristic is inherited. Examine your parents' and siblings' earlobes to see from which of your parents you acquired this characteristic.

Gregor Johann Mendel (Father of Genetics)

Many scientists in the past were captivated by heredity. An Austrian monk named Gregor Johann Mendel (1822–1884) worked on the difficult task of discovering the answers. He chose several pea plants, nurtured them year after year, gathered a lot of data, and for the first time, analysed and

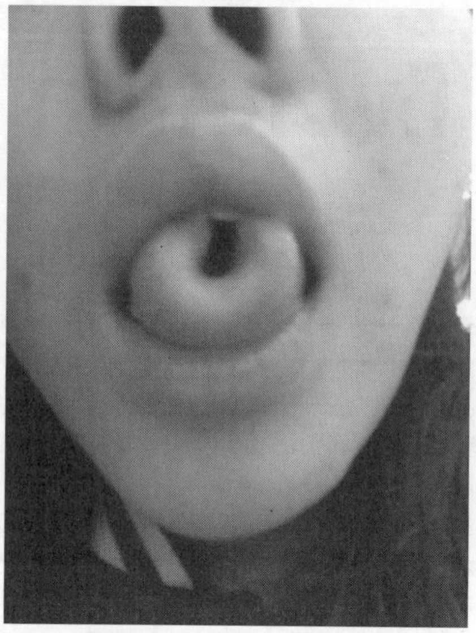

Fig. 7.1 Tongue rolling
(Depicts typical tongue roll, used for tongue rolling examples; Z33333t; 2017; https://commons.wikimedia.org/wiki/File:Tongue_rolled.jpg. Licensed under: CC-BY-SA-4.0)

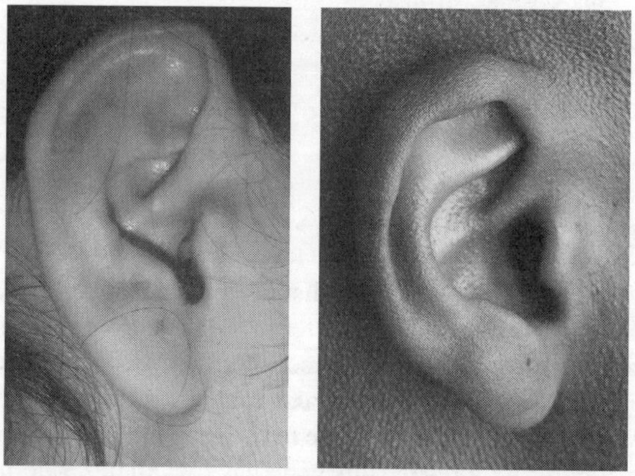

Fig. 7.2 Ear lobe, whether fixed (left) or free (right), is hereditary
(Ear of a twenty one year old Bengali female; Alif Ahmed; 2015; https://commons.wikimedia.org/wiki/File:Human-ear-female.jpg. Licensed under: CC-BY-SA-4.0 and Human right ear; Tomas Gunnarsson; 2016; https://commons.wikimedia.org/wiki/File:Human_right_ear_(cropped).jpg. Licensed under: CC-BY-SA-4.0)

proposed certain laws of inheritance. It was only years after his death, however, that his extraordinary work was recognised when Carl Correns, Erich von Tschermak and Hugo de Vries independently conducted experiments and came to the same conclusions as Mendel (Fig. 7.3).

Fig. 7.3 Gregor Johann Mendel
(Portrait of Gregor Johann Mendel, Garrison; Wellcome Library, London; 2014; https://commons.wikimedia.org/wiki/File:Portrait_of_Gregor_Johann_Mendel,_Garrison._Wellcome_M0002351.jpg. Licensed under: CC-BY-4.0)

Dominant and Recessive Alleles

Of the two alleles of a gene, one is dominant (super-ruling) and the other is recessive (subordinate or submissive).

Consider tongue rolling once more to understand this better. Many individuals may roll their tongue into a tube that extends from the mouth. See if you can accomplish it. The majority of you can, but others might not. On a certain set of chromosomes, there is a gene for tongue rolling. As a result, each person possesses a pair of these genes. It is customary in genetics to use a capital letter to denote the dominant gene, in this case 'R' for tongue rolling, and a smaller letter, such as 'r' for non-rolling, to denote the recessive gene. Consequently, a recessive gene is one that does not function when the opposing (dominant) gene is present.

Alleles occupy the same position (locus) on homologous chromosomes and affect the same characteristic but in different ways. For a pair of chromosomes showing four imaginary genes A, B, C and D:

(i) AA is homozygous (similar) dominant pair

(ii) bb is homozygous (similar) recessive pair

(iii) Cc and dD are heterozygous (dissimilar) pairs

The chromosome as a whole can be neither dominant nor recessive.

GENOTYPE AND PHENOTYPE

The three situations pertaining to any pair of genes, as for example in tongue rolling, can be as follows:

- RR (both dominant): tongue roller
- Rr (one dominant, one recessive): tongue roller
- rr (both recessive): not a tongue roller

As you can see, tongue rollers and non-rollers have different genetic constitutions. The genetic constitution (pertaining to the kinds of genes possessed) is called genotype and the expressed character is called phenotype ('*phene*' means 'to show'). There are two genotypes for every trait:

- Homozygous dominant, with a comparable pair of RR
- Heterozygous dominant, with a dissimilar pair Rr

The genotype of a non-roller is one single homozygous recessive pair, rr.

The following details can be seen in the pedigree (family) chart shown in (Fig. 7.4):

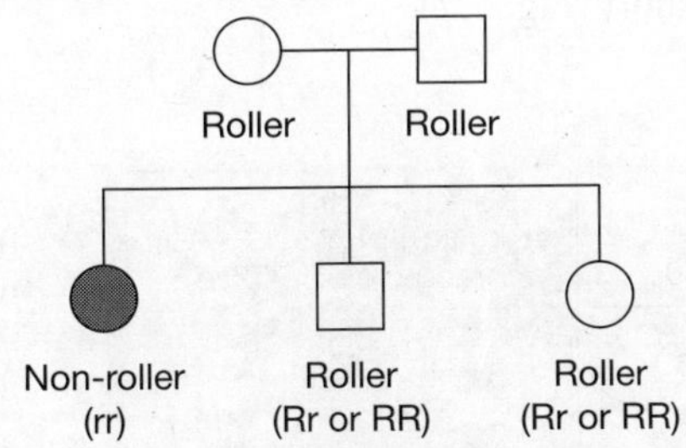

Fig. 7.4 A pedigree analysis showing rolling of tongue

- Females are represented by circles and males by squares. Both the mother and the father roll their tongues (hollow symbols represent the usual expressed character). Two of the three newborns can roll (hollow symbols), but only one cannot (solid symbol).
- Only the parents could have passed on the recessive trait (rr) of non-rolling to one of the children.
- The non-roller child in the family chart with genotype 'r' must have received one 'r' gene from one parent and the other 'r' gene from the other parent.
- Each of the two parents is heterozygous (Rr).
- If, on the other hand, one parent was homozygous (similar pair RR) dominant and the other heterozygous (dissimilar pair Rr) dominant, then every child would have received at least one dominant gene from the homozygous dominant parent through the sex cells. The situation can be schematically explained using a Punnett square (Fig. 7.5).

The punnett square below shows the genotypes of two parents for rolling tongue, which is a mendelian trait. The allele for the ability to roll the tongue is (R) is dominant over the allele for the inability to roll the tongue (r).

	(Mom) R	**R**
(Dad) R	RR	RR
r	Rr	Rr

Fig. 7.5 Punnett square

In a Punnett square, the various gametes (sex cells with the relevant feature) of one parent, the female, are arranged on one side of the square, while those of the other parent, the male, are arranged along the other. Next, the sub-squares contain the potential pairings (genotypes) of the opposing gametes. The genotypes can be written under the resulting phenotypes.

The common dominant and recessive forms of hereditary traits in humans are given in Table 7.1.

Table 7.1 Dominant and recessive forms of some common hereditary traits in humans

Character	Dominant trait	Recessive trait
Eye colour	Brown	Blue
Tongue	Rolling	Non-rolling
Hand	Right-handedness	Left-handedness
Ear lobe	Free	Attached
Lips	Thick	Thin
Rh Blood group	Positive	Negative
Colour vision (red green)	Normal	Colour blind
Cheek	Dimpled	Normal

Some rather uncommon hereditary traits:

- Polydactyly (extra fingers and toes)
- Dominant albinism (total absence of skin pigment)
- Recessive camptodactyly (inability to straighten the little finger)

MENDEL AND HIS WORK ON INHERITANCE

Mendel is known as the Father of Genetics. His laws of inheritance state that:

- Every trait or character (for example, the colour of flowers, the height of the plant, the colour and texture of the seed, the colour and texture of pods, and the location of the flower on the plant) is governed by a pair of factors. During gamete formation, the two factors in a pair segregate or separate. One component goes to one gamete and the other to the other gamete. The feature is expressed by a combination of factors after fertilisation.
- The dominant factor of the two elements governing a certain trait may express itself even if the other is present. The other factor, known as recessive, expresses itself only when the dominant factor is absent. For example, the pea plant's tallness factor is always expressed in the progeny, while dwarfness is expressed only if the tallness factor is absent.
- **Law of Independent Assortment:** Also known as Mendel's Second Law, it states that the alleles (alternative forms of a gene) for different traits segregate independently during the formation of gametes. In other words, the inheritance of one trait does not influence the inheritance of another trait. This law is based on Gregor Mendel's experiments with pea plants, where he observed that the segregation of alleles for one trait did not affect the segregation of alleles for another trait. As a result, the assortment of alleles for different traits is random and occurs independently of each other, contributing to the genetic diversity observed in offspring. This is a fundamental principle in the field of genetics and has significant implications for understanding patterns of inheritance and genetic variation.

Application of Mendel's Laws

A knowledge of the Mendelian principles gives us an idea about the new combinations in the progeny of hybrids and enables us to predict their frequency. Such knowledge is crucial for both plant and animal breeders in order to create better breeds. Hybridisation can create new plant varieties with novel combinations of valuable traits.

Mendel's Experiments

Mendel conducted his experiments on inheritance using the garden pea. The characteristics he studied are shown in (Fig. 7.6). To study one character (at a time), Mendel conducted a series of experiments in which he crossed pollinated plants.

Monohybrid cross: This is a cross between two plants with one set of contrasting traits. For example, plants that are tall or short (Fig. 7.7).

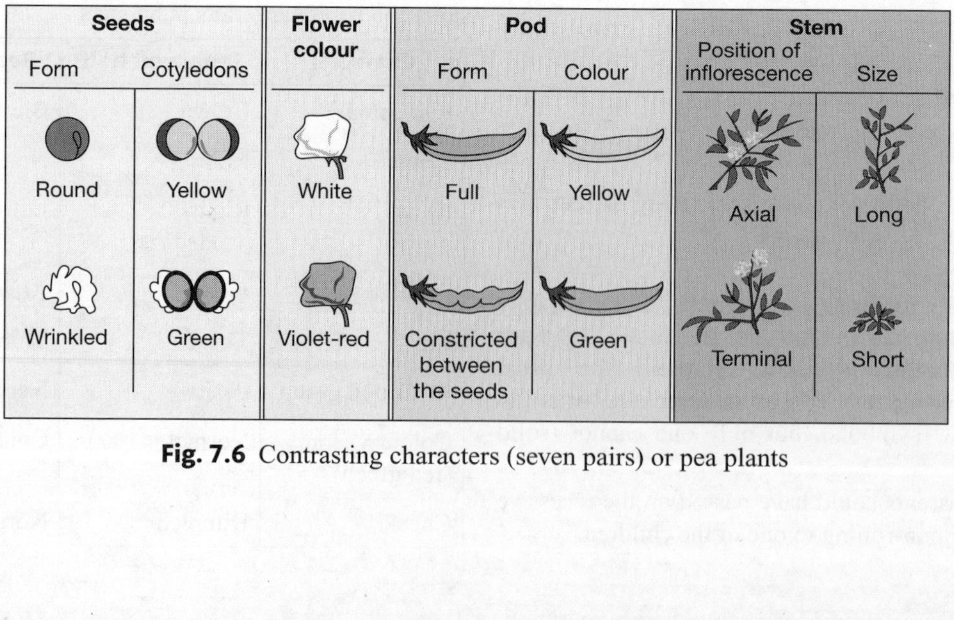

Fig. 7.6 Contrasting characters (seven pairs) or pea plants

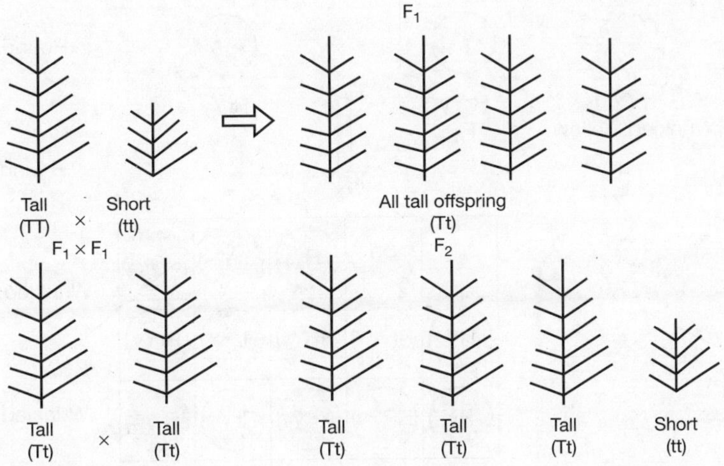

Phenotypic ratio 3 : 1
Genotypic ratio 1 : 2 : 1
TT – Both dominant gene
Tt – Both recessive gene
Tt – One dominant, one recessive gene

Pure condition | Homozygous | Heterozygous condition | Hybrid

Phenotypicratio: 3 : 1
Genotypic ratio: 1 : 2 : 1
Phenotype - Physical appearance [Tall or Short]
Genotype - Genetic make-up [TT, Tt or tt]

Fig. 7.7 Inheritance of traits over two generations

Observations:

1. All F progeny were tall (no medium height plant/ halfway characteristic)
2. One-fourth of F progeny were short
3. Phenotypic ratio F – 3 : 1
   Genotypic ratio F – 1 : 2 : 1

Conclusions:

1. TT and Tt are both tall plants, whereas tt is short.
2. A single copy of T is sufficient to make the plant tall, whereas both copies of t are required to make the plant short.
3. Characters/Traits such as 'T' are dominant (because they manifest themselves), whereas 't' is a recessive trait (because it is suppressed).

Dihybrid cross: This is a cross between two plants having two pairs of contrasting characters. For example, the shape and colour of seeds (Fig. 7.8).

Phenotypic ratio:
Round, yellow: 9
Round, green: 3
Wrinkled, yellow: 3
Wrinkled, green: 1

Genotypic ratio:
RRYY : 1
RRYy : 2
RrYY : 2
RRyy : 1
RrYy : 4
Rryy : 2
rrYY : 1
rrYy : 2
rryy : 1
RATIO: 1 : 2 : 2 : 1 : 4 : 2 : 1 : 2 : 1

Observations:

1. All Rr Yy round and yellow seeds were produced when RRYY was crossed with RRYY in the F_1 generation.
2. Self-pollination of F plants resulted in the parental phenotype plus two mixes (recombinants). Seed plants appeared in the ratio of 9 : 3 : 3 : 1 and were round wrinkled and green yellow.

Conclusions:

1. Round and yellow seeds are dominant characteristics.
2. The occurrence of new phenotypic combinations demonstrates that round and yellow seed genes are inherited separately.

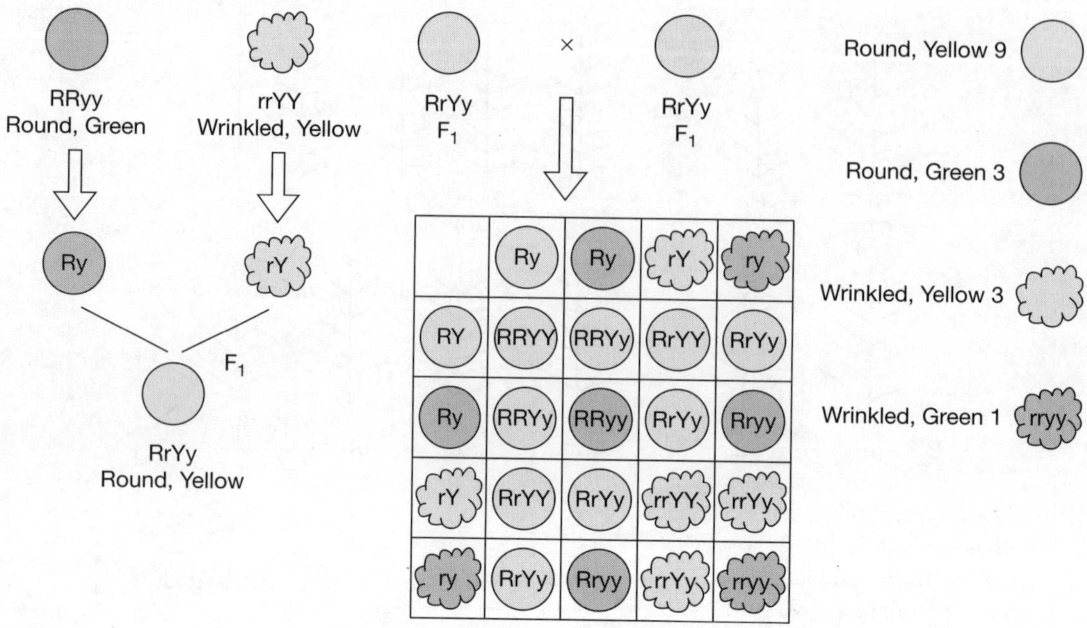

Fig. 7.8 Dihybrid cross/Independent inheritance of two separate traits, shape and colour of seeds

Expression of Traits

The information for generating proteins is contained in cellular DNA. A gene is a segment of DNA that supplies information for a certain protein; for example, the height of a plant is determined by the growth hormone, which is regulated by a gene. The plant will grow taller if the gene is efficient and more growth hormone is released. The plant will remain short if the gene for that particular protein is modified and less of it is secreted. During sexual reproduction, both parents contribute equally to the DNA of the future generation.

How can gametes, or germ cells, receive a single set of genes from parents with two copies?

Each gene set is present as separate independent pieces called chromosomes, rather than as a single lengthy thread of DNA. Each cell receives two chromosome copies, one from each parent. Because there is reductional division in the sex organs at the time of gamete formation, each germ cell or gamete has one copy. When fertilisation occurs, the number of chromosomes in the progeny is returned to normal, ensuring the species' DNA integrity.

Factors that Determine the Sex of Offspring

In some species, the temperature at which fertilised eggs are stored affects the sex of the offspring. Others, such as snails, can alter their sex, demonstrating that in these animals, sex is not genetically determined. In humans, the sex of an individual is determined genetically; that is, the offspring's sex is determined by genes inherited from the parents.

In humans, not all of the chromosomes are paired. There are 22 pairs of chromosomes (called autosomes), but one pair, the sex chromosomes, is unusual in that it does not have a perfect pair in males. Females have a perfect pair, represented by XX. Males, on the other hand, have a normal-sized X but a shorter Y, written as XY. The X chromosome is present in all gametes or ova produced by a homogenetic female. Males with heterogeneous chromosomes produce two types of sperms: one with the X chromosome and one with the Y chromosome. The baby's sex will be determined at fertilisation (Fig. 7.9).

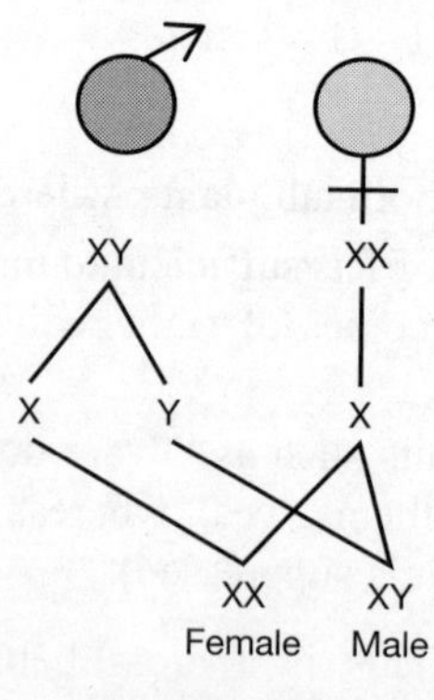

Fig. 7.9 Sex determination in human beings

If a sperm with the X chromosome fertilises an ovum with the X chromosome, the child will have the XX chromosomes and be female. If a sperm with a Y chromosome fertilises an ovum with an X chromosome, the child will be male and have XY chromosomes.

Mutation

A mutation is an abrupt alteration in one or more chromosomes, one or more genes, or both. Changes in specific characteristics or qualities are brought about by mutation, which modifies the hereditary material of an organism's cells. For instance:

- **Sickle-cell disease:** A gene mutation is the cause of this blood condition. Sickle-shaped RBCs are produced as a result of the mutation, which alters the DNA.
- **Radioactive radiations:** These also change the structure of genes, and their consequences are observable across many generations. A number of malformations in the bodies of plants and animals were caused by an atomic explosion that took place in Japan (Hiroshima, Nagasaki) during the Second World War in 1945. These deformities are still present today.

EVOLUTION

Every living thing on the planet has a common ancestry. Evolution is a series of incremental changes over millions of years that has resulted in the emergence of new species. Charles Robert Darwin was an English biologist who studied different types of life and proposed the concept of 'species evolution through natural selection'. He believed that a species' characteristics are passed down from its forefathers.

Acquired and Inherited Traits

An acquired trait is not passed down to the offspring. Germ cells are formed in the reproductive organs of sexually reproducing species, while somatic cells are found throughout the body.

Environmentally induced changes in somatic cells are not passed down to the progeny. This is due to the fact that a change in a somatic organ generated by a physiological response in the body does not become a change in the reproductive organs.

Inherited traits are those that are genetically inherited or passed down from generation to generation. Hugo de Vries demonstrated how heritable variants work. According to him, heritable mutations occur when the genes of the germplasm alter. He termed this mutation. When a trait spreads throughout a population, it indicates that it is favoured by natural selection.

Speciation

A species is a collection of individuals that can interbreed and produce fruitful offspring. Speciation is the process of dividing a population into distinct species that are unable to procreate with each other.

Genetic drift: This occurs as a result of random variations in the frequencies of specific genes. A specific environmental event can cause genetic variation in small populations.

Natural selection: This causes changes in individuals, which leads to the emergence of new species. For example, if the ecological parameters on the island differ slightly from those on the mainland, it leads to changes in the morphology and food choices of species across generations.

Gene flow: Migration causes gene flow, which is the transfer of genes from one group to another. For example, new gene combinations are introduced into a population when brown and green beetles breed.

All sexually reproducing organisms will be subject to speciation owing to inbreeding, genetic drift and natural selection, which will result in the emergence of a new species (Fig. 7.10).

Evolution and Classification

Characteristics are genetic qualities that are passed down from one generation to the next. These can be a specific form or function. They can be used to find out how closely organisms are related in evolutionary terms. The more similar features two species share, the more closely they are related. And the closer they are linked, the more recent a common ancestor they will have had.

A brother and sister, for example, are closely linked. They share common ancestors, mainly their parents, from the first generation before them. A girl and her first cousin are related as well, but not as closely as a girl and her brother. This is due to the fact that cousins share common ancestors – their grandparents in the second generation prior to them, rather than the first.

Tracing Evolutionary Relationships

Homologous and analogous traits: Organs with homologous traits have the same basic structure and origin but perform different functions. Because they acquired their limbs from a common ancestor, mammals, birds, reptiles and amphibians all have four limbs with the same basic limb pattern. However, these limbs have been modified to carry out different tasks (Fig. 7.11).

Analogous organs are those that have diverse architecture and origin but execute the same function. For example, the architecture of the wings of bat and bird are different but they are both used for flying (Fig. 7.12).

Fossils: When creatures die, their corpses usually degrade and disappear. However, some body pieces may fail to dissolve fully. They then harden and the material they are buried in retains the impression of the body part. Fossils

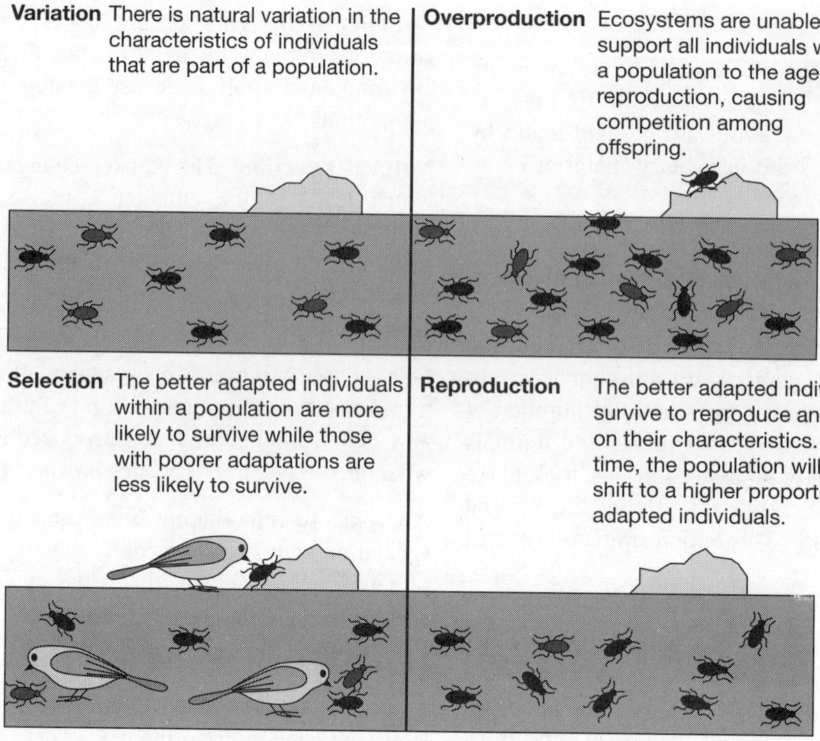

Fig. 7.10 Natural selection of beetles

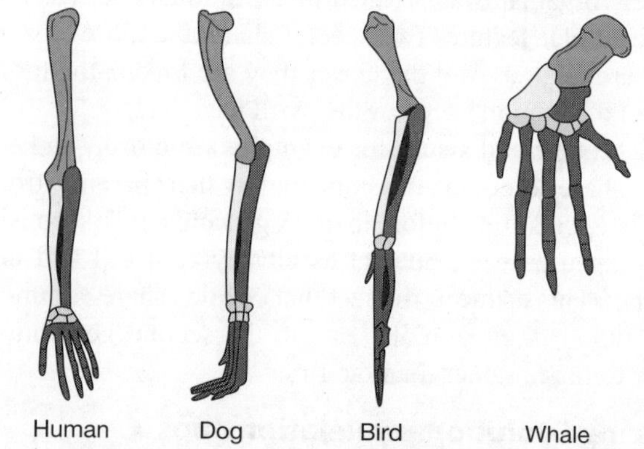

Fig. 7.11 Homologous organs

are thus the preserved remains of living things (Fig. 7.13). They are the remains or evidence of a plant or animal that lived during a previous geological epoch and which was retrieved from the soil. The process of converting an organism into a fossil is known as fossilisation. The study of fossils is known as palaeontology. The age of fossils can be determined in two ways. One method is to start digging and looking for fossils. It can be assumed that the fossils found in layers that are closer to the surface are more recent than those found deeper. The ratio of the different isotopes of elements found in the fossil material is the second method of dating fossils.

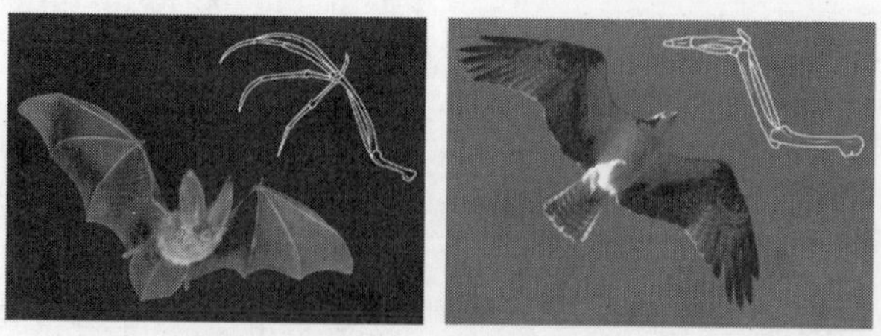

Fig. 7.12 Analogous organs such as the wing of a bat and the wing of a bird
(Biology is designed for multi-semester biology courses…; CNX OpenStax; 2016; https://commons.wikimedia.org/wiki/File:Figure_20_02_01.jpg. Licensed under: CC-BY-4.0)

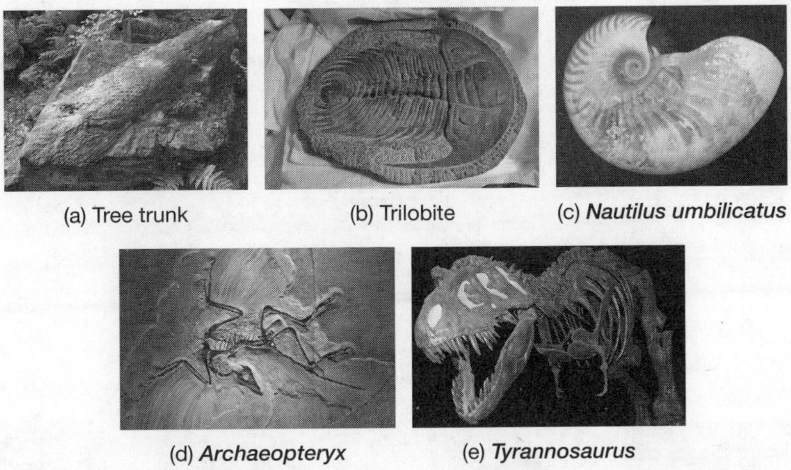

(a) Tree trunk (b) Trilobite (c) *Nautilus umbilicatus*

(d) *Archaeopteryx* (e) *Tyrannosaurus*

Fig. 7.13 Various kind of fossils

(Fossilised tree trunk in the old quarry…; OAnick; 2020; https://commons.wikimedia.org/wiki/File:Fossilised_tree_trunk_(Close).jpg. Licensed under: CC-BY-SA-4.0; Trilobite (Paradoxides sp.); Mike Peel; 2010; https://commons.wikimedia.org/wiki/File:BLW_Trilobite_(Paradoxides_sp.).jpg. Licensed under: CC-BY-SA-2.0-UK; Preserved specimen; Naturalis Biodiversity Center; 2015; https://commons.wikimedia.org/wiki/File:Naturalis_Biodiversity_Center_-_RMNH.MOL.310980_1_-_Nautilus_umbilicatus_Linnaeus,_1758_-_Nautilidae_-_Mollusc_shell.jpeg. Licensed under: CC0; Archaeopteryx lithographica…; James L Amos; 2014; https://commons.wikimedia.org/wiki/File:Archaeopteryx_fossil.jpg. Licensed under: CC0; View of the fossil/cast Tyrannosaurus Rex…; Pierre Camateros; 2010; https://commons.wikimedia.org/wiki/File:Fossil_Tyranausaurus_Rex_at_the_Royal_Tyrell_Museum,_Alberta,_Canada.jpg. Licensed under: CC-BY-SA-3.0)

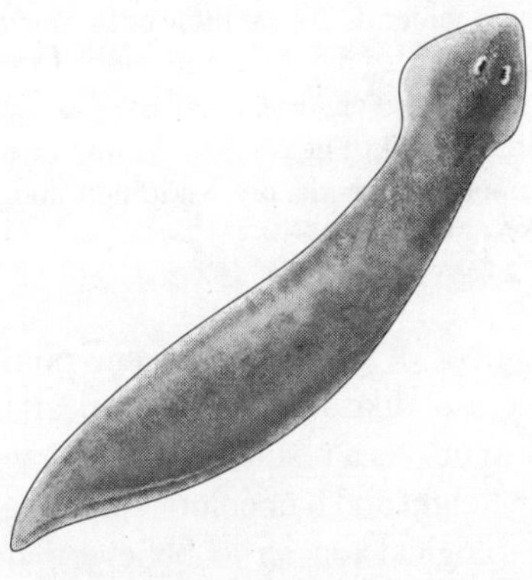

Fig. 7.14 *Planaria* eyespots to detect light
https://commons.wikimedia.org/wiki/File:202107_Planaria.svg

Biological convergence: Evolution is a slow process. Organisms gradually evolved complex organs. The eyes of the octopus and those of vertebrates evolved separately. Despite their varied beginnings, these structures are structurally similar, which is a classic example of biological convergence (Fig. 7.14).

Biological convergence is a phenomenon in which two unrelated species, if they have a common ancestor, become quite similar after a period of time and a few generations.

A change that was formerly advantageous for one attribute could now be useful for a completely different function. Long feathers, for example, were thought to provide insulation in cold weather. Some reptiles, such as dinosaurs, had feathers, but only a handful were able to fly. Birds use feathers for flight, which is an example of adaptation. It is a trait of a particular species that may be useful for performing a completely new role after evolution.

Examining the organ structure of fossils allows us to assess how far back evolutionary links can be traced. An excellent example is the wild cabbage plant. Artificial selection has produced broccoli, kohlrabi and kale from the common ancestor, the wild cabbage (Fig. 7.15).

Another method for tracing evolutionary ties is to look at how DNA has altered during reproduction. Comparing the DNA of different species should enable us to determine how much the DNA has altered during the evolution of new species. This approach is currently widely used to define evolutionary relationships.

Evolution and progress: Evolution is simply the process of generating diversity and influencing that diversity through environmental selection. The newly created species are not superior to the older ones in any way. Natural selection and genetic drift, as in the evolution of humans and chimps from a shared ancestor, have resulted in the formation of a population that is unable to breed with the original. The new forms that have evolved are more complicated than their ancestors. It is a species' ability to adapt to its environment, not its complexity, that ensures its survival.

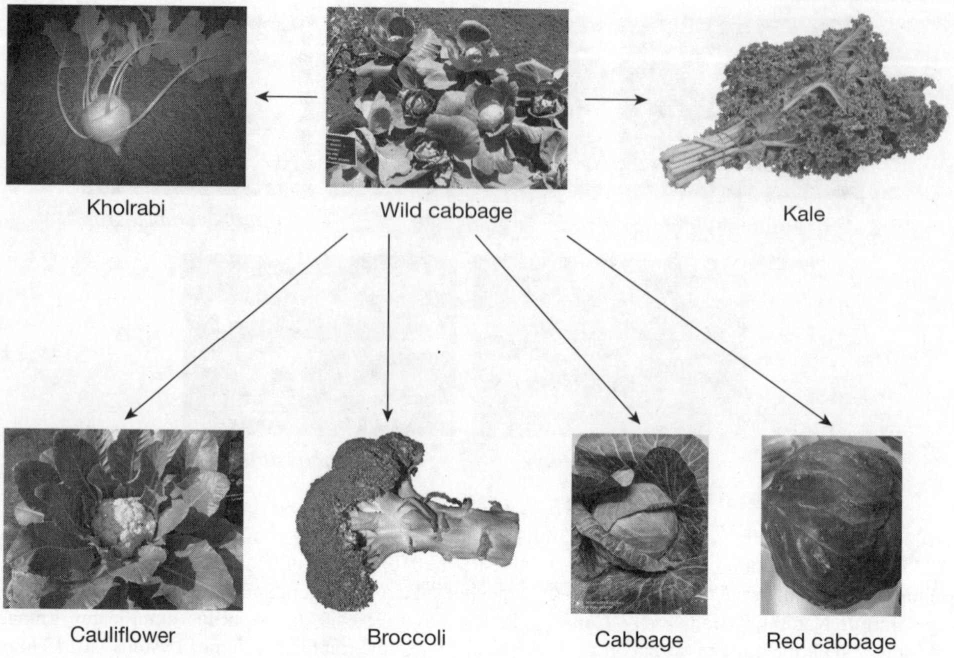

Fig. 7.15 Evolution of wild cabbage

(I grew this specimen in my backyard…; Tarunpant; 2012; https://commons.wikimedia.org/wiki/File:Kohlrabi_against_an_exotic_cushion,_Gurgaon,Haryana.jpg. Licensed under: CC-BY-SA-3.0; *Brassica oleracea*; Bibliotheque de l'Universite Laval; 2023; https://commons.wikimedia.org/wiki/File:Brassica_oleracea_15-p.bot-brassi.oler-14.jpg. Licensed under: CC-BY-SA-4.0; A bundle of kale; Evan-Amos; 2011; https://commons.wikimedia.org/wiki/File:Kale-Bundle.jpg. Licensed under: CC0; cauliflower from my home garden; Nidheesh.mp; 2013; https://commons.wikimedia.org/wiki/File:Cauliflower_in_home_garden.jpg. Licensed under: CC-BY-SA-3.0; Broccoli; Tia Monto; 2020; https://commons.wikimedia.org/wiki/File:Broccoli.png. Licensed under: CC-BY-SA-3.0; cabbage winter vegetable but now available all time; Pbhunia195; 2021; https://commons.wikimedia.org/wiki/File:Cabbage_2.jpg. Licensed under: CC-BY-SA-4.0; Cabbage (Red color), Phulbani; Skb4umama; 2016; https://commons.wikimedia.org/wiki/File:Cabbage_(Red_color%3F.jpg. Licensed under: CC-BY-SA-3.0)

Human evolution: Excavation, time-dating, fossil analysis and DNA sequence detection have all been used to trace the links in human evolution. All modern people, whether African or American, have the same gene pool, and so all modern humans are members of the same species, *Homo sapiens*. Individual variations, on the other hand, are caused by a huge number of genes in the gene pool. As a result, no two people are alike in terms of appearance, aptitude, conduct, and so on. As a result, human characteristics such as skin colour, height and hair colour vary greatly. However, there is no biological reason to believe that people with diverse characteristics belong to separate races.

SUMMARY

- Variations that occur during reproduction may be passed down through the generations. These changes might boost an organism's chances of surviving.
- Organisms that reproduce sexually have two copies of the same gene. The characteristic that manifests itself is referred to as the dominant trait and the other as the recessive trait, if the copies are not identical.
- It is possible for traits to be inherited individually, leading to novel combinations of traits in children produced through sexual reproduction.
- In many species, different factors influence sex. In humans, the paternal chromosome (X for girls, Y for boys) determines the child's sex.
- Variations within the species could offer advantages for survival or could just lead to genetic drift.
- Environmental changes that affect non-reproductive tissues are not inherited.
- When geographic isolation and variation are combined, speciation may occur.

- The classification of organisms traces the evolutionary links.
- By studying both fossils and contemporary organisms, evolution may be understood.
- Even intermediate phases had a survival advantage; complex organs may have evolved as a result.
- During evolution, organs or characteristics may be modified to serve new purposes. For instance, it is believed that feathers initially evolved for warmth before being modified for flying.

PRACTICE QUESTIONS

Analyse Your Concepts (School Exam Based)

Multiple Choice Questions

1. In humans, if gene B gives brown eyes and gene b gives blue eyes, what will be the colour of the eyes of a person having (i) Bb (ii) BB:
(a) (i) Brown (ii) Brown
(b) (i) Blue (ii) Blue
(c) (i) Blue (ii) Brown
(d) (i) Brown (ii) Blue

2. What is the probability that the progeny will be a boy?
(a) 50% (b) 56%
(c) 47.43% (d) It varies

3. Which of the following does Darwin's theory not include?
(a) Natural selection
(b) Survival of the fittest
(c) Evolution through inheritance
(d) Struggle for the existence of life

4. Which of the following is not a component of DNA?
(a) Adenine (b) Thymine
(c) Guanine (d) Uracil

5. What is a phenotype?
(a) The genetic makeup of an individual
(b) The physical and observable characteristics of an individual
(c) The process by which genes are expressed
(d) The inheritance of traits from parents

6. Which of the following is an example of a dominant trait?
(a) Blue eyes (b) Attached earlobes
(c) Cleft chin (d) Haemophilia

7. What is the process by which individuals with certain advantageous traits are more likely to survive and reproduce?
(a) Natural selection (b) Genetic drift
(c) Gene flow (d) Mutation

8. What is the mechanism by which new species arise?
(a) Gene flow (b) Mutation
(c) Genetic drift (d) Speciation

Match the Following

Each question contains statements given in two columns which have to be matched. Statements in column I have to be matched with statements in column II.

1. Match the following:

Column I	Column II
(i) Father of genetics	(a) Kale
(ii) *Origin of Species*	(b) Fossil
(iii) Wings of bat and butterfly	(c) Classification of organisms
(iv) Eyespots	(d) Kohlrabi
(v) Ammonite	(e) Homologous organs
(vi) Selection for larger leaves	(f) Gregor Mendel
(vii) Evolutionary relationships	(g) Analogous organs
(viii) Selection for swollen parts	(h) Charles Darwin
(ix) Limbs of frog and humans	(i) *Planaria*

2. Match the following:

Column I	Column II
(i) DNA replication	(a) RNA polymerase
(ii) Protein synthesis	(b) Ribosomes
(iii) Transcription	(c) DNA polymerase
(iv) Translation	(d) mRNA

3. Match the following:

Column I	Column II
(i) Speciation	(a) Charles Darwin
(ii) Natural selection	(b) Reproductive isolation
(iii) Mutation	(c) *Origin of Species*
(iv) Genetic drift	(d) Random events

4. Match the following:

Column I	Column II
(i) Monohybrid cross	(a) 9 : 3 : 3 : 1
(ii) Dihybrid cross	(b) Law of Dominance
(iii) Mendelian inheritance	(c) Law of Segregation
(iv) Phenotypic ratio	(d) Law of Independent Assortment

True or False

Read the following statements and write your answer as true or false.

1. An individual can pass on to its progeny the experiences of its lifetime.
2. All the variations in a species have equal chances of surviving in the environment in which they find themselves.
3. The sex of an infant is not case of inheritance of characteristics.
4. Gregor Mendel conducted experiments on pea plants to study the inheritance of traits.
5. The genotype refers to an individual's physical characteristics, while the phenotype refers to their genetic makeup.
6. A Punnett square is a diagram used to predict the possible genotypes and phenotypes of offspring from a cross between two parents.
7. Natural selection is a process by which individuals with advantageous traits are more likely to survive and reproduce, leading to a gradual change in a population over time.
8. Lamarck's theory of evolution proposed that acquired traits could be passed on to offspring, leading to evolutionary change.
9. Genetic drift refers to the change in allele frequencies in a population due to random events, such as natural disasters or chance mating.
10. Mutations are always harmful and can never result in new traits or adaptations.
11. Analogous structures are structures that have a similar evolutionary origin, but serve different functions in different organisms.

Fill in the Blanks

Complete the following statements with an appropriate word/term to be filled in the blank spaces.

1. The formation of new species is called _________.
2. The earliest members of *Homo sapiens* can be traced from _________.
3. Human beings and chimpanzees had a _________ ancestor a long time ago.
4. Artificial selection of wild cabbage for arrested flower development and for sterile flowers has generated _________ and _________, respectively.
5. Different structures but same function is a characteristic of _________ organs.
6. The more characteristics two species have in common, the more _________ they are related.
7. The notion of _________ provides diversity without any adaptation.
8. The sex of the child is largely _________ determined in the human beings.
9. The germ cells combine to restore the normal number of _________ in the progeny.
10. Charles Darwin put forth the idea of _________ selection for the evolution of species.

Assertion–Reason Questions

The following questions consists of two statements: Assertion (A) and Reason (R). Answer these questions selecting the appropriate option given below:

(a) Both A and R are true and R is the correct explanation of A

(b) Both A and R are true but R is not the correct explanation of A
(c) A is true but R is false
(d) A is false but R is true

1. Assertion (A): In humans, the type of chromosomes a child receives from his or her father will determine the sex of the child.
Reason (R): Girls (XX) inherit the X chromosome from their father, whilst boys (YY) inherit the Y chromosome from their father (XY).
2. Assertion (A): Geographic isolation cannot have a significant role in the speciation of an asexually reproducing creature.
Reason (R): Organisms that reproduce asexually do not need the assistance of other living things to procreate.
3. Assertion (A): The process of evolution is very sluggish.
Reason (R): Over the course of an organism's lifespan, new characteristics are gathered into it.
4. Assertion (A): Genetic variation is essential for the process of natural selection.
Reason (R): Genetic variation provides the raw material upon which natural selection acts, allowing for the differential survival and reproduction of individuals with advantageous traits.
5. Assertion (A): Mutations can introduce new genetic variation into a population.
Reason (R): Mutations are random changes in DNA that can occur spontaneously or due to exposure to radiation or chemicals.

One-word Answers

1. Who proposed the theory of evolution?
2. Which part of the DNA provides information for a protein?
3. What is the primary coordinating system in the human body?
4. What are the specialised cells that transmit electrical signals in the nervous system called?
5. What is the hormone produced by the pancreas that regulates blood sugar levels?
6. What is the neurotransmitter responsible for pleasure and reward in the brain?
7. What is the part of the brain responsible for regulating balance and posture?
8. What is the hormone produced by the adrenal gland that prepares the body for "fight or flight" response?
9. What is the hormone produced by the pineal gland that regulates sleep-wake cycles?
10. What is the junction between two neurons called?

Define the Term

1. Hereditary
2. Genetics
3. Clones
4. Allele
5. Evolution

Groups of Terms

Given below are groups of terms. In each group, the first pair indicates the relationship between the two terms. Rewrite and complete the second pair on a similar basis.

1. Adaptation: Evolution :: Phenotype : ________
2. Homozygous: Heterozygous :: Purebred : ________
3. Darwin: Wallace :: Theory of evolution : ________
4. Meiosis: Crossing over :: Mitosis : ________
5. Genetic drift: Bottleneck effect :: Natural selection : ________

Identify the Odd One Out

1. RNA, DNA, mRNA, tRNA
2. Analogous structures, Homologous structures, Vestigial structures, Mitochondrial structures
3. Charles Darwin, Gregor Mendel, Alfred Russel Wallace, Francis Crick
4. Genetic drift, Natural selection, Artificial selection, Hybridisation
5. Incomplete dominance, Codominance, Multiple alleles, Polygenic inheritance

Image-Based Questions

1. A plant with blue flowers denoted by BB is crossed with a plant with white flowers denoted by bb.
   (a) State the colour of the flower you would expect in F_1 generation plants.
   (b) What is the percentage of white flower plants in the F_2 generation if the flowers of F_1 plants are self-pollinated?

(c) State the expected ratio of the genotypes BB and Bb in the F_2 progeny.

2. (a) What is the scientific term of the organs shown here?

(b) How do these organs provide evidence in support of evolution?

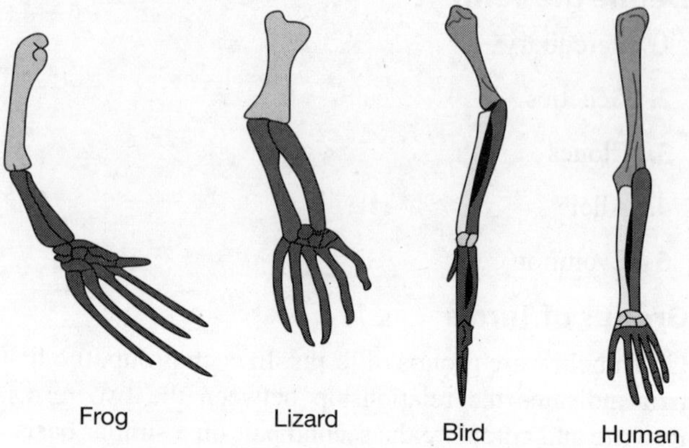

3. From the set of figures, make a pair of homologous and analogous organs each and give one reason in case of both to justify your answer.

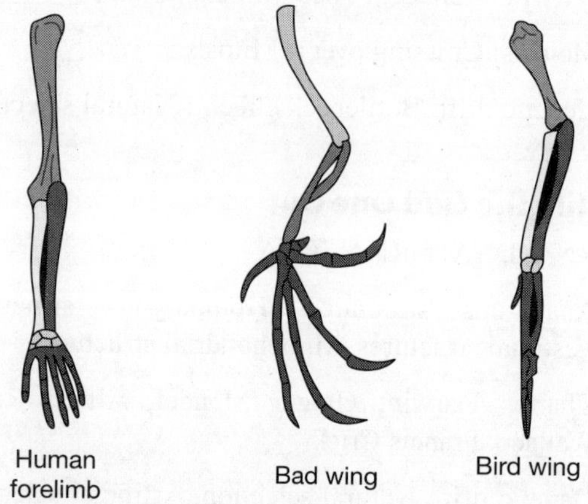

4. Fill in the blanks.

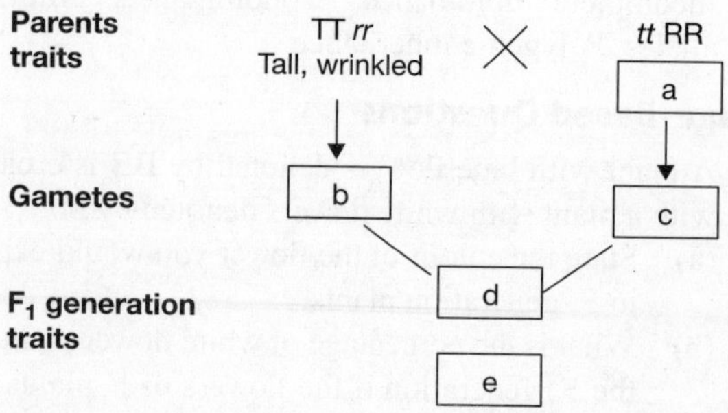

Very Short Answer Questions

1. Who coined the term genetics?
2. What is inheritance?
3. What is paleontology?
4. What is a gene?
5. What is an allele?
6. What is genetic variation?
7. What is the difference between a genotype and a phenotype?
8. What is natural selection?

Short Answer Questions

1. What is heredity?
2. Name the plant on which Mendel performed his experiments.
3. What is a gene?
4. What is genetics?
5. What is evolution?
6. What is natural selection?
7. What is genetic drift?
8. What is gene flow?
9. What is DNA?
10. What is a mutation?
11. What is genetic engineering?
12. What is a pedigree chart?

Long Answer Questions

1. All the variations in a species do not have equal chances of survival. Why?
2. "Only variations that confer an advantage to an individual organism will survive in a population." Justify this statement.
3. What connections exist between the scientific fields of classification and evolution?
4. Using examples, define comparable and homologous organs.
5. Explain the process of natural selection and its role in evolution.
6. What is genetic drift? How does it differ from natural selection?

7. What is gene flow? How does it contribute to genetic diversity?
8. Explain the process of genetic engineering and its applications.
9. What is a pedigree chart? How can it be used to study genetic traits and disorders?

Case Study-Based Question

1. Pea plants can produce seeds that are wrinkled or smooth. The dominance of one phenotype over the other is absolute. A farmer chooses to use pollen from a plant with wrinkled seeds to pollinate one blossom of a plant with smooth seeds. All of the seeds in the ensuing pea pod are smooth.
   (i) Which of the following results can be drawn?
   (1) The allele for smooth seeds dominates over that of wrinkled seeds.
   (2) The plant with smooth seeds is heterozygous.
   (3) The plant with wrinkled seeds is homozygous.
   (a) 1 only (b) 1 and 2 only
   (c) 1 and 3 only (d) 1, 2 and 3
   (ii) Which of the following crosses will give smooth and wrinkled seeds in the same proportion?
   (a) RR x rr (b) Rr x rr
   (c) RR x rr (d) rr x rr
   (iii) Which of the following crosses can be used to determine the genotype of a plant with dominant phenotype?
   (a) RR x RR (b) Rr x Rr
   (c) Rr x RR (d) RR x rr
   (iv) On crossing two heterozygous smooth-seeded plants (Rr), a total of 1000 plants were obtained in the F_1 generation. What will be the respective number of smooth and wrinkled seeds obtained in the F_1 generation?
   (a) 750, 250 (b) 500, 500
   (c) 800, 200 (d) 950, 50
   (v) The characters which appear in the first filial generation are called:
   (a) Recessive characters
   (b) Dominant characters
   (c) Lethal characters
   (d) Non-Mendelian characters
2. A couple is concerned about their child, who was born with sickle cell anaemia. They have several questions about the condition and the implications it will have for their child's future.
   (i) Sickle cell anaemia is a genetic disorder that affects the production of which protein in red blood cells?
   (a) Insulin (b) Haemoglobin
   (c) Collagen (d) Elastin
   (ii) Sickle cell anaemia is caused by a mutation in which gene?
   (a) BRCA1 (b) CFTR
   (c) TP53 (d) HBB
   (iii) How is sickle cell anaemia inherited?
   (a) Autosomal recessive
   (b) Autosomal dominant
   (c) X-linked dominant
   (d) X-linked recessive
   (iv) What are some common symptoms of sickle cell anaemia?
   (a) Fatigue and weakness
   (b) Jaundice and gallstones
   (c) Pain and swelling in joints and extremities
   (d) All of the above
   (v) What is a potential complication of sickle cell anaemia?
   (a) Stroke
   (b) Kidney failure
   (c) Vision loss
   (d) All of the above
   (vi) What is a common treatment option for sickle cell anaemia?
   (a) Blood transfusions
   (b) Bone marrow transplant
   (c) Gene therapy
   (d) All of the above
   (vii) What is the potential impact of sickle cell anemia on a person's future?
   (a) Increased risk of infections
   (b) Reduced life expectancy
   (c) Difficulty having children
   (d) All of the above
3. In a remote island, a population of finches with varying beak sizes lives in an environment where there are mainly two types of seeds available: small and large. Over time, it was observed that the average beak size of the finches increased. Which of the following statements best explains this observation?
   (a) The increase in beak size was due to a random genetic mutation that spread throughout the population.
   (b) The increase in beak size was due to mating preferences among the finches.
   (c) The increase in beak size was due to natural selection favoring individuals with larger beaks who were better able to survive and reproduce in the environment.
   (d) The increase in beak size was due to genetic drift caused by a small group of finches with larger beaks migrating to the island.

4. Two parents with blood types A and B have a child with blood type O. Which of the following statements is true regarding the inheritance of blood type?
   (a) The child inherited both the A and B alleles from the parents.
   (b) The child inherited the O allele from one parent and the A or B allele from the other parent.
   (c) The child inherited the O allele from both parents.
   (d) The child inherited the AB allele from one parent and the O allele from the other parent.
5. A family has a history of Huntington's disease, a genetic disorder that causes the progressive break-down of nerve cells in the brain. The father has the disease and each of his children has a 50% chance of inheriting the disease. Which of the following statements is true regarding the inheritance of Huntington's disease?
   (a) Huntington's disease is caused by a recessive allele and therefore, only one parent needs to have the disease for their children to inherit it.
   (b) Huntington's disease is caused by a dominant allele and therefore, both parents need to have the disease for their children to inherit it.
   (c) Huntington's disease is caused by a dominant allele and therefore, each child has a 50% chance of inheriting the disease regardless of whether the father or mother has the disease.
   (d) Huntington's disease is caused by a recessive allele and therefore, only children who inherit the disease allele from both parents will develop the disease.
6. A population of birds is separated into two groups by a physical barrier. Over time, the two groups evolve differently and become two separate species. This process is an example of:
   (a) Genetic drift
   (b) Gene flow
   (c) Natural selection
   (d) Allopatric speciation
7. Fossil evidence shows that a certain species of dinosaur existed for millions of years before becoming extinct. Which of the following statements is true regarding the process of evolution?
   (a) Evolution is a linear process that always leads to the development of more advanced organisms.
   (b) Evolution is a process of change that can result in the development of new species or the extinction of existing species.
   (c) Evolution only occurs over very short periods of time and is not observable in the fossil record.
   (d) Evolution occurs in response to environmental changes and is therefore a random process.

NCERT Corner

1. A normal pea plant bearing coloured flowers suddenly starts producing white flowers. What could be the possible cause?
2. Mention any two recessive traits of the garden pea.
3. What is the phylogenetic system of classification?
4. What will be the percentage of ab gametes produced by AaBb parents?
5. Mendel crossed a pure white recessive pea plant with a dominant pure red plant. What will be the first-generation hybrids?
6. Name the chemicals which are essential for the origin of life.
7. Why are males said to be heterogametic?
8. What is the percentage possibility of a couple having daughters?
9. Name two organisms in which sex determination is regulated by environmental factors.
10. Clones of sheep are copies of each other except for physical health. What type of variation is it?
11. Write down the basic difference between autosomes and allosomes.
12. What will be the sex of the embryo if an egg is fertilised by sperm having:
    (a) 22+x22+x
    (b) 22+y22+y
13. Mention two sources of variation.
14. What are Mendelian factors?
15. What happened when Mendel crossed two traits of a character in a pea plant?
16. Who provided experimental evidence to support the theory of the origin of life from inanimate matter?
17. Write down the differences between homologous organs and analogous organs. In which category would you place the wings of a bird and the wings of a bat? Explain your answer with reasons.

COMPETITION WINDOW (OBJECTIVE TYPE)

1. New species may be formed if:
   (i) DNA undergoes significant changes in germ cells
   (ii) Chromosome number changes in the gamete
   (iii) There is no change in the genetic material
   (iv) Mating does not take place
   (a) (i) and (ii) (b) (i) and (iii)
   (c) (ii), (iii) and (iv) (d) (i), (ii) and (iii)

2. Two pea plants, one with round green seeds (RRyy) and the other with wrinkled yellow (rrYY) seeds, produce F_1 progeny that have round, yellow (RrYy) seeds. When F_1 plants are selfed, the F_2 progeny will have a new combination of characters. Choose the new combination from the following:
   (i) Round, yellow (ii) Round, green
   (iii) Wrinkled, yellow (iv) Wrinkled, green
   (a) (i) and (ii) (b) (i) and (iv)
   (c) (ii) and (iii) (d) (i) and (iii)

3. A basket of vegetables contains carrot, potato, radish and tomato. Which of them represent homologous structures?
   (a) Carrot and potato
   (b) Carrot and tomato
   (c) Radish and potato
   (d) Radish and carrot

4. If the fossil of an organism is found in the deeper layers of earth, we can predict that:
   (a) The extinction of the organism occurred thousands of years ago
   (b) The extinction of the organism occurred recently
   (c) The fossil position in the layers of earth is not related to its time of extinction
   (d) The time of extinction cannot be determined

5. A trait in an organism is influenced by:
   (a) Paternal DNA only
   (b) Maternal DNA only
   (c) Neither paternal nor by maternal DNA
   (d) Both maternal and paternal DNA

6. Select the group which shares the maximum number of common characters:
   (a) Two genera of a family
   (b) Two individuals of a species
   (c) Two species of a genus
   (d) Two genera of two families

7. According to evolutionary theory, the formation of a new species is generally due to:
   (a) Sudden creation by nature
   (b) Accumulation of variations over several generations
   (c) Clones formed during asexual reproduction
   (d) Movement of individuals from one habitat to another

8. Select the statements that describe characteristics of genes:
   (i) Genes are specific sequence of bases in a DNA molecule
   (ii) A gene does not code for proteins
   (iii) In individuals of a given species, a specific gene is located on a particular chromosome
   (iv) Each chromosome has only one gene
   (a) (i) and (ii) (b) (i) and (iii)
   (c) (i) and (iv) (d) (ii) and (iv)

9. In peas, a pure tall plant (TT) is crossed with a short plant (tt). The ratio of pure tall plants to short plants in F_2 is:
   (a) 1 : 3 (b) 3 : 1
   (c) 1 : 1 (d) 2 : 1

10. Some dinosaurs had feathers although they could not fly, but birds have feathers that help them to fly. In the context of evolution, this means that:
    (a) Reptiles have evolved from birds
    (b) There is no evolutionary connection between reptiles and birds
    (c) Feathers are homologous structures in both organisms
    (d) Birds have evolved from reptiles

11. Human teeth and elephant's tusk are an example of which process?
    (a) Homologous organs
    (b) Vestigial organs
    (c) Analogous organs
    (d) Rudimentary organs

12. Independent inheritance of two separate traits, shape and colour of seeds in Mendel's cross of pea plants, resulted in a observable ratio of:
    (a) 3 : 1
    (b) 19 : 4 : 2 : 1
    (c) 1 : 1
    (d) 9 : 3 : 3 : 1

13. Which of these is not a vestigial organ in human beings?
(a) Appendix
(b) Wisdom tooth
(c) Gall bladder
(d) Nictitating membrane

14. Evolution of wild cabbage is an example of:
(a) Genetic drift (b) Natural selection
(c) Mutation (d) Artificial selection

15. The modern-day human species has evolved from its earliest ancestors, who lived in:
(a) Africa (b) Australia
(c) Asia (d) China

16. Two pink-coloured flowers on crossing resulted in 1 red, 2 pink and 1 white flower progeny. The nature of the cross will be:
(a) Double fertilisation (b) No fertilisation
(c) Cross fertilisation (d) Self pollination

17. Which of the following statements is incorrect?
(a) For every hormone, there is a gene.
(b) For every molecule of fat, there is a gene
(c) For the production of every enzyme, there is a gene.
(d) For every protein, there is a gene.

18. If a round, green-seeded pea plant (RR yy) is crossed with a wrinkled, yellow-seeded pea plant (rr YY), the seeds produced in the F_1 generation will be:
(a) Wrinkled and yellow (b) Round and green
(c) Wrinkled and green (d) Round and yellow

19. In human males, all the chromosomes are paired perfectly, except one. This/These unpaired chromosome(s) is/are:
(i) Large chromosome (ii) Small chromosome
(iii) Y chromosome (iv) X chromosome
(a) (i) and (ii) (b) (iii) only
(c) (iii) and (iv) (d) (ii) and (iv)

20. A zygote which has an X chromosome inherited from the father will develop into a:
(a) Girl
(b) Boy
(c) X chromosome does not determine the sex of a child
(d) Either boy or girl

21. The wing of a bird and the wing of an insect are:
(a) Homologous organs (b) Analogous organs
(c) Vestigial organ (d) Both (a) and (b)

22. A Mendelian experiment consisted of breeding tall pea plants bearing violet flowers with short pea plants bearing white flowers. The progeny all bore violet flowers but almost half of them were short. This suggests that the genetic make-up of the tall parent can be depicted as:
(a) TTWW (b) TTww
(c) TtWW (d) TtWw

23. An example of homologous organs is:
(a) Our arm and a dog's foreleg
(b) Our teeth and an elephant's tusks
(c) Potatoes and runners of grass
(d) All of the above

24. In evolutionary terms, we have more in common with
(a) A Chinese schoolboy
(b) A chimpanzee
(c) A spider
(d) A bacterium

Answer Keys

Multiple Choice Questions

1.(a) 2. (a) 3. (c) 4. (d) 5. (b)
6. (c) 7. (a) 8. (d)

Match the Following

1. (i) (f), (ii) (h), (iii) (g), (iv) (i), (v) (c), (vi) (a), (vii) (b), (viii) (d) (ix) (e)
2. (i) (c), (ii) (b), (iii) (a), (iv) (d)
3. (i) (b), (ii) (c), (iii) (d), (iv) (d)
4. (i) (c), (ii) (d), (iii) (b) (iv) (a)

True or False

1. F 2. F 3. F 4. T 5. F
6. T 7. T 8. T 9. T 10. F
11. F

Fill in the Blanks

1. Speciation
2. Africa
3. Common
4. Broccoli, cauliflower
5. Analogous
6. Closely
7. Genetic drift
8. Genetically
9. Chromosomes
10. Natural

Assertion–Reason Questions

1. (a) 2. (a) 3. (c) 4. (a) 5. (a)

One-word Answers

1. Charles Darwin
2. Coding region
3. Nervous system
4. Neurons
5. Insulin
6. Dopamine
7. Cerebellum
8. Adrenaline (or epinephrine)
9. Melatonin
10. Synapse

Define the Term

1. The transmission of characters or traits from parents to their offspring (children) is termed heredity.
2. Genetics is the study of heredity and other variations.
3. Those organisms that are identical copies of one another are termed clones. They are exact copies of each other.
4. An allele is one of the alternative forms of a gene, located at a specific position on a chromosome. Alleles arise through genetic mutations and can result in variations in a trait. For example, a gene that determines eye color may have different alleles, such as blue, brown, or green.
5. Evolution refers to the gradual change in the inherited characteristics of biological populations over successive generations. It occurs through the processes of variation, inheritance, and differential reproductive success, driven primarily by natural selection. Evolutionary processes lead to the development of new species, the adaptation of organisms to their environments, and the diversity of life on Earth.

Groups of Terms

1. Natural selection
2. Hybrid
3. Natural selection
4. Cell division
5. Adaptation

Identify the Odd One Out

1. DNA
2. Mitochondrial structures
3. Francis Crick
4. Hybridisation
5. Multiple alleles

Image-Based Questions

1.

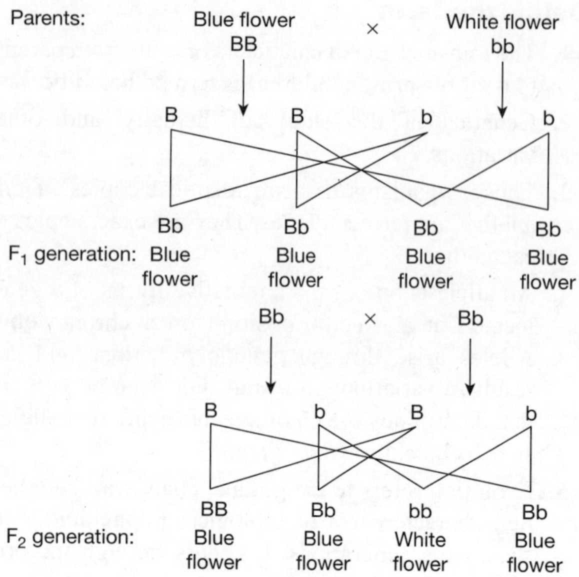

(a) The colour of all the flowers in the F_1 genration will be blue.

(b) Percentage of white flower plants in the F_2 generation will be 25.

(c) The ratio of genotype BB and Bb in F_2 progeny will be 1 : 2.

2. (a) Homologous organs

(b) They are fore limbs which have the same basic plan and anatomy but have undergone evolution and become modified to perform different functions.

3. (a) Analogous Organs: Wing of bat and wing of bird. In bird the fingers are reduced and the wing is covered by feathers.

(b) Homologous Organs: Fore limp of wing of bird/bat. They have the same body plan, being made of humerus, radioulna and hand.

4. (a) Dwarf round

(b) Tr

(c) tR

(d) TtRr

(e) Tall round

Very Short Answer Questions

1. William Bateson
2. The process of genetic transmission of characteristics from parent to offspring is called inheritance.
3. The study of fossils is known as paleontology.
4. A gene is a segment of DNA that contains the instructions for making a particular protein or RNA molecule.
5. An allele is one of the different versions of a gene that can exist at a single locus.
6. Genetic variation refers to the differences in DNA sequences between individuals of the same species.
7. A genotype refers to the genetic makeup of an individual, while a phenotype refers to the physical characteristics or traits that are produced as a result of the genotype.
8. Natural selection is the process by which certain traits become more or less common in a population over time due to their effects on survival and reproduction.

Short Answer Questions

1. The continuity of features from one generation to the next is known as heredity. It is also defined as the transmission of traits from parents to offspring.
2. Mendel performed his experiments on *Pisum sativum*, the garden pea.
3. A gene is a unit of heredity that is responsible for the transmission of traits from parents to offspring.
4. Genetics is the branch of biology that deals with the study of genes, heredity and variation in organisms.
5. Evolution is the process of gradual change in the genetic makeup of a population over time, leading to the development of new species from existing ones.
6. Natural selection is the process by which certain traits that are advantageous to an organism's survival and reproduction become more common in a population over time.
7. Genetic drift is the random fluctuation in the frequency of alleles in a population due to chance events, such as genetic mutations or changes in the environment.
8. Gene flow is the movement of genes from one population to another due to migration or interbreeding, which can lead to genetic diversity within a population.
9. DNA is the genetic material that contains the instructions for the development and function of all living organisms. It is composed of nucleotides and is organised into chromosomes in the cell nucleus.
10. A mutation is a change in the DNA sequence that can occur spontaneously or as a result of exposure to mutagens, and can lead to genetic variation and evolution.

11. Genetic engineering is the manipulation of an organism's genetic material to produce desired traits or outcomes, such as the development of disease-resistant crops or the production of insulin for medical use.
12. A pedigree chart is a diagram that shows the genetic relationships among members of a family over several generations, and is used to study the inheritance of traits and genetic disorders.

Long Answer Questions

1. Hint: Variations in a species arise through genetic mutations, but not all variations provide equal advantages in terms of survival and reproduction. Factors such as environmental conditions, competition for resources, and interactions with other organisms can influence the chances of survival for different variations. Elaborate on these points.
2. Hint: Natural selection favours variations that provide an advantage to an individual organism in its specific environment. These advantageous variations increase an organism's fitness, enabling it to survive and reproduce more successfully than individuals without those variations. Over time, these advantageous variations become more common in a population, leading to the concept of "survival of the fittest." Elaborate on these points.
3. Hint: Classification is the process of organising and categorising organisms based on their shared characteristics. Evolutionary relationships between organisms play a significant role in classification, as organisms that share a more recent common ancestor tend to have more similar characteristics. The study of classification provides insights into the patterns of evolutionary history and the relatedness of different species. Elaborate on these points.
4. Hint: Comparable organs: These are organs that have similar functions in different species but may have different structures. For example, wings of bats and wings of birds are comparable organs as they both serve the function of flying but have different anatomical structures.

   Homologous organs: These are organs that have a similar basic structure and origin but may have different functions. For example, the forelimbs of humans, whales, cats and bats all have the same basic bone structure, suggesting a common ancestor, even though they have different functions (such as, walking, swimming, grasping and flying). Elaborate on these points.
5. Hint: Natural selection is the process by which individuals with advantageous traits that are well-suited to their environment are more likely to survive and reproduce, passing on those traits to the next generation. It involves the differential survival and reproduction of individuals based on their inherited variations, resulting in the accumulation of favourable traits in a population over time. Natural selection is a driving force of evolution, leading to the adaptation and diversification of species. Elaborate on these points.
6. Hint: Genetic drift refers to random changes in the frequency of alleles in a population over time, primarily due to chance events. It is more pronounced in small populations and can lead to the loss of certain alleles or fixation of others. In contrast, natural selection is a non-random process that results from the differential reproductive success of individuals with advantageous traits. Elaborate on these points.
7. Hint: Gene flow is the movement of genes from one population to another through migration or the exchange of genetic material. It contributes to genetic diversity by introducing new alleles into a population or spreading existing alleles. Gene flow can occur between populations of the same species, reducing genetic differences between them. Elaborate on these points.
8. Hint: Genetic engineering is the deliberate modification of an organism's genetic material through biotechnology techniques. It involves manipulating DNA sequences to introduce desired traits or modify existing ones. Genetic engineering has various applications, including the production of genetically modified crops, development of medical treatments, and advancements in biotechnology research. Elaborate on these points.
9. Hint: A pedigree chart is a diagram that depicts the genetic relationships and inheritance patterns of a family over multiple generations. It shows the transmission of genetic traits and disorders through different family members. Pedigree charts can be used to study patterns of inheritance, determine the presence of genetic disorders, and analyse the likelihood of certain traits being passed on to future generations. Elaborate on these points.

Case Study-Based Questions

1. (i) c, (ii) b, (iii) d, (iv) a, (v) b
2. (i) b, (ii) d, (iii) a, (iv) d, (v) d, (vi) d, (vii) d
3. (c)
4. (c)
5. (c)
6. (d)
7. (b)

NCERT Corner

1. Mutation is responsible for the appearance of white flowers.
2. Dwarf (height of plant) and wrinkled seed.
3. Organisms are classified depending on their evolutionary relations.
4. 25%
5. All red
6. Proteins and nucleic acid
7. Males are called heterogametic because their Y chromosomes are different.
8. 50%
9. Turtle and lizard
10. Phenotypic variation
11.

Autosomes	Allosomes
These chromosomes are not involved in determining sex.	These chromosomes place a part in defining a person's sex.
Males and females have similar traits.	Males (XY) and females (XY) are dissimilar.
In humans, there are usually 44 autosomes or 22 pairs.	In humans, there are usually two allosomes or one pair.

12. (a) Female (b) Male
13. DNA copying errors and fertilisation at random.
14. Mendelian factors are heredity components that pass on a character from parents to offspring.
15. Mendel discovered that each feature was inherited separately from the others, resulting in its own 3 : 1 ratio.
16. Miller and Urey
17.

Homologous organs	Analogous organs
They have the same basic structural design	They have different basic structural design
They perform different functions	They perform similar functions
Their appearance is different	They have similar appearance

Wings of a bird and bat are analogous organs as they have similar appearance and perform similar functions but have different basic structural design.

Competition Window

1.(a) 2. (b) 3. (d) 4. (a) 5. (d)
6. (b) 7. (b) 8. (b) 9. (c) 10. (d)
11.(a) 12. (d) 13. (c) 14. (d) 15. (a)
16. (d) 17. (b) 18. (d) 19. (c) 20. (a)
21.(a) 22. (c) 23. (d) 24. (a)

CHAPTER

8 Control and Coordination

PREREQUISITES

You must have knowledge of:

- Human anatomy and physiology
- Psychology
- Basic mathematics, physics and chemistry
- Evolutionary biology

LEARNING OBJECTIVES

At the end of this chapter, you will be able to:

- Explain the role of the nervous system and hormones in controlling and coordinating various body activities
- Identify the main components of the nervous system and list their functions, emphasising their role in informed decision-making
- Explain the role of neurons in the transmission of nerve impulses
- Identify the location and functions of the spinal cord in evoking reflex actions
- Recognise the importance of reflex, voluntary and involuntary activities, as well as hormones, in the proper functioning of the body
- Examine the role of some endocrine glands in regulating growth and behaviour

INTRODUCTION

We pay attention to our bodies on a regular basis, yet few of us recognise what a well-tuned mechanism it is! Digestive juices are secreted when we eat food, but only when there is food in the food canal and only for as long as the food needs to be digested. Only when our muscles are activated, do they move. Even when the outside temperature changes, our body temperature remains constant. Can you explain how different organs work together to fulfil their duties correctly at the appropriate time? Are you aware of which organs are in charge of our thoughts, feelings, emotions and actions?

The neurological and endocrine systems ensure that the body functions in a coordinated and controlled manner. The brain, spinal cord, sense organs and nerves make up the nervous system, whereas the endocrine system is controlled by hormones, which are created by specific glands and released directly into the bloodstream. These two systems collaborate to communicate, integrate and coordinate the operations of our body's many organs and systems.

Some examples from our daily lives can help us grasp the complexity of the processes coordinated by the neurological and endocrine systems to assist us in performing a variety of simple and complex tasks.

Do you have any idea why we are hungry? Yes! You are correct. When our bodies want energy, we crave food. The eyes see the meal, the brain records it, and a sequence of synchronised activities are activated. Appropriate nervous system activity directs the relevant muscles in a person's hands and fingers to pick up and place the food in their mouth. When enough food has been consumed, signals from the brain's appetite centre indicate fullness, and the person stops eating.

When food reaches the alimentary canal, different digestive juices are secreted to aid digestion (for example, gastric juice, bile and pancreatic juice under the control of certain endocrine glands). Food is taken into the bloodstream after a series of digestive processes to provide the energy required. Some other digestive processes are not mentioned here. Many of these processes are invisible to us, but they are critical for digesting food and for producing energy for our daily activities.

The neurological system and the endocrine system work together as a team to govern and organise all of our activities, including physical actions, thought processes and emotional behaviour. It is worth noting that we may not even realise how important these systems are to our health and wellbeing. For example, we do not have to remember to breathe or digest food.

Another collection of actions, known as reflex actions, are frequently performed in response to stressful or dangerous scenarios. For example, if a hand comes in contact with a hot object, it should be removed immediately.

Thus, the neurological and endocrine systems are essential for the smooth functioning of our daily lives. Let us take a closer look at the structure and operation of these systems.

SCIENCE IN DAILY LIFE

Give an example of a process in your body that is brought about by the neurological and hormone systems working together.

THE NERVOUS SYSTEM

The nervous system perceives and responds to stimuli in the internal or external environment. A stimulus is an agent or a change in the environment that can cause the body to respond. There are many types of stimuli: physical (touch, prick, pressure), auditory, chemical, radiant (light), heat or cold, and electrical.

Neurons or nerve cells communicate with one another and other tissues to receive and transmit information throughout the body. Figure 8.1 (a) depicts the generalised structure of a neuron. It is divided into three parts:

- **Dendrites:** These are branched cytoplasmic extensions from the cell body. The dendritic tips of neurons receive impulses and trigger a chemical reaction that produces an electrical impulse, which is then conveyed to the cell body.
- **Cell body:** The nucleus of the cell body is highly defined and surrounded by cytoplasm. Like all other cells, it has organelles. The impulse from the dendrite is passed on by the cell body to the axon.
- **Axon:** This is a branch of the cell body that is unusually long in comparison to the others. The axon, or nerve fibre, is surrounded by an insulating fatty coating termed the neurilemma in most neurons. At intervals known as nodes of Ranvier, the fatty sheath is missing. The lack of a neurilemma aids the nodes of Ranvier in producing electrical activity and transmitting nerve impulses. The axon's end segments contain swellings that look like bulbs; they store chemicals called neurotransmitters. The axon bulbs of one neuron are close to the dendrites of another neuron. The joining of two neurons is called the synapse and the area between two neurons at the synapse is termed the synaptic cleft. Between the millions of nerve cells, there are many synapses.

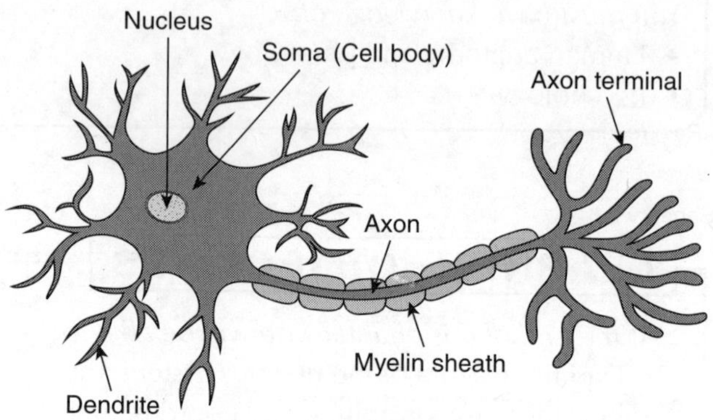

Fig. 8.1 (a) Structure of a neuron

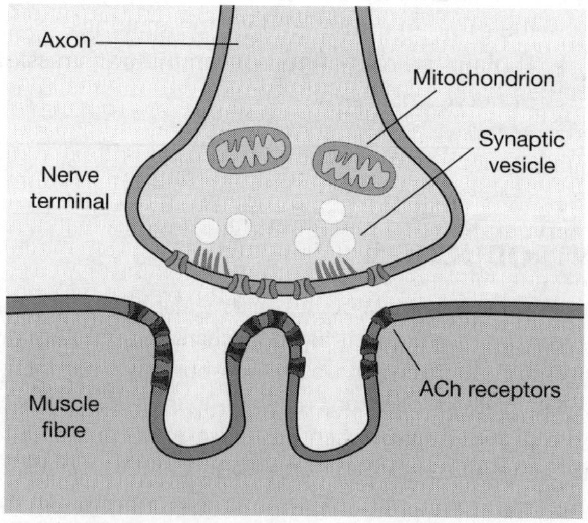

Fig. 8.1 (b) Neuromuscular junction

Impulses travel from one neuron to the next through the synapse. A neurotransmitter is produced in the synaptic cleft of the synapse when an impulse reaches the end of the first neuron. These molecules cross the gap, or synapse, and cause the next neuron to fire a similar electrical impulse. Finally, neurons send impulses to other cells, such as muscle cells (Fig. 8.1 (b)) or glands, in order to trigger the desired activity.

Neurons can be of three types (Fig. 8.2):

- **Sensory neuron:** An impulse is transmitted from the receptor (sense organ) to the central nervous system (either the brain or the spinal cord) through the sensory neurons.
- **Motor neuron:** This transmit signals from the central nervous system to the effectors.
- **Connecting neurons:** The sensory and motor neurons are interconnected by association (connecting) neurons, which are found in the brain and spinal cord.

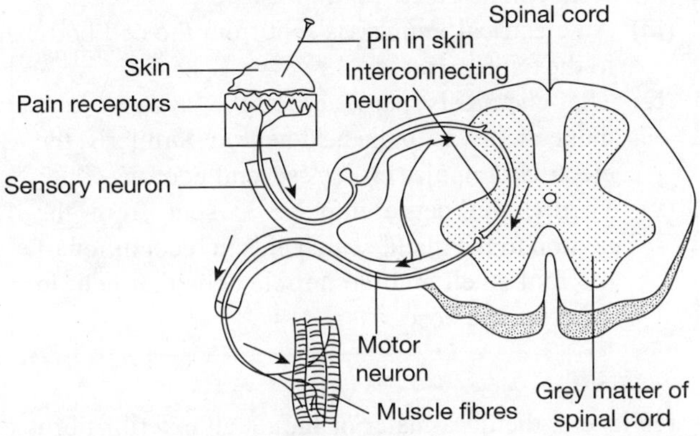

Fig. 8.2 Three types of neurons (sensory, motor and association), their synapse, and the direction of nerve signal transmission

Nerves

Nerves are thread-like structures that branch out from the brain and spinal cord to practically every area of the body. A nerve is made up of a tubular sheath encasing a bundle of nerve fibres (axons). It is similar to an underground electric line with multiple conducting wires that are all isolated from each other. The axon's medullary sheath works as an insulator, preventing impulse mixing between adjacent axons.

Nerves are of three types:

- **Sensory nerves** are made up of sensory fibres. The receptors (sense organs) send impulses to the brain or spinal cord through these nerves. The optic nerve, for example, originates in the eye and ends in the brain.
- **Motor nerves** are made up of motor fibres. These nerves transmit signals from the brain or spinal cord to effector organs such as muscles or glands. A nerve that originates in the brain and transmits impulses to the eye muscles is an example.
- **Mixed nerves** are nerves that have both sensory and motor fibres and perform a variety of functions. The spinal nerve is an example.

Fun Fact!

The nerves that emerge from the spinal cord are known as spinal nerves, while those that exit from the brain are known as cranial nerves.

Sense Organs

The stimulus is received by receptor organs such as the nose, eyes and ears, as indicated in Fig. 8.3. The input then travels through the sensory nerves to the spinal cord and brain, where it is processed. The motor neurons then convey the message to the appropriate organ (muscles or gland) for action. A response is elicited in this manner.

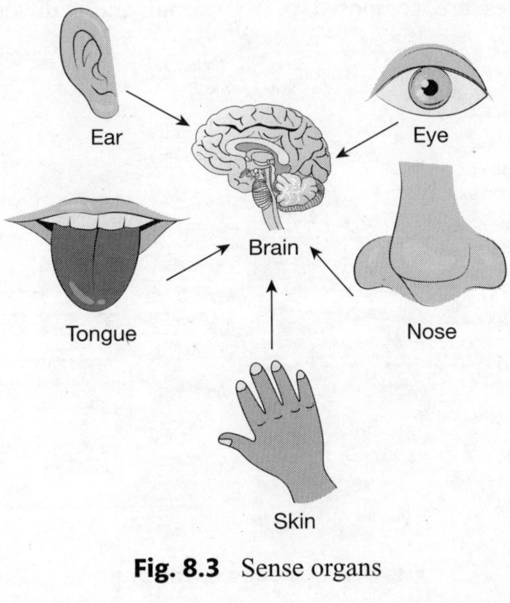

Fig. 8.3 Sense organs

Divisions of the Nervous System

The nervous system is divided into two parts: the central nervous system (CNS), which contains the brain and spinal cord, and the peripheral nervous system (PNS), which comprises the nerves arising from the brain and spinal cord Fig. 8.4.

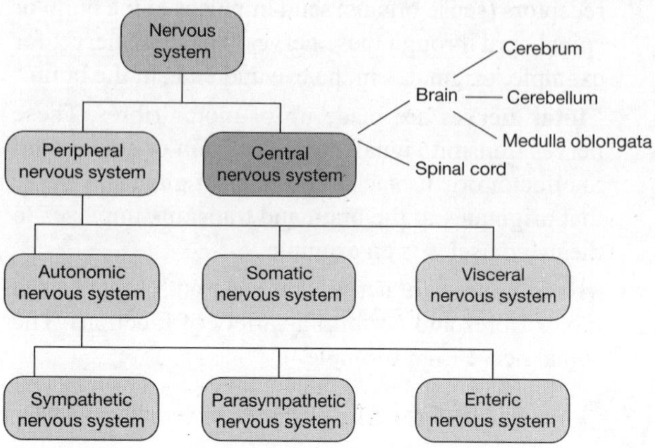

Fig. 8.4 Main divisions of the nervous system

SCIENCE IN DAILY LIFE

Consider how you would shield your nose if you were passing by a waste disposal place. Mark the events below to put them in a logical order.

To track the events that occur in the nervous system, number them from 1 to 5.

(i) Chemicals are released at the axon's end by an electrical impulse.

(ii) A neuron's dendritic cells respond to a stimulus by releasing chemicals. This chemical reaction results in an electrical impulse.

(iii) An electrical impulse is sent from the cell body to the axon.

(iv) The chemicals pass across the synapse and make their way to the next neuron. Similarly, the electrical impulse crosses several neurons.

(v) Finally, the nerve impulse is sent from the neuron to the gland, which aids in recognition of the foul smell and the muscle cells that help in covering the nose.

The Central Nervous System

The central nervous system (CNS) is known as the body's 'information processor'. The brain is located beneath the skull, while the spinal cord is located within the vertebral column.

The Brain

The brain is an extremely fragile organ. It is adequately protected within the bony cranium (brain box). It is further protected by three meninges (membranous coverings) that continue rearward over the spinal cord, as seen in Fig. 8.5. The meninges are composed of three layers: the dura mater or the tough exterior fibrous membrane; the arachnoid or the thin delicate intermediate layer that provides a web-like cushion; and the pia mater, which is the innermost highly vascular membrane that is generously supplied with blood.

A watery fluid called cerebrospinal fluid fills the area between the covering membranes, acting as a cushion to protect the brain from shocks.

The three key regions of the brain are as follows:

- **Cerebrum:** The cerebrum is the largest part of the brain and is divided vertically into the right and left

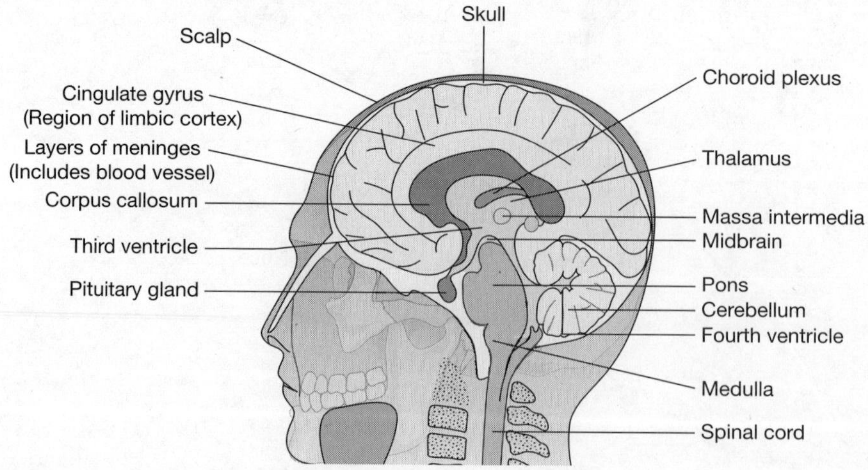

Fig. 8.5 Structure of the brain

Fun Fact!

Meningitis is a dangerous health disease caused by inflammation of the meninges, which you may have heard about. Microorganisms such as bacteria, virus, fungi and amoeba infect the meninges and cerebrospinal fluid around the brain and spinal cord, causing meningitis. Coughing, sneezing, kissing, sharing eating utensils, and using the same toothbrush can all spread meningitis. Good cleanliness can protect you from contracting the illness. Effective meningitis vaccines are also available.

hemispheres. The exterior surface is ridged, grooved and very convoluted. The cell bodies of neurons are found in the cerebrum's outer section, or cortex. It is termed grey matter as it is greyish in colour. The cortex is responsible for our ability to think, reason, invent, plan and memorise. The white matter that makes up the inner region of the cerebrum is primarily composed of axons, or nerve fibres, of the neurons. The cerebrum is the seat of intelligence, consciousness and willpower in general. It is in charge of all voluntary acts. It helps us make well-considered decisions. Figure 8.6 depicts some of the primary functions connected with various brain areas.

- **Cerebellum:** This lies beneath the cerebrum and is a considerably smaller part of the brain. It does not have any convolutions, but it does have many furrows. It also has a grey matter outer cortex and a white matter inner cortex. The cerebellum's main function is to keep the body in 'balance' and to coordinate muscle action. The cerebrum and cerebellum work closely together. When you stand up and walk, for example, the cerebrum generates the desire to do so. Walking necessitates the coordinated action of numerous muscles. The cerebellum is responsible for the proper coordination and timing of muscle contraction and relaxation.
- **Medulla oblongata:** Located near the base of the skull, this is the lowest region of the brain. It is generally triangular in shape and continues as the spinal cord behind it. Its main function is to keep track of the activity of visceral organs, such as the alimentary canal, movement, breathing, heartbeat, and a variety of other involuntary processes. Injury to the medulla usually leads to death because vital functions such as breathing and heartbeat are affected.

Fun Fact!

A person, when inebriated, generally walks clumsily. It is because under the effect of alcohol, the cerebellum is unable to coordinate muscular movements properly.

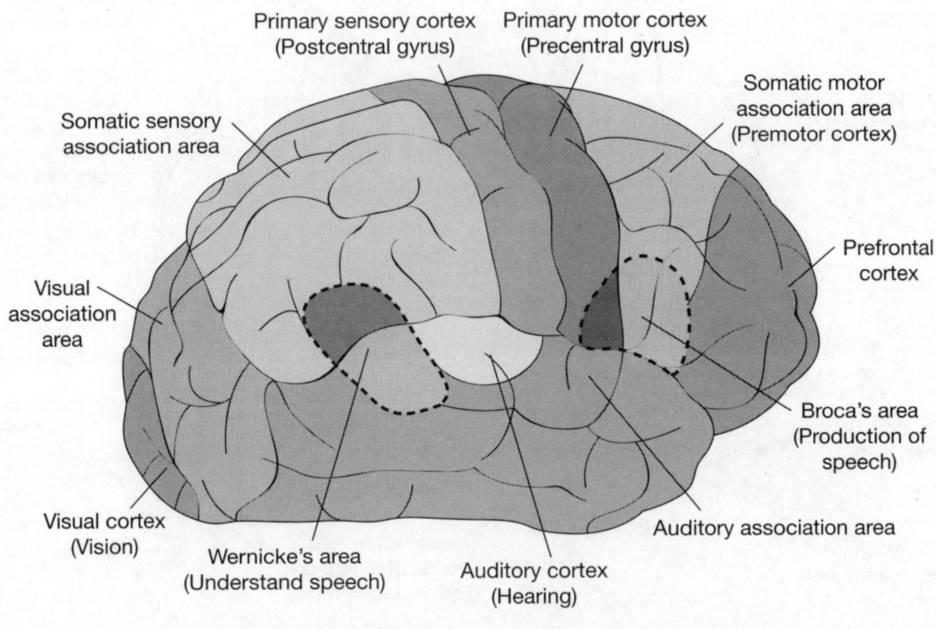

Fig. 8.6 Main functions of the different areas of the brain

The Spinal Cord

The spinal cord is a bundle of nerves that runs through the backbone. It is an essential component of the CNS. It begins at the medulla oblongata and runs almost all the way down the length of the backbone. It is located within the neural canal.

The backbone is also known as the vertebral column. The internal structure is depicted in Fig. 8.7. In the spinal cord, the grey and white matter are arranged differently than in the brain. Millions of bundles of axons (nerve fibres), which make up the white matter that lies underneath the grey matter cortex, connect the neurons in various regions of the brain to form functional circuits.

There is a short central canal that traverses the length of the brain and connects to the cavities. It is filled with cerebrospinal fluid, which functions as a shock absorber and serves as a medium for food, waste and breathing gases to be exchanged with neurons. Externally, the spinal cord is protected by the same three membranes that protect the brain: dura mater, arachnoid and pia mater.

The following **functions** are performed by the spinal cord:

- It is responsible for reflexes beneath the neck.
- It transmits sensory signals from the skin and muscles to the brain.
- It transmits motor responses from the brain to the trunk and limb muscles.

Reflex Action

We may be confronted with a life-threatening scenario that requires immediate action and does not allow us time to think and make a decision. Reflex action is used to achieve such responses. The term 'reflex' is used to describe a sudden and immediate reaction to anything. What do you do if there is a sudden dust storm? You instantly close your eyes to avoid dust particles from entering your eyes. When you touch a hot pan while cooking, how do you react? Do you remove your hand to avoid the hot pan? There is an instantaneous and automatic reaction in both of these circumstances.

Certain activities in our bodies are spontaneous and do not require processing by the brain. Reflex actions are acts or responses like these. Involuntary acts that occur without conscious mental processes are known as reflex actions. Consider the following scenarios: when particles fall into your eye, tears are immediately flushed to remove them (glandular secretion); if your hand unintentionally touches a hot pan, you immediately remove it (muscular movement) as shown in Fig. 8.8; you shiver (muscular contractions) when it is very cold or sweat (glandular secretion) when it is very hot. All of these are examples of reflex actions.

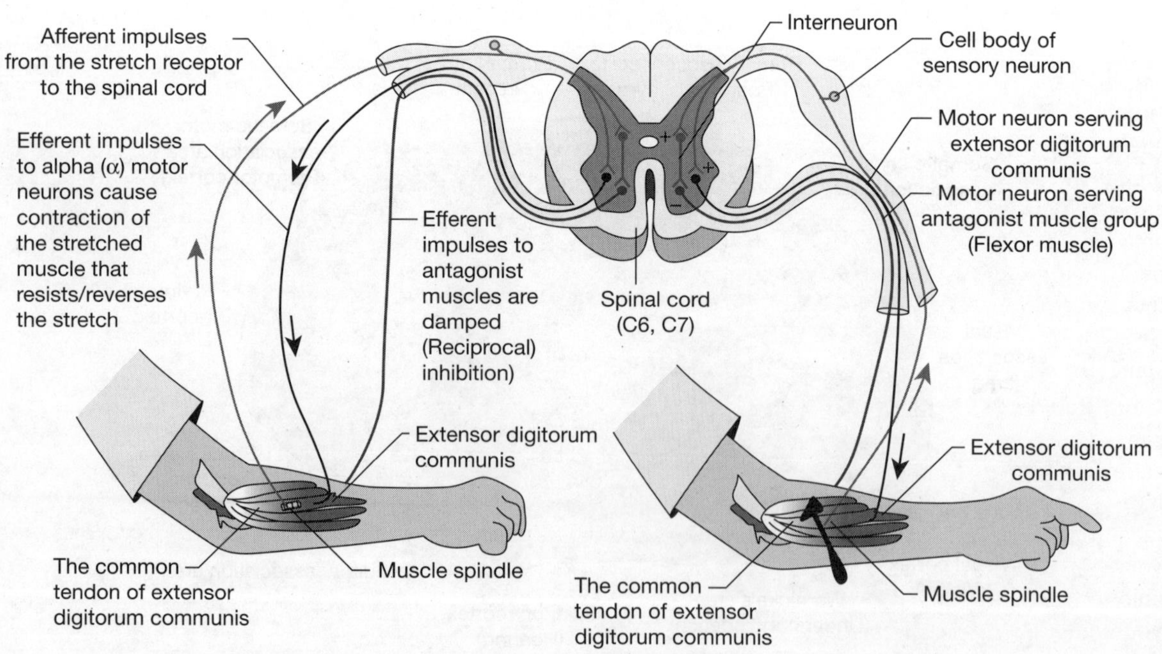

Fig. 8.7 Internal structure of the spinal cord and the nervous pathway

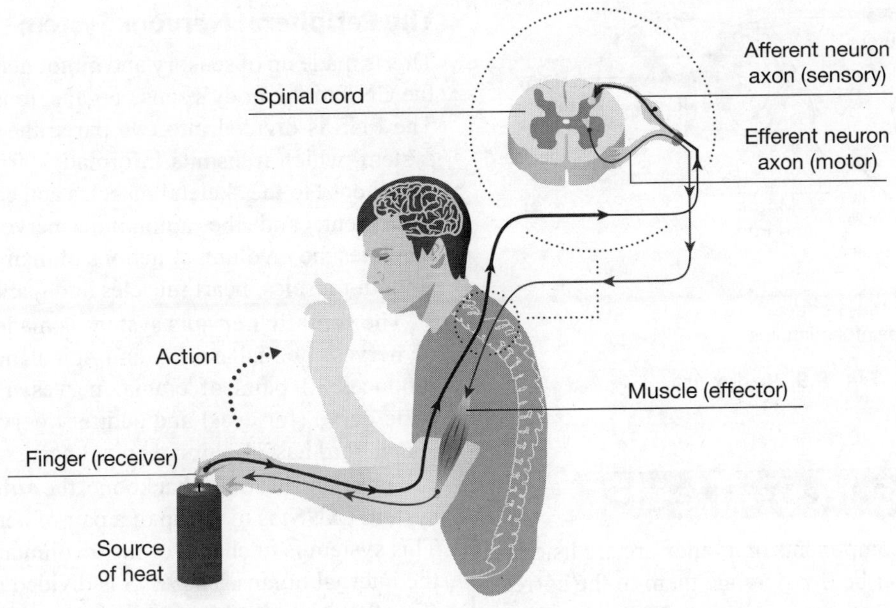

Fig. 8.8 Reflex of withdrawal of hand on touching a hot object brought about by the spinal cord

> ***Fun Fact!***
>
> A reflex action is a spontaneous, autonomic, and mechanical response to a stimuli that is controlled by the spinal cord without the brain's involvement. Sensory stimulations cause all involuntary responses or reflexes, which result in either a muscle activity or a glandular secretion.

Reflexes are of two types:

- **Inborn (natural) reflex:** Close your eyes and attempt to follow your body's rhythm. What are your thoughts? You may have the impression that you are breathing slowly. Natural reflexes refer to any activity that does not require any prior knowledge or experience. These reflexes are inborn, meaning they are passed down from generation to generation. Swallowing, coughing and blinking of the eyelids are some instances.
- **Conditioned reflex:** What happens when you can smell your favourite cuisine without actually eating it? In anticipation, your mouth starts to moisten (salivation). This is known as a conditioned (acquired) reaction. This phenomena is based on your previous experience, in which you have been able to link a distinct aroma to a specific dish that you enjoy. If you had never eaten the food before, the aroma would not have caused you to salivate.

  Conditioned reflexes are actions that evolve throughout the course of a person's life as a result of experience or learning.

  It is vital to reinforce the conditioned response on a regular basis in order to keep it alive. If the smell of a favourite food item is not followed by the food itself on several occasions, you will stop reacting to the smell with salivation after a certain time.

Reflex Arc

In the example mentioned earlier, the process of detecting a signal or input and responding to it with an output action can be accomplished rapidly if the nerves that detect heat were coupled to the nerves that move muscles in a more straightforward manner. A reflex arc is a term used to describe such a link (Fig. 8.9). Where should such reflex arc connections be made? On their way to the brain, nerves from the body connect in a bundle in the spinal cord.

Because the brain's reasoning process is not rapid enough, animals evolved reflex arcs. In fact, many animals lack the intricate neural network that is required for thinking. Thus, reflex arcs are likely to have evolved as an efficient way for immediate reactions.

A reflex arc may be represented as follows:

Stimulus → Receptor in the sense organ → Afferent (sensory) nerve fibre → CNS →

Efferent (motor) nerve fibre → Muscle (to contract)/ Gland (to secrete)

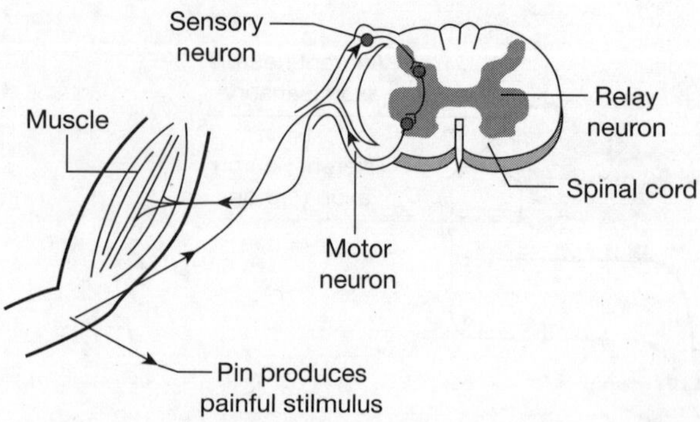

Fig. 8.9 Reflex arc

SCIENCE IN DAILY LIFE

The numerous components of a reflex arc are listed in a random manner below. Arrange them in the correct order:

Motor neuron, Sensory neuron, Effector, Stimulus, CNS, Receptor, Response

The Peripheral Nervous System

This is made up of sensory and motor nerves and connects the CNS to the body's sense organs, muscles and glands. The PNS is divided into two parts: the somatic nervous system, which transmits information from the brain and spinal cord to the skeletal muscles and controls voluntary movement; and the autonomic nerve system, which regulates the involuntary actions of many internal organs, smooth muscles, heart muscles and glands.

The **somatic nervous system** is made up of two types of nerves: cranial nerves and spinal nerves. The brain produces 12 pairs of cranial nerves. For example, the optic nerve (for eyes) and auditory nerve (for ears). The spinal cord has 31 pairs of spinal nerves.

On either side of the backbone, the **autonomic nervous system (ANS)** is made up of a pair of nerves and ganglia. This system is in charge of the involuntary movements of the internal organs. The ANS is divided into two sections (Fig. 8.10): sympathetic and parasympathetic.

During times of stress, the sympathetic nervous system becomes more active. It prepares the body for action. Its effects during the stress reaction include

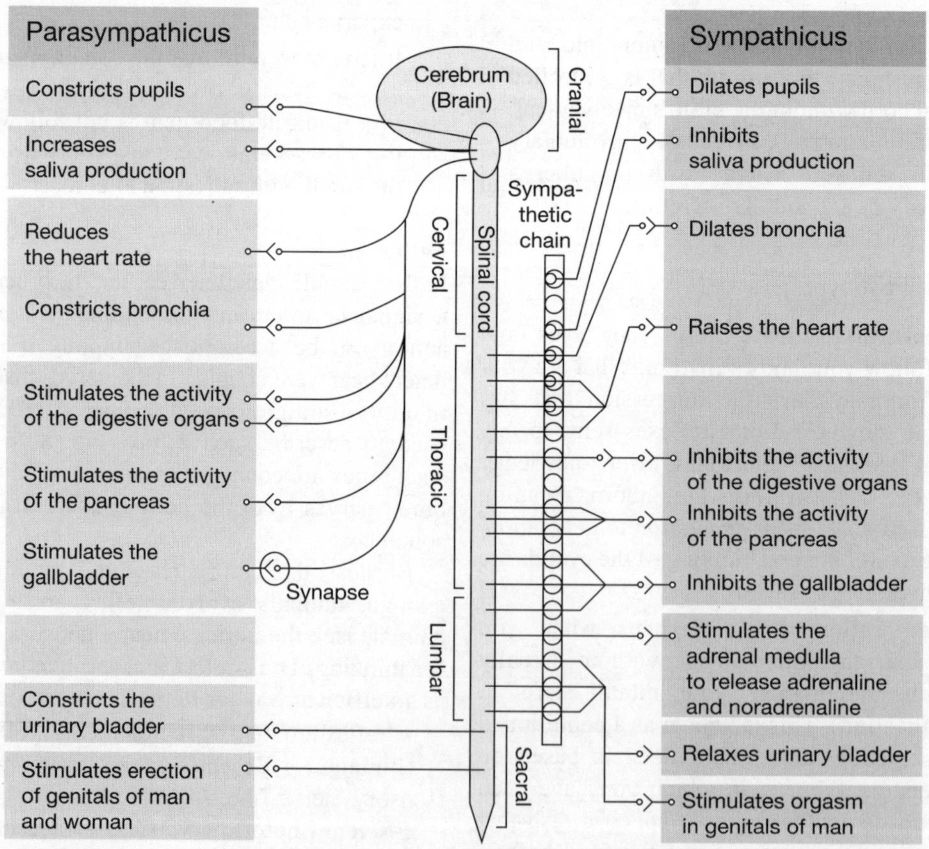

Fig. 8.10 Autonomic nervous system showing the opposing effects of the two parts, sympathetic and parasympathetic

the 'fight-or-flight response,' which is predominantly mediated by the hormone adrenaline.

The parasympathetic nervous system performs actions that do not require an immediate response, such as saliva and tear production, digestion, and so on. The functions of the sympathetic and parasympathetic nervous systems may appear to be diametrically opposed, yet they are actually complementary rather than hostile.

THE ENDOCRINE SYSTEM

From infancy to adulthood and old age, we have observed the changes and growth that occur in humans. During adolescence, you may notice more pronounced changes in height and weight, as well as very substantial development of secondary sexual traits. In truth, our bodies evolve throughout our lives. Endocrine glands, which are specific glands in our bodies, manage these changes.

The primary purpose of these glands is to produce chemical secretions known as hormones. Hormones play a critical role in the control, coordination and regulation of the body's tissues, organs and systems. To promote optimal functioning of the body, well-coordinated processes regulate the release of precise amounts of hormones.

The endocrine system is thus in charge of chemical coordination in the human body (Fig. 8.11).

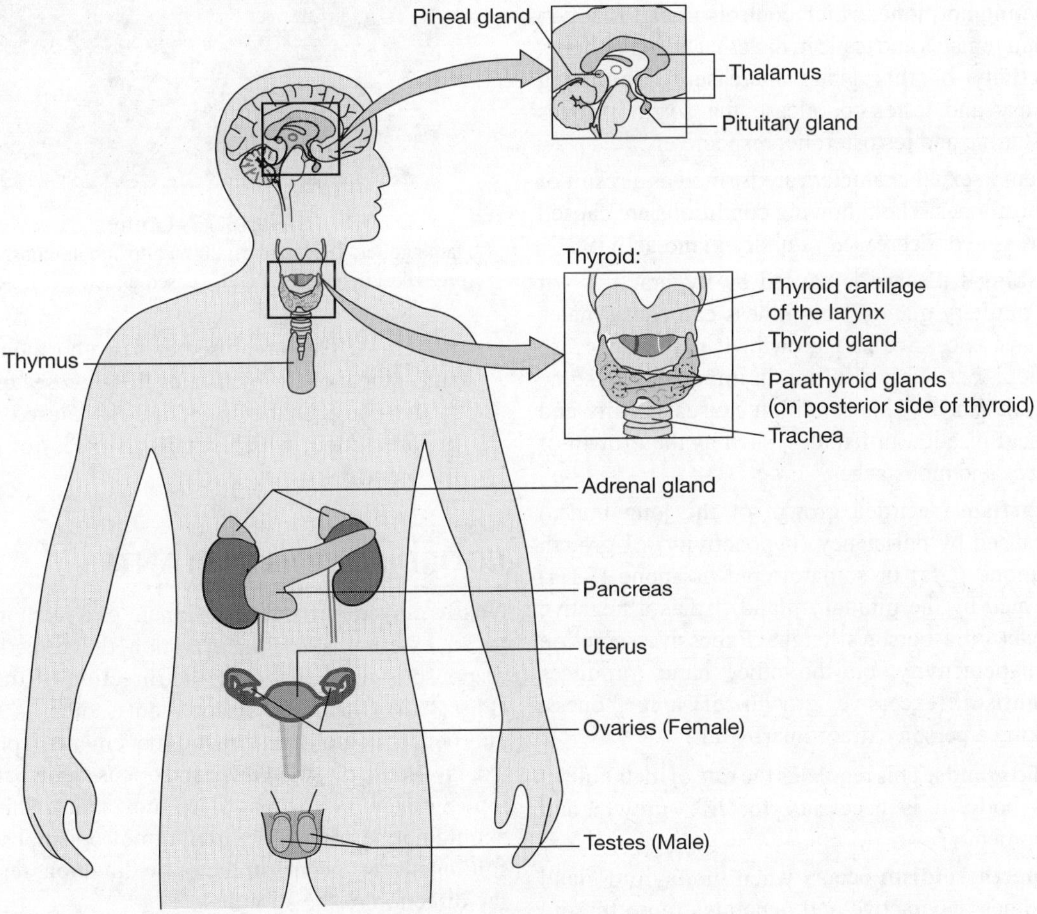

Fig. 8.11 The endocrine system

Fun Fact!

'Hormone' comes from the Greek '*hormaein*', which means 'to set in action' or 'to spur on'.

A hormone is a substance released by an endocrine gland and transported through the bloodstream to a target organ elsewhere in the body to induce a certain function.

Let us take a look at some of the most significant endocrine glands, the hormones they produce, and the effects they have on the body.

- **Pituitary gland:** This is a tiny gland found near the base of the brain. It is essential for a child's development from puberty to full reproductive maturity. The pituitary gland produces the gonad stimulating hormone, which controls gonad function (ovary in females and testis in males). During puberty, the activity of this gland increases, stimulating the ovary and testes to release the sex hormones progesterone and testosterone, respectively.

  Secondary sexual characters are formed as a result of these hormones. The following conditions are caused by increased or decreased pituitary gland activity:

  - **Cushing's disease** is caused by hyperactivity of the pituitary glands. This illness can cause males to have excessive hair growth. It may also cause testicular atrophy, resulting in impotency in some circumstances. This condition causes sterility and masculinisation in females, such as the growth of beards and moustaches.
  - **Dwarfism** (retarded growth of the long bones) is caused by deficiency (hypoactivity) of growth hormone (GH) or somatotropic hormone (STH) secreted by the pituitary gland. It has a negative impact on a person's height. Excessive secretion or hyperactivity, on the other hand, produces **gigantism** (excessive growth of long bones), making a person extraordinarily tall.
- **Thyroid glands:** This regulates the rate of metabolism in our body. It is necessary for life, growth and development.
  - **Hyperthyroidism** occurs when the thyroid gland becomes hyperactive and generates more thyroid hormone than is required for optimal function.
  - **Hypothyroidism** is a disorder in which the thyroid gland becomes underactive and generates less thyroid hormone than is required.
  - **Cretinism** is a disorder in which physical and mental growth is severely stunted due to an untreated congenital deficit of the thyroid hormone (hypothyroidism) or protracted iodine deficiency in the diet.
  - **Goitre** is a thyroid illness marked by an enlargement of the gland that manifests itself outwardly as a swelling in the front of the neck. A shortage of iodine in the diet causes simple goitre (Fig. 8.12).

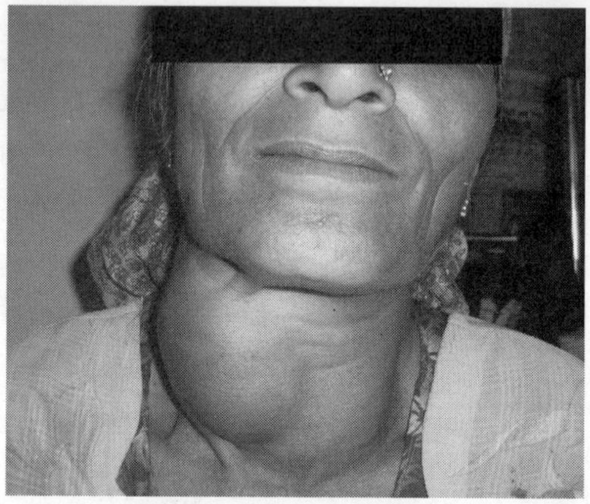

Fig. 8.12 Goitre

(A large goiter; JS Bhandari; 2009; https://commons.wikimedia.org/wiki/File:Goiter.JPG. Licensed under: CC-BY-SA-3.0)

- **Pancreas:** This organ secretes two hormones, insulin and glucagon, which aid in glucose metabolism in the body. Diabetes mellitus is caused by insulin hyposecretion, which results in excess of glucose in the blood.

COORDINATION IN PLANTS

Nastic movement is the movement of a plant in response to an external stimulus in which the direction of the response is not defined by the direction of the stimulus (Fig. 8.13). Flat sections of plants, such as the leaves and petals, demonstrate nastic movements. For example, a daisy bloom closes at dusk and opens again before dawn, a phenomenon known as sleep movement. This response should not be confused with thigmotropism, because leaf folding always occurs in the same direction, regardless of the direction of the stimulus.

There are two types of nastic movements:

- **Photonasty** is a nastic movement that corresponds to the day's light and dark phases. Primrose flowers, for example, bloom in the evening but close throughout the day.

Fig. 8.13 Giant sensitive tree
(*Mimosa pigra*, leaves and inflorescence…; Wibowo Djatmiko; 2008; https://commons.wikimedia.org/wiki/File:Mimo_pigra_080209-4579_ckoep.jpg. Licensed under: CC-BY-SA-3.0)

Fig. 8.14 Phototropism

- **Nyctinasty** is a movement triggered by darkness. The leaves and blossoms of a plant take on a different stance at night than they do during the day. For example, the leaves of the raintree fold by twilight.

Tropism is the movement of plant organs towards or away from a stimulus. Because tropic movements are slow, the stimulus must be applied for a longer period of time before the consequences become apparent.

The following are the forms of tropic motions in plants:

- **Phototropism** is the term used to describe the movement of plant components towards or away from light. Positive phototropism describes how most plants' shoots grow towards the source of light. Roots are negatively phototropic because they grow away from light (Fig. 8.14).
- **Geotropism** is the movement of plant organs in response to gravity. Because roots grow in the direction of gravity, they are positively geotropic. Negatively geotropic means that the shoot develops upwards, against gravity (Fig. 8.15).
- **Thigmotropism** is defined as the movement of plant organs in response to stimuli induced by physical contact with solid objects. Plants with weak stems employ twining stems and tendrils to climb on other plants/objects for support. As a result, twining stems and tendrils are thigmotropic.
- **Hydrotropism** is the movement of plant organs in reaction to water. Roots are positively hydrotropic, meaning they grow towards a source of moisture (Fig. 8.16).

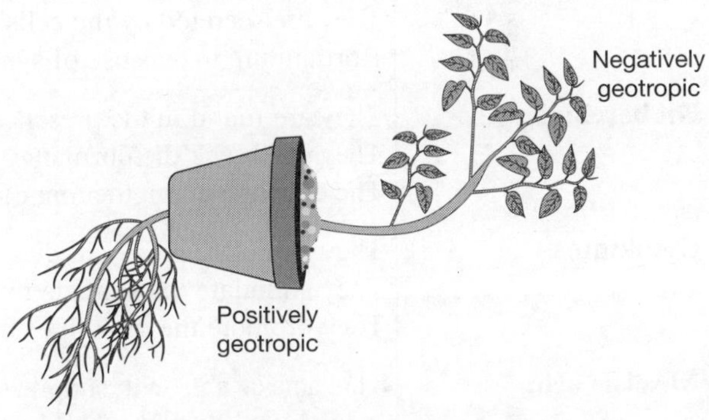

Fig. 8.15 Gravitropism

- **Chemotropism** refers to the movement of plant organs in response to a chemical stimulus. Negative chemotropism occurs when the plant organs develop away from the chemical response, while positive chemotropism occurs when plant parts grow towards the chemical response. Pollen tubes, for example, grow towards the sugary material released by the flower's stigma.

Plant Hormones

Plant hormones regulate cell division, cell expansion and cell differentiation, among other aspects of plant growth. The various plant hormones are listed in Table 8.1.

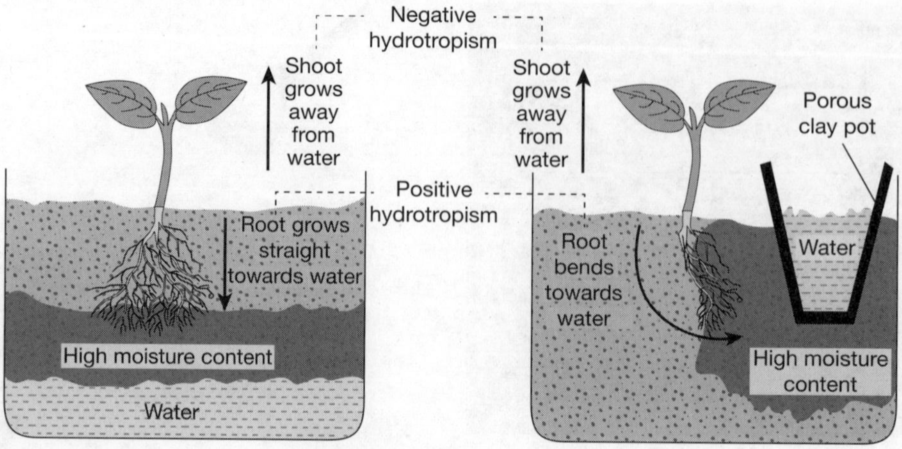

Fig. 8.16 Hydrotropism

Table 8.1 Plants hormones

Auxins	These are phytohormones. They are secreted by the cells at the tips of stems and roots, and they help plants grow. Horticulture makes use of synthetic auxins.
Gibberellins	They are found in the presence of auxins and promote cell differentiation. They can break the dormancy of seeds. The encourage lengthening of shoots.
Cytokinins	They stimulate plant cell division and delay the ageing of leaves. They stimulate the stomata to open. They promote the growth of fruit.
Abscisic acid	This acts as a growth suppressant. It encourages seed and bud dormancy. It promotes stomatal closure and encourages withering and falling of leaves. It causes detachment of flowers and fruits from the plants.

SUMMARY

- The nervous system and hormones perform the activities of control and coordination.
- The nervous system's reactions can be categorised as reflex, voluntary or involuntary actions.
- The nervous system functions through our muscles and transmits information through electrical impulses. It receives information from our sense organs.
- An organism's response to a change in the environment is known as a stimulus.
- Coordination is the methodical collaboration of several organs within an organism to achieve an appropriate response.
- Chemical coordination is present in both plants and animals.
- To create the intended impact, hormones released in one area of an organism travel to another area.
- The hormones' actions are regulated by a feedback mechanism.
- Phyto-hormones are the hormones found in plants.
- Auxin is a hormone that encourages cell expansion and growth in plants.
- Gibberellins are hormones that encourage cell differentiation and activate dormant seeds and buds.
- Cytokinin is a hormone that encourages cell division and the opening of stomata.
- Abscisic acid aids in slowing the plant's growth and encourages withering and the dropping of food and leaves.

PRACTICE QUESTIONS

Analyse Your Concepts (School Exam Based)

Multiple Choice Questions

1. Posture and balance of the body are controlled by:
   (a) Cerebellum
   (b) Medulla oblongata
   (c) Pons
   (d) Cerebrum

2. Breathing is controlled by which part of the brain?
   (a) Medulla oblongata (b) Cerebellum
   (c) Hypothalamus (d) Cerebrum

3. Which part of the nervous system controls the reflex activities of the body?
   (a) Spinal cord (b) Brain
   (c) Cerebrum (d) Cerebellum

4. Which of the following acts as an endocrine and exocrine gland?
   (a) Pancreas (b) Thyroid
   (c) Adrenal (d) Liver

5. Identify which of the following statements about thyroxine is incorrect:
   (a) The thyroid gland requires iodine to synthesize thyroxine.
   (b) Thyroxine is also called the thyroid hormone.
   (c) Iron is essential for the synthesis of thyroxine.
   (d) Thyroxine regulates protein, carbohydrate and fat metabolism in the body.

6. Which gland secretes the growth hormone?
   (a) Pituitary
   (b) Thyroid
   (c) Hypothalamus
   (d) Adrenal

7. The secretion of which hormone leads to physical changes in the body around the age of 10–12 years?
   (a) Testosterone from the testes and oestrogen from the ovary.
   (b) Oestrogen from the adrenal glands and testosterone from the pituitary gland.
   (c) Oestrogen from the testes and testosterone from the ovary.
   (d) Testosterone from the thyroid gland and oestrogen from the pituitary gland.

8. A diabetic patient suffers from deficiency of which hormone?
   (a) Thyroxine (b) Insulin
   (c) Oestrogen (d) Testosterone

Match the Following

Each question contains statements given in two columns which have to be matched. Statements in column I have to be matched with statements in column II.

1. Match the hormones given in Column I with their functions given in Column II:

Column I	Column II
(i) Thyroxine	(a) Causes breasts to develop in females
(ii) Adrenaline	(b) Causes males to start producing sperms
(iii) Insulin	(c) Prepares the body for an emergency
(iv) Oestrogen	(d) Controls the metabolic rate
(v) Testosterone	(e) Regulates the amount of sugar in blood

2. Match the following parts of the brain with their functions:

Column I	Column II
(i) Cerebrum	(a) Balance and coordination
(ii) Cerebellum	(b) Thinking and intelligence
(iii) Medulla	(c) Reflexes and involuntary actions

3. Match the following reflex actions with their stimuli:

Column I	Column II
(i) Sneezing	(a) Bright light
(ii) Blinking	(b) Dust or irritants
(iii) Salivation	(c) Foul taste or smell

4. Match the following glands with their hormones:

Column I	Column II
(i) Pituitary	(a) Growth hormone
(ii) Thyroid	(b) Thyroxine
(iii) Adrenal	(c) Adrenaline

5. Match the following sense organs with their stimuli:

Column I	Column II
(i) Ear	(a) Chemicals
(ii) Nose	(b) Sound waves
(iii) Tongue	(c) Light

True or False

Read the following statements and write your answer as true or false.

1. The highest coordinating centre in the human body is the brain.
2. Proper growth hormone in childhood leads to dwarfism in humans.
3. The main function of the cerebrum is thinking.
4. In human males, the testes lie in the scrotum, as it helps in the production of sperms.
5. The growth of the roots of plants towards water is called geotropism.

Fill in the Blanks

Complete the following statements with an appropriate word/term to be filled in the blank spaces.

1. Control and coordination are the functions of the nervous system and _______ in our body.
2. The nervous system uses _______ impulses to transmit messages.
3. The central nervous system consists of _______ and _______.
4. The largest part of the brain is _______.
5. Deficiency of the _______ hormone in childhood leads to dwarfism.
6. The growth of pollen tubes towards the ovules is the result of _______ movement.

Assertion–Reason Questions

The following questions consists of two statements: Assertion (A) and Reason (R). Answer these questions selecting the appropriate option given below:

(a) Both A and R are true and R is the correct explanation of A
(b) Both A and R are true but R is not the correct explanation of A
(c) A is true but R is false
(d) A is false but R is true

1. Assertion (A): We can detect the aroma of flowers.
   Reason (R): The sense of smell is detected by the phono receptors present in the nose, which send impulses to the brain.
2. Assertion (A): The signal from a neuron travels in both directions.
   Reason (R): The target tissue's entire cell population participates in the response.
3. Assetion (A): Flowering is caused by abscisic acid.
   Reason (R): It is classified as a growth inhibitor.
4. Assertion (A): Neurons send signals across the junction in the form of electrical impulses.
   Reason (R): The junction between two neurons is referred to as synapse. Neurotransmitters are released at the terminal ends.
5. Assertion (A): Positive phototropism refers to movement towards the direction of light.
   Reason (R): Auxin diffuses to the sunny side of a shoot when sunlight hits a certain part of the plant. Cell growth is induced by auxin concentration and results in a stem that bends towards the light.

One-word Answers

1. Which hormone helps in breaking dormancy?
2. What is the full form of IAA?
3. Name a gaseous plant hormone.
4. Name the stress hormone in plants.
5. What is the functional unit of the nervous system?
6. What is the role of dendrites in a neuron?
7. What is the gap between two neurons called?

8. What is the name of the nervous system that controls voluntary actions?
9. Which hormone is responsible for fight or flight response?
10. What is the name of the largest part of the brain?
11. What is the name of the hormone that regulates blood sugar levels?
12. What is the name of the gland that produces melatonin?
13. Which reflex action is responsible for closing of eyes due to bright light?
14. What is the name of the hormone that regulates water and salt balance in the body?

Define the Term

1. Coordination
2. Neuron
3. Stimuli
4. Receptor
5. Nerve impulse
6. Synapse
7. Synaptic cleft
8. Neurotransmitter

Groups of Terms

Given below are groups of terms. In each group the first pair indicates the relationship between the two terms. Rewrite and complete the second pair on a similar basis.

1. Insulin: Glucose uptake :: Glucagon: _______
2. Neuron: Nervous system :: Muscle fibre: _______
3. Cerebrum: Memory :: Cerebellum: _______
4. Photoreceptor: Light :: Thermoreceptor: _______
5. Adrenaline: Fight or flight :: Endorphins: _______
6. Sensory neuron: Afferent pathway :: Motor neuron: _______
7. Olfactory bulb: Smell :: Gustatory cortex: _______
8. Pupil: Iris :: Cornea: _______
9. Hormone: Target organ :: Neurotransmitter: _______
10. Pituitary gland: Antidiuretic hormone :: Adrenal gland: _______

Identify the Odd One Out

1. (a) Cerebrum (b) Cerebellum (c) Medulla oblongata (d) Spinal cord
2. (a) Insulin (b) Glucagon (c) Estrogen (d) Testosterone
3. (a) Cytokinin (b) Gibberellin (c) Abscisic acid (d) Bacteriophage
4. (a) Neuron (b) Nerve (c) Hormone (d) Neurotransmitter
5. (a) Pons (b) Thalamus C. Hypothalamus (d) Pituitary gland

Image-Based Questions

1. Label the parts (a), (b), (c) and (d) and show the direction of flow of electrical signals in the given figure.

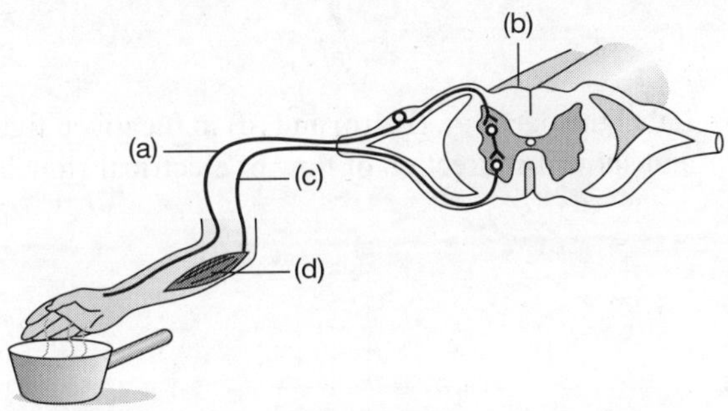

2. Label the endocrine glands in the given figure.

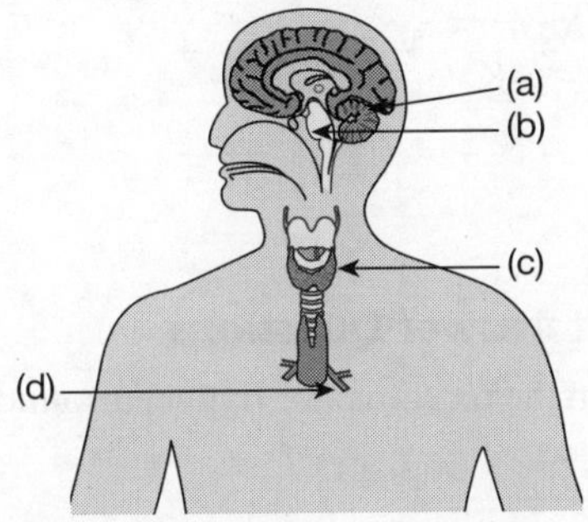

3. Label the parts of a neuron in the figure.

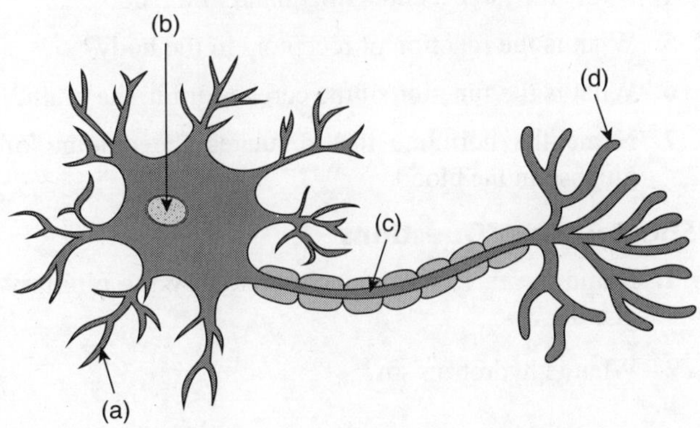

4. Label the parts (a), (b) and (c) in the given figure and show the direction of flow of information:

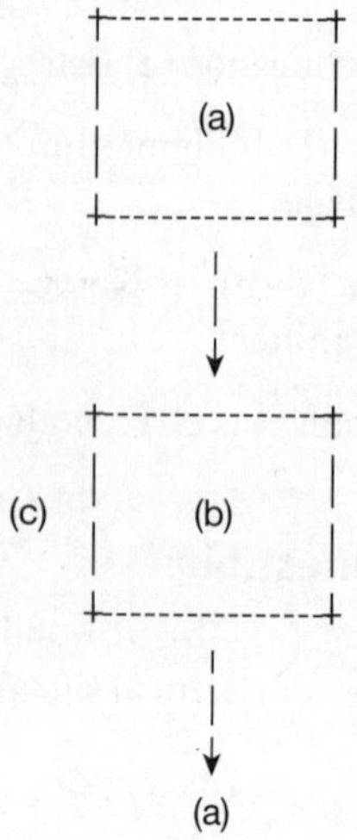

5. Label the parts (a), (b), (c) and (d) in the given figure and show the direction of flow of electrical signals:

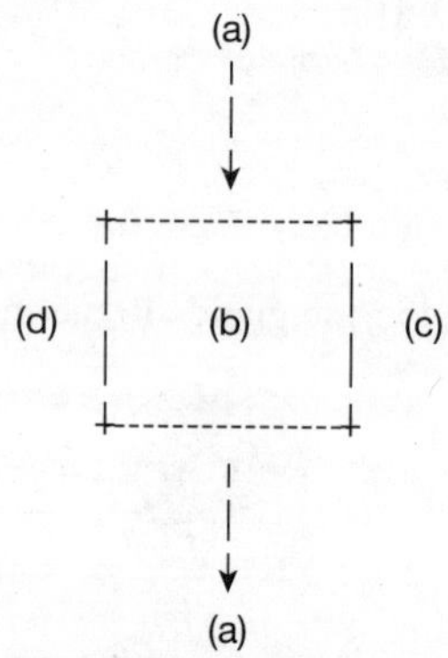

Very Short Answer Questions

1. Mention the receptors for light and sound in animals.
2. What is phototropism?
3. Define coordination.
4. Name the largest endocrine gland in the body.
5. What is the function of receptors in the body?
6. What is the function of the cerebellum in the brain?
7. Name the hormone that regulates the amount of glucose in the blood.

Short Answer Questions

1. Name the three hormones secreted by the pituitary gland.
2. What is hydrotropism?
3. What is coordination?
4. List the functions of testosterone and oestrogen.
5. What is the role of auxin hormone in phototropism?
6. What is the difference between voluntary and involuntary actions?
7. What is the difference between a hormone and a neurotransmitter?
8. What is the role of the hypothalamus in the endocrine system?

Long Answer Questions

1. Describe the phenomena of release and transport of neurotransmitters.
2. List the parts of the human forebrain along with their functions.
3. Write down the basic difference between reflex actions and involuntary actions.
4. Explain the difference between hormonal control and coordination and the nervous system in animals.
5. What is the difference between the manner in which movement takes place in a sensitive plant and the movement of our legs?
6. Name the major endocrine glands present in the human body. Which of these glands also function as exocrine glands?
7. (a) Write the names of five endocrine glands found in the human body. Name the hormones secreted by each gland.
   (b) How do hormones reach the organs they control?
   (c) Name the gland that controls the secretion of hormones from the pituitary.
   (d) How does our body respond when adrenaline is secreted in large amounts into blood?
   (e) Name the disease which occurs in adults due to deficiency of iodine in the diet. What is the main symptom of this disease?
8. Explain the mechanism of reflex action.
9. What are the different types of tropisms?
10. Explain the role of different parts of the human nervous system.
11. What is the role of hormones in the human body?

Case Study-Based Questions

1. The variety of tissues that may be directed to get ready for an activity would be constrained if a squirrel's body was designed only to respond to electrical impulses sent by nerve cells. On the other hand, if a chemical signal were also transmitted, it would reach every cell in the body and bring about the necessary broad alterations. Many animals, including humans, accomplish this by utilising adrenaline, which is released by the adrenal glands.
   (i) Adrenaline targets which hormone?
   (ii) The thyroid releases which hormone?
   (iii) What does thyroxine do?
   (iv) What are the three brain-based hormonal glands?

2. Some plants, like the pea plant, use their tendrils to cling to other plants or fences. These tendrils are delicate to the touch. The portion of the tendril in contact with any support grows less quickly than the portion away from the support. As a result, the tendril starts to wrap around the object and attach to it. More frequently, plants grow slowly in a certain direction in response to stimuli. The direction of this development gives the impression that the plant is moving.
   (i) What are the types of tropism shown by plants?
   (ii) Which of the following is exemplified by the touch-me-not plant?
   (iii) Which plant hormone is responsible for cell division?
   (iv) Name one growth-inhibiting hormone found in plants.

3. The body is made up of complex networks of carefully placed neurons. It is located at the front of the skull and receives messages from throughout the body. Before acting on those impulses, it considers them. Naturally, this thinking portion of the brain in the skull needs to be connected to nerves from various parts of the body in order to receive these impulses.
   (i) What is a reflex?
   (ii) What does the motor neuron do?
   (iii) The long cable that extends through the main cell in the neurons is called:
      (a) Dendrite
      (b) Axon
      (c) Node
      (d) Myelin sheath
   (iv) What are dendrites?

4. Animal neurological and muscular systems enable control and synchronisation. For humans, touching a hot object is a risky and urgent circumstance. We must recognise it and react to it. How can we tell if something is hot when we are touching it? The specialised tips of some nerve cells detect every bit of information from our surroundings. These receptors are typically found in the inner ear, nose, tongue and other sense organs. This means that whereas olfactory receptors pick up smell, gustatory receptors pick up taste.
   (i) Name the main parts of a neuron.
   (ii) Which part of the neuron receives the initial signal?
   (iii) What is the function of the myelin sheath?
   (iv) Which of the following is not a part of the nervous system?
      (a) Axon
      (b) Dendrites
      (c) Myelin sheath
      (d) Villi

5. Examples of voluntary behaviours that are based on choosing what to do next include writing, talking, shifting a chair and clapping at the conclusion of a programme. For this, the brain must communicate with the muscles. The peripheral nervous system, which is made up of cranial nerves that emerge from the brain and spinal nerves that emerge from the spinal cord, facilitates communication between the central nervous system and the various regions of the body. Thus, the brain enables us to think and act on the basis of that thinking.
   (i) Name the three main parts of the brain.
   (ii) The fluid present in our brain is called:
      (a) Cerebronasal fluid
      (b) Cerebrospinal fluid
      (c) Spinal fluid
      (d) Amniotic fluid
   (iii) What is the role of the hypothalamus?
   (iv) What does the medulla do?

NCERT Corner

1. Explain tropic movement with an example.
2. Explain the side effects, if intake of iodine in our diet is low?
3. Answer the following:
   (a) Name the hormone which is responsible for the changes noticed in females at puberty.
   (b) Dwarfism results due to deficiency of which hormone?
   (c) Blood sugar level rises due to deficiency of which hormone?
   (d) Name the hormone which requires iodine for its synthesis.
4. Answer the following:
   (a) Name the endocrine gland associated with the brain.
   (b) Which gland secretes digestive enzymes as well as hormones?
   (c) Name the endocrine gland associated with the kidneys.
   (d) Which endocrine gland is present in males but not in females?
5. What is a reflex action? Explain with an example.
6. How do auxins promote the growth of a plant?
7. What is the role of the pituitary gland in our body?
8. What is the difference between the voluntary and involuntary actions?
9. How is the movement of leaves of a sensitive plant different from the movement of a shoot towards light?

COMPETITION WINDOW (OBJECTIVE TYPE)

1. Any change in the environment to which an organism responds is called:
   (a) Response (b) Coordination
   (c) Stimulus (d) Hormone
2. Which plant hormone promotes dormancy in seeds and buds?
   (a) Auxin (b) Abscisic acid
   (c) Cytokinin (d) Gibberellin
3. The response of plant roots towards water is called:
   (a) Chemotropism (b) Phototropism
   (c) Geotropism (d Hydrotropism
4. Movement of a sunflower in accordance with the path of the sun is due to:
   (a) Geotropism (b) Chemotropism
   (c) Phototropism (d) Hydrotropism
5. Which plant hormone promotes cell division?
   (a) Cytokinin (b) Gibberellin
   (c) Auxin (d) Abscisic acid
6. The main function of abscisic acid in plants is:
   (a) To promote cell division
   (b) To increase the length of cells
   (c) To promote growth of the stem
   (d) To inhibit growth
7. Fall of mature leaves and fruits from plants is triggered by which of the following substances?
   (a) Abscisic acid (b) Cytokinin
   (c) Gibberellin (d) Auxin
8. The roots of plants are:
   (a) Positively phototropic
   (b) Negatively geotropic
   (c) Positively geotropic
   (d) None of these
9. The part of the body which responds to the instructions sent from the nervous system is called:
   (a) Effector (b) Receptor
   (c) Nerves (d) Muscles
10. The longest fibre in the cell body of a neuron is called:
    (a) Axon (b) Cytoplasm
    (c) Sheath (d) Dendrite
11. Which nerves transmit impulses from the central nervous system towards muscle cells?
    (a) Motor nerves
    (b) Sensory nerves
    (c) Relay nerves
    (d) Cranial nerves

12. A microscopic gap between a pair of adjacent neurons over which nerve impulses pass is called:
(a) Neurotransmitter (b) Synapse
(c) Axon (d) Dendrite

13. 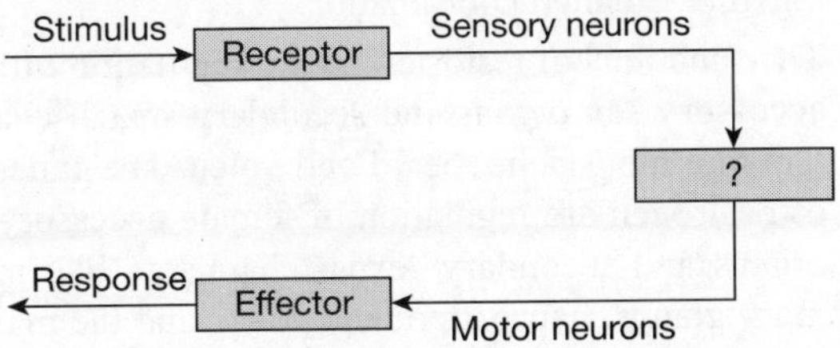

Give the missing term:
(a) Cranial nerves (b) Brain
(c) Spinal cord (d) Relay nerves

14. The highest coordinating centre in the human body is the:
(a) Brain (b) Heart
(c) Spinal cord (d) Kidney

15. The main function of the cerebrum is:
(a) Balancing (b) Hearing
(c) Memory (d) Thinking

Answer Keys

Multiple Choice Questions

1. (a) 2. (a) 3. (a) 4. (a) 5. (c)
6. (a) 7. (a) 8. (b)

Match the Following

1. (i) (d), (ii) (c), (iii) (e), (iv) (a), (v) (b)
2. (i) (b), (ii) (a), (iii) (c)
3. (i) (b), (ii) (b), (iii) (c)
4. (i) (a), (ii) (b), (iii) (c)
5. (i) (b), (ii) (a), (iii) (c)

True or False

1. T 2. F 3. T 4. T 5. F

Fill in the Blanks

1. Hormones 2. Electrical
3. Brain, spinal cord 4. Cerebrum
5. Growth 6. Chemotropic

Assertion-Reason Questions

1. (b) 2. (d) 3. (c) 4. (c) 5. (b)

One-word Answers

1. Auxin 2. Indole acetic acid
3. Ethylene 4. ABA
5. Neuron 6. Receive impulses
7. Synapse 8. Somatic
9. Adrenaline 10. Cerebrum
11. Insulin 12. Pineal
13. Photopupillary 14. Aldosterone

Define the Term

1. Coordination is the process by which two or more organs work together and complement one another's functions.
2. Neurons are extremely specialised cells that are capable of detecting, receiving and transmitting various stimuli.
3. An organism's action or response is triggered by a stimulus. Temperature, light, hunger, thirst and pain are examples of stimuli.
4. Our ability to perceive our surroundings is made possible by our sensory receptors. They react to external stimuli.
5. Electrical impulses are also referred to as nerve impulses. The information is carried by the electric current that flows from one end of the neuron to the other.
6. Synapse is the point at which two neurons connect and communicate.
7. Synaptic cleft is the space between two neurons at a synapse, where the nerve impulse is transmitted through certain chemical structures.
8. Neurotransmitters are chemical messengers that help in transmission of the nerve impulse between two neurons at a synapse. They are released by one neuron's nerve endings and received by the receptors present on the surface of the dendrites of another neuron.

Groups of Terms

1. Release of glucose
2. Spinal cord
3. Retina

4. Glucose
5. Response
6. Regulation
7. Impulse transmission
8. Synapse
9. Metabolism
10. Breathing

Identify the Odd One Out

1. (d) 2. (c) 3. (d) 4. (c) 5. (d)

Image-Based Questions

1. (a) Sensory neuron, (b) Spinal cord (CNS), (c) Motor neuron, (d) Effector (muscle in arm)
2. (a) Pineal gland, (b) Pituitary gland, (c) Thyroid, (d) Thymus
3. (a) Dendrite, (b) Cell body, (c) Axon, (d) Axon terminal
4. (a) Sensory receptor (b) Spinal cord (c) Brain

   Direction of flow of information: Sensory receptor → Spinal cord → Brain
5. (a) Sensory neuron (b) Interneuron (c) Motor neuron (d) Effector organ (muscle or gland)

   Direction of flow of electrical signals: Sensory neuron → Interneuron → Motor neuron → Effector organ

Very Short Answer Questions

1. Photoreceptor is the receptor for light and phono receptor is the receptor for sound in animals.
2. The movement of plants towards light is called phototropism.
3. Coordination refers to the harmonious functioning of different organs and systems in the body to perform a particular task.
4. The thyroid gland is the largest endocrine gland in the body.
5. Receptors detect changes in the external or internal environment and generate nerve impulses that are transmitted to the brain or spinal cord.
6. The cerebellum is responsible for coordinating voluntary movements and maintaining posture and balance.
7. Insulin is the hormone that regulates the amount of glucose in the blood.

Short Answer Questions

1. The three hormones secreted by the pituitary glands are growth hormone, adrenocorticotrophic hormone and follicle stimulating hormone (FSH).
2. Hydrotropism is a type of tropism where the plant shows movement in response to water.
3. An organ system cannot work independently. All the organs are linked and the way in which they work together is called coordination.
4. The functions of testosterone are regulation of male accessory sex organs and secondary sexual characters like moustache, beard and voice. The functions of oestrogen are regulation of female accessory sex organs and secondary sexual characters like mammary glands, hair pattern and voice, and the maintenance of pregnancy.
5. The auxin hormone is responsible for causing the stem to bend towards light during phototropism. It promotes the elongation of cells on the darker side of the stem and inhibits cell elongation on the lighter side, resulting in the bending of the stem towards the light.
6. Voluntary actions are those that are under conscious control and are initiated by the brain, while involuntary actions are those that occur automatically and are not under conscious control.
7. The human eye responds to changes in the amount of light entering it by adjusting the size of the pupil. In bright light, the pupil contracts to reduce the amount of light entering the eye, while in dim light, the pupil dilates to allow more light to enter.
8. Hormones are chemical messengers that are secreted by the endocrine glands and travel through the bloodstream to target cells, while neurotransmitters are chemical messengers that are released by neurons and transmit signals across synapses to target cells.
9. The hypothalamus is a part of the brain that controls the release of hormones from the pituitary gland. It receives input from other parts of the brain and responds by releasing hormones that either stimulate or inhibit the release of hormones from the pituitary gland.

Long Answer Questions

1. Hint: Neurotransmitters are released from the presynaptic neuron, cross the synaptic cleft, and bind to receptors on the postsynaptic neuron. Elaborate on this.
2. Hint: The forebrain consists of the cerebrum, thalamus, and hypothalamus. The cerebrum is responsible for higher cognitive functions, the thalamus relays sensory information, and the hypothalamus regulates various bodily functions. Elaborate on these points.

3. Hint: Reflex actions are involuntary and automatic responses to a stimulus, while involuntary actions encompass a broader range of involuntary movements and bodily functions. Elaborate on this.
4. Hint: Hormonal control and coordination involve the release of hormones into the bloodstream to regulate body functions, while the nervous system uses electrical impulses to transmit signals for rapid control and coordination. Elaborate on these points.
5. Hint: Sensitive plants exhibit movement due to changes in turgor pressure, while the movement of our legs is controlled by skeletal muscles and the nervous system. Elaborate on this.
6. Hint: Major endocrine glands include the pituitary, thyroid, adrenal, pancreas and gonads. The pancreas also functions as an exocrine gland. Elaborate on these points.
7. (a) Hint: Examples of endocrine glands are the pituitary gland (secretes growth hormone), thyroid gland (secretes thyroxine), adrenal glands (secrete adrenaline and cortisol), pancreas (secretes insulin and glucagon) and gonads (secrete oestrogen and testosterone).
   (b) Hint: Hormones reach their target organs through the bloodstream.
   (c) Hint: The hypothalamus controls the secretion of hormones from the pituitary gland.
   (d) Hint: Large amounts of adrenaline trigger the body's fight-or-flight response.
   (e) Hint: Deficiency of iodine in the diet can lead to a disease called goitre, with the main symptom being an enlarged thyroid gland.

   Elaborate on these points.
8. Hint: Reflex action involves a rapid, involuntary response to a stimulus mediated by a reflex arc, which includes a sensory receptor, sensory neuron, motor neuron and an effector organ. Elaborate on these points.
9. Hint: Tropisms include phototropism (response to light), gravitropism/geotropism (response to gravity) and thigmotropism (response to touch). Elaborate on these points.
10. Hint: Different parts of the human nervous system, such as the brain, spinal cord and peripheral nerves, play roles in sensory perception, motor control and information processing. Elaborate on these points.
11. Hint: Hormones act as chemical messengers in the body, regulating various physiological processes, including growth, metabolism, reproduction and stress response. Elaborate on these points.

Case Study-Based Questions

1. (i) Heart
   (ii) Thyroxine
   (iii) It controls the body's metabolism of carbohydrates, proteins and fats and encourages the ideal balance for growth.
   (iv) Pineal, pituitary and hypothalamus
2. (i) Phototropism, gravitropism, chemotropism, thigmotropism, thermotropism and hydrotropism.
   (ii) Thigmotropism
   (iii) Cytokinins
   (iv) Abscisic acid
3. (i) Sudden response to a stimulus
   (ii) Sends skeletal muscle impulses from the spinal cord
   (iii) (b)
   (iv) A neuron's dendrites are its outgrowths, where it receives signals (information) from other neurons.
4. (i) Dendrites, nucleus, axon, nerve ending and cell body.
   (ii) Dendritic tip
   (iii) Electrical impulses may move swiftly and effectively along nerve cells thanks to the myelin coating. These impulses decelerate if myelin is compromised.
   (iv) (d)
5. (i) Forebrain, midbrain and hindbrain
   (ii) (b)
   (iii) Regulates homeostasis, releases hormones
   (iv) All involuntary actions, including blood pressure, salivation and vomiting are under its control.

NCERT Corner

1. The directional growth movements of plants due to external stimuli are known as tropic movements. They can be towards the stimulus or away from it. For example, in phototropic movement, shoots respond by bending towards light while roots respond by bending away from light.
2. If iodine intake is low in our diet, release of thyroxine from the thyroid gland will be less. Therefore, protein, carbohydrate and fat metabolism will be affected. A person will suffer from goitre in case of iodine deficiency in the body.

3. (a) Oestrogen (b) Growth hormone (c) Insulin (d) Thyroxine
4. (a) Pituitary (b) Pancreas (c) Adrenal (d) Testes
5. A reflex action is an automatic response to a stimulus, which occurs without any conscious control. For example, if a person touches a hot object, the reflex action would be the immediate withdrawal of the hand from the hot object. This occurs without any conscious effort or decision-making, as it is a protective mechanism that helps to prevent injury.
6. Auxins are plant hormones that promote the growth of a plant. They are produced in the tips of shoots and roots and travel downwards to promote the growth of the plant in a specific direction. Auxins help to elongate cells by increasing their water uptake and also stimulate cell division. They also play a role in the formation of lateral roots and the development of fruit. Overall, auxins help to regulate the growth and development of plants.
7. The pituitary gland is a small gland located at the base of the brain, which is often called the "master gland" as it plays a key role in regulating the functions of other glands in the body. The pituitary gland produces and releases several hormones that control growth, reproduction, and metabolism. For example, it produces growth hormone, which stimulates the growth of bones and muscles, and luteinizing hormone, which triggers the production of sex hormones in the ovaries and testes. The pituitary gland also regulates the production of thyroid hormones and adrenal hormones, which are important for maintaining the body's energy levels and responding to stress.
8. Voluntary actions are those that are under our conscious control, and we can choose to perform them or not. For example, walking, speaking and writing are all voluntary actions. In contrast, involuntary actions are those that occur automatically and without our conscious control. Examples include breathing, blinking and digesting food. Involuntary actions are often reflexive in nature and help to maintain the body's internal environment.
9. The movement of leaves of the sensitive plant is different from the movement of a shoot towards light in several ways. The movement of the leaves of the sensitive plant is a rapid response to touch or other stimuli, and it involves the folding of the leaves inwards. This response is due to the movement of water in the plant cells, which causes the cells to shrink and the leaves to fold. In contrast, the movement of a shoot towards light is a slower response that occurs over a period of time. This response is due to the distribution of auxins in the plant, which causes the cells on the shaded side of the shoot to elongate more than those on the sunny side, resulting in the curvature of the shoot towards the light source.

Competition Window

1. (c)	**2.** (b)	**3.** (d)	**4.** (b)	**5.** (a)
6. (d)	**7.** (a)	**8.** (c)	**9.** (a)	**10.** (a)
11. (a)	**12.** (b)	**13.** (c)	**14.** (a)	**15.** (d)

CHAPTER

9 The Nervous System

PREREQUISITES

You must have knowledge of:

- Biology
- Anatomy and physiology
- Cell biology
- Biochemistry

LEARNING OBJECTIVES

At the end of this chapter, you will be able to:

- Describe the structure of each of the following:
  - Ganglion
  - Nerve
  - Grey matter
  - Nerve tract
  - White matter
- Describe the organisation of the nervous system
- Explain the functions of the principle components of the nervous system

INTRODUCTION

Every living thing must become aware of its surroundings in order to react appropriately and survive. In addition to acts that change the exterior environment, a lot of internal activity takes place of which we are ignorant. The timing and coordination of all such acts must be perfect. The neural system and the hormone system work together to coordinate this process.

The nervous system carries out the following key tasks:

- Keeps us updated on the outside world through our senses
- Makes it possible for us to recall, reflect and reason
- Regulates and synchronizes all voluntary muscular movements, such as holding a book in your hand while you read it or jogging
- Controls unconscious processes such as breathing and heartbeat

THE NEURON

The brain, spinal cord, sensory receptors and a large number of nerves make up our nervous system. Neurons, or nerve cells, are the building blocks of the brain and spinal cord.

The three main components of the neuron are the cell body, dendrites and axon (Fig. 9.1):

- The **cell body** (perikaryon or cyton) contains the same cell organelles as is present in other cells, with the exception of the centrosome, which is missing because nerve cells are not capable of dividing.
- **Dendrites** are branched cytoplasmic protrusions of the cell body. They transport nerve signals to the cyton/cell body.
- The **axon** can range in length from a few millimetres to more than one metre; it is a long process that begins from the cell body. In most neurons, a myelin (or medullary) sheath, which is a white insulating sheath, surrounds the axon. The myelin sheath is covered by an outer thin sheath called the neurolemma. Gaps known as nodes of Ranvier can be seen throughout the length of the myelin sheath. Collaterals are the lateral branches that some axons may have. The end of the axon contains swellings that resemble bulbs. This is where neurotransmitters, a class of chemicals, are stored. Axon terminals are in close proximity to the dendrites of one or more neurons, but they are not linked. The spaces between are known as synaptic clefts.

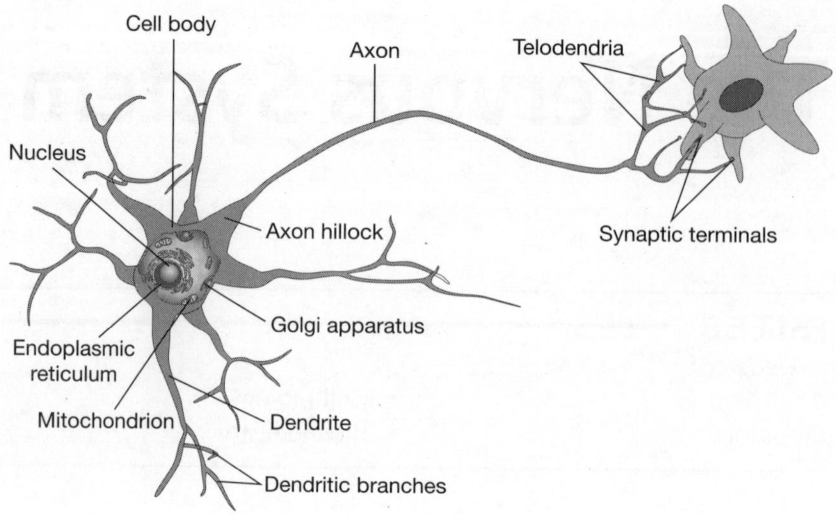

Fig. 9.1 Structure of the neuron

A few keywords related to nervous activity:

- A **stimulus** is a substance or a sudden modification of the internal or external environment that causes an organism or any of its constituent parts to alter.
- A **response** is a modification of an organism brought on by a stimulus.
- An **impulse** is a wave of agitation caused by an electrical disruption that impairs a nerve cell.
- **Receptors** are specific epithelial cells that, when stimulated, send waves of impulses in the direction of the central nervous system.
- **Effectors** are the muscles or glands that contract or produce substances in response to an impulse from the brain or spinal cord.

Nerve Impulse Conduction

The outer surface of the nerve fibre bears a positive (+) charge when it is in its normal (resting) state. This state is referred to as polarized. Polarization occurs because there are more Na+ ions outside the axon membrane. The axon membrane at that location becomes more permeable to Na+ ions upon stimulation (mechanical, electrical, chemical or heat); this causes these ions to migrate inward and results in loss of polarization (depolarization). This is called the **excited region** (Fig. 9.2).

The depolarization point serves as a stimulant for the membrane's next region, which also eventually depolarizes. While this takes place, the prior region is repolarized as a result of the active transport of Na+ ions once more to the outside. The sodium pump uses ATP as a source of energy to accomplish this transport. The depolarization waves that occur during nerve impulse conduction are followed by repolarization.

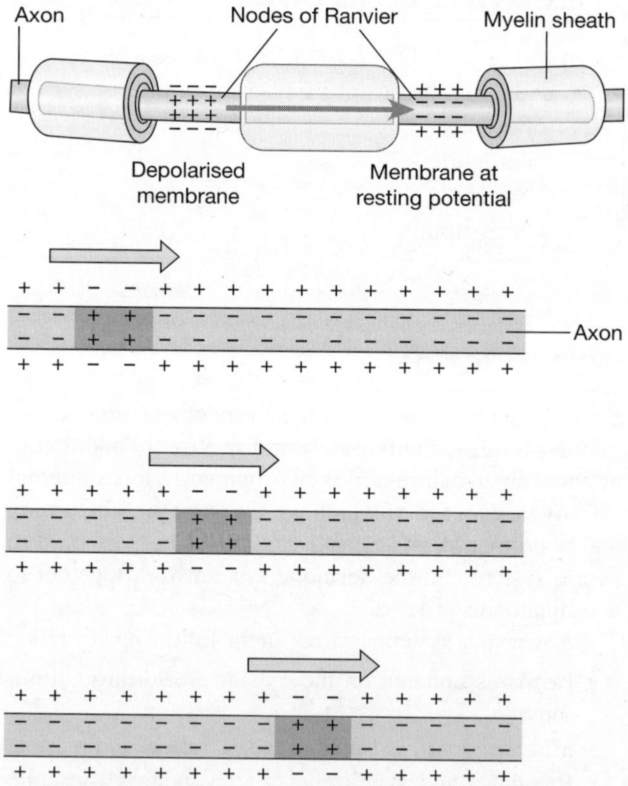

Fig. 9.2 Nerve impulse transmission

Synapse: The term synapse refers to the point of contact between the terminal branches of one neuron's axon and the dendrites of another neuron that have a fine gap. The neuron that follows receives the nerve impulse in this situation. This process uses chemicals. Acetylcholine is a substance that is released as the impulse reaches the terminal end of an axon. The dendrites of the next neuron receive a fresh impulse from this molecule. An enzyme quickly breaks down the chemical to prepare the synapse for subsequent transmission.

Types of Neurons

- **Sensory neurons** transmit the impulse to the central nervous system (the brain or spinal cord) from the receptor (sense organ) (Fig. 9.3).
- **Motor neurons** transfer the signal from the brain to an effector (muscle or gland).
- The brain and spinal cord include association **(connection) neurons**, which link the sensory and motor neurons.

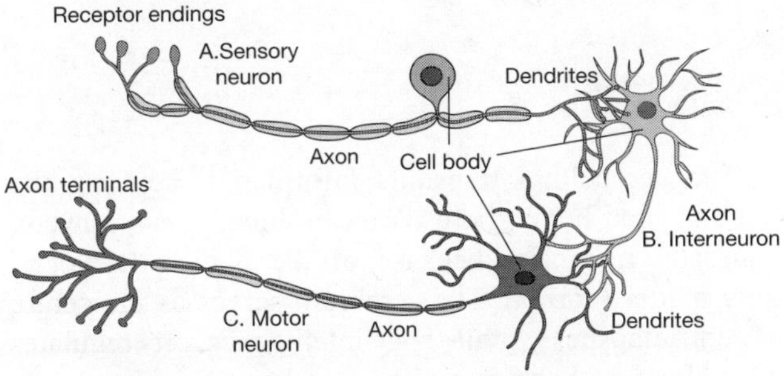

Fig. 9.3 Sensory, motor and interneurons

THE NERVE

The white structures that resemble threads and which arise from the brain and spinal cord to branch out to nearly every area of the body are called nerves. A nerve can be compared to an underground electric cable with several conducting wires that are all isolated from one another. In order to prevent impulses from mingling in adjacent axons, the myelin sheath of the axon functions as insulation.

There are three types of nerves:

- **Sensory nerves** are made up only of sensory fibres that send signals from receptors (sense organs) to the brain or spinal cord. For example, the optic nerve, which originates in the eye and terminates in the brain.
- The only components of **motor nerves** are motor fibres, which transmit impulses from the brain or spinal cord to the effector organs (muscles or glands) to activate them. For example, a nerve that emerges from the brain and feeds the muscles that rotate the eye.
- **Mixed nerves** are those that have fibres that are both motor and sensory. The spinal nerve is an example.

Clusters of nerve cells (cell bodies) called **ganglia** are the places from where nerve fibres may emerge or enter.

DIVISIONS IN THE NERVOUS SYSTEM

- The **central nervous system (CNS)** comprises the brain, which is located in the skull, and the spinal cord, which is housed within the vertebral column.
- The nerves that leave and enter the brain and spinal cord are a part of the **peripheral nervous system (PNS)**. This can be further divided into:
  - The **somatic nervous system (SNS)**, which delivers information to the skeletal (voluntary) muscles.
  - The **autonomic nervous system (ANS)**, which is made up of two networks of nerves and ganglia that regulate the involuntary movements of various internal organs' (smooth muscles, heart muscles and glands).

The Brain

The human brain is the largest among all animal brains when compared to the size of the body. Inside the brain box (cranium) of the skull, the brain is an extremely sensitive organ that is carefully protected. The adult brain weighs 1.35 kg, or about 2% of the body weight. About 80% of it is made up of water and it uses more than 25% (one-fourth) of the total oxygen the body receives.

The meninges, which extend backwards on the spinal cord, are three membranous coverings that protect the brain:

- The toughest outermost fibrous membrane is called the dura mater.
- The arachnoid is the delicate intermediate layer, which is thin and provides a cushion like a web.
- The piamater is the most vascular membrane in the body and is well-supplied with blood.

The swelling of the meninges is known as meningitis.

The cerebrospinal fluid, which is a liquid that fills the gap between the covering membranes, functions as a cushion to protect the brain from shocks. The core cavities (ventricles) of the brain and the spinal cord's central canal are both filled with the same liquid.

Regions of the Brain

The forebrain, midbrain and hindbrain are considered to be the three main regions of the brain (Fig. 9.4).

Forebrain

- **Cerebrum (cerebral hemispheres):** This is the part of the brain that controls intelligence, memory, consciousness, willpower and voluntary activities.
- **Diencephalon:** This consists of:
  - **Thalamus** (relays pain and pressure impulses to cerebrum)
  - **Hypothalamus** (controls the body temperature and pituitary)

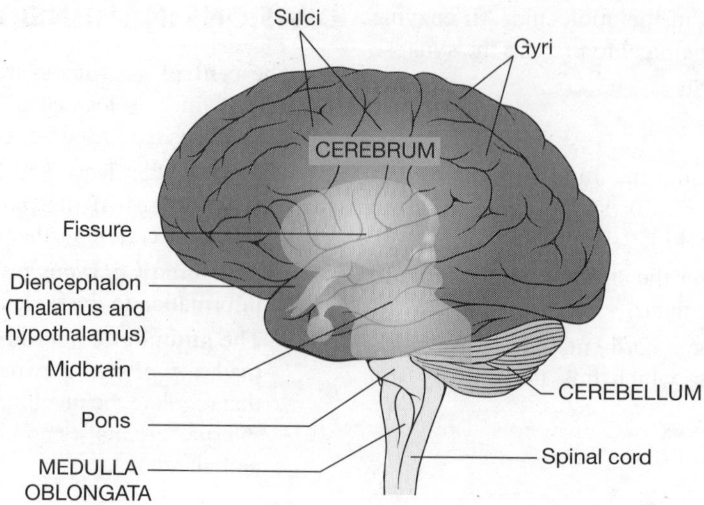

Fig. 9.4 The brain

Midbrain: This is a tiny tubular portion that controls the reflexes involving the eyes and ears.

Hindbrain

- The **cerebellum** controls body balance and muscle activity.
- The **pons** is situated in the cerebellum's shadow, at the centre of the brain. It regulates muscular actions on both sides of the body and transmits signals from one hemisphere of the cerebellum to the other.
- The **medulla oblongata** controls the activities of internal organs, heartbeat, breathing, etc.

The cerebrum, cerebellum and medulla oblongata are the three primary regions of the brain that are visible from the outside.

Cerebrum

The cerebrum is the largest part of the brain, and it is split into cerebral hemispheres (right and left). They have ridges and grooves all over their very wavy exterior. Each cerebral hemisphere is hollow inside, and its walls are divided into an outer cortex and an inner medulla.

The grey matter refers to the cortex, which houses the neuronal cell bodies and is greyish in colour. Convolutions are created by folding a layer of grey matter. The folds are referred to as gyri, and the grooves as sulci (creases). To accommodate more nerve cells, such a system expands its surface area. It is thought that the human brain has more convolutions because it contains more brain (nerve) cells, which leads to better intellect.

The white matter that makes up the inner part of the cerebrum primarily comprises the axons (nerve fibres) of the neurons. A sheet of fibres known as the corpus callosum, or hard body, connects the two hemispheres of the brain.

The cerebellum transmits information between the left and right hemispheres. We can think, reason, invent, plan and memorise because of the highly developed grey matter (cortex). The cerebrum serves as the centre of consciousness, will and intelligence. It regulates all voluntary behaviour. The cerebrum is also home to what is typically referred to as the subconscious or unconscious mind. More current sensations that rule conscious action obscure many earlier ones. Past events may be remembered in dreams or under hypnosis.

Cerebellum

A relatively smaller portion of the brain, the cerebellum, lies just underneath and at the base of the massive cerebrum. It has several furrows but no convolutions. This also features a grey matter-based outer cortex. It has white matter in the centre, which in median section resembles a branching tree. Maintaining the body's balance and coordinating muscle activity are the cerebellum's primary roles. The cerebrum is the origin of the urge to move a muscle. For instance, the impulse to stand up and walk originates in the cerebrum (conscious part). The coordinated use of numerous muscles is required for walking. The cerebellum is in charge of coordinating and timing their contraction and relaxation properly.

Medulla Oblongata

The lowest part of the brain, found at the base of the skull, is called the medulla oblongata. It roughly resembles a triangle and continues as the spinal cord behind. Its job is to regulate the operations of the body's internal organs, including the peristaltic movement of the alimentary canal, breathing motion, heartbeat and several other uncontrollable actions. Most often, death occurs when the medulla is injured.

The Spinal Cord

The spinal cord (Fig. 9.5) is contained within the neural canal of the vertebrae and extends from the medulla of the brain the whole length of the spine to the second lumbar vertebra. On the inner side is the grey matter, which contains the cell bodies of the motor (efferent) and association neurons, and on the outer side is the white matter. Axons that travel longitudinally to and from the brain as well as crossing from one side to the other can be found in the white matter. The centre has a small canal that spans its full length and connects with the ventricles of brain. Additionally, it contains cerebrospinal fluid, which serves as a shock-absorbing cushion and a conduit for the exchange of nutrients, waste and breathing gases with neurons. The duramater, arachnoid and piamater membranes that surround the brain are also present on the exterior of the spinal cord.

Functions

- Responsible for reflexes below the neck.
- Transports sensory signals from the muscles and skin to the brain.
- Transfers motor commands from the brain to the trunk and limb muscles.

Somatic Nervous System

This consists of two types of nerves: cranial and spinal. There are 12 pairs of cranial nerves and 31 pairs of spinal nerves.

- **Cranial nerves:** These consist of sensory nerves, such as olfactory (for the nose), optic (for the eyes) and auditory (for the ears); motor nerves, such as those leading to the eye and muscles; and mixed nerves, such as those that travel to and from the tongue and cheeks.
- **Spinal nerves:** These protrude from the spinal cord. Thirty-one pairs are present, including eight pairs in the neck, twelve in the thorax, five in the lumbar region, five in the sacral region and one pair in the coccygeal region. A typical spinal nerve has two roots: dorsal and ventral, which emerge from the spinal cord. There is an ovoid dorsal ganglion in each dorsal root. Each spinal neuron consists of both sensory and motor fibres, making it a mixed nerve. The sensory and motor fibres split at the intersection of the two roots; the motor fibres enter the ventral root and the sensory fibres the dorsal root. Both roots enter the spinal cord's grey matter and exit in the respective dorsal and ventral horns that protrude from the grey matter.

Autonomic Nervous System

On either side of the backbone, there are two chains of nerves and ganglia that make up the autonomic nervous system (ANS). The involuntary movements of internal organs are managed by this system. The ANS is divided into two sections: sympathetic and parasympathetic. Between the neck and waist, the sympathetic system nerves emerge from the spinal cord. The parasympathetic nervous system is located in the head and neck region at the front and the sacral region at the back at two places. In general, the actions of these two components are diametrically opposed. Adrenaline, a hormone released by the adrenal glands on the kidneys, stimulates the sympathetic nervous system.

Emotions like grief, wrath, fear and sexual excitation have a major impact on the ANS. Long-term emotional stress can lead to high blood pressure, stomach ulcers and other problems.

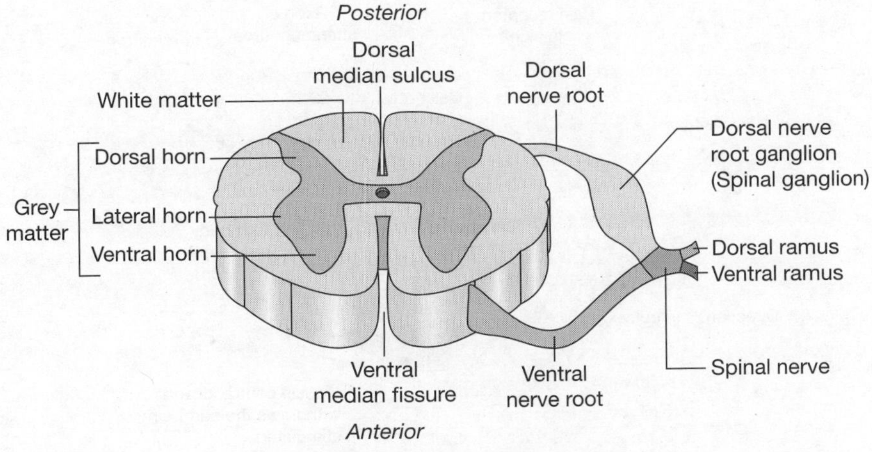

Fig. 9.5 The spinal cord

REFLEXES

Our body engages in two different types of actions: voluntary (done intentionally) and involuntary (performed unconsciously).

Examples of voluntary actions are:

- You turn on the TV and select a particular channel with the remote control in order to view a certain programme.
- You choose an apple and eat it.
- You raise your arm to check the time by looking at the watch on your wrist.

Unintentional (or involuntary) behaviours are also called reflexes. A reflex action (Fig. 9.6) is an automatic, swift and involuntary bodily response to a stimulus. Examples are:

- When a particle enters your eye, tears immediately flow to flush the particle out (glandular secretion).
- Quick removal of the hand when it unintentionally touches a hot iron (muscular movement).
- Muscle spasms from shivering in cold weather or sweating in heat (glandular secretion).
- Widening of the pupil when viewing in low light and vice versa (muscular movement).
- Moving food that has been swallowed along the food canal (muscular movement).
- Heartbeat that never stops (muscular movement).

Any sensory stimulation that results in a muscle motion or glandular secretion is the source of all involuntary actions or reflexes.

Types of Reflexes

Reflexes are of two types: natural (inborn) and conditioned (acquired).

- A **natural (inborn) reflex** is one that does not require any prior knowledge or training. These reflexes are innate or passed down from one's parents. Examples:
  - Blinking, coughing, sneezing: these are protective.
  - Salivation, swallowing, peristalsis: provide functional efficiency.
- A **conditioned (or acquired) reaction** is one that evolves over the course of a lifetime as a result of experience or learning. For instance, salivation is when your mouth waters at the mere sight or smell of a good food that you are accustomed to. This indicates that the response would not happen if you had not had that meal earlier. Salivation might also take place as a normal reaction. When you chew or eat food, saliva begins to descend.

Pavlov's experiment: The dog experiment (Fig. 9.7) conducted by Pavlov is a well-known illustration of conditioned reflex. This Russian researcher observed that no animal will secrete saliva in reaction to the sound of a bell or any other sound under normal circumstances. However, after a sufficient amount of training, the dog will begin to salivate whenever a bell is rung if food is served at that time. This behaviour continues even when food is not served at the time. In other words, a stimulus that did not previously trigger a salivary gland response in the organism does so now.

A dog will flee if it observes you merely crouching down while appearing to pick up a stone. In dogs, this conditioned reaction is one that is frequently seen. In a circus, wild animals are taught to perform a variety of tricks, mostly by conditioned reactions.

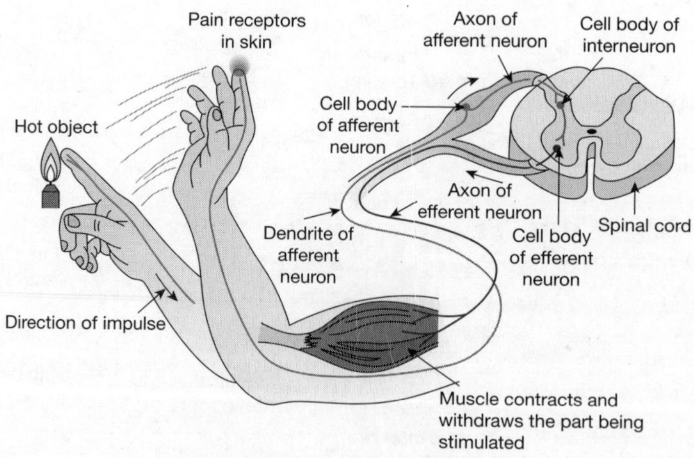

Fig. 9.6 The reflex action

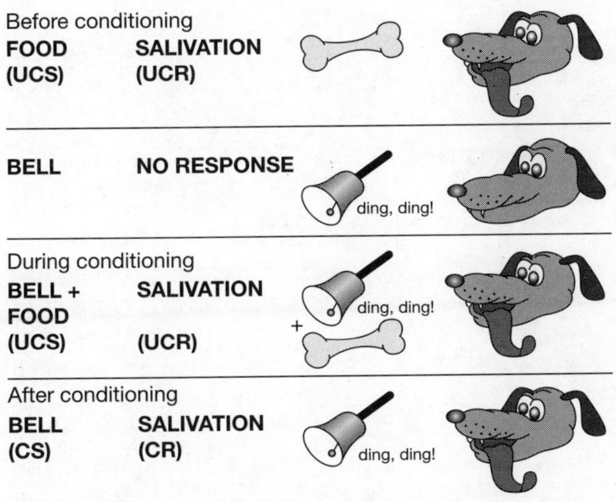

Fig. 9.7 Pavlov's dog experiment (UCS: Unconditioned Stimulus; UCR: Unconditioned Response; CS: Conditioned Stimulus; UR: Unconditioned Response)

Some Common Reflexes in Humans

Natural (Inborn)

- A knee-jerk occurs when one leg, which is relaxed and supported by the other leg while sitting, is abruptly tapped below the kneecap.
- Rapidly closing the eyelids in response to abrupt objects coming towards them or bright lights flashing nearby.
- Withdrawal of the hand or any other body part after being abruptly pricked.
- Peristaltic reflexes: Peristaltic waves move chyme (food) through the small intestine. When chyme causes a section of the intestine to become bloated, the intestinal wall stretches as a stimulus, and its contraction pushes the food forward as a response.
- The reflexive coughing that occurs when food eaten enters the windpipe.
- The reflexive sneezing that occurs when an irritant enters the nose.

Format like Natural (Inborn): The majority of daily routines and actions are conditioned reflexes.

- You stand up as the teacher enters the classroom.
- You tie lace your shoes while speaking, not knowing whether to place the right over the left or the other way around.
- Engaging in a musical performance.
- Using the keyboard's keys while using a computer.
- Automatically signalling to the right with your hand while turning your bike or car to the right.
- Braking your automobile or bicycle when someone suddenly passes in front of you.

Nervous Pathways in Reflexes

To be effective, a reflex action must be rapid. Therefore, the route for sending and receiving information must be short. In a basic reflex, impulses exit at the same level at which they enter the spinal cord rather than travelling up and down it. Two or three neurons may be involved in this. When two neurons are involved, they are a ventral motor neuron, which has its fibre in the ventral root, and a ventral sensory neuron, which is located in the ventral root of the grey matter of the spinal cord. The sensory (afferent meaning conveying towards) fibres of the dorsal root of the spinal nerve receive the impulse from the sense receptor, and as it reaches the motor neuron in the spinal cord, an immediate response impulse is transmitted to the effector muscle (or organ) to react. The reflex arc in the spinal cord that connects the sensory fibre and the motor neuron occasionally includes a third neuron, known as the intermediate, linking, association or relay neuron. Following the immediate response, the sensation is also transmitted to the brain through ascending neurons, which are longitudinal nerve fibres that link the spinal cord.

Neurons at various levels of the spinal cord are involved in complex reflex activity. If you trip on a coil of rope while walking in the dark on a road, you instinctively jump to the side out of fear that it might be a snake. This reaction involves almost all of the skeletal muscles.

THE SENSE ORGANS

Our body's principal sensory organs are our eyes, ears, tongue, nose and skin, which are each sensitive to different types of light, sound, taste, smell and touch. Balance, bodily motions, hunger and thirst, pain, etc. are additional sensations. The sensory cells in these organs that are categorised as receptors that sense the actual feeling.

Any specialised tissue or cell that is responsive to a certain stimulation is a receptor.

- **Mechanoreceptors** are touch and skin pressure receptors that respond to mechanical change.
- **Chemoreceptors** affect the taste and smell receptors in the tongue and nose.
- **Photoreceptors**, which respond to light, are rods and cones in the retina of the eye.
- **Thermoreceptors** are skin-based heat and cold receptors that respond to temperature changes.

The Eyes

- On the front of the head, the two eyes are housed in large receptacles called **orbits**. Six muscles allow for the rotation of each eye, which is shaped like a ball (Fig. 9.8).

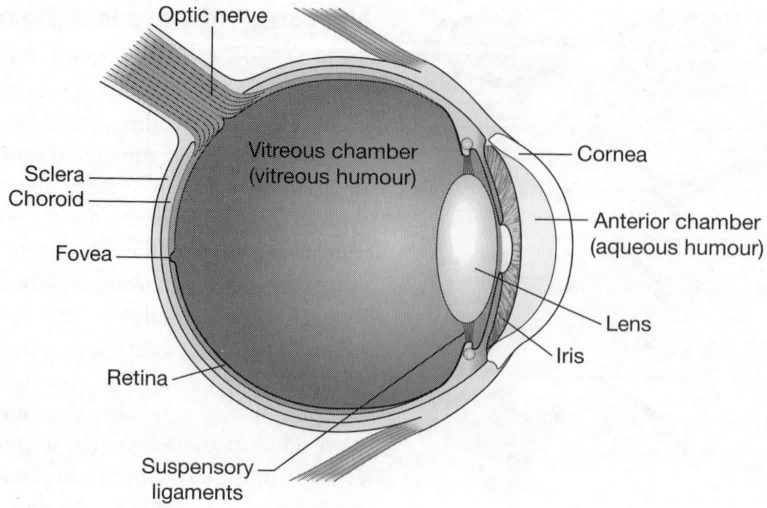

Fig. 9.8 The eye

- The upper and lower moveable **eyelids** exclude light and protect the front (external) surface of the eyes. **Eyelashes** on each eyelid are curled outwards to stop bigger objects from falling into the eye.
- Although the **eyebrows** are not a component of the eye, they serve as a barrier to keep water or sweat from entering the eyes.
- The top sideward region of the orbit is home to the **lacrimal glands**, which produce tears. The gland's secretion is released over the front surface through 6–12 ducts.

  The tears also keep the front surface of the eye clean by washing away dust particles. The tears are disseminated by the motions of the eyelids (blinking). Due to the lysozyme enzyme, which destroys bacteria, they also have an antimicrobial quality. The tear ducts empty their contents into a sac that is located at the inner angle of the eye. The secretion is transported into the nasal cavity by a nasolacrimal duct. The tear glands release a lot of fluid that 'waters the eyes' or overflows as 'tears' when they are irritated or in certain emotional situations. Both great delight and despair may cause you to cry.

  - **Functions** of tears:
    - Lubricate the surface of the eye
    - Wash away dust particles
    - Help in killing germs
    - Communicate emotions

- A thin membrane called the **conjunctiva** covers the entire front of the eye. It is continuous with the inner lining of the eyelids. It is reduced to a single transparent epithelial layer over the cornea. Conjunctivitis is a condition in which a viral infection causes this outermost layer to turn red.
- The eye is protected by the bony orbital socket, brows, tear ducts and the conjunctiva in different ways.

Structure of the Eyeball

Three concentric layers make up the eyeball's wall:

- **The outer sclerotic layer:** The stiff fibrous tissue that makes up the sclerotic layer, often known as the sclera, is white in colour. The sclerotic layer, which may be seen through the conjunctiva, makes up the white area on the front of the eye. In the front, where it covers the coloured portion of the eye, known as the cornea, it protrudes out and turns translucent. Sometimes a patient's cornea becomes opaque (white) and dysfunctional. In such circumstances, a healthy cornea from a donor eye can be used to replace the damaged one.
- **The middle choroid layer:** Blood vessels are abundantly present in the choroid layer, which nourishes the eye. It has a dark, black pigment called melanin that stops light from reflecting and dispersing inside the eye. The ciliary body is formed in the front of the eye by the choroid expanding (containing circular muscles). The ciliary body's smooth muscles modify the lens's shape.

  The lens is partially covered by the iris, which is an outgrowth of the choroid and has a circular aperture in the centre, known as the pupil. The colour of the iris determines whether an eye is blue, brown or black. The iris has circular muscles to narrow the pupil and radial muscles to broaden it. The amount of light entering the eye is controlled by this modification of the pupil's size. By shining a torchlight into your eyes while gazing in a mirror, you can quickly see this. The pupil dilates in low-light conditions while constricting in bright light. Because each person's iris muscles have an own pattern and arrangement, they serve as another means of identification.

- **The innermost layer:** The retina is light-sensitive. It has two different types of sensory cells:
  - The **rod cells**, which have ends that resemble rods, are sensitive to low light but are insensitive to colour. They contain the visible purple pigment rhodopsin. The retina's rod cells are found almost everywhere.
  - **Cones** (inner ends conical) are responsible for colour vision and are sensitive to intense light. Cones have different visual pigments. The yellow area is where the cone cells are predominantly located.

Rods and cones are not distributed equally. At the rear of the eye, almost at the centre of the eyeball's horizontal axis, is an area known as the macula lutea, also known as the **yellow spot** or fovea centralis. The greatest concentration of sensory cells, especially cones, is found here. As a result, this is where colour vision and the brightest vision are located. Fewer cones and more rods are seen throughout the rest of the retina.

The **blind area** is located near the nasal side and lateral to the yellow spot. This area lacks sensory cells, making it the point of no vision. The optic nerve emerges from the eyeball at this location where the nerve fibres from all of the retina's sensory cells converge and unite.

Just behind the pupil, there is a translucent biconvex crystalline substance that serves as the **lens**. It has fibres for translucent lenses (long thin cells which have lost their nuclei). The suspensory ligament, a group of fibres that connects the lens to the ciliary body, collectively holds the lens in place. The ciliary body, which is also a component of the choroid, is located where the choroid and iris meet. The lens is fairly elastic and has muscles in the ciliary body, which adjust the shape of the lens to view objects at various distances.

The lens splits the eyeball's inner cavity into two chambers: the aqueous chamber is in front of the lens, and the vitreous chamber is behind the lens.

- The front chamber between the lens and the cornea is known as the aqueous chamber. It contains aqueous humour, a transparent, watery liquid. The aqueous humour has two purposes:
  - Prevents physical shock while keeping the lens wet
  - Refracts light
- The bigger ocular cavity located below the lens is called the vitreous chamber. It is filled with vitreous humour, a transparent jelly-like thick fluid. The vitreous humour has the following two purposes:
  - It aids in maintaining the eyeball's shape
  - It safeguards the retina and nerve terminals

How We See

Following are the four main steps in 'seeing' an object:

1. Through the transparent framework (conjunctiva, cornea, aqueous humour, lens, vitreous humour), light rays from the object enter the eyes.
2. To begin with, the cornea's curvature bends light rays somewhat, and then the lens bends them more to generate an image on the retina. The retinal image is a real, reversed image.
3. The image's light energy causes chemical alterations in the sensitive cells (rods and cones). These alterations cause nerve impulses to be produced, which pass through the optic nerve and the visual region of the brain, where they produce the perception of sight.
4. Even though the image created in the eye is inverted, our brain interprets the image in a variety of ways, such as by 'seeing' the items upright.

An object's picture needs to be in fine focus in both eyes for it to be seen clearly. **Accommodation** is the term used to describe the process of focusing the eye to perceive objects at various distances. This is mostly caused by a shift in the elastic lens's curvature, which makes it either thinner or fatter. The lens is more flattened or narrower for distance vision. The lens becomes more convex or rounded for near vision (closer than 6 m). The ciliary muscles cause these alterations in the lens's shape. The lens is less convex and stretched by the suspensory ligaments in the normal state (ciliary muscles relaxed), making it ideal for viewing distant objects.

The ciliary muscles, which are circular, tighten when we look at adjacent objects and have a tendency to drag the ciliary body slightly forward. Because of its flexibility, the lens thickens and becomes more rounded or convex as the suspensory ligament tension is released, rendering it loose.

Only the cones of the retina, which are stimulated only by intense light, allow for colour vision. On a moonlit night, only the rods work, making it impossible to distinguish red, violet or purple flowers in a garden.

The Ear

The human ear helps us in balance and in hearing sounds (Fig. 9.9). It is divided into the outer ear, middle ear and inner ear.

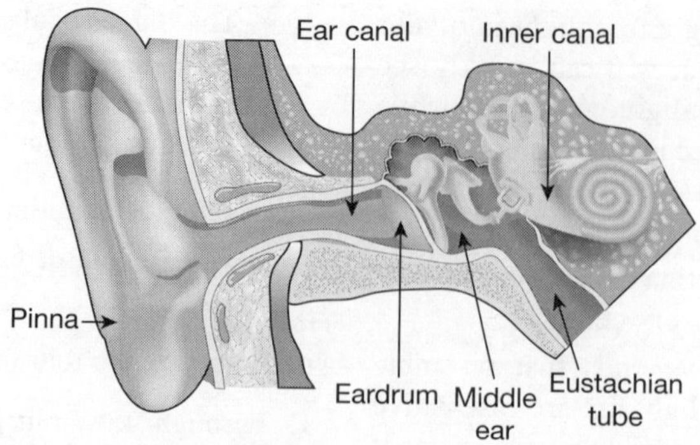

Fig. 9.9 The ear

- The auditory canal and the projecting pinna, often known as the auricle, make up the outer ear (or tympanum).
- In addition to a Eustachian tube that connects the middle ear chamber to the throat, the middle ear contains three microscopic bones: the malleus, incus and stapes, or hammer, anvil and stirrup in common usage. The three bones are collectively to as the ear ossicles. The inner surface of the eardrum is where the hammer bone's handle is attached. The stirrup is coupled to its opposite end by the anvil, which is also connected to it. The oval window, a membrane-covered entrance to the inner ear, is where the flat portion of the stirrup attaches. The inner and middle ears are connected by a second hole called the round window, which is also shielded by a thin membrane.
- The cochlea and semicircular canals are the two main components of the inner ear, often known as the membranous labyrinth. The spiral-shaped cochlea resembles the shell of a snail. There are 2.5 turns in total. Three parallel canals, separated by membranes, make up its interior winding hollow. Endolymph is present in the median (cochlear) canal, whereas perilymph is present in the other two canals. The middle canal is home to sections with sensory cells and the organ of Corti, a spiral organ. The auditory nerve is joined by the nerve fibres that emerge from these cells. On the basilar membrane are the sensory cells. The remaining portion of the inner ear is made up of three semicircular canals that are placed in three separate planes, one of which is horizontal and the other two vertical, at right angles to one another. Each canal has a wider end that creates an ampulla with sensory cells for dynamic balancing while the body is moving and nerve fibres that connect to the auditory nerve from these cells.

  The utriculus and sacculus are seen on the short stem that connects the bases of semicircular canals to the cochlea. Additionally, while the body is motionless, such as when standing, these areas include sensory cells for static balance.

Functions of the Ear

The internal ear is involved in two sensory functions: hearing and balance.

Hearing: The external auditory canal receives sound waves that are collected by the pinna. They make contact with the vibrating eardrum. The Eustachian tube balances the air pressure on either side of the eardrum, enabling unrestricted vibration. The three ossicles begin to vibrate as a result of the eardrum's movement. The first two ossicles function like a lever, amplifying the vibration of the last ossicle (stirrup). The fluid within the cochlear canals also vibrates as a result of the vibrations transmitted from the vibrating stirrup to the membrane of the oval window. The auditory nerve transmits the impulses to the brain through the cochlea's spiral organ, the cochlea's fluid-driven vibrations stimulating the sensory cells' hair-like processes. The cochlear canal is designed to accommodate sounds of various pitches in its regions. The majority of sounds we experience are a synthesis of vibrations occurring at numerous different rates of speed, or pitches. Vibrations at all speeds are not detectable by us (frequencies). Only noises between 20 and 20,000 Hertz can be detected by our sensory receptors; however, those between 1000 and 4000 Hz are the ones that are most clearly audible. Higher frequency sounds can be heard by dogs.

Balance: The fluid inside the semicircular canals is disturbed as the head is rotated in various directions. The auditory nerve sends the nerve impulse through the nerve fibres linked to the sensory hair cells that are being pushed against by the moving fluid in the canals to the brain. The sensory cells of the semicircular canals are focused on dynamic balance, or when the body is moving. The utriculus and sacculus contain similar sensory patches that register the static (positional) balance with regard to gravity.

The Sense of Taste

The taste buds on the tongue are where the sensation of taste is located. An ovoid collection of sensory and supporting cells is a taste bud. Nerve fibres extend from the bases of the sensory cells, which contain hair-like projections at their ends. The taste hairs extend into the outer taste pore, which is situated on the epithelium's surface. These holes allow substances in solution to enter, stimulating the sensory hairs. The taste buds are primarily found on the tongue's upper surface, while some are also be found in the mouth and throat.

The Sense of Smell

The nasal chamber contains delicate epithelial layers where the sense of smell is located. The smell cells have projections that resemble hairs, and they react to particles dispersed in the nasal mucus discharge. The olfactory nerve carries the impulse generated by these cells to the brain.

Flavour is a combination of taste and scent when consuming food or beverages. Both grated apple and grated onion have a similar flavor, which is slightly sweet, if you hold your nose. Food loses flavour and aroma when you have a cold and a blocked nose, which has the same effect.

The Sense of Touch

Our body's largest sense organ and the most frequently employed sensory instrument is its sense of touch. Our body is covered in skin, which serves as the sensory organ for touch. In daily life, we identify a variety of substances mostly through touch.

We perceive temperatures through touch, from touching ice to experiencing cold chills to receiving an abrupt shock when touching a hot pan. A soft cotton ball can be distinguished from a rough jute bag through touch. As a result, we can comprehend smoothness or roughness.

We can detect a touch on any region of our body, and every touch sends a unique signal to our brain, allowing us to understand where we are being touched.

SUMMARY

- The fundamental component of the neurological system is the neuron. It is made up of a cell body with a nucleus, dendritic filaments, and an axon, a long fibre that is frequently encased in a myelin sheath.
- A nerve is a collection of axons.
- Brain cells and axons, respectively, make up the majority of the grey and white matter in the brain and spinal cord.
- Nerve impulses travel along axons. Synapses allow adjacent neurons to communicate with one another.
- Reflexes are automatic responses to stimuli. The path taken by nerve impulses involved in reflexes is a reflex arc.
- The nervous system is divided into the peripheral nervous system, which consists of nerves connecting to the brain and spinal cord, and the central nervous system, which is made up of the brain and spinal cord.
- The peripheral nervous system is thought to include the autonomic nervous system.
- The brain consists of three major regions: the forebrain which includes the cerebral hemispheres (or cerebrum), hypothalamus and pituitary gland; the hindbrain or brain stem containing the medulla oblongata and the cerebellum. The brain and spinal cord are encased in protective membranes called meninges.
- The optic, olfactory, acoustic and vagus nerves are among the 12 pairs of cranial nerves.
- The spinal cord is a meninge-encased wire of nerve tissue that extends from the brain to the end of the tail.
- The autonomic nervous system regulates bodily processes that are unconscious of conscious control. The sympathetic nervous system is responsible for the flight or fight reaction, which includes increased heart rate, bronchial dilation, pupil dilation, and decreased stomach activity. The parasympathetic nervous system is linked to slowed heart rate, constricted pupils and more active gut.

PRACTICE QUESTIONS

Analyse Your Concepts (School Exam Based)

Multiple Choice Questions

1. The neurological system of our body is essential to its operation. What are its primary duties?
   (a) It increases awareness about the exterior world with the help of sense organs
   (b) It renders support in harmonising and regulating all voluntary muscular activities
   (c) It contributes in controlling involuntary activities
   (d) All of the above

2. The main components of the nervous system are:
   (a) Brain and spinal cord
   (b) Spinal cord and numerous nerves
   (c) Brain, spinal cord, sense receptors and numerous nerves
   (d) Spinal cord

3. Of what organ are the main components neurons or nerve cells?
   (a) Heart (b) Brain
   (c) Lungs (d) Liver

4. The three crucial portions of which cell body are dendrites, axon and cell membrane?
   (a) Neuron (b) Kidney
   (c) Hormones (d) Spleen

5. Perikaryon, also known as cyton, is another name for?
   (a) Blood cell (b) Dendrites
   (c) Axon (d) Cell body

6. All of the following are cell organelles, with the exception of:
   (a) Chromosomes (b) Diplosome
   (c) Centrosome (d) Kinetochore

7. Dendrites are essential for carrying nerve impulses to the:
   (a) Axon (b) Cyton
   (c) Neuron (d) Cell body

8. White matter is where the majority of axons are covered in this white insulating sheath:
   (a) Myelin sheath (b) Neurolemma
   (c) Nodes of Ranvier (d) Synaptic clefts

9. When the myelin sheath is shielded by an external, thin coating, it is referred to as:
   (a) Bulbs (b) Collaterals
   (c) Neurotransmitters (d) Neurolemma

10. Nodes of Ranvier are defined as:
    (a) The gaps enclosed by myelin sheath
    (b) The closure of gaps portrayed by myelin sheath
    (c) Gaps depicted by myelin sheath all over the length
    (d) None of the above

Match the Following

Each question contains statements given in two columns which have to be matched. Statements in Column I have to be matched with statements in Column II.

1.

Column I	Column II
(a) Cerebrum	(p) Controls the pituitary
(b) Cerebellum	(q) Controls vision and hearing
(c) Hypothalamus	(r) Relays information for vision and hearing
(d) Midbrain	(s) Regulates digestion
	(t) Maintains body posture

2.

Column I	Column II
(a) Frontal lobe	(p) Controls balance and coordination
(b) Parietal lobe	(q) Controls emotions and social behaviour
(c) Occipital lobe	(r) Controls touch and spatial awareness
(d) Temporal lobe	(s) Controls vision and object recognition
(e) Cerebellum	(t) Controls voluntary movements

3.

Column I	Column II
(a) Axon	(p) Receives messages from other neurons
(b) Dendrite	(q) Carries messages away from the cell body
(c) Cell body	(r) Connects the neuron to other neurons
(d) Myelin sheath	(s) Increases the speed of nerve impulse transmission
(e) Synapse	(t) Site of communication between two neurons

4.

Column I	Column II
(a) Central Nervous System	(p) Includes the brain and spinal cord
(b) Peripheral Nervous System	(q) Controls voluntary movements
(c) Somatic Nervous System	(r) Regulates involuntary functions
(d) Autonomic Nervous System	(s) Transmits sensory information
(e) Sensory Nervous System	(t) Carries messages to and from the brain

True or False

Read the following statements and write your answer as true or false.

1. Sometimes, medicines dropped in the eyes enter the nose and throat.
2. Ciliary muscles regulate the size of the pupil.
3. Yellow spot of the retina is the region of colour vision.
4. The auditory nerve is used only for perceiving sound.
5. The malleus, incus and stapes are collectively called the ear ossicles.
6. Flavour and taste are one and the same thing.
7. Short-sightedness and hyperopia are one and the same thing.
8. Blind spot is so-named because no image is formed on it.

Fill in the Blanks

Complete the following statements with an appropriate word/term to be filled in the blank spaces.

1. The central nervous system consists of ______ and ______.
2. ______ is absent in neurons.
3. The central space of the brain is ______.
4. ______ fluid is present inside and outside the brain.
5. The nervous system is responsible for the ______ of all body activities.
6. The nervous system is divided into two main parts ______.
7. The brain is the main organ of the central nervous system and is responsible for ______ information from the body.
8. The spinal cord is a long, cylindrical structure that extends from the brain and is responsible for ______ between the brain and the rest of the body.
9. Neurons are the basic functional units of the nervous system and are responsible for ______ between different parts of the body.
10. The dendrites of a neuron receive signals from other neurons or sensory receptors, while the axon ______ signals to other neurons or effectors.
11. ______ is a fatty layer that surrounds the axon of some neurons and helps to speed up the transmission of electrical signals.
12. The ______ is the tiny gap between two neurons where signals are transmitted.
13. The peripheral nervous system consists of two main types of nerves: ______.
14. The autonomic nervous system is responsible for regulating ______ functions such as heart rate, digestion and breathing.

Assertion–Reason Questions

The following questions consists of two statements: Assertion (A) and Reason (R). Answer these questions selecting the appropriate option given below:

(a) Both A and R are true and R is the correct explanation of A

(b) Both A and R are true but R is not the correct explanation of A
(c) A is true but R is false
(d) A is false but R is true

1. Assertion (A): The medulla oblongata controls vital functions such as breathing and heart rate.
   Reason (R): The medulla oblongata is part of the brainstem, which is responsible for controlling basic life-sustaining functions.
2. Assertion (A): Damage to the hippocampus can result in memory loss.
   Reason (R): The hippocampus is responsible for the formation and storage of long-term memories.
3. Assertion (A): Multiple sclerosis is a disease that affects the myelin sheath surrounding nerve cells.
   Reason (R): The myelin sheath helps to transmit electrical signals along the length of a nerve cell.
4. Assertion (A): The sympathetic nervous system is responsible for the fight or flight response.
   Reason (R): The fight or flight response is a physiological reaction to stress or danger, and is controlled by the sympathetic nervous system.
5. Assertion (A): The cerebellum is responsible for coordination and balance.
   Reason (R): The cerebellum receives input from sensory systems in the body and sends output to motor systems, allowing for precise movements.
6. Assertion (A): Parkinson's disease is caused by the degeneration of dopaminergic neurons in the substantia nigra.
   Reason (R): Dopaminergic neurons in the substantia nigra are responsible for producing dopamine, a neurotransmitter that is important for movement control.
7. Assertion (A): The corpus callosum is a bundle of nerve fibres that connects the two hemispheres of the brain.
   Reason (R): The corpus callosum allows for communication and coordination between the two hemispheres of the brain.

One-word Answers

1. Mention the basic unit of the brain.
2. What is the protective membranous covering of the brain?
3. The fluid that surrounds and fills the brain.
4. The intersection of two nerve cells.
5. What is the basic functional unit of the nervous system?
6. What is the main organ of the central nervous system?
7. Which part of the neuron receives signals from other neurons or sensory receptors?
8. Which part of the neuron sends signals to other neurons or effectors?
9. What is the fatty layer that surrounds the axon of some neurons and helps to speed up the transmission of electrical signals?
10. What is the tiny gap between two neurons where signals are transmitted?
11. Which part of the brain controls vital functions such as breathing, heart rate and blood pressure?
12. Which part of the brain is responsible for the coordination and control of voluntary movements?
13. What is the name of the division of the nervous system that regulates involuntary functions such as heart rate, digestion and breathing?
14. What is the name of the long, cylindrical structure that extends from the brain and is responsible for relaying messages between the brain and the rest of the body?

Define the Term

1. Neuron
2. Nerve
3. Stimulus
4. Synaptic cleft
5. Reflex action
6. Corpus callosum

Groups of Terms

Given below are certain groups of terms. In each group the first pair indicates a relationship between the two terms. Rewrite and complete the second pair on a similar basis.

1. Neuron: Axon :: Muscle: ___________
2. Synapse: Neurotransmitter :: Receptor: ___________
3. Central Nervous System: Brain :: Peripheral Nervous System: ___________
4. Cerebellum: Motor Control :: Hippocampus: ___________

5. Dendrite: Incoming Signals :: Axon: __________
6. Cones: Iodopsin :: Rods: __________

Identify the Odd One Out

1. Cerebrum, cerebellum, brainstem, liver
2. Neuron, axon, myelin sheath, hormone
3. Synapse, neurotransmitter, dendrite, ribosome
4. Spinal cord, autonomic nervous system, brain, muscle
5. Motor neuron, sensory neuron, interneuron, hormone
6. Brainstem, midbrain, hindbrain, spleen
7. Cerebellum, medulla oblongata, pons, pancreas
8. Central nervous system, peripheral nervous system, somatic nervous system, digestive system
9. Sympathetic nervous system, parasympathetic nervous system, motor nervous system, autonomic nervous system
10. Forebrain, midbrain, hindbrain, pituitary gland

Image-Based Questions

1. Given below is the diagram of a neuron. Name the parts numbered A to D.

1. NEURON

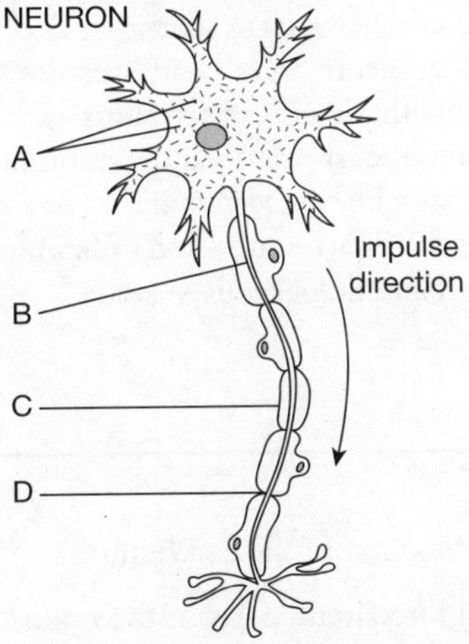

2. A heated object has been touched by you. Draw a diagram that shows the steps that result in a response, such as a rapid hand pullback.

Very Short Answer Questions

1. Which is the brain's outermost layer?
2. Which ear ossicle attached to the tympanum?
3. What is the photosensitive pigment present in the rods of the retina?
4. Which is the nerve which transmits impulses from the ear to the brain?
5. What is the eye defect caused due to shortening of the eyeball from front to back?
6. What is the tube which connects the cavity of the middle ear with the throat?
7. What are the photoreceptors found in the retina of the eye?

Short Answer Questions

1. What is the lacrimal gland?
2. How do the yellow spot and blind spot vary from each another?
3. Name an eye defect caused by ageing. What causes it?
4. What is meant by power of accommodation of the eye?
5. Describe the features of the image that appears on the retina of the eye.
6. What is meant by optical illusion? Give an example.
7. Name the four principal tastes and the regions of the tongue concerned with their perception.

Long Answer Questions

1. Explain how you focus the image of a distant object in your eye after finishing a book by raising your head.
2. Occasionally, a clear image from a dream that you saw comes to mind. How did this event affect your eyes?
3. If you gently close your eyes and push them against your hands, you might be able to see some brilliant light particles. When there is no light entering your eyes, how do you see this?
4. Differentiate between the following pairs:
   (a) Cerebrum and cerebellum (function)
   (b) Sympathetic nervous system and parasympathetic nervous system (overall effect on body)
   (c) Sensory nerve and motor nerve (direction of impulse carried)
5. Explain the following terms:
   (a) Motor nerve
   (b) Autonomic nervous system
   (c) Conditioned reflex
   (d) Association neuron

6. What are the advantages of having a nervous system?

7. What is the difference between a reflex action and voluntary action?

Case Study-Based Questions

1. Two hungry boys (A and B) enter a restaurant and see the menu. Boy B starts salivating but not A. What is the initial response of Boy A and B upon seeing the menu at the restaurant?
   (a) Both boys start salivating
   (b) Boy A starts salivating but not B
   (c) Boy B starts salivating but not A
   (d) Neither boy starts salivating

2. What is the physiological process behind Boy B's salivation upon seeing the menu?
   (a) The sight of food triggers the release of digestive enzymes in Boy B's mouth
   (b) The sight of food triggers the release of saliva from the salivary glands in Boy B's mouth
   (c) The sight of food triggers the production of hunger hormones in Boy B's stomach
   (d) The sight of food triggers the contraction of Boy B's stomach muscles

3. What type of neural pathway is involved in Boy B's salivation response to the menu?
   (a) Sensory neurons → Motor neurons
   (b) Motor neurons → Sensory neurons
   (c) Association neurons → Sensory neurons
   (d) Sensory neurons → Association neurons → Motor neurons

4. A person is walking in a park and suddenly hears a loud noise. What effective change will occur in the body, and which part of the autonomic nervous system brings it about?
   (a) Increased heart rate and respiratory rate; sympathetic nervous system
   (b) Decreased heart rate and respiratory rate; parasympathetic nervous system
   (c) Increased heart rate and respiratory rate; parasympathetic nervous system
   (d) Decreased heart rate and respiratory rate; sympathetic nervous system

5. A person is sleeping and their bladder is full. What effective change will occur in the body, and which part of the autonomic nervous system brings it about?
   (a) Contraction of the bladder; parasympathetic nervous system
   (b) Relaxation of the bladder; sympathetic nervous system
   (c) Contraction of the bladder; sympathetic nervous system
   (d) Relaxation of the bladder; parasympathetic nervous system

6. A person is exercising vigorously. What effective change will occur in the body, and which part of the autonomic nervous system brings it about?
   (a) Decreased heart rate and respiratory rate; sympathetic nervous system
   (b) Increased heart rate and respiratory rate; parasympathetic nervous system
   (c) Increased heart rate and respiratory rate; sympathetic nervous system
   (d) Decreased heart rate and respiratory rate; parasympathetic nervous system

NCERT Corner

1. Name the types of neurons based on the function they perform.

2. What is the name of the junction between two neurons?

3. What is the function of the cerebellum?

4. What are the three main parts of the brain?

5. What is the function of medulla oblongata?

COMPETITION WINDOW (OBJECTIVE TYPE)

1. The insulating sheath covering the axon is called:
   (a) Plasmalemma
   (b) Neurolemma
   (c) Dura mater
   (d) Pia mater

2. Which one of the following pairs of brain parts and its function is not correctly matched?
   (a) Cerebrum - memory
   (b) Cerebellum - the balance of body
   (c) Medulla oblongata - controls activities of internal organs
   (d) Pons - consciousness

3. A mixed nerve is one which:
   (a) Carries sensation from two or more sense organs
   (b) Contains both sensory and motor fibres
   (c) Has a common root but branches into two or more nerves to different organs
   (d) Has two or more roots from different parts of the brain

4. Reflex action is controlled by:
   (a) Brain
   (b) Spinal cord
   (c) Autonomic nervous system
   (d) Peripheral nervous system

5. Which one of the following is responsible for controlling the body temperature?
   (a) Liver (b) Lung
   (c) Hypothalamus (d) Spinal cord

6. Which of the following is absent in a neuron?
   (a) Ribosome (b) Centrosome
   (c) Chromosome (d) Lysosome

7. The basic structural and functional unit of the brain is:
   (a) Nephron
   (b) Nerve
   (c) Spinal cord
   (d) Neuron

Answer Keys

Multiple Choice Questions

1. (d) **2.** (c) **3.** (b) **4.** (a) **5.** (d)
6. (c) **7.** (b) **8.** (a) **9.** (d) **10.** (c)

Match the Following

1. (a) (q) (b) (t) (c) (p) (d) (r)

2.

Column I	Column II
(a) Frontal lobe	(t) Controls voluntary movements
(b) Parietal lobe	(r) Controls touch and spatial awareness
(c) Occipital lobe	(s) Controls vision and object recognition
(d) Temporal lobe	(q) Controls emotions and social behavior
(e) Cerebellum	(p) Controls balance and coordination

3.

Column I	Column II
(a) Axon	(q) Carries messages away from the cell body
(b) Dendrite	(p) Receives messages from other neurons
(c) Cell body	(r) Connects the neuron to other neurons
(d) Myelin sheath	(s) Increases the speed of nerve impulse transmission
(e) Synapse	(t) Site of communication between two neurons

4.

Column I	Column II
(a) Central Nervous System	(p) Includes the brain and spinal cord
(b) Peripheral Nervous System	(q) Controls voluntary movements
(c) Somatic Nervous System	(r) Regulates involuntary functions
(d) Autonomic Nervous System	(s) Transmits sensory information
(e) Sensory Nervous System	(t) Carries messages to and from the brain

(*Continued*)

True or False

1. T **2.** F **3.** T **4.** F **5.** T
6. T **7.** F **8.** T

Fill in the Blanks

1. Brain, Spinal cord
2. Centrioles
3. Cerebrospinal fluid (CSF)
4. CSF
5. Control and coordination
6. Central nervous system and peripheral nervous system
7. Receiving and processing
8. Relaying messages
9. Transmitting electrical and chemical signals
10. Sends
11. Myelin sheath
12. Synapse
13. Sensory nerves and motor nerves
14. Involuntary

Assertion–Reason Questions

1. (a) **2.** (a) **3.** (a) **4.** (a) **5.** (a)
6. (a) **7.** (a)

One-word Answers

1. Neuron/Nerve cell
2. Meninge
3. Cerebrospinal fluid
4. Synapse
5. Neuron

6. Brain
7. Dendrite
8. Axon
9. Myelin sheath
10. Synapse
11. Brain stem
12. Cerebellum
13. Autonomic nervous system
14. Spinal cord

Define the Term

1. The structural and basic building blocks of the neurological system are neurons. They are trained to send electrical impulses carrying information to various parts of the body.
2. Nerves are composed of axons, or nerve fibres, and are distinct from neurons that are encased in tubular sheaths.
3. A stimulus is an agent or a sudden alteration of the internal or external environment that affects an organism or any of its parts.
4. Synaptic clefts are spaces between the axon terminals and dendrites of one or more neurons.
5. Reflex action is a spontaneous, rapid or instantaneous involuntary bodily response to a stimulus.
6. A sheet of fibre known as the corpus callosum connects the two hemispheres of the brain.

Groups of Terms

1. Fibres
2. Ligand
3. Nerves
4. Memory
5. Outgoing Signal

Identify the Odd One Out

1. Liver
2. Hormone
3. Ribosome
4. Muscle
5. Hormone
6. Spleen
7. Pancreas
8. Digestive system
9. Motor nervous system
10. Pituitary gland

Image-Based Questions

1. A - Cyton

   B - Axon

   C - Neurolemma

   D - Node of Ranvier

2.

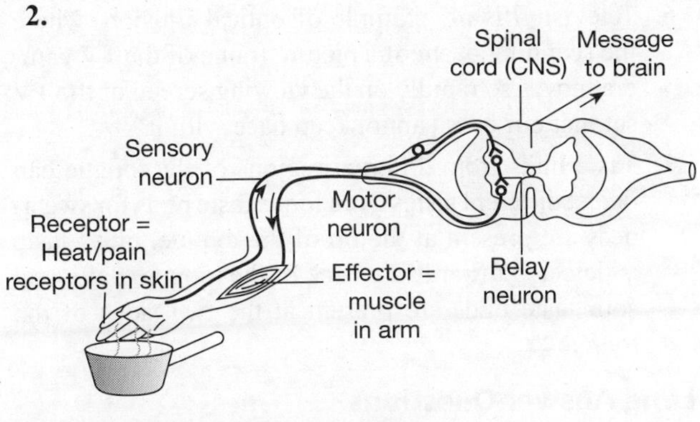

Very Short Answer Questions

1. The brain's outermost layer is the duramater.
2. The ear ossicle attached to the tympanum is the malleus.
3. The photosensitive pigment present in the rods of the retina is rhodopsin.
4. The nerve which transmits impulses from the ear to the brain is the auditory nerve.
5. The eye defect caused due to shortening of the eyeball from front to back is hyperopia.
6. The tube which connects the cavity of the middle ear with the throat is the Eustachian tube or pharyngotympanic tube.
7. The photoreceptors found in the retina of the eye are rods and cones.

Short Answer Questions

1. A gland that secretes tears. The lacrimal glands are found in the upper, outer part of each eye socket.
2. The yellow spot is present in the posterior pole of the eye, lateral to the blind spot. The blind spot is present at the point of origin of the optic nerve.
3. Presbyopia is a condition in which the eye lens loses its ability to accommodate owing to ageing. Here, the person cannot see objects that are close by clearly due to decrease in ocular accommodation power.
4. Power of accommodation is the ability of the eye lens to focus near and far objects clearly on the retina by adjusting its focal length. In this process ciliary muscles help in changing the shape of lens by compressing and elongating it. Power of accommodation of the eye is limited.
5. The image created on the retina is real and inverted, and the optic nerve sends chemical signals to the brain, which converts them into upright images that can be seen with the eyes.

6. Television is an example of optical illusion, where the scanning beam of a picture frame of the TV camera moves so rapidly on the viewing screen of the TV set that our eyes cannot keep pace with it.
7. Taste buds from different regions of the tongue can detect different tastes from food. Taste buds for sweetness are present at the tip of the tongue, bitter taste is detected towards the back of the tongue. Salty and sour taste buds are present at the two sides of the tongue.

Long Answer Questions

1. Hint: The process of focusing on a distant object involves the adjustment of the lens in your eye. Elaborate on this point.
2. Hint: This question is about the impact of a clear image from a dream on your eyes. Elaborate on this point.
3. Hint: This question is about the phenomenon of seeing light particles when there is no light entering the eyes. Elaborate on this point.
4. Hints:
   (a) Function: One is responsible for higher cognitive functions, while the other is involved in coordination and balance.
   (b) Overall Effect on Body: One prepares the body for stressful situations, while the other promotes relaxation and restoration.
   (c) Direction of Impulse Carried: One carries sensory information from the body to the brain, while the other carries motor commands from the brain to the muscles.

   Elaborate on these points.
5. Hints:
   (a) Type of nerve involved in controlling muscle movements.
   (b) Part of the nervous system responsible for involuntary bodily functions.
   (c) Learned response to a specific stimulus.
   (d) Neuron that connects sensory and motor neurons in the nervous system.

   Elaborate on these points
6. Hint: Consider the ability to respond to stimuli, coordinate body functions, and adapt to the environment. Elaborate on this point.
7. Hint: Reflex Action: Involuntary and automatic response to a stimulus without conscious thought. Voluntary Action: Deliberate and intentional action initiated by conscious decision-making. Elaborate on this point.

Case Study-Based Questions

1. (c) 2. (b) 3. (d) 4. (a) 5. (d)
6. (c)

NCERT Corner

1. Motor neurons make the connection between the brain and muscles throughout the body. Sensory neurons are neurons that let us feel sensation. Interneurons are the nerve cells that connect motor neurons to other motor neurons and sensory neurons to other sensory neurons.
2. The junction between two neurons is called as a synapse.
3. The function of the cerebellum is to maintain body posture, muscle tone, coordination and equilibrium of the body.
4. Frontal lobe: It is associated with parts of speech, planning, reasoning, problem-solving and movements.

   Parietal lobe: Helps in movements, the perception of stimuli and orientation.

   Occipital lobe: It is related to visual processing.
5. The medulla oblongata helps regulate breathing, heart and blood vesselfunction, digestion, sneezing and swallowing. This part of the brain is a centre for respiration and circulation. Sensory and motor neurons (nerve cells) from the forebrain and midbrain travel through the medulla.

Competition Window

1. (b) 2. (d) 3. (b) 4. (b) 5. (c)
6. (b) 7. (d)

CHAPTER

10 The Circulatory System

PREREQUISITES

You must have knowledge of:

- Anatomy of the heart
- The components of blood, including red blood cells, white blood cells, platelets and plasma
- Exchange of oxygen and carbon dioxide between the lungs and blood
- Immune system and its functions
- Basic principles of nutrition and how nutrients are transported

LEARNING OBJECTIVES

At the end of this chapter, you will be able to:

- Explain the structure and functions of the heart
- Describe the different types of blood vessels and their functions
- Describe lymph and the lymphatic system
- Describe tissue fluid

INTRODUCTION

The bodies of almost every organism, including humans, contain many types of fluids. Such fluids serve as both a distributing system (to deliver substances) and a collecting system (to gather substances) to and from the various parts of the body, including the farthest cell.

Every organ in our body requires the involvement of the circulating body fluids. For example:

- The respiratory system pulls in air, and the oxygen taken up from it in the lungs needs to be transmitted to all the regions of the body.
- The digestive system digests and absorbs nutrients required to be transported to every body cell. Similarly, the CO_2 gathered from the entire body must be transported to the lungs for release.
- All the surplus water, salt and nitrogenous waste such as urea must be removed from various regions of the body and transported to the excretory system for elimination.
- The circulating blood must transport hormones secreted by the endocrine system throughout the body.

All such functions that need transport are performed by the two circulating fluids, blood and lymph. Besides transport, these fluids perform some other functions also.

BODY FLUIDS

There are three principal fluids in our body:

- **Blood,** contained in the heart and in the blood vessels (arteries, veins and capillaries) of the circulatory system
- **Tissue fluid,** occupying spaces between cells in the organs
- **Lymph,** contained within lymph vessels and lymphatic organs such as the spleen and the tonsils

Closed vascular system is the term used to describe this kind of blood circulation. In contrast, blood primarily circulates through open areas in other species, such as insects; this is referred to as an **open circulatory system**. The heart pumps blood into vascular-free areas of human tissue.

Other non-circulating fluids include vitreous **humour** in the eyes, synovial fluid in the cavities of the skeletal joints, and others that are found in specific organs.

BLOOD

Blood is always in motion, from the heart to the arteries and back through the veins. When drawn from an artery, the blood is a rather thick liquid that is bright red in colour; when drawn from a vein, it is dark crimson. An average adult human has 5–6 litres of blood by volume in the body. Perhaps we have all 'tasted' our blood when there is a cut in the tongue or bleeding from the gums and found it to be salty? Blood is slightly alkaline with a pH of 7.3–7.45 (7 is neutral, neither acidic nor alkaline).

Functions of Blood

The main functions of the blood in our body can be categorised under two broad heads: transportation and protection.

Transportation

- Moving digested food from the digestive tract to the tissues. Simple sugars like glucose, amino acids, vitamins, mineral salts, and others are among these compounds.
- Delivery of oxygen to the tissues from the lungs. This process is carried out by red blood cells working in conjunction with haemoglobin to create the unstable molecule oxyhaemoglobin, which disintegrates when it reaches the tissues to deliver oxygen.
- Transporting carbon dioxide from the tissues and to the lungs. This is carried out by blood plasma in conjunction with haemoglobin.
- Moving waste products from the tissues to the liver, kidney or skin for removal or to make them harmless.
- Distribution of hormones secreted by special glands (endocrine glands) directly into blood.
- Distribution of heat. Blood helps in keeping the temperature of the body uniform by distributing heat.

Protection

- Blood forms a clot wherever there is a cut in a blood vessel. The clot serves to prevent further loss of blood and the entry of disease-causing germs.
- Its white blood corpuscles protect the body from diseases by engulfing bacteria which may have entered the body.
- It generates antibodies and antitoxins that neutralise toxins or eradicates the germs that enter the body.

Composition of Blood

Blood consists of:

- Plasma (fluid part), constituting about 55–60%
- Cellular elements (red and white cells and platelets), constituting about 40–45%

Plasma

Plasma is a light-yellow-coloured, alkaline liquid. It mainly consists of:

- Water, 90–92%
- Proteins, 7–8%
- Inorganic salts, 1%
- Other substances, traces

The inorganic salts are mainly sodium chloride and sodium bicarbonate. Among other substances in the plasma are glucose, amino acids, fibrinogen, hormones, urea, etc. The plasma from which the protein fibrinogen has been removed is called **serum**.

Cellular Elements

The formed or cellular elements (shaped structures visible under magnification) of the blood are of three types: red blood cells (erythrocytes), white blood cells (leukocytes) and blood platelets (thrombocytes) (Fig. 10.1).

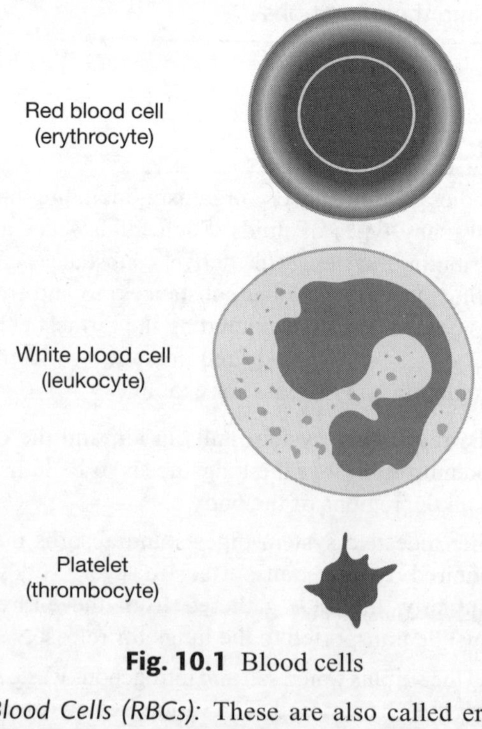

Fig. 10.1 Blood cells

Red Blood Cells (RBCs): These are also called erythrocytes. They are minute, biconcave, disc-like structures, flat in the centre and thick and rounded at the periphery.

They have a diameter of only 7 microns (one micron is equal to one thousandth of a millimetre). They are particularly effective at absorbing oxygen due to their small size and the concavities on either side, which give them a wide surface area. Due to their small size, RBCs may pass through the body's tiny capillaries, where they move in a single file. An adult human male has about 5 million RBCs per cubic mm of blood and an adult female has about 4.5 million RBCs per cubic mm of blood.

RBCs have a colourless spongy body or stroma which contains a respiratory pigment haemoglobin (Hb). A protein component and an iron-containing component (haemin) make up haemoglobin. It may easily react with oxygen to generate oxyhaemoglobin, an unstable substance that swiftly releases oxygen to the tissues in need of it. A very small amount of carbon dioxide can be carried by haemoglobin in the form of carbamino haemoglobin.

Is it correct to say that veins carry CO_2 and no oxygen? Oxygen is also carried through the veins, though slightly less so than by arteries. The oxygen saturation of haemoglobin increases from 97% to 99% inside the lungs. Only around 23% of the oxygen in Hb is released inside respiring tissues, while 75% of the oxygen is still present in the blood returning through the veins.

Haemoglobin produces the stable molecule carboxyhaemoglobin due to its high affinity for carbon monoxide (HbCO). As a result, the blood's capacity to deliver oxygen is reduced, which can occasionally result in fatalities. For instance, burning furnaces without enough ventilation in small rooms can cause carbon monoxide poisoning, which can be fatal.

Life and death of RBCs

- In adults, RBCs are produced in the marrow of the long bones, especially in the ribs, breastbone and ilium of the hip girdle.
- They are produced in the liver and spleen of an embryo.
- RBCs are produced in the bone marrow of every bone in children until the age of five.
- Red blood cells lack nuclei when they are mature, but when they are produced, they have one. They become enucleated as they mature.
- An RBC has a lifespan of approximately 120 days.
- The spleen, liver and bone marrow destroy old and weak red blood cells; the liver retains the iron portion, while the remaining portion is expelled as a bile pigment (bilirubin).
- In a normal adult, approximately 20,000,000 RBCs are destroyed every minute. In other words, about 1% of the total erythrocytes in the body are destroyed every day.

Mammalian red blood cells: Mature mammalian red blood cells circulate in the blood system and lack specific organelles. They have no nucleus, mitochondria or endoplasmic reticulum. Thus, though deficient of the above organelles, the mature red blood cells, in reality, are more efficient in carrying out their task of picking up and delivering oxygen. The factors making them more efficient in this work are as follows:

- Loss of nucleus causes red blood cells to become biconcave, increasing their surface area to volume ratio and allowing them to absorb more oxygen. The space in between is increased, so more RBCs can be accommodated in the same space.
- Red blood cells cannot utilise oxygen without mitochondria (cellular respiration occurs in mitochondria). Thus, the oxygen that the lungs take in is completely delivered to the tissues without being used up. Secondly, as there are no mitochondria, all of the glucose that the RBCs do not use is transferred to the blood plasma.
- Without the endoplasmic reticulum, RBCs are more flexible and can fit through tiny capillaries.
- The concentration of RBCs in newborns is higher that in adults, averaging 6–7 million per cubic millimetre (1 mm^3). The RBC count decreases by 5% as you sleep. It increases after physical activity, pregnancy and emotional stress. People who live at altitudes of 4,200 m or higher have about 30% higher RBC levels. Polycythaemia is the medical term for an excessively high RBC count, while erythropenia is the medical term for an abnormally low RBC count.

White Blood Cells (WBCs): These are also known as leukocytes (*leukos*: white) because they spend most of their time (90%) in tissue fluid or in the lymph and very little (only about 10%) in the blood.

The red bone marrow, lymph nodes and occasionally even the liver and spleen, all produce WBCs. Their lifespan is around two weeks. A neutrophil has a lifespan of only a few hours and 125 billion of them are created daily. RBCs and WBCs are destroyed in the same way.

There are five main types of white blood cells (WBCs) and their functions are as follows:

- **Neutrophils:** These are the most abundant type of WBCs and play a key role in fighting bacterial infections. They can engulf and destroy bacteria through a process called phagocytosis.

- **Lymphocytes:** There are two main types of lymphocytes, B-cells and T-cells. B-cells produce antibodies, which help to recognize and destroy invading pathogens, while T-cells can directly attack infected cells or abnormal cells.
- **Monocytes:** These are the largest type of WBCs and can differentiate into macrophages, which are specialised cells that can engulf and destroy pathogens, dead cells, and other debris.
- **Eosinophils:** These WBCs play a role in the immune response to parasitic infections and can also contribute to allergic reactions.
- **Basophils:** These WBCs are involved in the immune response to allergens and can release histamine, which causes inflammation.

Leukaemia or a malignancy of the tissue is the excessive production of WBCs at the expense of RBCs. Usually, it is a deadly condition. The only available treatment is blood transfusion. Leucopenia is an abnormal reduction in the WBC count.

Functions

- **Phagocytosis:** In this process, the majority of WBCs – particularly neutrophils – adopt solid particles – primarily bacteria – into their bodies. This is a mechanism of disease defence. An abnormal increase in WBC count of up to 50,000 or more per cubic millimetre is a sign of a physical infection.
- **Inflammation:** The body's reaction to injury and localised invasion of microorganisms result in inflammation. A number of symptoms, such as increased localised heat, redness, swelling and discomfort are present in the inflamed area. Leukocytes, in particular neutrophils and monocytes, travel through the blood arteries by diapedesis and fight the dangerous bacteria there. They also get rid of the damaged cells by phagocytosis. The majority of pus is composed of tissue cells and dead, bacterially degraded white blood cells.
- **Antibody production:** White blood cells, particularly lymphocytes, produce antibodies that either eradicate or neutralise infections or the toxins they secrete. The injection of pathogens or pathogen components that have been weakened through vaccination stimulates the creation of particular antibodies that, if the pathogens were to enter the body, would later wipe them out.

  Toxins, which are poisonous substances, are released by pathogens when they get into the bloodstream. In response, the lymphocytes in blood produce chemical substances called antibodies that freely circulate in the blood plasma. Antibodies are specific to the poisons for which they are made.

Blood Platelets: Blood platelets are tiny, non-nucleated, round or oval entities that float in the blood. In a cubic millimetre of blood, an adult has between 200,000 and 400,000 of them.

The megakaryocytes, which are enormous cells seen in the red bone marrow, are where the platelets are generated. Platelets are entirely enclosed by a membrane after being budded off from the megakaryocytes. Their life expectancy is 3–5 days, and they are primarily eliminated in the spleen (a lymphatic organ located in the abdomen). They play a crucial role in blood clotting. The blood clotting process begins at the site of injury when platelets break down and release the chemical thrombokinase.

When a blood vessel is cut, blood escapes from it. But soon a clot is formed on the wound (Fig. 10.2) and the flow of blood is stopped. If it were not so, the injured person would bleed to death. Clotting (also called coagulation) occurs in a series of steps as follows:

1. Thrombokinase is released at the site of a wound by the wounded tissue cells and platelets that disintegrate (also called thromboplastin). (Factor X or Stuart factor are more recent names for it.)
2. The thrombokinase functions as an enzyme, converting the plasma's prothrombin (inactive) into thrombin with the aid of the calcium ions present (active).
3. In the presence of calcium ions, thrombin interacts with plasma's soluble fibrinogen to transform it into insoluble fibrin. Fibrin is a material that solidifies to form threads. These tiny, sticky fibrin strands create a network (mesh) at the wound.
4. Blood cells are trapped in the network of the fibrin; the network then shrinks and squeezes out the rest of the plasma which is in the form of a clear liquid, the serum. The solid mass which is left behind is called clot (or thrombus).

Blood Transfusion and Blood Groups

It may occasionally be essential to inject blood directly into a patient's body while they are having surgery. This is known as a blood transfusion. Blood is administered from a healthy donor through one of the patient's veins. However, in order to do so, the kind or type of blood being transfused must match (or be compatible) with the recipient's blood.

Karl Landsteiner discovered the blood grouping system. Human RBCs have certain proteins on their

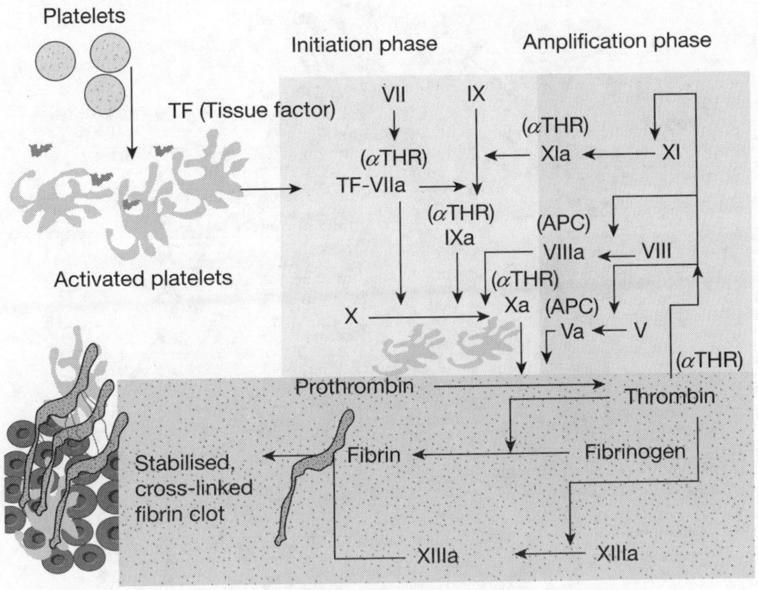

Fig. 10.2 Blood clotting

surface, called antigens. The blood's plasma contains complimentary antibodies. There are two categories of antigens in humans: antigen A and antigen B. Depending on the presence or absence of these antigens, there are four types of blood groups: A, B, AB and O.

There are several systems of blood grouping. However, two of them, ABO and Rh, are the most important.

ABO system: According to this system, human blood is classified into four types: A, B, AB and O. Besides the transfusion of one's own type of blood (A to A, AB to AB, and so on), the other possible transfusions are as follows:

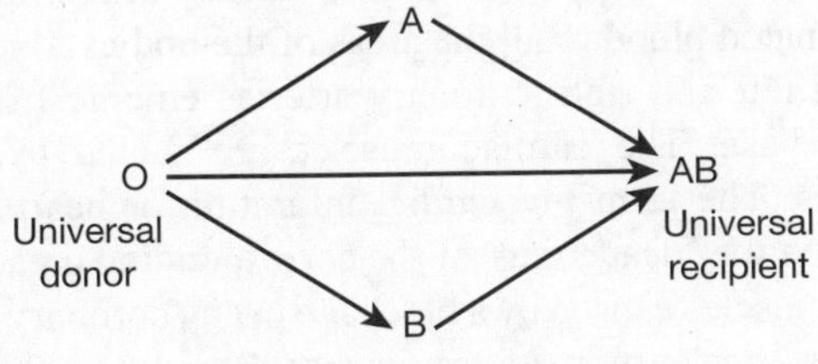

Fig. 10.3 Blood typing

- O type blood can be given to persons of all types of blood, that is, O, A, B and AB. Hence a person with O type is called universal donor (Fig. 10.3).
- A person with AB type of blood can receive blood from all types, that is, from AB, A, B and O, and is, therefore, called universal recipient.
- A person with A type can receive blood from A and O types and a person with B type from B and O types only.

Rh system: Most people have a chemical called Rh factor in their blood. (Rh stands for Rhesus, the domesticated monkey, in which the factor was first identified). Within two weeks of transfusion, the recipient's blood becomes sensitised to the Rh material when the blood of such a person (Rh-positive) is given to someone without it (Rh-negative). Now, if a second transfusion is given to a person who is Rh-negative, the antibody created by the previous transfusion creates a response with the transfused blood that could even be fatal.

This is comparable to how an allergy manifests itself.

Pregnancy and the Rh factor: If a woman is Rh-negative and has a Rh-positive child, she may become sensitive (when the husband is Rh-positive). The first Rh-positive child will be healthy, but if the mother becomes sensitized, the second positive child, if conceived shortly after, may experience complications that could sometimes result in foetal death and abortion.

Blood Circulatory System

The circulatory system consists of the heart, arteries, veins and capillaries.

The Heart

The heart is located between the two lungs and directly above the diaphragm. The narrow end of the roughly triangular heart is directed toward the left (Fig. 10.4),

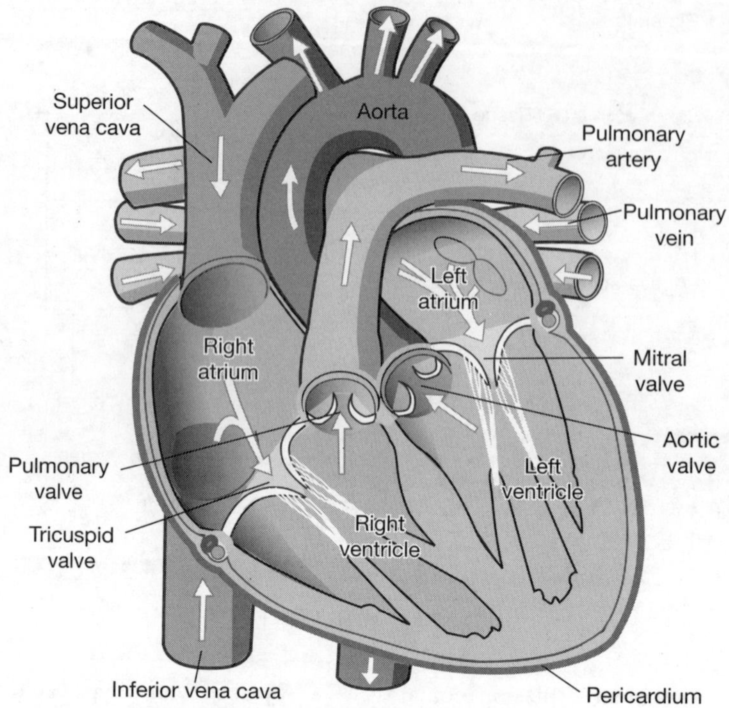

Fig. 10.4 Human heart

and when the heart is contracting, this end experiences the strongest contraction, creating the impression that the heart is on the left.

The adult heart is roughly the size of a closed fist, measuring 12 cm in length and 9 cm in width. The pericardium, a double-walled membrane, guards the heart. It contains pericardial fluid that acts as a lubricant to prevent mechanical damage and lessen friction during heartbeats.

The heart has two upper atria (one atrium) and two lower ventricles, making up its four chambers. Because their primary duty is to absorb blood from the body and pump it into the ventricles, the atria (also known as auricles) have thinner walls (Fig. 10.5). The ventricles must pump blood over great distances; so they have robust, muscular walls. Only the oxygenated blood required by the lungs is pumped by the right ventricle. The left ventricle pumps blood against gravity to the extreme ends of the body, including the brain and the toes of the feet. As a result, its walls are thicker.

Blood vessels entering the heart: The anterior vena cava and posterior vena cava are two significant arteries that enter the right atrium. Deoxygenated blood from the anterior or top of the body, such as the head, chest and arms, is transported through the anterior vena cava, also known as the superior vena cava or precaval. The posterior (or inferior) vena cava receives blood from the lower parts of the body, including the legs and abdomen. Four pulmonary veins are received by the left atrium (two from each lung). Oxygenated blood is delivered by the pulmonary veins.

Blood vessels leaving the heart: Arising from the ventricles are two large blood vessels. The right ventricle gives rise to the pulmonary artery, which delivers deoxygenated blood to the lungs for oxygenation. The left ventricle gives rise to the aorta, which delivers oxygenated blood to all the areas of the body.

Six left and right coronary arteries emerge from the aortic base. The cardiac muscles are fueled by these arteries. The term myocardial infarction or heart attack refers to the 'deadening' of the corresponding area of the heart muscles caused by a blockage in any coronary artery or in any one or more of its branches. Blood from the heart walls is gathered by the cardiac veins, which then drain it into the right auricle. Chest pain due to insufficient supply of blood to the heart muscle is known as angina pectoris.

There are four **valves** in the heart:

- The opening between the right auricle and the right ventricle is where the right atrio-ventricular valve is situated. It is also known as the tricuspid valve because it has three thin, triangular cusps that resemble leaves. The tendinous cords (chordae tendinae) that originate from the papillary muscles,

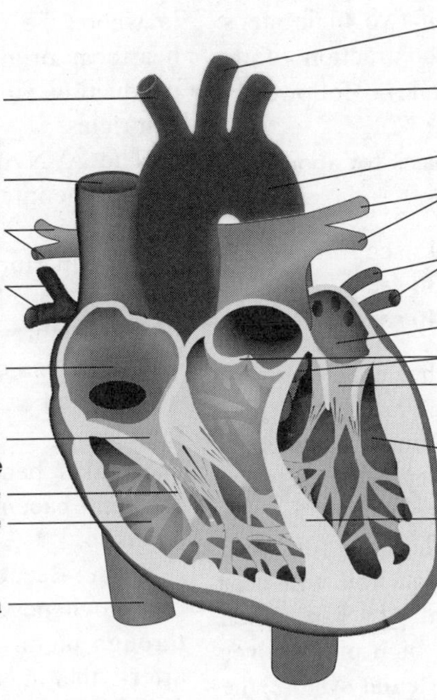

Fig. 10.5 Internal structure of the human heart

which are muscular protrusions of the ventricular wall, hold the flaps' apices (plural of apex) in place.

- Similarly situated on the left side of the heart is the left atrioventricular valve. It is known as the bicuspid (or mitral) valve because it contains two cusps.
- At the point where the right ventricle empties into the pulmonary artery are the pulmonary semilunar valves. There are three of these pocket-shaped items.
- Aortic semilunar valves are located at the point of origin of the aorta from the left ventricle. These are also three in number and pocket shaped.

Atrial Systole

1. The atrial (auricular) muscles contract.
2. The openings of the vena cava and pulmonary vein close.
3. Blood enters the ventricles by crossing through the tricuspid valve and bicuspid valve.
4. Semilunar valves at the roots of the pulmonary artery and aorta are closed, producing the sound 'dup' to prevent flow of blood back into the ventricles.

Ventricular Systole

1. The ventricular muscles contract.
2. The tricuspid valve and bicuspid valve close with a jerk, producing the 'lubb' sound.
3. Blood passes into the aorta and pulmonary artery through the semilunar valves.
4. The atria draw in blood through the openings of the vena cava and pulmonary vein.
5. Chordae tendinae hold the valves in position, preventing their upturning due to the pressure exerted by the contracting ventricles.

Circulation of Blood in the Heart

1. The two atria contract first. At this point, the ventricles are empty and are relaxed (or dilated). As a result, blood from the atria flows freely into the ventricles.
2. Then the atria relax and the ventricles contract. Blood under pressure from the ventricles has a tendency to return to the atria, but the flaps of the two cuspid valves tighten and puff up, blocking the channel and preventing blood from returning. The flaps of the valves are held in place by the chordae tendinae, which also prevent them from turning into the atria.
3. The ventricular blood then enters the pulmonary artery (from the right ventricle) and the aorta (from the left ventricle). The mouths of the pocket-like valves at the bases of these two blood vessels face away from the ventricles. Therefore, the blood leaving the ventricles presses the valves flat and obtains a clear passage in between.
4. When the ventricles dilate, blood from the pulmonary artery and the aorta tend to return; the blood fills the pockets of the valves and closes the passage.

Heartbeat: Each heartbeat consists of two main steps: the atria contract first followed by contraction of the ventricles. The contraction phase (systole) is followed by a relaxing expansion phase (diastole).

Each full beat of the human heart lasts for about 0.85 seconds.

Systole of atria (auricles)	0.15 sec
Systole of ventricles	0.30 sec
All chambers in relaxed state	0.40 sec

The ventricles begin to relax at the conclusion of ventricular systole (ventricular diastole). For a brief time, the atria (auricles) and ventricles are both in a relaxed condition (joint diastole). The **cardiac cycle** refers to the entire series of actions that make up a heartbeat.

Heart sounds: By placing a stethoscope on your chest, you may listen to your heartbeat. You will hear two sounds, 'lubb' and 'dup', followed by a gap. When the atrio-ventricular (tricuspid and bicuspid) valves close abruptly at the beginning of ventricular systole, the first sound, lubb', is made. The second sound 'dup' is produced when, at the beginning of ventricular diastole, the semilunar valves at the roots of aorta and pulmonary artery are closed.

The rate of heartbeat varies among different species and even in individuals of the same species. Some of the variations in beats per minute (at rest) in different species are (approximately) as follows:

Whale	15
Elephant	25
Horse	40
Adult men	64–72
Adult women	72–80
Cat	120
Newborn infant	140
Rat	250
Sparrow (in vigorous activity)	800–900

Do you find any relationship between the size of the body and the rate of heartbeat? The smaller the size, the faster the heart rate. This is because smaller animals lose more body heat due to the higher surface to volume ratio, and therefore, the increased heart rate distributes body heat faster. Secondly, the young growing stages, such as the human baby, show higher metabolism for body growth, and therefore, the faster heart rate keeps the 'supply' and 'take off' of the metabolic substances in the correct quantities.

Pacemaker: The pacemaker (sino-atrial node or SAN), which is found in the walls of the right auricle, is where the impulse or command that initiates the heartbeat originates. Through the use of specialised conducting fibres, this impulse is transmitted to the ventricles.

The AVN node (atrio-venticular node) is located next to the interauricular septum, close to the tricuspid valves, while the pacemaker (SAN) is situated close to the superior vena cava opening. From the AVN, a group of muscle fibres known as the bundle of His or Purkinje fibres travels to the interventricular septum. The branches of these fibres run along the ventricle's wall. Together, they constitute a system that sends an impulse to every area of the heart. Sometimes, the pacemaker becomes faulty, causing heart trouble. An artificial pacemaker may be fixed in the heart of such a person.

Pulse: Because of their elastic muscle walls, arteries dilate whenever the heart contracts, forcing blood through them. By gently placing your finger over an artery that has protruded superficially, such as the radial artery of the wrist on the side of the thumb, you may easily feel this distension. The heartbeat is indirectly counted when the pulse is taken. Exercise and even certain emotions can cause an increase in heart rate.

Blood pressure: The force that blood exerts on the artery walls as it flows through them is known as blood pressure. The upper limit (systolic pressure), which occurs when new blood is pushing through the artery as a result of the heart's ventricular contraction, is one of two limitations to this pressure. Diastolic pressure is the lower limit that is measured after the wave has passed. In adults, the normal blood pressure ranges from 60–80 mmHg to 100–140 mmHg. Hypertension is a rise in blood pressure above 140/90.

The Blood Vessels

The blood vessels are branched tubes extending from the heart to all the parts of the body. They are of three types: arteries, veins and capillaries (Fig. 10.6).

- An **artery** is a vessel that carries blood away from the heart towards any organ. It has thick muscular walls, a narrow lumen (the central bore) and the blood in it flows in spurts which correspond to the ventricular contractions of the heart.
- A **vein** is a vessel that carries blood from the various organs to the heart. It has thin muscular walls, a wider lumen and the blood in it flows uniformly. Veins contains thin, pocket-shaped valves whose openings face the direction of the heart; these valves prevent backflow of blood.

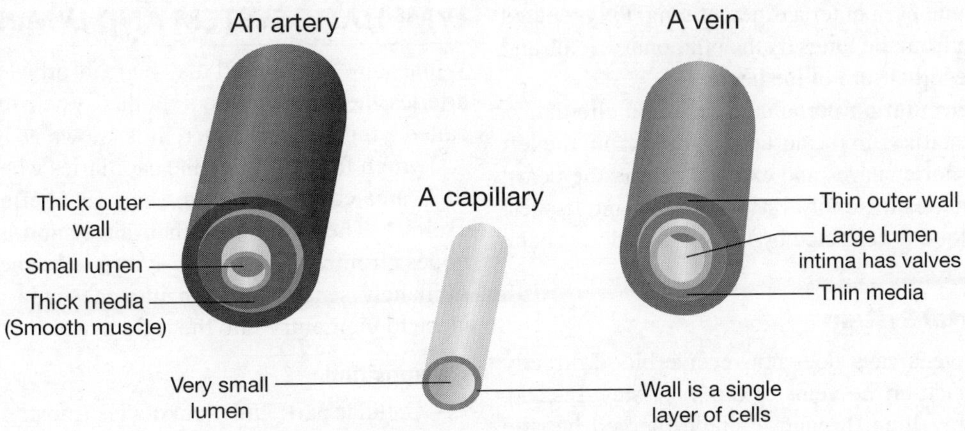

Fig. 10.6 Arteries, veins and capillaries

- The smallest or the final branch of an artery is called an **arteriole**. Arterioles are highly muscular and can change their diameter many times. Arterioles break up into capillaries.
- A **capillary** is a very narrow tube (about 8 micrometres in diameter); its wall consists of a single layer of squamous epithelial cells (endothelium), and it has no muscles. The human body's blood capillaries could stretch 100,000 kilometres if they were lined up end to end!

  Capillaries serve the following purposes:
  - They permit carbon dioxide from the intercellular fluid to diffuse inward.
  - They permit the flow of molecules such as glucose, amino acids, urea and hormones both inside and outward.
  - They enable the amoeboid migration of leukocytes (WBCs) to push them past the capillary walls. Capillaries can dilate (vasodilation) and contract (vasoconstriction) which, respectively, increases or decreases the blood supply to the body parts. Evidence of these changes in capillaries is seen in the colour of the skin. When you walk in the hot sun, your face turns pink (increased blood flow); when it is very cold, your face turns bluish (reduced blood flow).
- The capillaries gradually reunite and increase in size, assuming the same three layers (connective tissue layer, muscular layer and endothelium) as in the arteries and veins. The smallest united common branch is called a **venule**. The venules join to form larger veins. Compared with arterioles, the venules are larger with a much weaker muscular coat.

Pulmonary and Systemic Circulation

Before one full rotation, blood passes through the heart twice: once for the short pulmonary (lung) circulation and once for the long systemic (whole body) circulation. Because of this, the circulation of blood throughout the human body is also referred to as **double circulation** (Fig. 10.7).

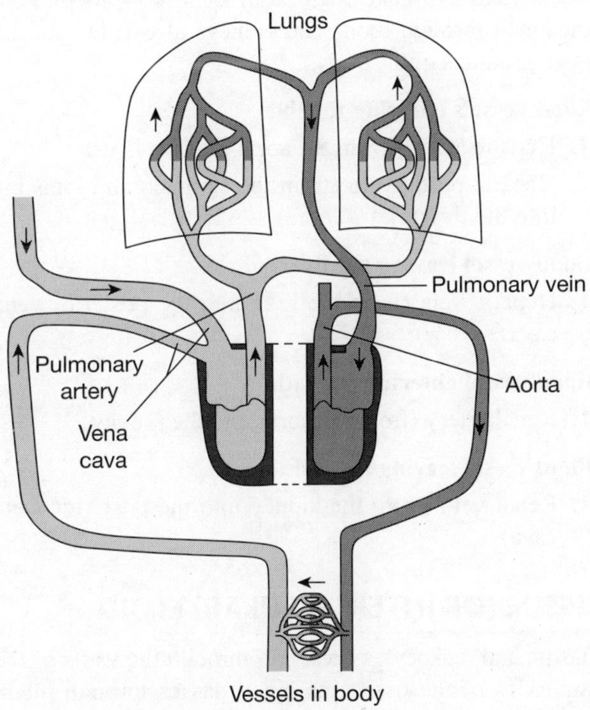

Fig. 10.7 Double circulation

The lungs are involved in **pulmonary circulation**. It begins in the right ventricle's pulmonary artery, which splits into two branches that each enter a different lung. Oxygenated blood is drawn from the lungs by the pulmonary veins and returned to the left atrium of the heart.

Systemic circulation pertains to the main circulation in the body. It starts with the aorta that arises from the left ventricle. The aorta curves and extends behind the heart, sending off arteries to supply various organs and tissues. From there, blood is collected by the veins and sent back into the heart.

Hepatic Portal System

The posterior vena cava does not receive blood directly from the stomach or the veins of the intestines. Instead, it begins by travelling through a single merged hepatic portal vein. Contrary to the usual characteristic of veins, it disintegrates into capillaries at that location, and their reunion forms a new vein known as the hepatic vein that connects to the posterior vena cava. So, by definition, a portal vein is one that originates from capillaries and terminates in them.

The liver works as a store to control the amount of nutrients going into general blood circulation; for instance, extra sugar is stored as glycogen after being absorbed by the stomach and intestines. Additionally, the liver detoxifies (makes harmless) some poisons that are consumed through food. The kidneys also filter out the waste products of the body.

Blood vessels entering the liver

1. Hepatic artery (from the aorta into the liver)
2. Hepatic portal vein (from the stomach and intestine into the liver)

Blood vessel leaving the liver

1. Hepatic vein (from the liver into the posterior vena cava)

Blood vessel entering the kidney

1. Renal artery (from the aorta into the kidney)

Blood vessel leaving the kidney

1. Renal vein (from the kidney into the posterior vena cava)

TISSUE (OR INTERCELLULAR) FLUID

Plasma and leukocytes 'leak out' through the walls of the capillaries of the tissues as blood passes through them. This fluid, also known as tissue fluid, intercellular fluid or extracellular fluid, bathes the cells. The cells take in oxygen and other necessary components from this fluid and release waste products like carbon dioxide back into it.

LYMPHATIC SYSTEM AND LYMPH NODES

While some tissue fluid may be reabsorbed into the blood arteries, the majority enters another system of tiny channels called lymph vessels, where it is known as lymph.

Lymph flows through the capillaries when the muscles in an area contract. This is a beneficial effect of physical exercise. The lymph vessels drain lymph into the lymph nodes from where fresh lymph channels arise and ultimately send the lymph into the main anterior veins close to their entry into the right auricle.

Composition

- Cellular part: Only leukocytes (mostly lymphocytes) (no blood platelets)
- Non-cellular part: water 94% and solids (proteins, fats, carbohydrates, enzymes, antibodies) 6%

Functions

- **Nutritive:** It supplies nutrition and oxygen to those parts where blood cannot reach.
- **Drainage:** It drains away excess tissue fluid and metabolites and returns proteins to the blood from tissue spaces.
- **Absorption:** Fats from the intestine are absorbed through lymphatics (lacteals located in the intestinal.
- **Defence:** Lymphocytes and monocytes of the lymph function to defend the body. The lymphatics also remove bacteria from the tissues. You may have experienced painful swellings in your groins or in the axils of arms when you get a boil or injury in the limbs. This is a protective sign. The lymph nodes tend to localise the infection and prevent it from spreading to the body as a whole. The tonsils on the sides of the neck are also lymph glands.

The Spleen

The spleen is a large lymphatic organ, about the size of a clenched fist, reddish brown in colour and situated in the abdomen, behind the stomach and above the left kidney.

Functions

- It acts as a blood reservoir. In an emergency such as haemorrhage, physical or emotional stress, or in carbon monoxide poisoning, the spleen releases the stored blood into the bloodstream.
- It produces lymphocytes.
- It destroys worn-out red blood cells (sharing this function with the liver).
- In an embryo, it produces RBCs.

SUMMARY

- The organ system that moves materials to and from all of the body's cells is the cardiovascular system. The heart, blood arteries and blood are the cardiovascular system's primary components.
- The heart is a muscular organ in the chest that mostly consists of cardiac muscles. Through repetitive, rhythmic contractions, the cardiac muscles pump blood through the blood arteries. Blood circulates through the four chambers of the heart and is restricted to one direction by valves.
- Blood is transported all over the body by blood vessels. The three primary types of blood vessels are capillaries, veins and arteries, which primarily transport blood away from the heart (which exchanges substances between the blood and cells of the body).
- There are two interrelated circulations in the cardiovascular system. Between the heart and the lungs, where blood is oxygenated, is the pulmonary circuit. Between the heart and the rest of the body, where it distributes oxygen, is the systemic circuit.
- Blood is a viscous connective tissue that travels through blood vessels and throughout the body. It consists of cells including erythrocytes, leukocytes and thrombocytes, as well as a liquid component called plasma that has a variety of dissolved chemicals.
- Blood pressure is an important factor in the circulatory system, and it can be influenced by factors such as heart rate, blood volume and the elasticity of blood vessels.
- The heart is composed of four chambers: the right atrium, the right ventricle, the left atrium and the left ventricle.
- The right side of the heart pumps deoxygenated blood to the lungs, while the left side of the heart pumps oxygenated blood to the rest of the body.
- Diseases and disorders of the circulatory system can have serious health consequences, including heart attacks, strokes and blood clots.

PRACTICE QUESTIONS

Analyse Your Concepts (School Exam Based)

Multiple Choice Questions

1. The nearest organ to which the heart supplies oxygenated blood is:
(a) Lung (b) Stomach
(c) Intestine (d) Heart itself

2. When a doctor is recording your pulse, he is pressing on your wrist exactly on a:
(a) Vein (b) Capillary
(c) Artery (d) Arteriole

3. The valve present between the right atrium and the right ventricle is the:
(a) Tricuspid valve (b) Bicuspid valve
(c) Semi-lunar valve (d) Mitral valve

4. The blood vessel supplying blood to the kidney is the:
(a) Renal vein (b) Renal artery
(c) Dorsal aorta (d) Hepatic vein

5. Angina pectoris is caused by:
(a) Defective nutrition
(b) Inadequate supply of oxygen to the heart muscle
(c) Defective functioning of mitral valve
(d) Infection by a virus

6. The chief function of lymph nodes is to:
(a) Produce WBCS
(b) Produce hormones
(c) Destroy old RBCs
(d) Destroy pathogens

7. Heart sounds are produced due to:
(a) Closure of tricuspid and bicuspid valves
(b) Rushing of blood through valves producing turbulence
(c) Closure of atrioventricular and semilunar valves
(d) Entry of blood into auricles

Match the Following

Each question contains statements given in two columns which have to be matched. Statements in column I have to be matched with statements in column II.

Column I	Column II
(1) SA node	(p) Plasma
(2) Defective haemoglobin in RBC	(q) Serum
(3) Muscle fibres located in heart	(r) Pacemaker
(4) The liquid squeezed out of blood during clotting	(s) Sickle cell anaemia
(5) Never tires, keep on contracting and relaxing	(t) Purkinje fibres
(6) Cardiac cycle	(u) Cardiac muscle
(7) Liquid part of the blood without corpuscles	(v) 0.85 seconds

True or False

Read the following statements and write your answer as true or false.

1. The heart of a normal human adult beats more than one lakh times per day.
2. WBCs contain haemoglobin.
3. RBCs are of several kinds whereas WBCs are of one kind.
4. Leukocytes show amoeboid movement.
5. The average life of red blood cells in our body is about 120 hours.
6. Serum is an immunological preparation of blood (without cells and fibrinogen) containing antibodies and antitoxins against specific diseases.
7. Blood group AB is a universal donor.
8. The walls of auricles are thicker than those of ventricles.

Fill in the Blanks

Complete the following statements with an appropriate word/term to be filled in the blank spaces.

1. The blood vessel that begins and ends in capillaries is the ________
2. A blood vessel which has a small lumen and a thick wall is ________
3. The valve which prevents back flow of blood in the veins and lymph vessels ________.

	Substance	From	To
4.	________	Lungs	Whole body
5.	Carbon dioxide	________	________
6.	Urea	________	________
7.	Digested carbohydrates	Intestine	________
8.	________	________	Target organs

Assertion–Reason Questions

The following questions consist of two statements: Assertion (A) and Reason (R). Answer these questions selecting the appropriate option given below:

(a) Both A and R are true and R is the correct explanation of A
(b) Both A and R are true but R is not the correct explanation of A
(c) A is true but R is false
(d) A is false but R is true

1. Assertion (A): Diapedesis causes WBCs to gather at the site of wounds.
   Reason (R): Leukocytes are being squeezed out of the endothelium, which is the cause.
2. Assertion (A): The majority of animal RBCs are nucleus-free.
   Reason (R): A complex molecule called haemoglobin, which contains iron, gives the entire cytoplasm of red blood cells (RBCs) a red colour.
3. Assertion (A): In the lymphatic system, lymph is known as tissue fluid.
   Reason (R): It comprises plasma proteins, RBCs and WBCs.
4. Assertion (A): In healthy blood arteries, blood clots.
   Reason (R): Heparin, an anticoagulant, is released by healthy blood vessels.
5. Assertion (A): Fibrins are produced by the conversion of inactive fibrinogens in the plasma, in the presence of the enzyme thrombin.
   Reason (R): Plasma without fibrinogen and blood corpuscles is called serum.

One-word Answers

1. Any vein that originates from an organ and travels to an organ other than the heart.
2. The kind of blood vessels which have no muscular walls.

3. An artery that conveys blood that has lost oxygen.
4. The kind of blood cells which can squeeze out through the walls of one category of blood vessels.
5. The smallest common blood vessels formed by the union of capillaries.
6. The blood vessel which starts from capillaries and ends in capillaries.
7. The phase of the cardiac cycle in which the auricles contract.
8. The valve present between the chambers on the right side of the human heart.
9. The phase of the cardiac cycle in which the ventricles are filled with blood from the atrium.
10. The fluid found between the membranes of the heart.

Define the Term

State the chief functional activity of each of the following:

1. Blood platelets
2. Neutrophils
3. Erythrocytes
4. Lymphocytes
5. Bone marrow
6. Venule

Groups of Terms

Given below are certain groups of terms. In each group the first pair indicates a relationship between the two terms. Rewrite and complete the second pair on a similar basis.

1. Lubb: Atrioventricular valves:: Dup: __________
2. Coronary artery: Heart :: Hepatic artery: __________
3. Capillary: Exchange :: Bridge: ____
4. Vein: Return :: Drain: ____
5. Aorta: Main pipeline :: Main road: ____

Identify the Odd One Out

1. (a) Addison's disease (b) Cushing's syndrome (c) Acromegaly (d) Leukemia
2. Thymine, Cytosine, Adenine, Pepsin
3. 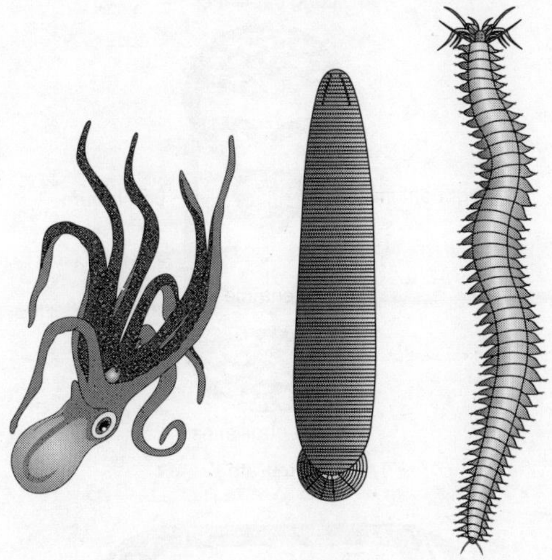
4. Thoracic duct, aorta, pulmonary vein, vena cava

Image-Based Questions

1. Label the parts and answer the following questions

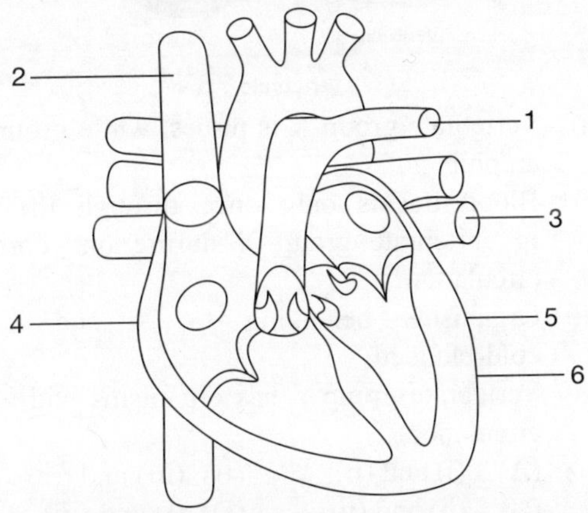

(a) Which type of blood is carried by the blood vessel marked 2?
(b) Name the main artery which takes the blood from the heart to different parts of the body?
(c) Which chamber of the heart receives deoxygenated blood from the body?
(d) Which chamber of the heart receives oxygenated blood from the lungs?

2. The image shows the path that blood takes through the circulatory system in two vertebrate groups X and Y, respectively. Observe the images carefully and identify the correct option:

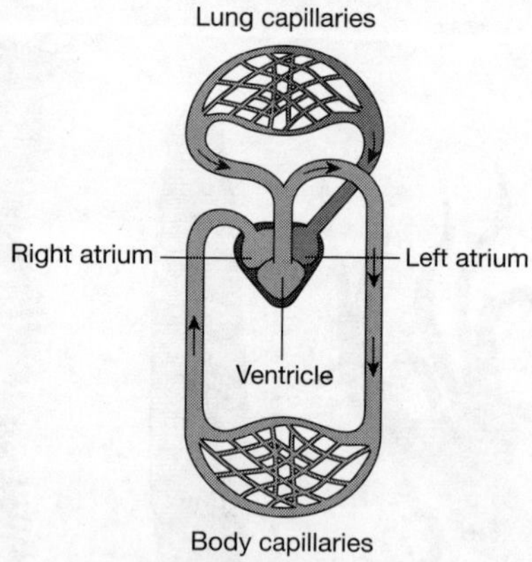

Vertebrate X

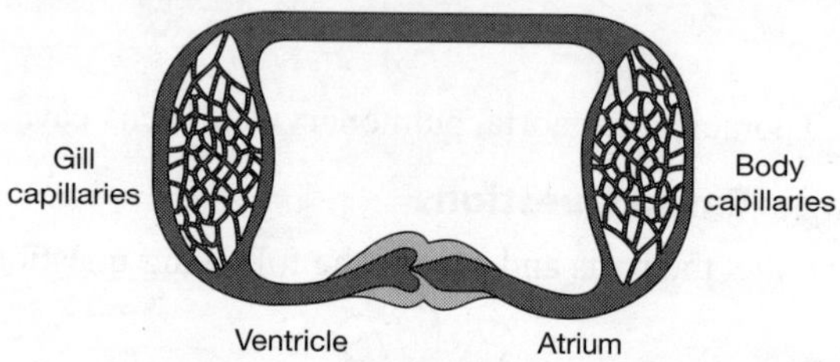

Vertebrate Y

(i) Vertebrate group X is pisces, while group Y is amphibian.
(ii) Blood travels only once through the heart in vertebrate group X during one complete circulation.
(iii) Organisms belonging to X and Y are cold-blooded.
(iv) Vertebrate group Y has organisms with single circulation.

(A) (i) and (ii) (B) (iii) and (iv)
(C) (ii) and (iii) (D) (i) and (iii)

3. Given below are the diagrammatic sketches of two kinds of blood vessels.

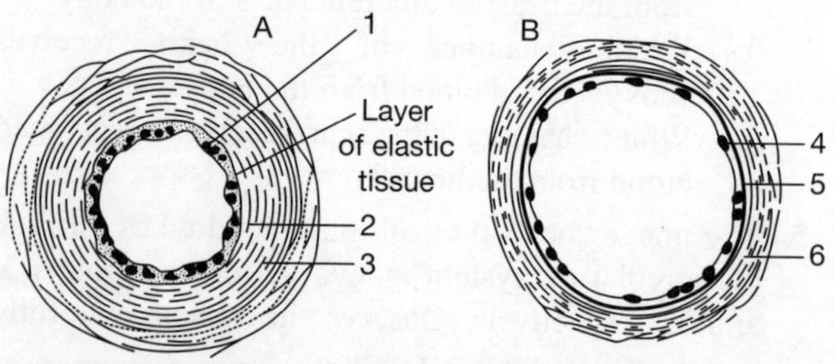

(a) Identify the blood vessels A and B.
(b) Name the parts numbered 1 to 6.
(c) Mention any two main differences between A and B.

4. Given below is a highly schematic diagram of the human blood circulatory system.

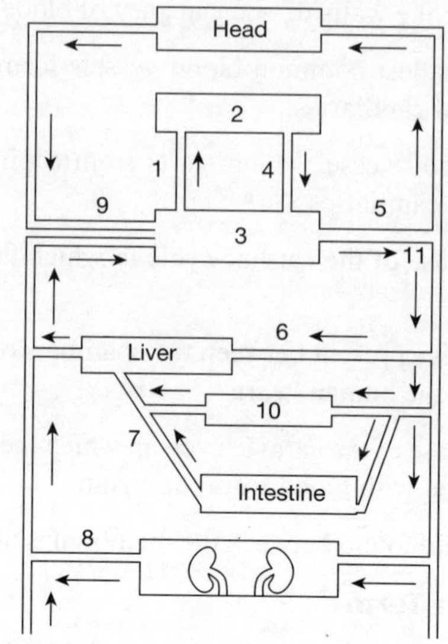

(a) Which part (state the number) represents the heart? Give reasons in support of your answer.
(b) Which numbers represent the following?

Aorta	Hepatic portal vein
Pulmonary artery	Superior vena cava
Renal Vein	Stomach
Dorsal aorta	

5. The diagram below shows part of the capillary bed in an organ of the human body. Some of the blood arriving at the capillaries at points labelled A, moves out into the spaces between the tissue cells. Study the diagram and answer the questions that follow:

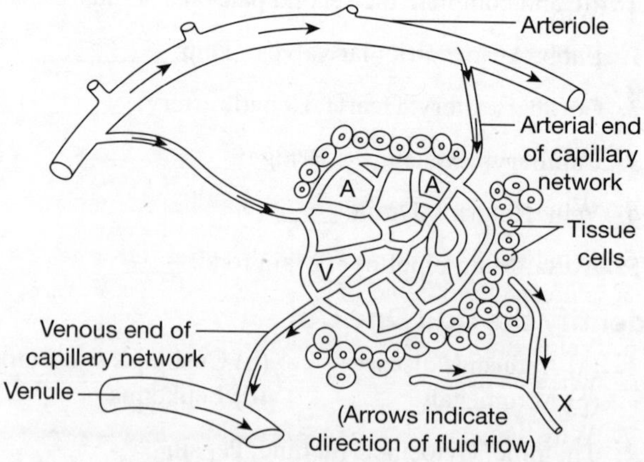

(a) When the liquid from the blood surrounds the cells, what is it called?

(b) Name any one important component of the blood which remains inside the capillaries and fails to move out into the spaces.
(c) Some of the liquid surrounding the cells does not pass directly back into the blood but eventually reaches it by another route through vessel X. Name the fluid present in vessel X.
(d) State two important functions performed in our body by the fluid present in the vessel X.

6. The diagram given alongside represents the human heart in one phase of its activity. Study the same and then answer the questions that follow:

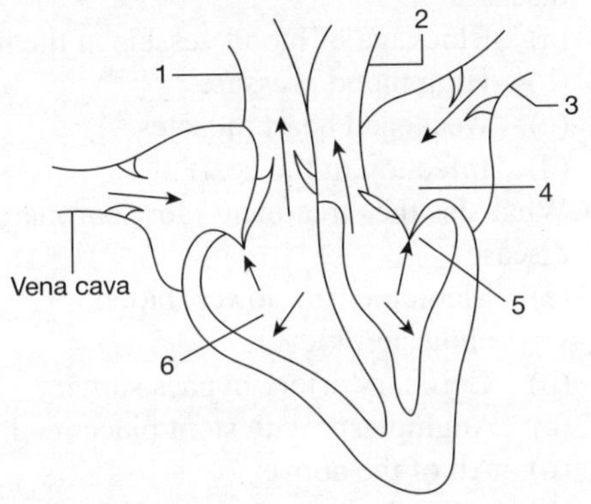

(a) Name the phase.
(b) Which part of the heart is contracting in this phase? Give a reason to support your answer.
(c) Name the parts numbered 1 to 6.
(d) What type of blood flows through the parts marked '1' and '2,' respectively?
(e) How many valves are closed in this phase?

Very Short Answer Questions

1. Why is the circulatory system also known as a transport system?
2. Define diapedesis.
3. Why are capillaries thin walled?
4. What is blood pressure? How is it measured?
5. What is the value of systolic BP and diastolic BP of a normal human adult?
6. Veins have valves at intervals in their inner lining whereas the arteries do not have valves. Explain.
7. Why does the number of leukocytes increase during infection?
8. In what ways does the blood entering the kidney differ from that leaving the kidney?
9. Which are the chambers of the heart that receive blood?
10. Which are the chambers of the heart that pump blood?

Short Answer Questions

1. What does 'double circulation' refer to?
2. Give reasons why a mature mammalian erythrocyte lacks nucleus and mitochondria?
3. How do the tonsils and spleen function?
4. What are three functions of the cardiovascular system?
5. Valves prevent blood from flowing backwards in the cardiovascular system. Why do you think this is important?

Long Answer Questions

1. When are the sounds 'lubb' and 'dup,' respectively, created during a heartbeat?
2. What is the origin of the widespread misconception that the heart is on the left side of the chest?
3. Differentiate between the individuals in each of the following pairs in light of the question in brackets:
   (a) Erythrocytes and leucocytes (function)
   (b) Artery and vein (direction of blood flow)
   (c) Artery and vein (type of blood primarily flowing through)
   (d) Tricuspid and bicuspid valves (location)
4. Describe the process of blood clotting in detail.
5. Explain the role of the heart in the circulatory system.

Case Study-Based Questions

1. When Alex visits the doctor, the doctor informs him that his blood pressure is 135/90 mm Hg. Please respond to the questions below on his blood pressure.
   (i) What is Alex's blood pressure according to the doctor?
   (a) 90/135 mm Hg (b) 135/90 mm Hg
   (c) 90 mm Hg (d) 135 mm Hg
   (ii) What is the normal range for blood pressure in adults?
   (a) 100/60 mm Hg
   (b) 120/80 mm Hg
   (c) 140/90 mm Hg
   (d) 160/100 mm Hg
   (iii) Which of the following factors can contribute to high blood pressure?
   (a) Lack of physical activity
   (b) Excessive salt intake
   (c) Stress
   (d) All of the above

(iv) What is the condition called when blood pressure remains consistently high over time?
(a) Hypotension (b) Hypertension
(c) Atherosclerosis (d) Thrombosis

(v) What are the potential health risks associated with untreated high blood pressure?
(a) Heart attack (b) Stroke
(c) Kidney damage (d) All of the above

2. Ramesh, a 45-year-old man, has been experiencing chest pain and shortness of breath for the past few weeks. He also has a family history of heart disease.

(i) What is the likely cause of Ramesh's symptoms?
(a) Hypertension (b) Atherosclerosis
(c) Thrombosis (d) Hypotension

(ii) How does atherosclerosis lead to chest pain and shortness of breath?
(a) By narrowing the blood vessels that supply oxygen to the heart
(b) By causing blood clots to form in the lungs
(c) By increasing the amount of carbon dioxide in the blood
(d) By decreasing the amount of oxygen in the blood

(iii) What lifestyle changes can Ramesh make to reduce his risk of heart disease?
(a) Eating a healthy diet and exercising regularly
(b) Quitting smoking and reducing alcohol intake
(c) Managing stress and getting enough sleep
(d) All of the above

(iv) If Ramesh's condition is left untreated, what is the potential outcome?
(a) Heart attack (b) Stroke
(c) Kidney damage (d) All of the above

3. Sarah is a 55-year-old woman who has a family history of heart disease. She experiences shortness of breath and chest pain during physical activity. Upon visiting the doctor, she is diagnosed with coronary artery disease.

(i) Which of the following is a risk factor for coronary artery disease?
(a) Lack of physical activity
(b) Family history of heart disease
(c) High intake of fruits and vegetables
(d) All of the above

(ii) What is the main cause of coronary artery disease?
(a) Blockage of blood vessels in the heart
(b) High blood pressure
(c) Weakened heart muscles
(d) Infection in the heart

(iii) What is the treatment for coronary artery disease?
(a) Medicines to lower blood pressure and cholesterol levels
(b) Coronary artery bypass surgery
(c) Angioplasty with stent placement
(d) All of the above

(iv) What are the symptoms of coronary artery disease?
(a) Shortness of breath and chest pain
(b) Nausea and vomiting
(c) Fatigue and weakness
(d) All of the above

(v) How can coronary artery disease be prevented?
(a) Maintaining a healthy weight and diet
(b) Quitting smoking
(c) Regular exercise
(d) All of the above

NCERT Corner

1. Define double circulation. Why is it necessary?
2. Describe the human circulatory system with the help of a neat labelled diagram.
3. Name the components of blood and explain their functions.
4. What is the difference between the transport of materials in xylem and phloem?
5. How are waste products removed from the body?
6. What is blood pressure? Explain its measurement.
7. What is a heart attack? How is it caused?
8. How is the rate and rhythm of heartbeat regulated in the human body?
9. Explain the role of valves in the human heart.
10. What is the difference between arteries and veins?

COMPETITION WINDOW (OBJECTIVE TYPE)

1. Which one of the following is responsible for digesting and absorbing nutrients that are required for transportation within the body?
   (a) Digestive system
   (b) Respiratory system
   (c) Excretory system
   (d) Hormones

2. The respiratory system plays an essential role in the domain of:
   (a) Eliminating body waste
   (b) Digesting nutrients
   (c) Breathing and exchange of gases
   (d) Controlling body processes

3. The main role of the excretory system is to flush out:
   (a) Excessive body waste (b) Nutrients
   (c) Fluids (d) All of the above

4. Hormones performs a significant function in:
   (a) Breathing process
   b) Controlling body mechanism
   (c) Managing body composition
   (d) Absorbing nutrients

5. Blood and lymph are two circulating fluids that contributes in the process of:
   (a) Absorption (b) Elimination
   (c) Transportation (d) Stagnation

6. The system of blood circulation in our body is called:
   (a) Open blood circulatory system
   (b) Partially closed circulatory system
   (c) Closed blood circulatory system
   (d) Partially closed blood circulatory system

7. Which of the following fluid(s) in our body does not depict the ability to circulate?
   (a) Tissue fluid
   (b) Blood
   (c) Lymph
   (d) Synovial fluids and vitreous humour

8. In comparison to red blood cells, white blood cells do not possess:
   (a) Nucleus (b) Protein
   (c) Haemoglobin (d) None of the above

9. A drastic reduction in the quantity of white blood cells is termed:
   (a) Leukemia (b) Anaemia
   (c) Lymphomas (d) Leukopenia

10. The process through which most white blood cells, and particularly neutrophils, exhibit the capacity to engulf germs is known as:
    (a) Pinocytosis (b) Phagocytosis
    (c) Endocytosis (d) Exocytosis

11. Enhanced local heat, swelling, pain and redness are some of the features of:
    (a) Inflamed spot (b) Allergy
    (c) Tomour (d) Infection

12. The entrance of disease-causing germs into the blood vessels might lead to the formation of poisonous substances which is called:
    (a) Enzymes (b) Virus
    (c) Toxins (d) Anti-toxins

13. Having inadequate blood clotting proteins, a hereditary disease that prevents blood from properly clotting, is known as:
    (a) Thalassemia (b) Sickle cell anaemia
    (c) Colour blindness (d) Haemophilia

14. Who played an important part in discovering the concept of blood grouping ?
    (a) Alister Hardy (b) Karl Landsteiner
    (c) Carl Linnaeus (d) Erik Acharius

15. Two different types of antigens are found in the majority of human beings. Which two types are they?
    (a) Antigen A and Antigen B
    (b) Antigen C and Antigen D
    (c) Antigen E and Antigen F
    (d) None of the above

16. The term 'antibodies' literally means:
    (a) Lack of complementary proteins in plasma of the blood
    (b) Plasma of the blood
    (c) Complementary proteins contained in the plasma of the blood
    (d) Complementary proteins contained in the lymph

17. The circulatory system is primarily comprised of:
(a) Heart
(b) Blood
(c) Blood vessels
(d) All of the above

18. ABO system and Rh system are considered as two of the most significant mechanisms of:
(a) Blood clotting
(b) Blood transfusion
(c) Blood genotype
(d) Blood grouping

19. Non-granular WBCs are:
(a) Lymphocytes and monocytes
(b) Lymphocytes and basophils
(c) Eosinophils and basophils
(d) Eosinophils and monocytes

20. The nearest organ to which the heart supplies oxygenated blood is the:
(a) Lung (b) Stomach
(c) Intestine (d) Heart itself

21. The chief function of lymph nodes is to:
(a) Produce WBCs
(b) Produce hormones
(c) Destroy old RBCs
(d) Destroy pathogens

22. Heart sounds are produced due to:
(a) Closure of tricuspid and bicuspid valves
(b) Rushing of blood through valves producing turbulence
(c) Closure of atrioventricular and semilunar valves
(d) Entry of blood into auricles

Answer Keys

Multiple Choice Questions

1. (d) **2.** (c) **3.** (a) **4.** (b) **5.** (b)
6. (d) **7.** (c)

Match the Following

Column I	Column II
(1) SA node	(p) Pacemaker
(2) Defective haemoglobin in RBC	(q) Sickle cell anaemia
(3) Muscle fibres located in the heart	(r) Purkinje fibres
(4) The liquid squeezed out of blood during clotting	(s) Serum
(5) Never tires, keep on contracting and relaxing	(t) Cardiac muscles
(6) Cardiac cycle	(u) 0.85 seconds
(7) Liquid part of the blood without corpuscles	(v) Plasma

True or False

1. T **2.** F **3.** F **4.** T **5.** F
6. T **7.** F **8.** F

Fill in the Blanks

1. Portal veins 2. Arteries
3. Semi-lunar valves 4. Oxygen
5. Whole body, Lungs 6. Liver, Kidney
7. Liver
8. Hormones, Endocrine glands

Assertion–Reason Questions

1. (b) **2.** (b) **3.** (c) **4.** (d) **5.** (c)

One-word Answers

1. Portal vein 2. Capillaries
3. Pulmonary artery 4. Leukocytes
5. Venules 6. Arterioles
7. Atrial systole 8. Tricuspid valve
9. Ventricular diastole 10. Pericardial fluid

Define the Term

1. Blood platelets, also called thrombocytes, are small, irregularly shaped cell fragments that are produced in the bone marrow. Platelets play a critical role in the blood clotting process, which helps to prevent excessive bleeding in response to injury or damage to blood vessels.

2. Neutrophils are a type of white blood cell, also known as leukocytes, that play a critical role in the immune system's response to infection and inflammation. Neutrophils are produced in the bone marrow and circulate in the bloodstream, but they are also found in other tissues, such as the lungs and the gastrointestinal tract.
3. Erythrocytes, also known as red blood cells, are the most common type of blood cell in the human body. Their main function is to transport oxygen from the lungs to the tissues and to carry carbon dioxide from the tissues to the lungs for elimination. Erythrocytes are produced in the bone marrow and circulate in the bloodstream for about 120 days before being removed from circulation and replaced with new cells.
4. Lymphocytes are a type of white blood cell, also known as leukocytes, that play a critical role in the immune system's response to infection and disease. Lymphocytes are produced in the bone marrow and mature in the thymus gland (in the case of T lymphocytes) or in the bone marrow itself (in the case of B lymphocytes). They circulate in the bloodstream and lymphatic system and are found in various tissues, such as the spleen, lymph nodes, and tonsils.
5. Bone marrow is a soft, spongy tissue that is found in the centre of most bones in the body, such as the hips, ribs, sternum, and long bones of the arms and legs. It is responsible for producing blood cells, including red blood cells, white blood cells and platelets.
6. A venule is a small blood vessel that allows blood to flow from capillaries to larger veins in the body. Venules are part of the circulatory system and are responsible for transporting deoxygenated blood back to the heart.

Groups of Terms

1. Semilunar valves
2. Liver
3. Connection
4. Flow
5. Traffic flow

Identify the Odd One Out

1. (d)
2. Pepsin
3. Leech
4. Thoracic duct

Image-Based Questions

1. (a) 1. Left pulmonary artery; 2. Superior vena cava; 3. Left pulmonary vein; 4. Right auricle; 5. Left auricle; 6. Left ventricle

   (b) Deoxygenated blood is carried by the blood vessel marked 2.

   (c) Aortic arc (Aorta) is the main artery which carries oxygenated blood from heart to different parts of the body.

   (d) Right atrium in the chamber of the heart receives deoxygenated blood from the body.
2. (b)
3. (a) A represents Artery, B represents Vein

   (b)

   1 represents the external layer made of connective tissue

   2 represents the lumen

   3 represents the middle layer of smooth muscles and elastic fibres

   4 represents the endothelium

   5 represents tunica media

   6 represents tunica externa

   (c)

   Arteries (A)

   1 Progressively branched, decreasing in size.

   2 Smallest artery breaks into arterioles.

   3 Have thick and more muscular walls.

   Veins (B)

   1 Progressively unite, increasing in size.

   2 Smallest vein arises from venules.

   3 Have thin and less muscular walls.
4. (a) Structure 3 represents the heart. It forms the centre of double circulation. Also blood circulation (indicated by 1) begins with flow from the heart to the lungs.

   (b) Aorta 5

   Hepatic portal vein 7

   Pulmonary artery 1

   Superior vena cava 9

   Renal vein 8

   Stomach 10

   Dorsal aorta 11
5. (a) The liquid that surrounds the cells is called tissue fluid.

   (b) The important component of blood which remains inside the capillaries and fails to move out into the spaces is red blood cells.

   (c) Lymph is the fluid that is present in vessel X.

   (d) (i) Nutritive: Supplies nutrition to those parts where blood cannot reach.

(ii) Drainage: It drains away excess tissue fluid and metabolites and returns proteins to the blood from tissue spaces.

(iii) Absorption: Fats from the intestine are absorbed through lymphatics.

6. (a) The given diagram represents ventricular systole.

(b) Both the heart ventricles are contracting in this phase since the semilunar valves are open.

(c) (i) Pulmonary artery

(ii) Aorta

(iii) Pulmonary vein

(iv) Left atria

(v) Semilunar valves

(vi) Right ventricle

(d) Deoxygenated and oxygenated blood flows through the pulmonary artery and pulmonary vein, respectively.

(e) Bicuspid and tricuspid valves are closed in this phase to prevent backflow of blood into the atria.

Very Short Answer Questions

1. Circulatory system is also called transport system because it transports food, water, hormones, enzymes, electrolytes, antibodies and respiratory gases to or away from the body tissues.
2. The process in which the leukocytes or white blood cells squeeze out through the walls of blood capillaries at the site of injury to fight against pathogens.
3. The capillaries are thin walled, because they help in the exchange of gases and diffusion of materials into the cells. This diffusion is possible due to the thinness of its walls.
4. The blood flowing through the arteries exert some pressure on its walls. This pressure is known as blood pressure. It is measured by an instrument named as sphygmomanometer.
5. Systolic BP is 100–140 mmHg and diastolic BP is 60–80 mmHg.
6. In arteries, blood flows with pressure but in veins, the pressure falls considerably and backflow of blood is prevented by the semilunar valves.
7. The number of leukocytes increases during infection to fulfill the defence demand of the body.
8. The blood entering the kidneys contains excretory substances such as urea, uric acid, etc. These substances are removed from the blood in the kidneys and the blood is free from the waste products of metabolism when it leaves the kidneys.
9. The right and left atria
10. The right and left ventricles

Short Answer Questions

1. 'Double circulation' refers to the two separate circulatory loops that are present in mammals and birds. One loop is responsible for carrying oxygen-poor blood from the heart to the lungs (pulmonary circulation), where it is oxygenated, and then returning oxygen-rich blood to the heart. The second loop is responsible for carrying the oxygen-rich blood from the heart to the body's tissues (systemic circulation) and then returning oxygen-poor blood to the heart.
2. A mature mammalian erythrocyte, or red blood cell, lacks a nucleus and mitochondria for a few key reasons:

   Nucleus: The nucleus is responsible for controlling gene expression and directing cellular functions, but in the case of erythrocytes, this is not necessary. The primary function of red blood cells is to transport oxygen from the lungs to the body's tissues, and having a nucleus would take up valuable space that could be used to carry more haemoglobin, the protein that actually binds to oxygen.

   Mitochondria: Mitochondria are organelles responsible for producing energy in the form of ATP. However, erythrocytes primarily rely on anaerobic metabolism to produce ATP since they do not have access to oxygen while in circulation. Additionally, the lack of mitochondria allows erythrocytes to maintain their characteristic shape, which is crucial for their function in circulation.
3. The tonsils are a pair of lymphatic tissue masses located at the back of the throat. They act as a first line of defense against infections that enter the body through the nose and mouth. The tonsils contain immune cells, such as lymphocytes and macrophages, that can trap and destroy bacteria and viruses that enter the body. The tonsils can become inflamed and swollen when they are fighting off an infection, which can result in a sore throat while the spleen is an organ located in the upper left part of the abdomen. It is responsible for filtering blood and removing old or damaged red blood cells and platelets. The spleen also plays a role in fighting infections by producing and storing immune cells, such as lymphocytes and monocytes. In addition, the spleen can help to remove foreign substances, such as bacteria and viruses, from the bloodstream.

4. The three functions of the cardiovascular system are to transport oxygen and nutrients to cells in the body, remove waste products and defend the body against infection.
5. The body's cells require the transportation of oxygen and nutrients, as well as the removal of carbon dioxide and other wastes, through the cardiovascular system. For this to function, there must be a one-way flow of blood from the heart to the body's cells and back again. Therefore, it is crucial to avoid reverse flow because, if it did, deoxygenated blood would stay close to the body's cells rather than going ahead to get oxygenated again.

Long Answer Questions

1. Hint: "Lubb" is created at the beginning of ventricular contraction (systole), while "dup" is created at the beginning of ventricular relaxation (diastole). Elaborate on this.
2. Hint:The misconception that the heart is on the left side of the chest is due to the fact that the left ventricle, which is the main pumping chamber of the heart, is positioned slightly more towards the left side.Elaborate on this.
3. Hints:
   (a) Erythrocytes function in carrying oxygen and carbon dioxide, while leucocytes function in immune response.
   (b) Arteries carry blood away from the heart, while veins carry blood back to the heart.
   (c) Arteries primarily carry oxygenated blood, while veins primarily carry deoxygenated blood.
   (d) The tricuspid valve is located between the right atrium and right ventricle, while the bicuspid (mitral) valve is located between the left atrium and left ventricle.

   Elaborate on these points.
4. Hint: Blood clotting, or coagulation, is a complex process involving platelets, clotting factors and fibrin. In response to injury, platelets aggregate at the site and release chemicals that initiate a cascade of reactions, leading to the formation of fibrin strands. These strands create a meshwork that traps blood cells and forms a clot, preventing excessive bleeding. Elaborate on these points.
5. The heart acts as a pump in the circulatory system, pumping oxygenated blood to the body's tissues and organs through arteries. It also receives deoxygenated blood from the body's tissues and pumps it to the lungs for oxygenation through veins. Elaborate on this.

Case Study-Based Questions

1. (i) b, (ii) b, (iii) d, (iv) b, (v) d
2. (i) b, (ii) a, (iii) d, (iv) d
3. (i) b, (ii) a (iii) d, (iv) a (v) d

NCERT Corner

1. Double circulation is a type of circulation in which the blood passes through the heart twice in one complete cycle. It is necessary because it ensures efficient supply of oxygenated blood to all parts of the body. The first circulation (pulmonary circulation) carries deoxygenated blood from the heart to the lungs, where it is oxygenated, and then brings oxygenated blood back to the heart. The second circulation (systemic circulation) carries oxygenated blood from the heart to the rest of the body and brings deoxygenated blood back to the heart.
2. The human circulatory system consists of the heart, blood vessels, and blood. Here is a labelled diagram:

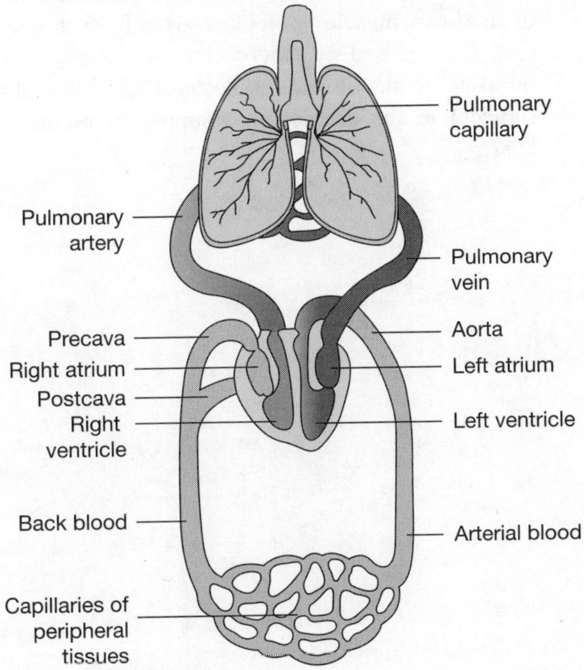

3. Blood consists of plasma, red blood cells, white blood cells, and platelets. Plasma carries nutrients, waste products, hormones, and proteins. Red blood cells carry oxygen from the lungs to the body's tissues and remove carbon dioxide from the body. White blood

cells fight infection and diseases. Platelets help in blood clotting to prevent excessive bleeding.

4. Xylem transports water and minerals from the roots to the leaves through the process of transpiration. Phloem transports food (sugar) from the leaves to the rest of the plant through the process of photosynthesis.
5. Waste products are removed from the body through the excretory system, which includes the kidneys, lungs, skin and liver. The kidneys filter waste products from the blood and excrete them in the form of urine. The lungs remove carbon dioxide from the body through exhalation. The skin excretes sweat, which contains waste products. The liver breaks down waste products and excretes them in the form of bile.
6. Blood pressure is the force exerted by the blood against the walls of the blood vessels. It is measured using a sphygmomanometer. The measurement consists of two numbers – the systolic pressure (when the heart contracts) and the diastolic pressure (when the heart relaxes).
7. A heart attack occurs when the blood flow to a part of the heart muscle is blocked, usually by a blood clot. It is caused by atherosclerosis, which is the build-up of plaque (fat, cholesterol and other substances) in the arteries that supply blood to the heart.
8. The rate and rhythm of heartbeat are regulated by the sinoatrial (SA) node, which is the natural pacemaker of the heart. The SA node sends electrical signals that cause the heart to contract and relax in a coordinated manner.
9. Valves in the human heart prevent the backflow of blood. The tricuspid valve and the mitral valve separate the atria (upper chambers) from the ventricles (lower chambers). The pulmonary valve and the aortic valve separate the right ventricle from the pulmonary artery and the left ventricle from the aorta, respectively.
10. Arteries carry oxygenated blood away from the heart to the rest of the body. They have thick, muscular walls to withstand the high pressure of blood flow. Veins carry deoxygenated blood back to the heart from the body's tissues. They have thinner walls and valves to prevent backflow of blood

Competition Window

1. (a)	**2.** (c)	**3.** (a)	**4.** (b)	**5.** (c)
6. (a)	**7.** (d)	**8.** (a)	**9.** (d)	**10.** (b)
11. (a)	**12.** (c)	**13.** (d)	**14.** (b)	**15.** (a)
16. (c)	**17.** (d)	**18.** (d)	**19.** (a)	**20.** (d)
21. (d)	**22.** (a)			

CHAPTER

11 The Excretory System

PREREQUISITES

You must have knowledge of:

- Structure and functions of major organs and systems in the human body
- The concept of homeostasis, which refers to the body's ability to maintain a stable internal environment

LEARNING OBJECTIVES

At the end of this chapter, you will be able to:

- Explain the functions and components of the excretory system
- Describe the role of the kidneys in filtering waste products from blood and producing urine
- Describe the mechanisms of urine storage and elimination, including the role of the bladder and urethra
- Explain the importance of a healthy excretory system for overall health and well-being

INTRODUCTION

The body's metabolic processes result in the formation of a sizable number of waste products. The break-down of proteins, lipids and carbohydrates releases significant volumes of CO_2 and H_2O. Proteins and other complex nitrogenous substances are the starting point for the formation of nitrogenous wastes such ammonia, urea and uric acid. These nitrogenous waste products are expelled with the aid of the excretory system because they become toxic or hazardous if retained in the body.

Excretion is the process through which chemical wastes, primarily nitrogenous ones, are eliminated from the body. Excretion is crucial for keeping the organism in a homeostatic (steady state) condition. The excretory system is made up of organs that are involved in the production, retention and removal of urine.

Excretion should not be confused with defecation, which is the act of releasing faeces, or the undigested remains of food from the intestine. Similarly, exhaling CO_2 through the lungs is a component of respiration, not excretion. Instead of the excretory system, the term urinary system may be more applicable for humans.

A variety of chemicals that are regularly generated in our bodies or ingested through food must be eliminated in order for them not to become hazardous. These include the following:

- **Carbon dioxide and water:** Every living organism releases energy through the oxidation of glucose, which results in the creation of carbon dioxide and water. The lungs are responsible for eliminating carbon dioxide. The water combines with the body's other water molecules.
- **Nitrogenous metabolic waste:** These include ammonia, urea and uric acid. These are mostly created in the liver from the dead protein remnants of other tissues. Amino acids, which are byproducts of protein

breakdown, cannot be stored by the body in excess. They are broken down in the liver to create excretable urea and glucose that can be used and stored.

- **Salts and vitamins:** It is necessary to dispose of extra salts such as common salt (NaCl) and some extra water-soluble vitamins (B and C). The kidneys are the principal organs that secrete salt.
- **Water:** With food and beverages, substantial amounts of water are consumed. The extra water is eliminated, which also serves the practical purpose of dissolving the dangerous substances and transporting them out.
- **Bile:** The break-down products of the haemoglobin of deceased RBCs include bile pigments, primarily yellow bilirubin. This is drawn out of the bloodstream by the liver cells, that then secrete it into bile juice, which is emptied into the duodenum through the common bile duct. These pigments undergo intestinal modification to become the pigments that give faeces their characteristic yellowish-brown hue. Urine contains some of these colours that are eliminated.

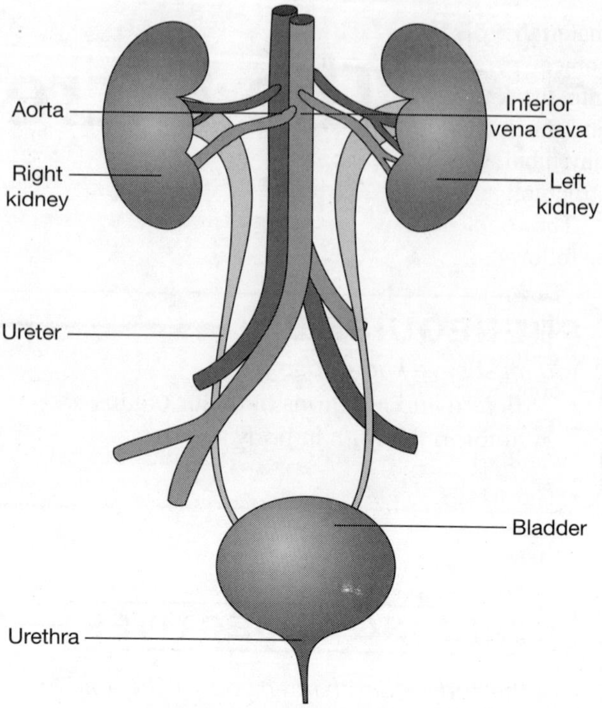

Fig. 11.1 The urinary system

THE EXCRETORY ORGANS

The main excretory organs, the kidneys, release excretory products, primarily urea, in the form of urine. Sweat gland excretion is accidental. The main function of these glands is to regulate body temperature (thermoregulation). Nitrogenous wastes are carried in trace amounts by the sweat produced. Actually, sweat glands are not excretory because they only release perspiration when it is necessary for cooling. Carbon dioxide is excreted by the lungs and discharged in the exhaled air.

The Kidneys (Urinary System)

The kidneys (Fig. 11.1) are two bean-shaped organs that are roughly 10 cm long and 6 cm wide. They are situated on either side of the backbone and are shielded from the surrounding tissue by the last two ribs. The kidney on the right side is a little lower than the one on the left. The median surface of each kidney has a notch (hilum) from which a tube called the ureter emerges.

The ureter links behind with the urine bladder, which is located in the lower region of the abdomen. The kidney is termed the pelvis, and the front end of the ureter is somewhat enlarged into it. Continuously passing through the ureters and accumulating in the urinary bladder is the urine produced by the kidneys. The urinary bladder periodically empties its contents into the external environment through the urethra. When the bladder contracts to empty, the apertures of the ureters (into the bladder) operate as valves to stop the urine from flowing backwards. A sphincter, which is a circular muscle, protects the passage of the urine into the urethra and only relaxes during micturition in response to brain stimulus.

Internal Structure

An outer dark cortex and an interior lighter medulla are the two primary sections visible in a longitudinal slice of the kidney (Fig. 11.2). The medulla is made up of a

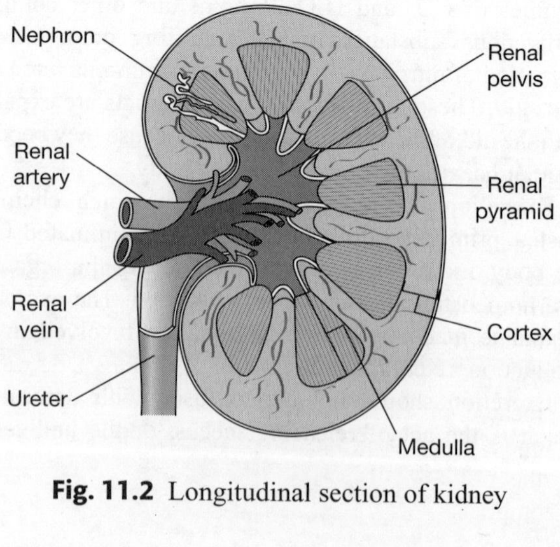

Fig. 11.2 Longitudinal section of kidney

material with fine stripes that is organised into many conical pyramids. Each pyramid's tip (papilla) extends into the kidney's pelvis. Uriniferous tubules, nephrons and renal or kidney tubules are only a few of the many tiny tubules that make up the kidney. These are both the structural and operational components of the kidney.

The components of each kidney tubule (Fig. 11.3) are as follows:

- **Bowman's capsule:** It resembles a hollow ball that has been squeezed on one side and has a thin-walled (single-cell-thick epithelium) cup. The tubule continues to a hollow interior region.
- **The glomerulus**, or knot-like collection of blood capillaries, is lodged in the cup's outer concavity.
- The **Malpighian capsule**, or renal capsule, is the term used to describe Bowman's capsule and the glomerulus collectively.

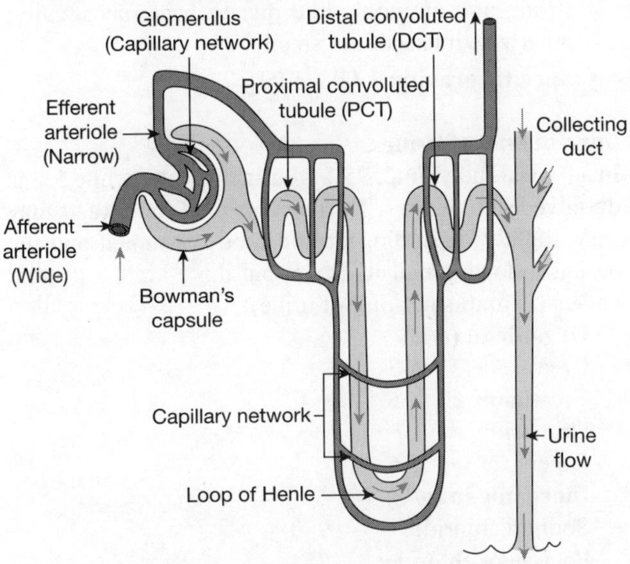

Fig. 11.3 Structure of a kidney tubule

- The first convoluted section of the tubule is called the **proximal convoluted tubule (PCT)**. In the slice, the cortex appears dotted because Bowman's capsule and the proximal convoluted region are both located here. The middle portion of the U-shape (the **loop of Henle**) looks like a hairpin and is not convoluted. It travels into the medulla before returning to the cortex to continue into the tangled section of the tubule.
- The **distal convoluted tubule (DCT)**, which is also located in the cortex, is where the kidney tubule ends. It opens into a **collecting duct**, which gathers the urine produced by several kidney tubules and empties it into the kidney's pelvis.

There are approximately 2 million uriniferous tubules (in both kidneys put together). Each tubule is 4–5 cm long. The total length of all the tubules is more than 60 km. This great length provides a huge surface for reabsorption of usable substances, especially water, as the contents move through them. About 1 litre of blood flows through the kidneys per minute. The glomerular filtrate produced in 24 hours is 160 litres, while the urine produced from the glomerular filtrate after reabsorption is 1.2 litre per day.

Blood Supply to the Kidney Tubules

The dorsal aorta gives rise to two renal arteries, each of which enters one kidney. Each renal artery divides and redivides multiple times to produce arterioles, and each of these arterioles enters Bowman's capsule as an afferent arteriole. This divides into several capillaries, forming a mass that resembles a knot and fits snugly inside Bowman's capsule. The efferent arteriole is made up of the glomerulus' reconnecting capillaries. After leaving Bowman's capsule, the efferent arteriole travels a short distance before fragmenting into the secondary vasa recta capillary network that surrounds the renal tubule and reunites to form a vein. The renal vein, which exits the kidney at the median surface and empties blood into the posterior vena cava, is ultimately formed by joining repeatedly with other kidney veins. All the blood in the body passes through the kidneys about 350–400 times a day.

Urine

The formation of urine occurs in two main steps:

1. **Ultrafiltration:** When blood passes through the glomerulus, greater pressure than that found in other capillaries is applied to it. This is because the efferent (outgoing) arteriole is smaller than the afferent (incoming) arteriole. The liquid portion of blood is forced into the renal tubule by this high pressure (hydrostatic pressure). Ultrafiltration is the term used to describe filtration under extremely high pressure. The plasma, together with most of the blood's organic and inorganic components, such as urea, glucose and amino acids, exits the glomerulus during ultrafiltration and enters Bowman capsule's cavity, which resembles a funnel. The term **glomerular filtrate** refers to the liquid that enters the renal tubule. Water, urea, salts, glucose and other plasma solutes are present in the glomerular filtrate. The two types of corpuscles, proteins and other big molecules that make up the thicker portion of the blood, still present in the glomerulus after ultrafiltration are transported by the efferent arteriole.

As a result, the blood leaving the glomerulus is fairly thick (Fig. 11.4).

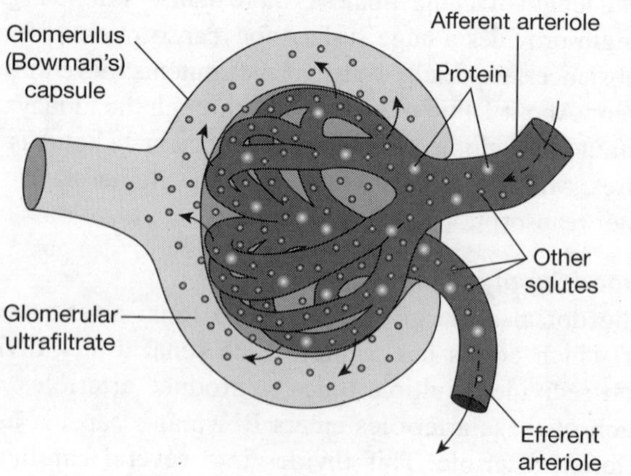

Fig. 11.4 Glomerular filtration

2. **Reabsorption:** The substance that passes through the glomerulus and into the renal tubule is not yet urine. Glucose and other salts, such as sodium, are among the many useful substances present in this very diluted solution. Quite a bit of water is reabsorbed along with the useful materials when the filtrate travels down the tubule. The extent of their reabsorption, however, depends on how little the blood's regular concentration is impacted. This is called **selective absorption**. The DCT is where many foreign compounds, including medicines like penicillin, as well as some naturally occurring molecules like potassium (K^+), are transported into the developing urine. The term **tubular secretion** refers to this route because it involves the activity of the cells that line the tubular wall.

Steps in Urine Formation

1. Glomerulus: Ultrafiltration.
2. Bowman's capsule: Receives the glomerular filtrate.
3. Proximal convoluted tubule: Reabsorbs most of the water (about two-thirds) and much of the glucose and sodium and chloride ions.
4. Loop of Henle: Absorbs some water and sodium ions
5. Distal convoluted tubule: Reabsorbs the remaining chlorides and some water. The walls secrete potassium and foreign chemicals such as penicillin and other drugs into the urine that is formed.

Excretion of Urine

Urine travels into the collecting ducts in the pelvis and through the ureter into the urinary bladder by ureteral peristalsis (waves of constriction in the ureters) and gravity. When the sphincter muscles at the urinary bladder's opening into the urethra relax in response to a nervous system impulse, urine is discharged from the urinary bladder into the urethra (in the penis in men and directly in females). This is known as **micturition**. The energy required for each of these processes that contribute to the production of urine makes the kidneys' oxygen requirement six to seven times greater than that of muscles.

Physical Properties of Urine

- **Colour**: Clear yellow (due to urochrome); colour varies with the diet.
- **Volume:** 1–1.5 litres per day; pH: 5–8, that is, usually it is slightly acidic (pH = 6). A protein diet makes it more acidic while a vegetable diet makes it more alkaline.
- **Odour**: On standing, the smell of urine becomes strong and ammonia-like due to bacterial activity; otherwise it has a faint smell.
- **Specific gravity:** 1.003–1.035

Constituents of Urine

In a typical individual, 95% of urine is water while 5% is dissolved solid wastes. The percentage of solid waste may vary slightly depending on the food consumed and the period following consumption, but these are the general values (in grams per litre of urine):

Organic in (g/L)

Urea:	2–3
Creatinine:	1–5
Uric acid:	0.7
Others:	2–6

Inorganic in (g/L)

Sodium chloride:	9–0
Potassium chloride:	25
Ammonia:	0–6
Others:	2–5

In addition to the typical components, urine may also expel some hormones and medicines including antibiotics and extra vitamins.

Abnormal Constituents of Urine

- **Blood cells/Haematuria**: Here, blood passes with urine, due to infection in the urinary tract, kidney stone or tumour.
- **Glucose/Glycosuria**: Excess glucose passes with urine due to diabetes mellitus (sugar diabetes).

Taste of Urine: Urine has a salty flavour. When it is sweet, it may be due to diabetes. When it is tasteless, it is a symptom of diabetes insipidus, a disorder marked by abundance of water in the urine (due to insufficient antidiuretic hormone (ADH)).

If albumin is present, it may be due to high blood pressure or increased permeability of the Bowman's capsule membrane on account of bacterial infections. Bile pigments may be seen in the case of anaemia, hepatitis (jaundice) or liver cirrhosis.

Regulation of Urine Output

Diuresis (increased urine output) and ADH, which is secreted by the posterior lobe of the pituitary gland, regulates how concentrated the urine becomes as a result of water reabsorption. Diuresis is the increased output of urine that results from a decrease in ADH secretion. Diuretics are substances that stimulate the production of urine, such as liquid diets, coffee and alcohol.

Gout and Kidney Stones

Due to uric acid's relative lack of water solubility, it has the potential to crystallise and accumulate in the joints, resulting in gout. When the kidneys are not functioning correctly, excessive uric acid and certain salts, such as calcium oxalate, may be the cause of kidney stones.

Osmoregulation

The kidney controls the blood's composition, or the amount of water and salts, in addition to filtering out waste such as urea. The term osmoregulation refers to a process that controls the blood's osmotic pressure. Getting adequate water from food or direct consumption aids the kidneys in performing as they should. We drink a lot of water in the summer in tropical countries. However, we urinate less frequently in the summer than in the winter and the urine is typically heavier. This is because during summer, we sweat a lot and the kidneys have to reabsorb more water from the urine, which causes the urine to become more concentrated.

A patient with cholera experiences watery bowels and vomiting. His intestines are unable to allow water into the blood. As a result, his kidneys reabsorb practically all of the water from the urine in the renal tubules, along with urea. In the end, the patient can suffocate from poisoning brought on by the build-up of large amounts of urea (uraemia). Restoring water to the blood with a glucose-saline drip or oral rehydration solution (ORS) is the initial treatment.

Artificial Kidney

If one kidney is damaged or removed for any reason, the other kidney can meet all the excretory requirements of the body, allowing the individual to continue living a normal life. However, if both kidneys fail, it is fatal. A dialysis machine is an artificial kidney that draws blood from the patient's radial artery, filters out urea and extra salts, and then returns the cleansed blood to a vein in the same arm. Dialysis should be carried out twice a week for around twelve hours in cases where the kidneys have suffered permanent damage.

SUMMARY

- Excretion is the process by which the body expels waste and extra water. It is an integral part of all living beings and a key mechanism through which the human body maintains homeostasis.
- The excretory system includes the kidneys, ureters, bladder and urethra.
- The paired kidneys are frequently regarded as the primary excretory organs. Their main function is to produce urine, which serves as a means of removing extra water and waste from the bloodstream.
- The bladder stores urine until it is eliminated from the body through the urethra.
- Sweat glands on the skin contribute to excretion by producing sweat. A tiny amount of urea, produced by protein catabolism, as well as extra water and salt are expelled through sweating.
- Gaseous wastes, principally carbon dioxide from cellular respiration in cells throughout the body, are expelled by the lungs.
- Disorders of the excretory system, such as kidney failure and urinary tract infections, can have serious consequences for overall health.

PRACTICE QUESTIONS

Analyse Your Concepts (School Exam Based)

Multiple Choice Questions

1. Maximum amount of water from the glomerular filtrate is reabsorbed in the:
   (a) Proximal convoluted tubule
   (b) Descending limb of loop of Henle
   (c) Ascending limb of loop of Henle
   (d) Distal convoluted tubule
2. Excretion primarily involves:
   (a) Removal of all byproducts during catabolism
   (b) Removal of byproducts during anabolism
   (c) Removal of nitrogenous wastes
   (d) Removal of excess water
3. Which one of the following is not an excretory activity?
   (a) Giving out carbon dioxide
   (b) Passing out faecal matter
   (c) Sweating
   (d) Removal of urea
4. In humans, urea is formed in the:
   (a) Ureter (b) Liver
   (c) Spleen (d) Kidney
5. What is the primary function of the excretory system?
   (a) To transport oxygen to the body's cells
   (b) To eliminate waste products from the body
   (c) To produce hormones that regulate body functions
   (d) To maintain the body's temperature
6. Which of the following is an organ of the excretory system?
   (a) Liver (b) Spleen
   (c) Lungs (d) Kidneys
7. What is the waste product that is eliminated by the excretory system?
   (a) Carbon dioxide (b) Oxygen
   (c) Urea (d) Glucose
8. What is the tube that carries urine from the kidney to the bladder called?
   (a) Urethra (b) Ureter
   (c) Nephron (d) Glomerulus
9. Which of the following is a common disorder of the excretory system?
   (a) Asthma (b) Diabetes
   (c) Cystitis (d) Meningitis

Match the Following

Each question contains statements given in two columns which have to be matched. Statements in column I have to be matched with statements in column II.

1.

Column I	Column II
(1) Ultrafiltration	(a) Henle's loop
(2) Concentration of urine	(b) Ureter
(3) Transport of urine	(c) Urinary bladder
(4) Storage of urine	(d) Malpighian corpuscle
(5) Urination	(e) Urethra

2.

Column I	Column II
(1) Kidneys	(a) Removal of solid waste
(2) Lungs	(b) Removal of excess water
(3) Skin	(c) Removal of carbon dioxide
(4) Liver	(d) Removal of urea and other nitrogenous wastes

True or False

Read the following statements and write your answer as true or false.

1. Blood is forced under enormous pressure through the glomerulus.
2. Glomerular filtrate is made up of a variety of elements, including water, salts, glucose and white blood cells.

3. The renal tubule completely reabsorbs the sodium chloride present in the glomerular filtrate.

4. In addition to the typical components, urine may also excrete extra vitamins but not antibiotics.

5. Kidney stones can be caused by high levels of uric acid and urine.

Fill in the Blanks

Complete the following statements with an appropriate word/term to be filled in the blank spaces.

1. The urinary bladder is muscular and is under ______ control.
2. The function of the ureter is ______ of urine.
3. Urine is carried from the kidneys by the ______ to the urinary bladder.
4. The inner wall of the urinary bladder is composed of ______.
5. The ______ are the primary organs of the excretory system.
6. The functional unit of the kidney is the ______.
7. The tube that carries urine from the bladder to the outside of the body is called the ______.
8. The waste product that is eliminated by the excretory system is called ______.
9. The process by which the kidneys filter waste products from blood is called ______.
10. The urinary system is responsible for maintaining the body's ______ balance.
11. The condition characterised by the presence of blood in urine is called ______.
12. The medical term for inflammation of the bladder is ______.
13. The process by which the kidneys reabsorb useful substances from the filtrate is called ______.
14. The condition characterised by the presence of protein in urine is called ______.

Assertion–Reason Questions

The following questions consist of two statements: Assertion (A) and Reason (R). Answer these questions selecting the appropriate option given below:

(a) Both A and R are true and R is the correct explanation of A
(b) Both A and R are true but R is not the correct explanation of A
(c) A is true but R is false
(d) A is false but R is true

1. Assertion (A): Aquatic mammals like whales and seals are said to be ureotelic.
   Reason (R): It is because their main nitrogenous waste product is urea.

2. Assertion (A): Sharks are said to be ammonotelic.
   Reason (R): Sharks have the ability to retain considerable amounts of ammonia.

3. Assertion (A): Ammonia should be removed from the body as rapidly as it is formed.
   Reason (R): Ammonia is insoluble in water.

4. Assertion (A): Both ammonia and urea are excreted by earthworms.
   Reason (R): Excretion in earthworm depends on the environment.

5. Assertion (A): In birds and reptiles, the main excretory product is a combined form of urine and faeces.
   Reason (R): Birds and reptiles do not have separate chambers for the excretion of urine and faeces.

One-word Answers

1. The tube arising from the notch of the kidney on the median side and connecting behind with the urinary bladder.
2. The tube that passes urine to the outside.
3. The inner lighter coloured region of the kidney.
4. The structural and functional unit of the kidney.
5. The blood vessel which: (a) enters the Malpighian capsule, (b) leaves the Malpighian capsule
6. The organ which produces urea.
7. The outer region of the kidney containing Bowman's capsule.
8. The tuft of capillaries inside Bowman's capsule.
9. The part of kidney tubules where the term urine is first used for the fluid in it.
10. The vein in which urea concentration is maximum.

Define the Term

1. Ultrafiltration
2. Micturition
3. Renal pelvis
4. Urea
5. Osmoregulation
6. Selective absorption

Groups of Terms

Given below are certain groups of terms. In each group the first pair indicates a relationship between the two terms. Rewrite and complete the second pair on a similar basis.

1. Sweat glands: Perspiration :: Kidneys: ______.
2. Liver: Detoxification :: Lungs: ______.
3. Ureters: Transport :: Urethra: ______.
4. Nephron: Filtration :: Loop of Henle: ______.
5. Renal arteries: Blood supply :: Renal veins: ______.
6. ADH: Water reabsorption :: ANP: ______.

Identify the Odd One Out

1. Kidneys, Ureters, Lungs, Urinary bladder
2. Nephrons, Renal artery, Renal vein, Ovaries
3. Urea, Urine, Urinary tract, Urethra

Image-Based Questions

1. Answer the questions based on the figure:

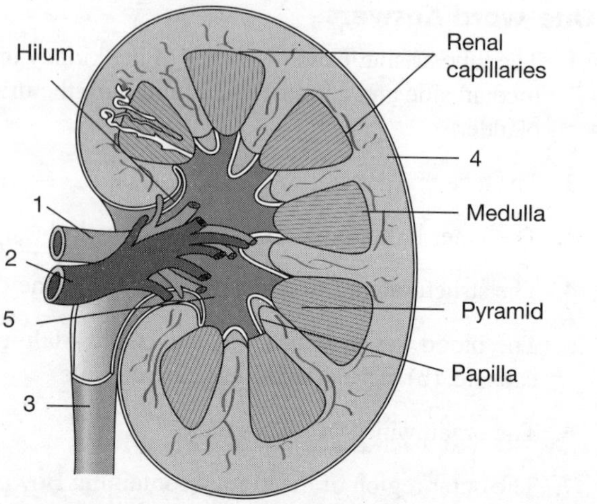

(a) Is it a longitudinal section or a cross-section?
(b) Name the parts numbered 1–5.
(c) Which area/part (give its name or the number) contains the following:
  (i) Malphigian capsule
  (ii) Blood vessel with least/no nitrogenous waste
  (iii) Freshly collected urine

2. Answer the questions based on the figure:

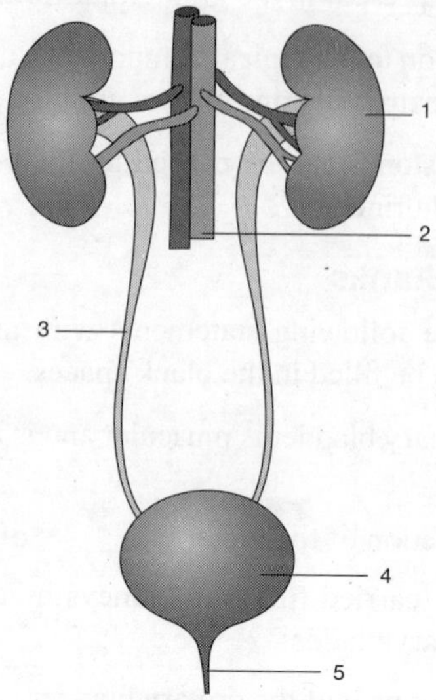

(a) Name all the organ-systems shown completely or even partially.
(b) Name the parts numbered 1 to 5.
(c) Name the structural and functional unit of the part marked '1'.
(d) Name the two main organic constituents of the fluid that flows down the part labelled '3'.
(e) Name the two main steps involved in the formation of the fluid that passes down the part labelled '3'.

3. The following diagram represents a mammalian kidney tubule (nephron) and its blood supply.

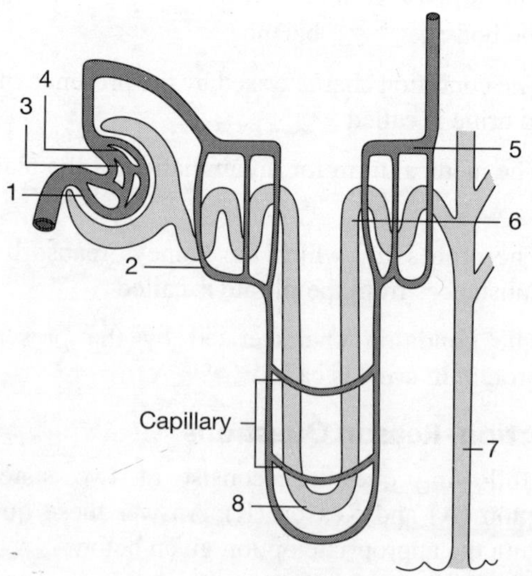

Parts indicated by 1 to 8 are as follows:

1. Afferent arteriole from renal artery
2. Efferent arteriole
3. Bowman's capsule
4. Glomerulus
5. Proximal convoluted tubule with blood capillaries
6. Distal convoluted tubule with blood capillaries
7. Collecting tubule
8. U-shaped loop of Henle

Study the diagram and answer the questions below:

(a) Where does ultrafiltration take place?
(b) Which structure contains the lowest concentration of urea?
(c) Which structure contains the highest concentration of urea?
(d) Which structure (normally) contains the lowest concentration of glucose?
(e) Where is the most water reabsorbed?
(f) State the reason for the high hydrostatic pressure in the glomerulus.
(g) Name the part of the nephron which lies in the renal medulla.

Very Short Answer Questions

1. What is excretion?
2. Why is urine yellow in colour?
3. What is the role of the urethra?
4. Why is glucose absent in the urine of a healthy person?
5. Name the end products of metabolism removed from the body?
6. What happens if waste products are not eliminated from the body?
7. What are ureters?
8. What is the function urinary bladder?
9. What are the two main parts of the uriniferous tubule?
10. Name the vein that carries deoxygenated blood from the kidneys.
11. What is gout?

Short Answer Questions

1. Name the main nitrogenous metabolic waste excreted out by mammals including humans.
2. What is a uriniferous tubule? How does it function?
3. What is dialysis? Under what conditions is it carried out?
4. Give the functions of the kidneys.

Long Answer Questions

1. Why is excretion necessary? Name the common excretory substances in our body.
2. Why is it necessary to maintain the normal osmotic concentration of blood?
3. If you donate one kidney to a needy patient, would it cause any harm to you? Give reasons.
4. In summer, urine is slightly thicker than in winter. Explain why.
5. Differentiate between the following pairs of terms:
   (a) Bowman's capsule and Malpighian tubule
   (b) Renal cortex and renal medulla
   (c) Renal pelvis and renal papilla
   (d) Urea and urine
   (e) Excretion and catabolism

Case Study-Based Questions

1. A patient visits the doctor with complaints of fatigue, nausea and a decrease in urine output. The patient reports feeling bloated and having swelling in the ankles. The doctor suspects a possible kidney problem.
   (i) Which of the following is a function of the excretory system?
   (a) Digestion
   (b) Transport of oxygen to cells
   (c) Regulation of blood pressure
   (d) Production of hormones
   (ii) Which organ is primarily responsible for filtering waste from the blood and producing urine?
   (a) Liver (b) Lungs
   (c) Kidneys (d) Stomach
   (iii) What is the main nitrogenous waste product produced by the kidneys?
   (a) Ammonia (b) Urea
   (c) Carbon dioxide (d) Glucose

(iv) What is the name of the functional unit of the kidney responsible for filtering blood?
(a) Nephron (b) Neuron
(c) Axon (d) Synapse

(v) What condition could the patient be suffering from?
(a) Diabetes (b) Hypertension
(c) Kidney stones (d) Renal failure

2. A patient visits the doctor with symptoms of back pain, blood in the urine and difficulty urinating. The doctor suspects a possible kidney stone.

(i) Which of the following structures is responsible for transporting urine from the kidneys to the bladder?
(a) Urethra (b) Ureter
(c) Prostate gland (d) Seminal vesicle

(ii) Which of the following would be an effective diagnostic tool to detect the presence of a kidney stone?
(a) X-ray (b) CT scan
(c) MRI (d) Ultrasound

(iii) What is the most common type of kidney stone?
(a) Calcium oxalate
(b) Uric acid
(c) Struvite
(d) Cystine

(iv) What is the name of the hormone that regulates the reabsorption of water in the kidneys?
(a) Aldosterone (b) Adrenaline
(c) Insulin (d) Growth hormone

(v) What is the medical term for the condition where the kidneys stop working properly and are unable to remove waste from the body?
(a) Kidney stones
(b) Renal failure
(c) Glomerulonephritis
(d) Nephrotic syndrome

NCERT Corner

1. Name the major nitrogenous excretory compound in humans. State how it is formed in the body.
2. Explain how the kidney regulates the volume and concentration of urine produced by it.
3. What are the functional units of the kidney? What is their structure and function?
4. Differentiate between ultrafiltration and selective reabsorption.
5. How does the skin help in the excretion of waste products in humans?
6. Why do freshwater fish excrete more water as urine, while saltwater fish excrete very little urine?
7. Explain why the loop of Henle is essential for urine concentration in the kidneys.
8. Explain the process of micturition in humans.
9. What is the role of the liver in excretion in humans?
10. What are the major excretory products of the lungs and skin?
11. What is the significance of urea recycling in the excretory system?
12. Explain how the excretory system helps in maintaining pH balance in the body.
13. Name two common nitrogenous waste products excreted by animals other than humans.
14. Explain how the countercurrent mechanism in the nephron helps in urine concentration.

COMPETITION WINDOW (OBJECTIVE TYPE)

1. The main excretory product in human beings is:
(a) Urea
(b) Ammonia
(c) Uric acid
(d) Ammonium chloride

2. The nephron in human beings is a part of the system for:
(a) Respiration (b) Nutrition
(c) Excretion (d) Circulation

3. Blood is purified by the __________.
(a) Kidney (b) Lungs
(c) Heart (d) Intestine

4. Which is the excretory product of earthworm?
(a) Ammonia and uric acid
(b) Urea and amino acid
(c) Amino acid
(d) Ammonia, urea and traces of creatinine

5. There are __________ uriniferous tubules in a kidney.
(a) About 10,000 (b) About 5,000
(c) Numerous (d) About 1.0×10^6

6. Conversion of excess amino acids into urea is performed in the:
(a) Lungs (b) Large intestine
(c) Liver (d) Cloaca

7. Columns of Bertin are found in the:
(a) Testes (b) Ovaries
(c) Kidney (d) Liver

8. Humans are:
(a) Ammonotelic
(b) Ureotelic
(c) Uricotelic
(d) None of the above

9. The yellow colour of urine is due to:
(a) Uric acid (b) Urea
(c) Urochrome (d) Melanin

10. The human kidney's __________ is regarded as its fundamental functioning unit:
(a) Exon (b) Nephron
(c) Cilia (d) Neuron

11. A series of metabolic processes known as the Krebs–Henseleit cycle occur in the:
(a) Brain (b) Liver
(c) Urinary bladder (d) Lungs

12. Bowman's capsule is located in the:
(a) Cortex (b) Henle's loop
(c) Bladder (d) None of the above

13. The junction of two or three main renal calyces is known as the:
(a) Renal pelvis
(b) Urethra
(c) Bowman's capsule
(d) None of the above

14. Which of the following is the largest exocrine gland in the human body?
(a) Pancreas (b) Liver
(c) Gall bladder (d) Thyroid

15. The extension of the ureter's forepart edge into the kidney is called the:
(a) Cervix (b) Hip
(c) Pelvis (d) Abdomen

16. Tears, sweat, saliva, milk and insulin are examples of:
(a) Excretion (b) Reabsorption
(c) Substitution (d) Secretion

17. The kidney is made up of a large quantity of tubules known as:
(a) Proximal convoluted tubule
(b) Distal convoluted tubule
(c) Uriniferous tubules
(d) Liver lobule

18. What is the full form of PCT?
(a) Proximal convex tubule
(b) Proximal convoluted tubule
(c) Primary convoluted tubule
(d) Proximal convoluted tissue

19. Bowman's capsule and the glomeruli are together referred to as:
(a) Malpighian capsule
(b) Glomerulus
(c) Loop of Henle
(d) Bowman's capsule

20. DCT stands for:
(a) Distal conclave tubule
(b) Double convoluted tubule
(c) Distal convex tubule
(d) Distal convoluted tubule

21. How many litres of blood pass through the kidneys every minute?
(a) 2 litres (b) 4 litres
(c) 1 litre (d) 3 litres

22. How many times a day does all the blood in the body flow through the kidneys?
(a) 350–400 times (b) 200–250 times
(c) 100–150 times (d) 450–500 times

23. The essential steps in the creation of urine include:
(a) Ultrafiltration (b) Reabsorption
(c) Tubular secretion (d) All of the above

Answer Keys

Multiple Choice Questions

1. (a) 2. (c) 3. (c) 4. (d) 5. (b)
6. (d) 7. (c) 8. (b) 9. (c)

Match the Following

1. (1) (d) (2) (a) (3) (b) (4) (c) (5) (e)
2. (1) (d) (2) (c) (3) (b) (4) (a)

True or False

1. T 2. T 3. F 4. T 5. T

Fill in the Blanks

1. Nervous
2. Passage
3. Ureters
4. Transitional epithelium
5. Kidneys
6. Nephron
7. Urethra
8. Urea
9. Filtration
10. Fluid and electrolyte
11. Hematuria
12. Cystitis
13. Reabsorption
14. Proteinuria

Assertion–Reason Questions

1. (a) 2. (b) 3. (b) 4. (c) 5. (c)

One-word Answers

1. Ureter
2. Urethra
3. Renal medulla
4. Nephron
5. Afferent arteriole, efferent arteriole
6. Liver
7. Renal cortex
8. Glomerulus
9. Distal convoluted tubule
10. Renal vein

Define the Term

1. Ultrafiltration is a process of filtration in which fluid is passed through a semi-permeable membrane under pressure, separating larger particles or molecules from smaller ones.
2. Micturition is the medical term for the process of urination or voiding. It is the act of expelling urine from the bladder through the urethra and out of the body.
3. The renal pelvis is a funnel-shaped structure that serves as a collecting reservoir for urine within the kidney.
4. Urea is an organic compound that is produced in the liver of mammals and some fish as a byproduct of the metabolism of amino acids.
5. Osmoregulation is the process by which living organisms regulate the concentration of water and solutes within their cells and tissues to maintain homeostasis. It is a physiological process that helps to balance the uptake and loss of water and ions, and to maintain the proper concentration of electrolytes, such as sodium, potassium and chloride ions, in the body fluids.
6. Selective absorption, also known as reabsorption, is the process by which substances that are necessary for the body are selectively reabsorbed from the filtrate back into blood. This occurs in various segments of the nephron, including the proximal convoluted tubule, the loop of Henle and the distal convoluted tubule. Substances such as glucose, amino acids and ions such as sodium, potassium and chloride are selectively reabsorbed based on the body's needs, while waste products such as urea and excess ions are excreted in the urine.

Groups of Terms

1. Urine
2. Oxygenation
3. Elimination
4. Reabsorption
5. Blood drainage
6. Sodium excretion

Identify the Odd One Out

1. Lungs
2. Ovaries
3. Urinary tract

Image-Based Questions

1. (a) It is the longitudinal section of the kidney.
   (b) The parts numbered 1–5 are:
   1 → Renal Artery
   2 → Renal Vein
   3 → Ureter
   4 → Cortex
   5 → Pelvis
   (c) Given below is the area/part containing the mentioned structure:
   (i) Cortex (4)
   (ii) Medulla
   (iii) Pelvis (5)
2. (a) Excretory system and Circulatory system.
   (b) The parts numbered 1–5 are:
   1 → Kidney
   2 → Renal Artery
   3 → Ureter
   4 → Urinary Bladder
   5 → Urethera
   (c) Nephron
   (d) Urea and ammonia
   (e) Ultrafiltration and selective absorption
3. (a) Ultrafiltration takes place in the glomerulus (4 in the diagram).
   (b) Efferent arteriole (2 in the diagram) contains the lowest concentration of urea.
   (c) Afferent arteriole (1 in the diagram) from the renal artery contains the highest concentration of urea.
   (d) Collecting tubule (7 in the diagram) contains the lowest concentration of glucose.
   (e) Most water is reabsorbed in the proximal convoluted tubule with blood capillaries (5 in the diagram).
   (f) The efferent arteriole is narrower than the afferent arteriole. This creates high hydrostatic pressure in the glomerulus.
   (g) Loop of Henle is the part of the nephron which lies in the renal medulla.

Very Short Answer Questions

1. Excretion is the process by which metabolic waste and other toxic substances are eliminated from the body. It is a vital function of the excretory system, which includes the kidneys, ureters, bladder, urethra and other organs such as the skin and lungs.
2. Urine is yellow in colour due to the presence of the pigment urochrome formed due to break-down of the haemoglobin of old RBCs.
3. The urethra is a tube that carries urine from the bladder to the outside of the body during urination. It is part of the excretory system.
4. Glucose is absent in the urine of a healthy person because it is completely reabsorbed from the glomerular filtrate.
5. Metabolic wastes or nitrogenous waste.
6. If waste products are not removed from the body, they build up inside the body and could become toxic and deadly.
7. The ducts that carry urine from the kidney to the urinary bladder or cloaca are called ureters.
8. The urinary bladder stores the urine till it is released through the urethra.
9. The two main parts of the uriniferous tubule are Malpighian corpuscle and renal tubule.
10. Renal vein.
11. Uric acid is relatively less soluble in water and may crystallise and get deposited in the joints of the body, resulting in a condition called gout.

Short Answer Questions

1. The main nitrogenous metabolic waste excreted out by mammals, including humans, is urea. Urea is produced in the liver during the break-down of amino acids, and is transported to the kidneys for excretion through urine. Other nitrogenous waste products that may be excreted by mammals include ammonia and creatinine, but these are typically present in much smaller quantities compared to urea.
2. The uriniferous tubule, also known as a nephron, is the functional unit of the kidney responsible for the formation and excretion of urine. Each kidney contains millions of uriniferous tubules. The uriniferous tubule is a long, coiled tube consisting of several parts, including the renal corpuscle, proximal convoluted tubule, loop of Henle, distal convoluted tubule and collecting duct. The uriniferous tubule functions to filter, modify and concentrate the blood to produce urine and maintain the body's fluid and electrolyte balance.
3. Dialysis is a medical treatment that is used to remove waste products and excess fluid from the blood when the kidneys are no longer able to perform this

function. It is a life-saving treatment for individuals with end-stage renal disease (ESRD), which is a condition in which the kidneys have lost almost all of their function. Dialysis helps to prevent the build-up of waste products and excess fluid in the body, which can lead to serious health problems such as heart failure, electrolyte imbalances, and even death.

Dialysis is usually carried out when the kidney function falls below 10–15% of normal, or when the patient experiences symptoms of kidney failure such as fatigue, swelling and shortness of breath. It is also used as a temporary measure for individuals who are waiting for a kidney transplant or for those who are unable to undergo transplant surgery.

4. (a) Blood is filtered by the kidneys.
   (b) Urine is made by the kidneys.
   (c) Osmoregulation is the process of maintaining or controlling the blood's osmotic pressure (water and salt concentration)
   (d) The kidneys control the blood's water, electrolyte and acid-base levels.
   (e) The kidneys eliminate foreign chemicals like pigments and medicines.
   (f) By eliminating compounds that are present in excess of what is immediately required, the kidneys control the chemical composition of bodily fluids.

Long Answer Questions

1. Hint: Explain the significance and need for excretion. The common excretory substances in our body include carbon dioxide, urine, sweat, faeces and bilirubin. Elaborate on these points.
2. Hint: Maintaining normal osmotic concentration of blood is important for cellular function and shape, fluid balance and hydration, kidney function and waste removal, nerve and muscle function, pH balance and acid-base regulation, and enzyme activity and metabolic processes. Elaborate on these points.
3. Hint: Elaborate on the following: adaptation of the remaining kidney; extensive evaluation and screening; surgical techniques and expertise; long-term studies and follow-up; and psychological and emotional support.
4. Hint: The key factor contributing to the difference in urine thickness between summer and winter is related to water balance and hydration levels in the body. Elaborate on this.
5. Hints:
   (a) Bowman's capsule is found in the mammalian kidney, while Malpighian tubules are found in insects and other arthropods. Bowman's capsule is involved in the filtration of blood and the formation of urine, while Malpighian tubules are responsible for the excretion of metabolic waste and regulation of water and ion balance.
   (b) The renal cortex and renal medulla are two distinct regions within the kidney. The renal cortex is located on the outermost region of the kidney, while the renal medulla is located deeper within the kidney.
   (c) The renal pelvis and renal papilla are both structures found in the kidney. The renal pelvis is a large, funnel-shaped area that collects urine from the renal pyramids.
   (d) Urea and urine are related but have distinct purposes and functions. Focus on their composition and roles in the body.
   (e) Excretion and catabolism are related to the elimination of waste products but involve different processes and concepts. Consider the processes involved and the nature of the substances being eliminated.

   Elaborate on these points.

Case Study-Based Questions

1. (i) c, (ii) c, (iii) b, (iv) a, (v) d
2. (i) b, (ii) b, (iii) a, (iv) a, (v) b

NCERT Corner

1. The main nitrogenous excretory compound in humans is urea. It is formed in the liver from excess amino acids.
2. The kidney regulates the volume and concentration of urine produced by it through the process of selective reabsorption and tubular secretion. By selectively reabsorbing substances like water, glucose, amino acids and ions, and secreting others like hydrogen ions and potassium ions, the kidney is able to regulate the composition and volume of urine produced.
3. The functional units of the kidney are called nephrons. Each nephron consists of a renal corpuscle (made up of the glomerulus and Bowman's capsule) and a renal tubule (made up of the proximal convoluted tubule, loop of Henle and distal convoluted tubule). The glomerulus filters blood to form urine,

and the renal tubule reabsorbs and secretes substances to regulate the composition of urine.

4. Ultrafiltration is the process by which blood is filtered in the glomerulus to form a filtrate. Selective reabsorption is the process by which substances like glucose, amino acids and ions are selectively reabsorbed back into the blood from the renal tubule.
5. The skin helps in the excretion of waste products in humans by sweating. Sweat contains water, salts and small amounts of urea and other wastes.
6. Freshwater fish excrete more water as urine to get rid of excess water they take in through osmosis. Saltwater fish excrete very little urine because they need to conserve water to compensate for the high salt concentration of their environment.
7. The loop of Henle is essential for urine concentration in the kidneys because it creates a concentration gradient in the renal medulla that allows for water reabsorption.
8. Micturition is the process of urination. It is initiated by the contraction of the muscular wall of the urinary bladder, which forces urine out of the body through the urethra.
9. The liver plays a role in excretion by producing bile, which helps in the elimination of waste products like bilirubin, cholesterol and drugs.
10. The main excretory products of the lungs are carbon dioxide and water vapour. The main excretory product of the skin is sweat.
11. Urea recycling is significant in the excretory system because it allows for the reabsorption of water and urea in the loop of Henle, which helps in the concentration of urine.
12. The excretory system helps in maintaining the pH balance in the body by selectively reabsorbing and secreting ions like hydrogen ions and bicarbonate ions.
13. Two common nitrogenous waste products excreted by animals other than humans are ammonia and uric acid.
14. The countercurrent mechanism in the nephron helps in urine concentration by creating a concentration gradient in the renal medulla that allows for water reabsorption. This is achieved by the flow of filtrate and blood in opposite directions in the loop of Henle.

Competition Window

1. (a)	2. (c)	3. (a)	4. (d)	5. (d)
6. (c)	7. (c)	8. (b)	9. (c)	10. (b)
11. (b)	12. (a)	13. (a)	14. (b)	15. (c)
16. (d)	17. (c)	18. (b)	19. (a)	20. (d)
21. (c)	22. (a)	23. (d)		

CHAPTER 12

The Endocrine Glands – The Producers of Chemical Messengers

PREREQUISITES

You must have knowledge of:
- Hormones
- Human anatomy and physiology
- Positive and negative feedback mechanisms

LEARNING OBJECTIVES

At the end of this chapter, you will be able to:
- List the types of hormones and describe how they contribute to maintaining homeostasis
- Describe the functions of hormones and how hormone production is controlled
- Explain the functions of various glands in the endocrine system
- Describe how the glands cooperate to preserve equilibrium

INTRODUCTION

As our body's functions are extremely complicated, they must be tightly controlled to ensure that each one occurs at the appropriate time and in the correct order. For instance, only when food is present in the food canal should gastric juice, bile and pancreatic juice be secreted into it. Although the neurological system is partially responsible for this type of regulation, hormones also play a role.

Hormones are secretions from specific cells or glands, and blood carries them to all the parts of the body, but their effects are only felt in one or more particular parts (target organ or cells). The majority of hormones are produced by specialised glands termed **endocrine glands**, sometimes known as ductless glands since their secretions are released directly into the circulation without passing through any particular duct. Some of these glands or body regions, like the stomach and duodenum, which ordinarily serve different fundamental functions, also produce some hormones.

Multiple glands and glandular cells make up the endocrine system, which is responsible for the body's overall shared function of chemical coordination. Almost all endocrine glands function in unison. A system is described as a collection of organs working together to accomplish a common goal.

GENERAL PROPERTIES OF HORMONES

- Hormones are released into the blood directly from their source (and not into lymph).
- They use chemical methods to control physiological processes. The body's enzyme systems are affected by them.
- They typically act on target organs or cells far from their source.
- Hormones produced in one species typically have a similar impact on other species.
- They are created in very small quantities and have high biological activity. For instance, adrenaline

is effective even at a concentration of 1 part per 300,000,000 parts.

- Chemically, some hormones are peptides (proteins like insulin), others are amines (derived from amino acids like adrenaline) and are water-soluble, and some are steroids produced from cholesterol like testosterone, which are lipid-soluble.
- Their over-production (hypersecretion/oversecretion) or under-production (hyposecretion/undersecretion) may have severe repercussions.
- Hormones are eliminated from the body and not retained in it.

ENDOCRINE GLANDS

Figure 12.1 depicts the main endocrine (hormone-producing) glands in the human body. These are listed below:

- Adrenal
- Pituitary
- Gonads
- Pancreas
- Parathyroid
- Thyroid
- Thymus

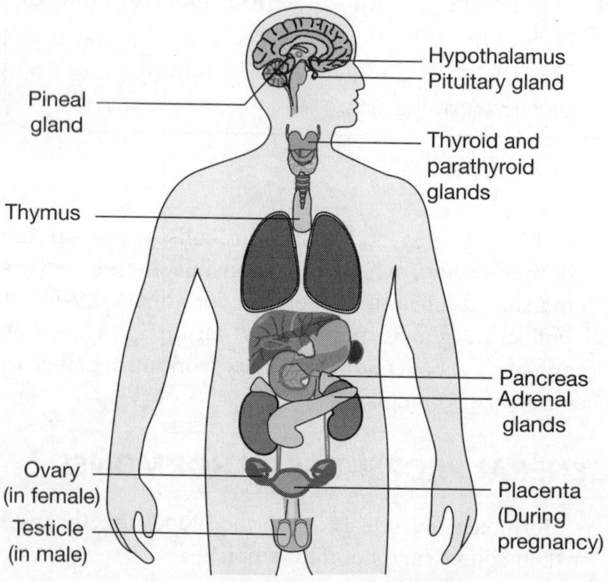

Fig. 12.1 Endocrine glands

Adrenal Glands

These are found on top of the kidneys (Fig. 12.2). They resemble caps. A central medulla and a peripheral cortex make up each adrenal gland.

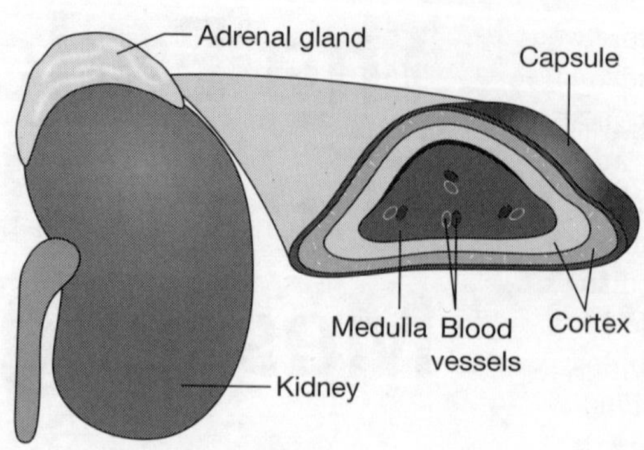

Fig. 12.2 Adrenal glands

The adrenal medulla releases adrenaline (also called epinephrine and noradrenaline). Adrenaline is a hormone that primes the body for 'fight', or facing danger, or 'flight', or escaping it, in the event of an emergency. The body receives additional energy and strength under the circumstance. When under emotional stress, extra hormones are released into the blood. Our adrenal glands release a lot of adrenaline when we are enthusiastic or irritated. The autonomic nervous system's nerve terminals stimulate the gland.

The **effects of adrenaline** are listed below:

- It causes an increase in blood pressure and an increase in heart rate, while lowering blood flow to the skin and visceral organs.
- The liver increases the amount of glucose secreted into blood.

The most well-known hormone secreted by the adrenal cortex is cortisone (which suppresses inflammation). The cortical hormones are divided into three groups:

- **Mineralocorticoids:** Control the metabolism of minerals, particularly Na^+ and K^+ ions.
- **Glucocorticoids:** Control the metabolism of proteins, carbohydrates and fats. The cortical hormones generally:
  - Increase the level of blood glucose
  - May have an impact on protein and fat metabolism
  - Control the body's salt and water balance
  - Help the body adjust to stresses such excessive cold or heat, burns, infections, etc.
- **Sex hormones:** Some cortical hormones exhibit sex hormone-like behaviours. Both sexes have these hormones. In young children, expansion of the cortex causes early onset of sexual maturity.

A mature woman may develop certain male traits, such as a beard, moustache and a deep male voice, if her adrenal cortex is over-developed. This condition is referred to as **adrenal virilism**. Mature men who experience overgrowth of the adrenal cortex may start to exhibit feminine traits like enlarged breasts.

Addison's disease is brought on by adrenal cortical hyposecretion. Loss of energy, changes in skin colour, weight loss, nausea, hypoglycemia (low blood sugar), sensitivity to pain and cold and susceptibility to infections are some symptoms.

Cushing's syndrome is caused by the adrenal cortex secreting too much hormone. Obesity, hyperglycemia (increased blood sugar), osteoporosis, weakness, and salt and water retention are symptoms of this condition.

Pancreas

The pancreas (Fig. 12.3) is a duct gland as well as a ductless gland. As a duct gland, it secretes pancreatic juice, which is emptied into duodenum. As a ductless gland, it has distinct clusters of hormone-secreting cells known as islets of Langerhans that are dispersed throughout the entire gland. The beta, alpha and delta cells that make up the islet cells produce the three hormones insulin, glucagon and somatostatin.

Insulin, which is released by beta cells, prevents a rise in blood sugar levels. This is accomplished primarily in two ways:

- It encourages the body cells to use glucose.
- It encourages the liver and muscles to store additional blood glucose as glycogen.

Diabetes is brought on by insufficient insulin secretion (more correctly diabetes mellitus or hyperglycemia). The term '*mellitus*', which refers to the transit of sugar (glucose) in urine, means 'honey'.

A diabetic person has high blood sugar levels, excretes a lot of sugar-laden urine, experiences thirst because of the water lost through excessive urination, loses weight and weakens with time. The person may occasionally lose their eyesight. In most cases, injecting insulin is not a cure but a way to provide the hormone that the pancreas is unable to create.

Hypoglycemia is a condition where the blood sugar level becomes very low and the person may become unconscious. If a diabetic patient receives too much insulin also, the patient may lose consciousness. A quick piece of sugary cookies or sweets is beneficial in cases of hypoglycemia, also known as insulin shock.

Thyroid

The thyroid is a bilobed, or butterfly-shaped, organ found in front of the neck, directly below the larynx (Fig. 12.4). The isthmus, which connects the two lobes, is narrow (interconnection). Thyroxine and calcitonin are two hormones secreted by it.

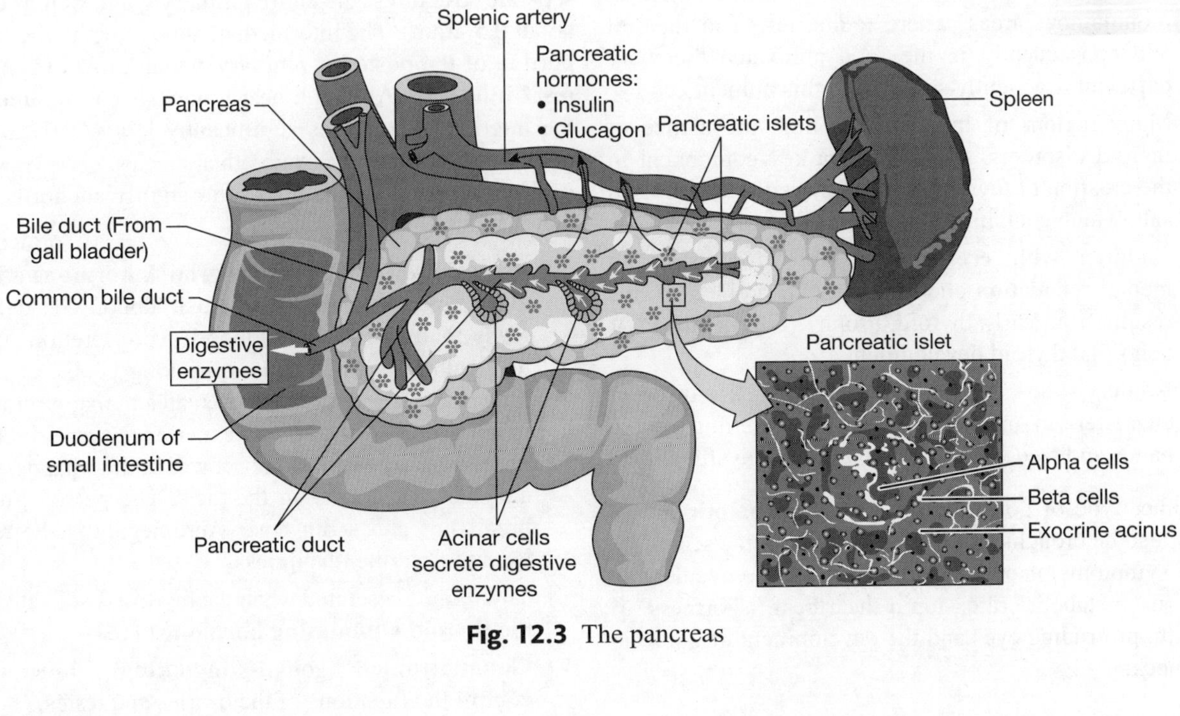

Fig. 12.3 The pancreas

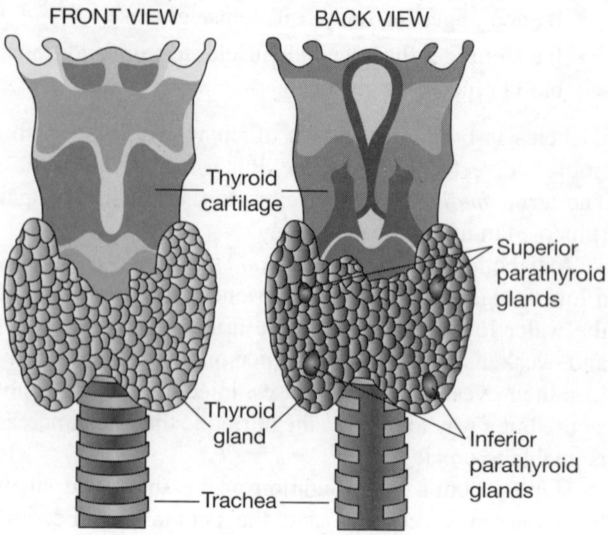

Fig.12.4 The thyroid gland

Thyroxine controls the rate of cellular oxidation, which generates heat when the body is at rest. A rise in secretion boosts metabolism, while a fall in secretion slows it down. Additionally, it affects bone ossification, body temperature, brain development and general body growth.

Three **disorders** can result from inadequate thyroxine secretion:

- A simple **goitre** is an enlargement of the thyroid that manifests as a neck swelling. Lack of iodine in meals is to blame for this. People who reside in mountainous areas, where iodine levels in the soil and consequently in the food cultivated there are deficient, frequently suffer from this ailment.

  Many regions of India have a high prevalence of thyroid disorders. As iodine is a key component in the creation of thyroxine, it is advised to use iodised salt, which contains iodine, in the diet.
- Children with **cretinism** exhibit dwarfism and mental retardation and have growth problems. This results from early thyroid atrophy (degeneration) or abnormal thyroid development.
- Adults who have **myxoedema**, a disorder characterised by sluggishness and swelling of the hands and face, are affected by thyroid dysfunction.

Another type of goitre known as **exophthalmic goitre** may also be brought on by excessive thyroxine secretion. The symptoms of over-secretion include a considerable rise in metabolic rate, rapid heartbeat, shortness of breath, protruding eyes and the development of goitre in the neck.

Parathyroid Glands

The necks of humans and other tetrapods contain small endocrine glands called parathyroid glands. The four parathyroid glands that are typically present in humans are dispersed around the back of the thyroid gland. When there is deficiency of calcium in blood, the parathyroid gland generates and secretes the parathyroid hormone, which is essential for controlling the levels of calcium in the bones and blood.

Two pairs of parathyroid glands are found behind the left and right thyroid lobes. The higher-positioned parathyroid glands on each side are known as the superior parathyroid glands, while the lower-positioned parathyroid glands are known as the inferior parathyroid glands. Each gland is a flat, yellowish-brown oval that is roughly 6 mm long, 3–4 mm broad, and 1–2 mm anteroposteriorly. It resembles a lentil seed.

A small protein called parathormone—also known as parathyroid hormone—plays a role in the regulation of calcium and phosphate balance and bone physiology. Calcitonin's effects are counterbalanced by those of the parathyroid hormone.

Pituitary

The pituitary gland (Fig. 12.5) is a tiny projection at the base of the midbrain, just below the hypothalamus, roughly the size of a pea. As it seems to regulate almost all other endocrine glands, it is commonly referred to as the master gland. The anterior and posterior pituitary lobes are two separate structures (the entire pituitary gland weighs only about 0.5 gram). The intermediate lobe refers to the front portion of the posterior pituitary, which differs from the rest of the lobe. Although nearly nonexistent in humans, the intermediate lobe is significantly larger and carries out more functions in certain other species.

The following list includes some significant hormones that the anterior lobe produces:

- Normal growth requires the **growth hormone** (GH), also known as somatotropin. Childhood GH deficit leads to dwarfism. Childhood over-secretion leads to gigantism, which causes the long bones to grow longer than they should and creates human up to 2.7 m tall. Adults who abruptly experience over-secretion of GH see excessive bone growth in their hands, feet and face, especially in the jaws. The person grows broad lips and a big nose. Acromegaly is the term used to describe this illness.
- Thyroxine is secreted when the thyroid is activated by the **thyroid stimulating hormone** (TSH).
- **Gonadotropic (gonad-stimulating) hormones** control the functions of the ovaries and testes.

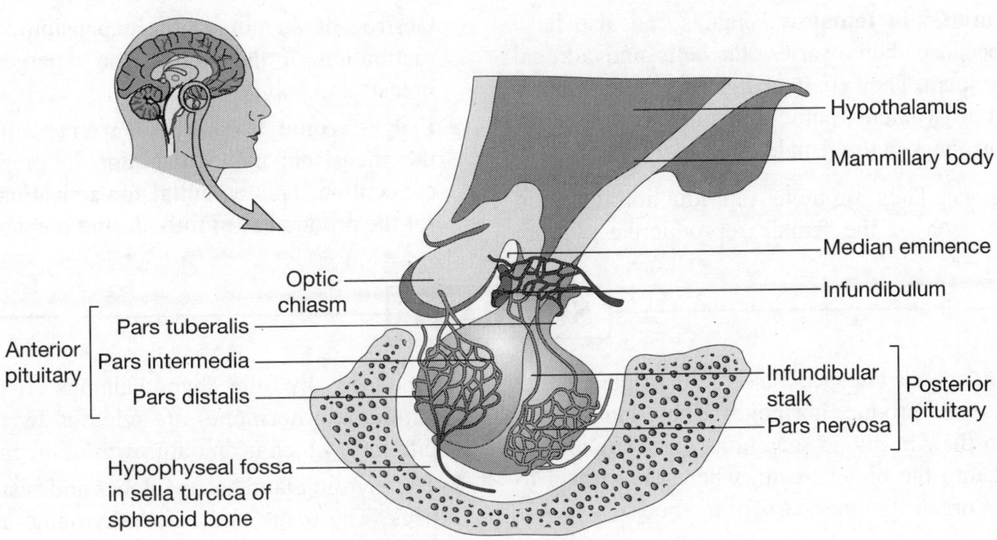

Fig. 12.5 The pituitary gland

- The activity of the adrenal cortex is regulated by the **adrenocorticotropic hormone** (ACTH). Tropic hormones are those that encourage the production of a specific hormone by another endocrine gland, such as the anterior pituitary's gonadotropic hormones and the gonads' production of specific hormones. The term tropic refers to alternation of the function of the designated organ, such as the thyroid's thyrotropic hormone or the adrenal cortex's adrenocorticotropic hormone.

Two hormones are produced by the posterior lobe: oxytocin and vasopressin. **Vasopressin**, another name for antidiuretic hormone (ADH), narrows blood vessels as blood pressure rises. Additionally, it affects the kidneys, causing the kidney tubules to absorb more water. ADH deficiency results in diabetes insipidus (also known as water diabetes, when urine is frequent and profuse), which causes the body to lose water and makes the sufferer thirsty. **Oxytocin** promotes uterine contractions in pregnant women, which results in the birth of the child. It also encourages milk ejection.

Thymus

The lymphatic system includes a small gland called the thymus. The thymus gland is situated in the upper chest, behind the breastbone (sternum). It is located in the mediastinum, a region of chest between the lungs. The thymus gland's main function is to create T-lymphocytes, or T-cells, which are specialised white blood cells. The bone marrow and thymus are the destinations of lymphocytes, or white blood cells. The thymus is where the lymphocytes develop into highly specialised T-cells.

Gonads

The gonads, which are also known as the male and female reproductive organs, are a component of the endocrine system. The male and female gonads, ovaries and testes, respectively, are in charge of manufacturing the sex hormones in our bodies. The gonads are the most important hormone-secreting organ in the human body

The functions of the gonads are listed below:

- Secreting the sex hormones
- Generating gametes
- Assisting in meiosis

Both in males and females, they are responsible for the maturation of secondary sex traits.

Gonad hormones in males: The male hormones or androgens are in charge of the development of the male reproductive system. The main hormones in the male gonads are:

- **Testosterone:** This hormone, which is made by the testes, is crucial for the development of bone, muscle and body hair. It causes the penis to expand, voice to deepen and facial hair to emerge.
- **Follicle-stimulating hormone**, or FSH, controls sperm cell growth and regulation. Its release is prevented by the compound inhibin.
- **Androstenedione:** This stimulates and controls the body's production of oestrogen and testosterone.

Gonad hormones in females: Females can also have androgens because their ovaries, fat cells and adrenal glands create them. They are in charge of promoting hair development in women's pubic and underarm regions. The main hormones in the female gonads are:

- **Oestrogens:** They promote reproduction and aid in the growth of the female reproductive system. Oestrogens encourage the expansion of the uterus, maturation of the vagina and enlargement of the breasts and pelvis.
- **Progesterone:** It controls a woman's uterus during the menstrual cycle and aids in preparing it for conception. It is essential for activating glands and for the production of milk during pregnancy.

SUMMARY

- Endocrine glands are specialised cells or glands that synthesise and produce hormones. These glands lack ducts, so their hormone secretions are immediately released into the bloodstream, where they travel to the target organs in other parts of the body.
- Adrenal glands, pancreatic, thyroid and pituitary glands are the main endocrine glands. The functions of the tissues are stimulated by hormones, which also regulate metabolism, growth and development.
- They synthesise, store and use chemicals, such as glucose, and they regulate metabolism. Nearly all of the endocrine glands work in unison. Together, they function as the endocrine system, a group of organs.
- When hormones attach to receptors on or in target cells, biological changes result. In response to hormone activity, a target cell's number of receptors may rise or fall.
- The main mechanism for regulating hormone levels is negative feedback, in which a hormone's production is inhibited as its level rises.
- The pituitary gland is situated in the base of the brain. The hypothalamus sends signals to the anterior pituitary, which creates six hormones. The hormones produced by the hypothalamus (oxytocin and antidiuretic hormone) are released by the posterior pituitary, which is an outgrowth of the brain.
- The thyroid gland has two lobes and is situated in the neck. Thyroxine and triiodothyronine are produced by the thyroid. Calcitonin is also made by the thyroid.
- The adrenal cortex and adrenal medulla make up the adrenal glands, which are situated on top of the kidneys. Corticosteroids, glucocorticoids and mineralocorticoids are all made by the adrenal cortex. Epinephrine and norepinephrine are produced by the adrenal medulla, which is the inner portion of the adrenal gland.
- In the abdomen, the pancreas is situated between the small intestine and the stomach. The islets of Langerhans, which are collections of endocrine cells in the pancreas, contain beta cells that release insulin and alpha cells that release glucagon.
- The thymus produces hormones that support the immune system's growth.
- In males and females, the gonads generate steroid hormones such as testosterone, oestrogen and progesterone. Leptin is produced by adipose tissue and encourages satiety signals in the brain.

PRACTICE QUESTIONS

Analyse Your Concepts (School Exam Based)

Multiple Choice Questions

1. A gland having endocrine as well as exocrine functions is:
   (a) Pituitary (b) Thyroid
   (c) Pancreas (d) Adrenal
2. Exophthalmic goitre is caused due to overactivity of:
   (a) Thymus (b) Thyroid
   (c) Parathyroid (d) Adrenal cortex
3. The secretion of ADH (anti-diuretic hormone) is inhibited by:
   (a) Emotion and stress (b) Nicotine
   (c) Alcohol (d) Morphine
4. Which of the following glands is responsible for producing insulin?
   (a) Adrenal gland (b) Pituitary gland
   (c) Thyroid gland (d) Pancreas
5. Which hormone is produced by the adrenal glands and is responsible for regulating the body's response to stress?
   (a) Cortisol (b) Thyroxine
   (c) Insulin (d) Melatonin
6. Which of the following is not a function of the endocrine system?
   (a) Regulating growth and development
   (b) Regulating metabolism
   (c) Regulating the body's response to stress
   (d) Regulating the body's voluntary movements

Match the Following

Each question contains statements given in two columns which have to be matched. Statements in column I have to be matched with statements in column II.

Column I	Column II
1. Pituitary gland	A. Testosterone
2. Sulphur dioxide	B. Calcium
3. Seminiferous tubules	C. Growth hormone
4. Clotting of blood	D. Acid rain
5. Gonad	E. Sperm

True or False

Read the following statements and write your answer as true or false.

1. The pituitary is of the size of a lemon.
2. The pituitary has three lobes: anterior, posterior and intermediate.
3. Over-secretion of the growth hormone in adults causes gigantism.
4. Vasopressin secreted by the anterior lobe controls reabsorption of water from the kidney tubules.
5. Hormones are chemical messengers that are transported by the nervous system.
6. The hypothalamus is a gland that produces hormones.
7. The adrenal glands are located on top of the kidneys.
8. The pancreas produces insulin and glucagon, hormones that regulate blood sugar levels.
9. Hormones only affect the organs or tissues they are secreted by.
10. The pituitary gland is often referred to as the master gland because it controls the functions of all other endocrine glands.
11. Hormone levels in the body are regulated solely by negative feedback mechanisms.
12. Hormones are only produced by the endocrine system.

Fill in the Blanks

Complete the following statements with an appropriate word/term to be filled in the blank spaces.

1. Regulation of activities inside the body is brought about by the ___________ system.
2. Hormones are secreted directly into ___________.
3. Chemically, some hormones are ___________, some steroids and some ___________.
4. Both ___________ and deficiency of hormones can lead to serious consequences.

5. Hormones produced in one species usually show ___________ influence in other species.

Assertion–Reason Questions

The following questions consist of two statements: Assertion (A) and Reason (R). Answer these questions selecting the appropriate option given below:

(a) Both A and R are true and R is the correct explanation of A
(b) Both A and R are true but R is not the correct explanation of A
(c) A is true but R is false
(d) A is false but R is true

1. Assertion (A): The thyroid gland produces hormones that regulate metabolism.
Reason (R): Metabolism is the process by which the body converts food into energy.
2. Assertion (A): The pituitary gland produces a hormone called prolactin that stimulates milk production in the breasts.
Reason (R): Prolactin is a steroid hormone that is also involved in regulating other reproductive functions.
3. Assertion (A): The hormone insulin is produced by the pancreas and helps regulate blood sugar levels.
Reason (R): In type 1 diabetes, the body does not produce enough insulin, leading to high blood sugar levels.
4. Assertion (A): The adrenal glands produce hormones that help the body respond to stress.
Reason (R): One of the hormones produced by the adrenal glands is cortisol, which helps regulate blood pressure and blood sugar levels.
5. Assertion (A): The hormone oxytocin is involved in inducing labour during childbirth.
Reason (R): Oxytocin is also known as the "love hormone" because it is released during social bonding activities such as hugging and kissing.

One-word Answers

1. The hormones that the pancreas makes.
2. The hormone that the adrenal medulla produces.
3. The condition brought on by excessive insulin secretion.
4. The hormone that the Langerhans islets' (beta) cells secrete.
5. The hormone which increases blood pressure.
6. The hormone causing more urine formation.
7. The hormones that activates the sympathetic nervous system as a whole.

Define the Terms

1. Endocrine system
2. Hormones
3. Myxoedema
4. Endocrine gland
5. Hormone
6. Target cell
7. Receptor
8. Feedback mechanism

Groups of Terms

Given below are certain groups of terms. In each group the first pair indicates a relationship between the two terms. Rewrite and complete the second pair on a similar basis.

1. Adrenal Cortex: Adrenal Medulla :: Endocrine Gland Layers : ____
2. Oxytocin: Uterine Contraction :: (Hormone) : ____
3. Thyroxine: Metabolic Rate :: (Hormone) : ____
4. Insulin: Glucose Uptake :: (Hormone) : ____
5. Hypothalamus: Hormone Regulation :: (Brain Structure) : ____

Identify the Odd One Out

1. Which of the following glands is not present in females: thyroid, adrenal, pituitary, prostate
2. The conditions cretinism, myxoedema, goitre, scurvy
3. The hormones insulin, glucagon, thyroxine
4. The hormonal sources adrenal cortex, adrenal medulla, cortisone, pituitary
5. Which of the following is the odd one out in terms of hormonal regulation of calcium levels in the body?
(a) Parathyroid hormone
(b) Calcitonin
(c) Vitamin D
(d) Adrenaline

Image-Based Questions

1. Study the diagram given below and then answer the questions that follow:

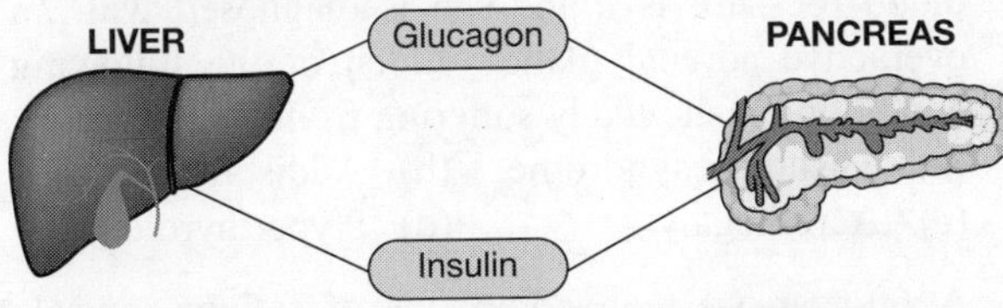

   (a) Name the cells of the pancreas that produce (1) glucagon (2) insulin.
   (b) Why is insulin not given orally but is injected into the body?
   (c) State the main function of (1) glucagon and (2) insulin.

2. The image below shows a gland responsible for secreting a hormone that regulates the sleep–wake cycle. Name the gland and the hormone it secretes.

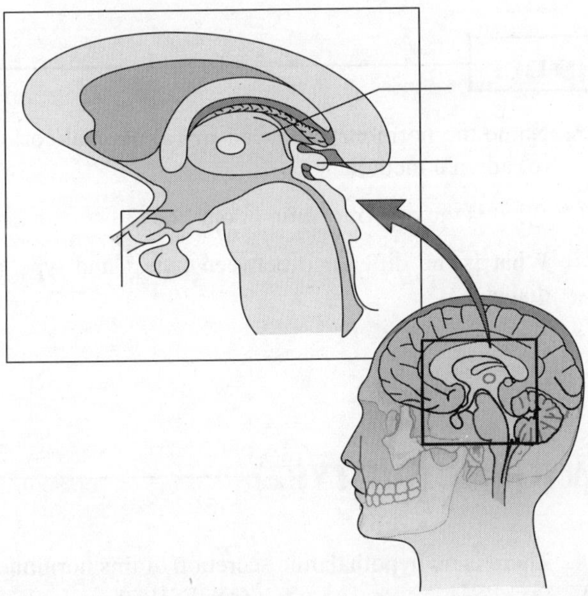

3. Identify the endocrine gland depicted in the image below.

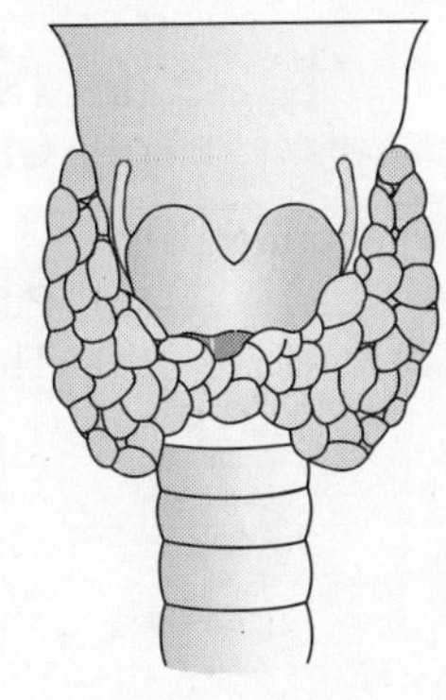

Very Short Answer Questions

1. Why is insulin not given orally but injected into the body?
2. What is the technical term for the cells of the pancreas that produce endocrine hormones?
3. Where in the body is the pancreas located?
4. What are the hormones secreted by the islets of Langerhans of the pancreas?
5. What are the two types of cortical hormones?
6. What are the hormones secreted by the posterior pituitary?
7. Which is the emergency hormone?

Short Answer Questions

1. What distinguishes endocrine glands from other glands?
2. List any two distinctions between an enzyme and a hormone.
3. Why is iodine necessary for our bodies?
4. Your mouth gets dry and your heart rate goes up when you stand up to give your first speech in front of a big crowd. What causes these changes?
5. What are the six hormones secreted by the anterior pituitary?

Long Answer Questions

1. Compare the pace, transmission and overall nature of the changes caused by hormonal reactions with the nervous response.
2. List three key distinctions between the actions of hormones and nerves in the body's regulatory system.
3. What is the endocrine system, and what are its functions?
4. What are the main endocrine glands, and what hormones do they produce?
5. What is the difference between endocrine and exocrine glands?
6. What is the role of hormones in the body?

Case Study-Based Questions

1. A 30-year-old man complains of excessive thirst and frequent urination. His blood glucose levels are high, and he has lost a significant amount of weight in a short time. Which of the following conditions is he likely suffering from?

(a) Cushing's syndrome (b) Addison's disease
(c) Diabetes mellitus (d) Acromegaly

2. A 12-year-old girl has a visible lump in her neck, and she complains of difficulty swallowing and breathing. She is diagnosed with an enlarged thyroid gland. Which of the following conditions is she likely suffering from?
(a) Hypothyroidism (b) Hyperthyroidism
(c) Goitre (d) Thyroid cancer

3. A 50-year-old man presents with excessive sweating, tremors, and a fast heartbeat. He has lost weight and has been feeling anxious and irritable. Blood tests show high levels of thyroid hormone. Which of the following conditions is he likely suffering from?
(a) Hypothyroidism (b) Hyperthyroidism
(c) Cushing's syndrome (d) Addison's disease

4. A 25-year-old woman presents with a hump on her back, a round face, and a protruding belly. She has been experiencing unexplained weight gain, and her blood pressure is high. She is diagnosed with an overactive adrenal gland. Which of the following conditions is she likely suffering from?
(a) Cushing's syndrome (b) Addison's disease
(c) Acromegaly (d) Hyperthyroidism

5. A 60-year-old man complains of fatigue, muscle weakness and joint pain. His blood pressure is low, and he has lost weight in a short time. Blood tests show low levels of cortisol. Which of the following conditions is he likely suffering from?
(a) Cushing's syndrome (b) Addison's disease
(c) Acromegaly (d) Hyperthyroidism

NCERT Corner

1. Name the hormone produced by (a) testes (b) ovaries.
2. What is the role of thyroxine in our body?
3. What is goitre? How can it be prevented?
4. Define diabetes. What are the common symptoms of diabetes?
5. Name the hormone produced by (a) adrenal cortex (b) adrenal medulla.
6. What is the role of insulin in our body?
7. What is the difference between type I and type II diabetes?

COMPETITION WINDOW (OBJECTIVE TYPE)

1. A gland that performs both endocrine and exocrine functions is:
(a) Pituitary (b) Thyroid
(c) Pancreas (d) adrenal

2. Exophthalmic goitre is brought on by excessive activity of:
(a) Thymus (b) Thyroid
(c) Parathyroid (d) Adrenal cortex

3. ADH deficiency results in:
(a) Diabetes mellitus (b) Diabetes insipidus
(c) Dwarfism (d) Acromegaly

4. There is no hypothalamic secretion of this hormone:
(a) PRH (b) FSH
(c) CRH (d) TRH

5. The anterior pituitary produces this hormone most frequently:
(a) LH (b) TSH
(c) ACTH (d) GH

6. This is not an endocrine gland:
(a) Adrenal (b) Pituitary
(c) Lacrimal (d) Thyroid

7. Secretion of cortisone hormone occurs in:
   (a) Medulla of adrenal (b) Cortex of adrenal
   (c) Pancreas (d) Thyroid

8. Which hormone among the following turns extra glucose into glycogen?
   (a) Glucagon (b) Thyroxine
   (c) Insulin (d) Adrenaline

9. Which of the following hormones promotes the conversion of glycogen to glucose in the liver:
   (a) Insulin
   (b) Adrenaline
   (c) Glucagon
   (d) Thyroxin

10. Which pituitary hormone causes monthly follicular development in females and the production of sperm in male testicles?
    (a) Prolactin
    (b) Growth hormone
    (c) Follicle-stimulating hormone
    (d) Luteinizing hormone

11. The 'fight-or-flight' reaction is brought on by these hormones:
    (a) Thyroxine and melatonin
    (b) Insulin and glucagon
    (c) Epinephrine and norepinephrine
    (d) Oestrogen and progesterone

12. What is the difference between endocrine and exocrine glands:
    (a) Endocrine glands release hormones, exocrine glands release waste
    (b) Endocrine glands are interconnected, exocrine glands are totally independent
    (c) Endocrine glands are formed by epithelial tissue, exocrine glands are connective tissues primarily
    (d) Endocrine glands are ductless, exocrine glands release secretions into ducts or at the surface of the body

13. The release of hormones is thought to be very important since it provides support in:
    (a) Controlling body activities
    (b) Regulating blood pressure
    (c) Minimising diabetes level
    (d) All of the above

14. Precise timing and accurate order must be maintained for all of the activities in order for this to happen:
    (a) Proper functioning of the nervous system
    (b) Regulation of the digestive system
    (c) Appropriate regulation of body activities
    (d) Effective maintenance of hormonal balance

15. The term endocrine glands is used frequently for:
    (a) External glands (b) Internal glands
    (c) Duct glands (d) Ductless glands

16. When a general customary function is executed by a set of organs, it is known as:
    (a) Process (b) System
    (c) Method (d) Organisation

17. Hormones play a key role in modulating body functions through which processes?
    (a) Physiological processes
    (b) Psychological processes
    (c) Biological processes
    (d) Neurological processes

18. A certain species' production of hormones provides information about:
    (a) Different effect in other species
    (b) Equivalent effect in other species
    (c) Identical effect in other species
    (d) Null effect in other species

19. Hormone secretion can be excessive or insufficient, which can lead to:
    (a) Minimal effect
    (b) Moderate effect
    (c) Neutral consequences
    (d) Serious consequences

Answer Keys

Multiple Choice Questions

1. (c) 2. (b) 3. (c) 4. (d) 5. (a)
6. (d)

Match the Following

1. (c) 2. (d) 3. (e) 4. (b) 5. (a)

True or False

1. F 2. F 3. F 4. F 5. F
6. T 7. T 8. T 9. F 10. T
11. F 12. F

Fill in the Blanks

1. Endocrine
2. Blood
3. Proteins, amino acid derivates
4. Excess
5. Similar

Assertion–Reason Questions

1. (a) 2. (c) 3. (a) 4. (a) 5. (c)

One-word Answers

1. Insulin and glucagon
2. Epinephrine (adrenaline) and norepinephrine (noradrenaline)
3. Type 2 diabetes
4. Insulin
5. Erythropoietin
6. Antidiuretic hormone (ADH)
7. Epinephrine and norepinephrine

Define the Term

1. The body's chemical processes are coordinated by the endocrine system, which is made up of several endocrine glands and glandular cells.
2. Blood circulates throughout the body carrying hormones, which are physiological secretions that solely affect one or more particular body parts. Particular cells or glands create them (target organ or cells).
3. Patients with hypothyroidism who develop myxoedema, a condition that makes them lethargic and results in swelling of the hands and face.
4. Endocrine gland is a gland that produces and secretes hormones directly into the bloodstream, which then travel to target tissues to regulate various bodily functions.
5. Hormone is a chemical substance produced by an endocrine gland that regulates the activity of specific cells or organs in the body.
6. Target cell is a cell that has receptors for a specific hormone and is capable of responding to its signals.
7. Receptor is a specialised protein on the surface of a cell that binds to a specific hormone or other signaling molecule, initiating a chain of biochemical events that lead to a cellular response.
8. Feedback mechanism is a regulatory system in which the output of a process or pathway feeds back to inhibit or stimulate the activity of the input, helping to maintain homeostasis. In the endocrine system, feedback mechanisms are often used to regulate hormone levels in the body.

Groups of Terms

1. Subdivision of Adrenal Gland
2. (Function)
3. (Effect)
4. (Function)
5. (Function)

Identify the Odd One Out

1. Prostate
2. Scurvy
3. Glucagon
4. Cortisone
5. (d)

Image-Based Questions

1. (a) The cells of the pancreas that produce glucagon are alpha cells, while the cells that produce insulin are beta cells.

   (b) Insulin is not given orally because it gets broken down in the stomach and small intestine by digestive enzymes. This means that it would not be able to reach the bloodstream and have its desired effect on glucose regulation. Instead, it is injected into the body to ensure direct absorption into the bloodstream.

(c) The main function of glucagon is to increase blood glucose levels by promoting the breakdown of glycogen in the liver and releasing glucose into the bloodstream. On the other hand, the main function of insulin is to decrease blood glucose levels by promoting the uptake of glucose by cells and the storage of glucose as glycogen in the liver and muscle tissue.

2. The gland is the pineal gland and the hormone it secretes is melatonin.
3. The endocrine gland depicted in the image is the thyroid gland.

Very Short Answer Questions

1. Insulin cannot be administered orally because the digestive process renders it useless.
2. The technical term for the cells of the pancreas that produce endocrine hormones is islets of Langerhans.
3. The pancreas is located in the abdomen behind the stomach.
4. The pancreatic islets of Langerhans secrete insulin, somatostatin and glucagon.
5. The two cortical hormones are glucocorticoids and mineralocorticoids.
6. The hormones released by the posterior pituitary are vasopressin and oxytocin.
7. Adrenaline is the emergency hormone.

Short Answer Questions

1. Endocrine glands are ductless glands, which implies that they release their hormone secretions directly into the bloodstream as opposed to other glands, which are exocrine glands and have ducts.
2. By regulating and controlling a variety of bodily processes, hormones and enzymes play a crucial part in human health. Enzymes are catalysts because they speed up reactions, but hormones are known as chemical messengers since they start many different processes.
3. The thyroid hormones thyroxine and triiodothyronine, which help in the production of proteins and enzyme function as well as controlling normal metabolism, are made from iodine.
4. The hormone known as adrenaline gets the body ready for any emergency event. Heart rate increases as a result of adrenaline. Moreover, it encourages the digestive system's arterioles to constrict, which reduces the blood flow to the digestive system.
5. Growth hormone (somatostatin), adreno-corticotropic hormone (ACTH), prolactin, follicle-stimulating hormone (FSH), thyroid-stimulating hormone (TSH) and luteinizing hormone (LH)

Long Answer Questions

1. Hint: Focus on the speed of response, method of transmission and duration of effects for each system. Elaborate on this.
2. Hint: Explain their mode of transmission, target cells and speed of response. Elaborate on this.
3. Hint: Consider the glands involved, their secretions (hormones) and their role in maintaining homeostasis. Elaborate on this.
4. Hint: Consider the well-known endocrine glands such as the pituitary gland, thyroid gland and adrenal glands and their respective hormones. Elaborate on this.
5. Hint: Consider the mode of secretion and target of secretion for each type of gland. Elaborate on this.
6. Hint: Explain how hormones act as chemical messengers and regulate various physiological processes. Elaborate on this.

Case Study-Based Questions

1. (c) 2. (c) 3. (b) 4. (a) 5. (b)

NCERT Corner

1. (a) Testosterone (b) Oestrogen
2. Thyroxine regulates the metabolism of carbohydrates, fats and proteins in the body.
3. Goitre is an abnormal enlargement of the thyroid gland. It can be prevented by consuming iodised salt.
4. Diabetes is a metabolic disorder in which the body is unable to produce or properly utilise insulin. The common symptoms of diabetes are frequent urination, increased thirst, hunger, weight loss, blurred vision and fatigue.
5. (a) Cortisol (b) Adrenaline
6. Insulin regulates the uptake and utilisation of glucose by cells in the body.

7. Type I diabetes is an autoimmune disorder in which the body's immune system attacks and destroys the cells that produce insulin, whereas type II diabetes is a metabolic disorder in which the body is unable to properly utilise insulin.

Competition Window

1. (c)	**2.** (b)	**3.** (b)	**4.** (d)	**5.** (c)
6. (c)	**7.** (b)	**8.** (c)	**9.** (c)	**10.** (c)
11. (c)	**12.** (d)	**13.** (d)	**14.** (c)	**15.** (d)
16. (b)	**17.** (a)	**18.** (b)	**19.** (d)	

CHAPTER 13

How Do Organisms Reproduce?

PREREQUISITES

You must have knowledge of:

- Cell division and mitosis
- Structure and functions of male and female reproductive organs
- Sexual and asexual reproduction
- Stages of development in humans and other animals
- Genetics and heredity
- DNA and the process of DNA replication
- Evolution and natural selection

LEARNING OBJECTIVES

At the end of this chapter, you will be able to:

- Recognise that reproduction is a necessary aspect for the survival of a species
- List the modes of reproduction
- Explain the types of asexual reproduction
- Identify the sex organs and describe sexual reproduction
- Recognize and emphasise the changes that occur in the human body as a result of puberty
- Describe the prevention and transmission of sexually transmitted diseases (STDs) and reproductive tract infections (RTIs)

INTRODUCTION

You must have heard from your parents what a happy occasion your birth was! Your parents, grandparents or family friends may have expressed their delight at seeing you take your first steps! They would undoubtedly recall numerous incidents from the past as you grew older.

Growing up is a natural process that all organisms go through. Microbes, plants and animals need to mature to the point where they can reproduce their own species. As a result, the species survives from generation to generation. Reproduction is the biological process that ensures the survival of a species.

Reproduction can take place in two ways: asexual and sexual. Single individuals can reproduce asexually to create bacterial and protozoan progeny. In sexual reproduction, a male gamete fuses with a female gamete and a new individual is born. In plants and animals, this is the more prevalent way of reproduction.

Let us study these two modes of reproduction in detail.

ASEXUAL REPRODUCTION

In smaller creatures such as bacteria and protozoa, as well as some algae, single-celled reproduction occurs in a variety of ways. Asexual reproduction in plants is accomplished through vegetative propagation. Sponges and *Hydra* are examples of animals that reproduce both asexually and sexually.

There are several types of asexual reproduction:

- **Binary fission:** Here, a cell divides into two individuals. This occurs in *Amoeba* and bacteria (Fig. 13.1).

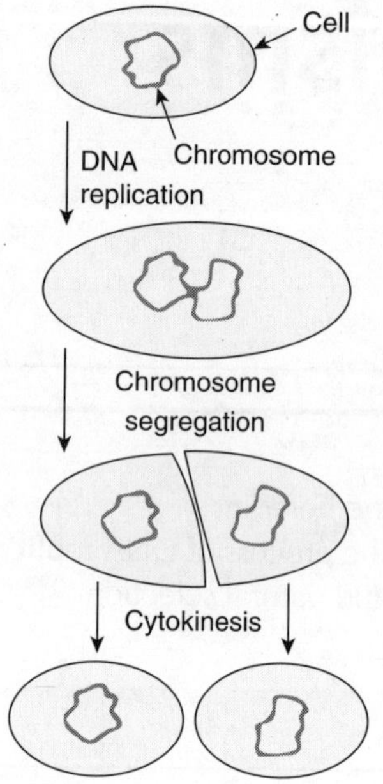

Fig. 13.1 Binary fission

- **Budding:** This occurs when a bud forms from the mother cell's body and remains linked to it. The parental nucleus elongates before splitting into two halves, one of which moves into the bud.

  In multicellular animals such as sponges and *Hydra*, a bud emerges from some portion of the body. It enlarges and then detaches from the parent body once all of its body parts have been produced (Fig. 13.2).

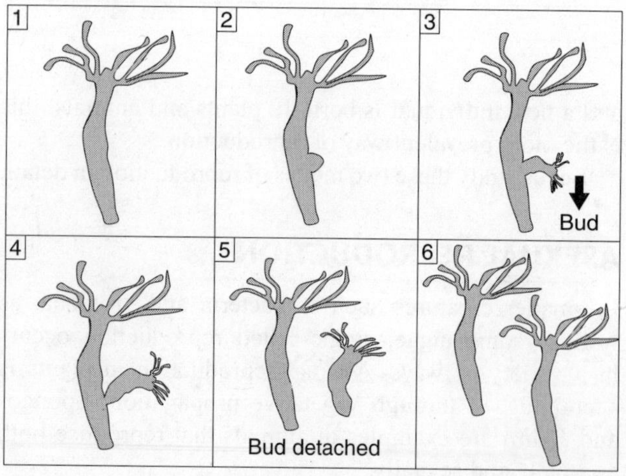

Fig. 13.2 Budding in *Hydra*

- **Spore formation:** Algae such as *Chlamydomonas* divide their cytoplasm and nucleus in order to produce 4–8 spores. Fungi, moss and ferns all produce spores for reproduction. Spores are single cells that develop into new individuals after being released from the parent plant (Fig. 13.3).

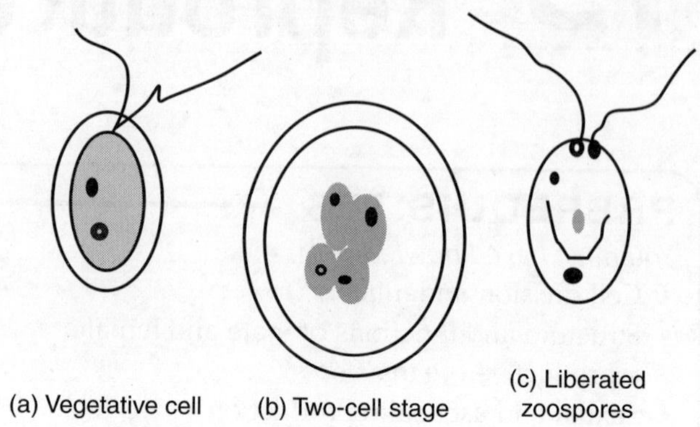

Fig. 13.3 Spore formation in *Chlamydomonas*

Asexual Reproduction in Plants

New plants can grow in nature from vegetative elements such as the root, stem or leaves. This form of asexual reproduction in plants is termed vegetative propagation (Fig. 13.4).

Humans have adapted the natural vegetative propagation methods to artificially propagate a large number of plants. The various artificial propagation methods are:

- **Cutting:** A branch of a plant is cut and planted in the ground. A new plant emerges as a result of the formation of roots.
- **Layering:** A branch of a plant is placed on the ground, with a section of it buried in the dirt. The plant's shoot which will be utilised as stock is chopped off. The stem is sliced vertically, and the scion is placed and taped into the stock. This component takes hold and sprouts a new plant. Table 13.1 lists the modes of reproduction and parts involved of different plants.

Other Methods of Asexual Reproduction

Researchers have used tissue culture to raise offspring from a single parent in the laboratory. Dolly, the cloned sheep, was a replica of her mother.

- **Micropropagation:** Tissue culture procedures have been standardised by researchers. Every live cell or portion of a plant has been discovered to be totipotent, meaning that it has the ability to produce new plants. An individual's cells come from a single cell, the zygote, and hence all the cells have the same

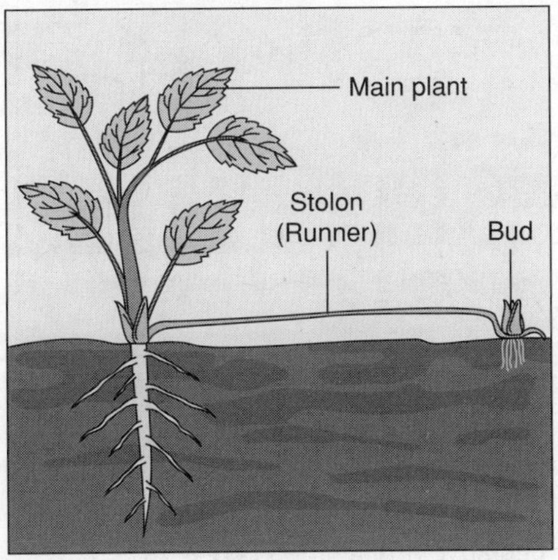

(i) Runner

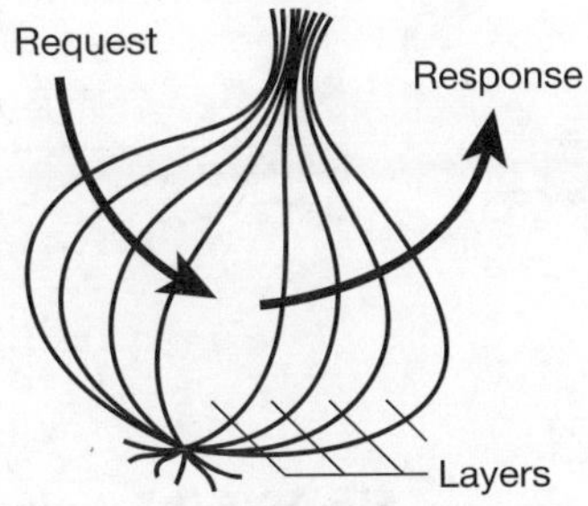

(ii) Bulb

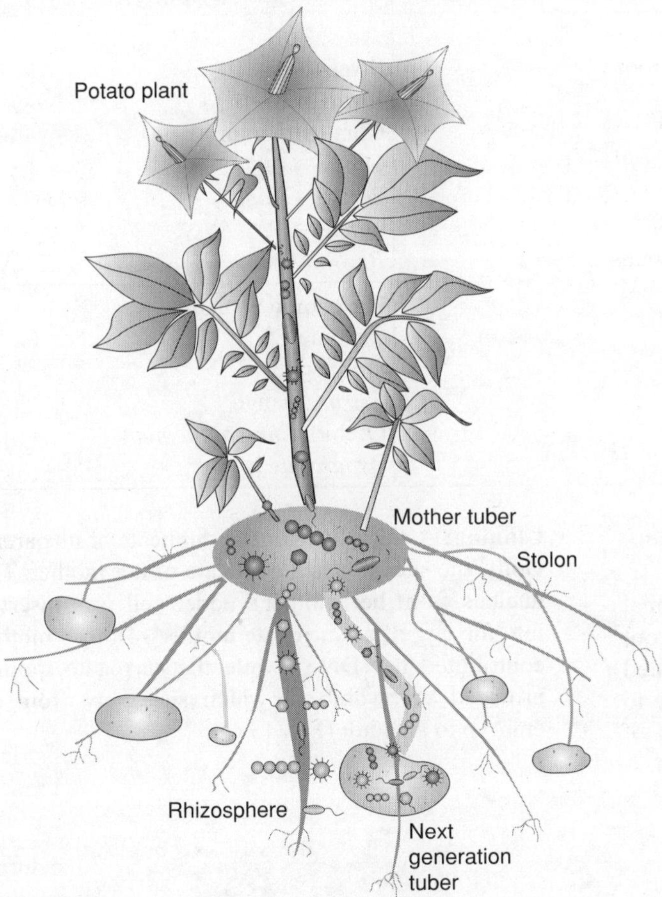

(iii) Potato tuber

(iv) Rhizome of ginger
(Ginger rhizome; Mk2010; 2018; https://commons.wikimedia.org/wiki/File:Ginger_rhizome.jpg. Licensed under: CC-BY-SA-4.0)

(v) Leaf of *Bryophyllum* with buds
(Closeup of a *Bryophyllum daigremontianum*; CrazyD; 2005; https://commons.wikimedia.org/wiki/File:Bryophyllum_daigremontianum_nahaufnahme2.jpg.Licensed under:CC-BY-SA-3.0-migrated)

Fig. 13.4 (a) Natural vegetative propagation (i)–(v)

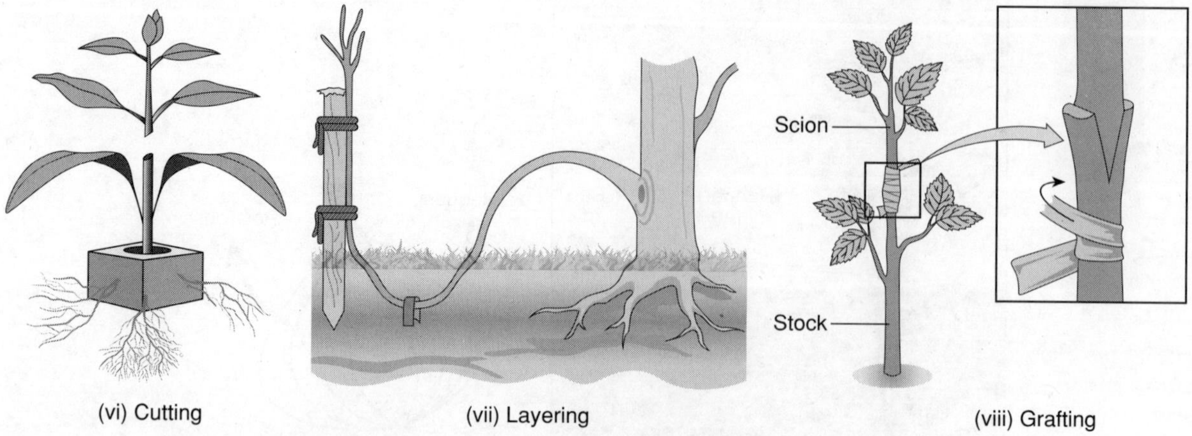

Fig. 13.4 (b) Artificial vegetative propagation (vi)–(viii)

Table 13.1 Modes of reproduction in different plants

Mode of reproduction	Plant part involved	Example
Natural Method		
(a) Root	Adventitious root	Sweet potato
(b) Stem	Runner	Lawn grass
	Sucker	Chrysanthemum
	Bulb	Onion
	Tuber	Potato, *Canna*
	Rhizome	Ginger
	Adventitious buds	*Bryophyllum*
Artificial Method	Cutting	Rose, money plant
	Layering	Jasmine
	Grafting	Grapevine
		Citrus, mango
		Orchid, *Chrysanthemum*, *Asparagus*

genes. Genes play an important role in activities, such as growth, development and reproduction.

Cells may be cultivated in a sufficient nutritional solution from a piece of a plant, such as the root or a leaf of a carrot, to form an undifferentiated clump of cells called a callus. This can give rise to new plantlets. Micropropagation is the process of propagating plants using tissue culture (Fig. 13.5).

- **Cloning:** A clone is a genetic duplicate of its parent. Dolly, the sheep, was a duplicate of her mother. The nucleus from her mother's udder cell was inserted into the egg of a 'surrogate mother'. Dolly's mother contributed her DNA, while the surrogate mother provided the womb in which she grew from an embryo to an adult (Fig. 13.6).

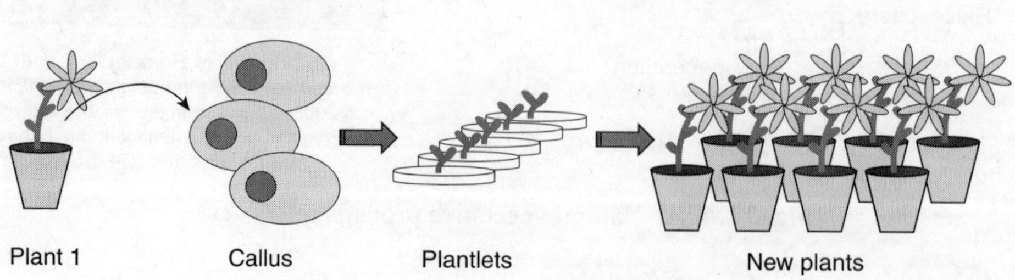

Fig. 13.5 Steps in micropropagation

Fig. 13.6 Dolly, the cloned sheep
(Dolly the Sheep National Museum of Scotland; Sgerbic; 2019; https://commons.wikimedia.org/wiki/File:Dolly_the_Sheep_National_Museum_of_Scotland.jpg. Licensed under: CC-BY-SA-4.0)

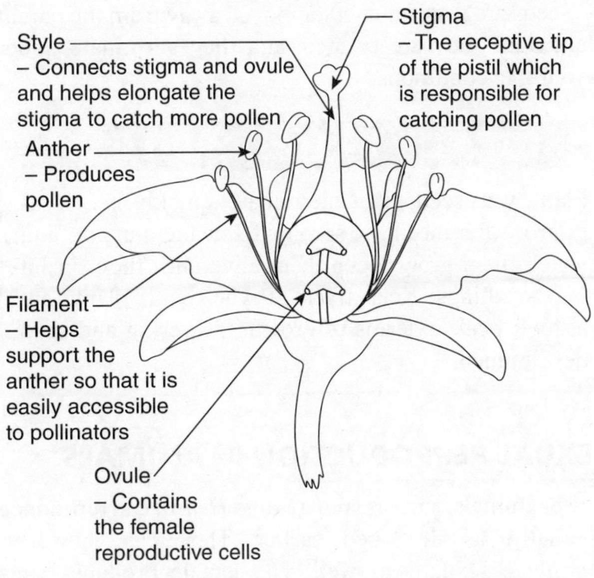

Fig. 13.7 Longitudinal section of a flower

SEXUAL REPRODUCTION IN PLANTS

Sexual reproduction necessitates the joining of male and female gametes. We now have a better understanding of how blooming plants reproduce sexually.

The flower is a flowering plant's reproductive organ (Fig. 13.7). The male portion is made up of stamens (androecium), which produce pollen. Male sex cells can be found in the pollen grains. Each flower may have a number of stamens. There are two components to each stamen: the pollen-bearing anther is the top section of the flower and it is held on the filament, which is the lower component.

The female component of the pistil (gynoecium) is the ovary, which contains eggs, ovules or the female sex cells. The stigma, style and ovary are the components of the pistil. In most plants, every flower has both male and female parts. They are referred to as bisexual. Male flowers with just the androecium and female flowers with only the gynoecium are found in some plants. They are said to be monogamous or unisexual.

Pollination and Fertilisation

The pollen grains and the ovule nuclei are brought together by a variety of forces, including wind, water and insects. Pollination is the transport of pollen grains from the anther to the stigma of a flower. When pollen from a flower fall on its own stigma and fertilises the ovule, it is known as self-pollination. Cross-pollination occurs when pollen from one flower falls on the stigma of another flower from the same species, fertilising the ovule of that flower.

SCIENCE IN DAILY LIFE

Search for the stamen and pistil in a wilted flower. Identify the various sections, then look up the terms for these pieces in your book's images.

As the pollen tube pushes through the pistil, each pollen grain produces a pollen tube, and the pollen grain nucleus reaches the ovule through this tube (Fig. 13.8). Fertilised ovules develop into seeds, which can germinate and grow into seedlings and new plants (Fig. 13.9)

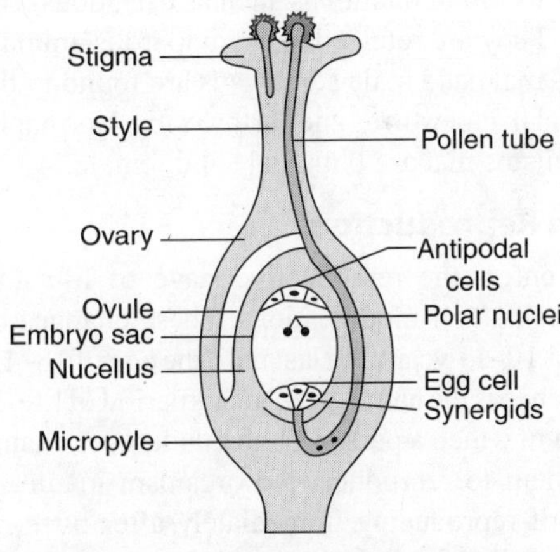

Fig. 13.8 Formation of pollen tube

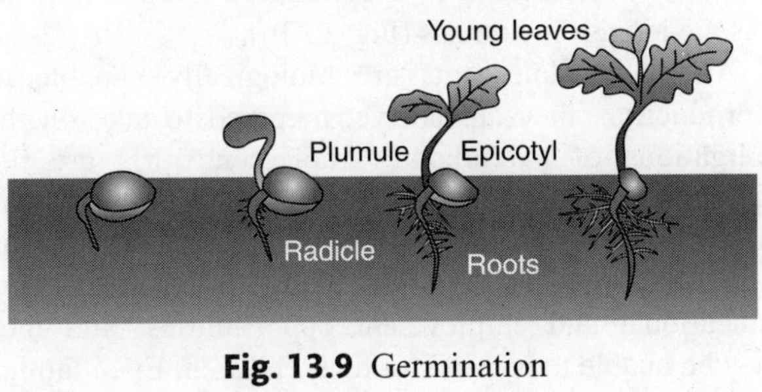

Fig. 13.9 Germination

Seeds are disseminated or moved away from the parent plant once they are created, and they germinate under favourable conditions.

SCIENCE IN DAILY LIFE

Plant wet seeds that germinate quickly (such as gram). After they have sprouted, sow them in pots and watch them grow. Keep them alive until they sprout into seedlings. Keep track of when your plants are at their peak in terms of growth, flowering and seed development.

SEXUAL REPRODUCTION IN ANIMALS

Lower animals, such as sponges and *Hydra*, can reproduce asexually, as described earlier. They can, however, reproduce sexually as well. The female produces eggs and the male generates sperms in all animals. The zygote is formed when an egg and a sperm combine to produce a zygote, which subsequently develops into an embryo, and the embryo becomes a fully formed person. Inside the egg, development may take place partially or totally. Oviparous creatures are those that lay eggs, such as fish, frogs, reptiles and birds. The infant develops inside the mother's womb in mammals such as cats, dogs, cows and humans. They are referred to as viviparous animals.

The female and male sex organs are found in the same individual in tapeworms and earthworms. Hermaphrodites are organisms that are both male and female.

Human Reproduction

As one enters the reproductive stage of life, the body undergoes natural modifications. These changes begin at the age of 10–11 years and last until the age of 18–19 years. Adolescence is the name given to this period of life. Puberty is the era in which a person's body undergoes changes that enable them to reproduce. No organism matures and is capable of reproducing immediately after birth; instead, maturity and adulthood are required. In animals, the time between birth and maturity is quite short. It is possible that it is the longest in humans (Fig. 13.10).

Although adolescents are biologically capable of reproduction, they are not yet prepared to take on the obligations of parenthood. Adolescent girls are not biologically mature for childbearing, and problems can arise for both the adolescent mother and her infant. Adolescent parents are less likely to have adequate educational and employment opportunities, and they may be unable to provide for their children. Even though the legal age of marriage in India is 18 for females and 21 for males, child marriage remains a problem. According to data from the National Family Health Survey carried out in 2005–06, 27% of girls and 3% of boys in the age range of 15–19 years were married at the time of the survey.

Psychological changes: Mental, emotional and intellectual maturity comes in stages:

- Adolescents can reason abstractly.
- They go through mood swings.
- They develop a sense of self-consciousness.
- The importance of one's self-image and identity grows.
- They place a high value on friendship.
- They begin to develop sexual attraction to one another.
- Peer opinions that differ from their personal convictions can cause tension and anxiety because it is crucial for them to fit in with the group standards.

Female Reproductive System

The female reproductive system is found in the lower abdomen. Table 13.2 lists the organs of the female reproductive system and their role. All female animals have an ovary and oviduct, while those that do not lay eggs but give birth to offspring have a uterus.

Figures 13.11 (a) and (b) depict the human female reproductive system and a section of the female reproductive tract, with the egg released from the ovary/fertilised in the fallopian tube and the zygote undergoing development until it reaches the uterus and implants in its wall for further development.

Male Reproductive System

The male reproductive organs are shown in Fig. 13.12. The functions of its parts are listed in Table 13.3.

Fun Fact!

The testes are found within in the scrotal sac, which is located outside the body. This is to ensure that the temperature at which sperms are created is 2°C lower than body temperature, which is necessary for the sperms to survive.

Fertilisation and Embryonic Development

Figure 13.13 depicts the stages of fertilisation. The sperm is formed in the testis (spermatogenesis) and the egg or ovum is produced in the ovary. A zygote is formed when the nuclei of the sperm and egg combine inside the egg. Fertilisation takes place in the fallopian tube (the oviduct).

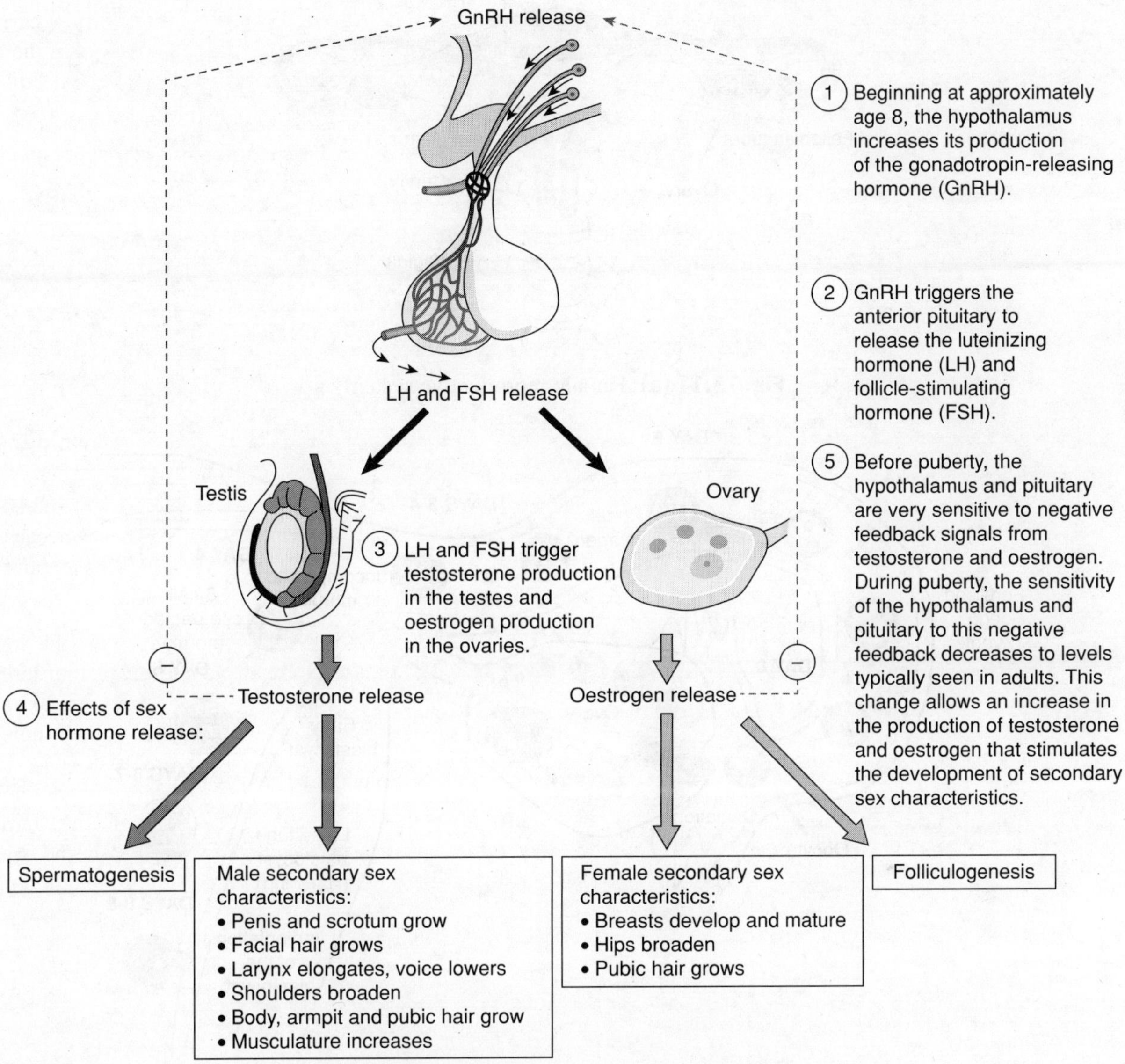

Fig. 13.10 Changes at puberty and adolescence

Table 13.2 The human female reproductive organs and their functions

A pair of ovaries	Produce ova
One cervix	The uterus's opening
One uterus	The womb, the place where the embryo grows and develops
Two fallopian tubes	The tubes through which eggs move from the ovaries to the uterus
One vagina	An opening

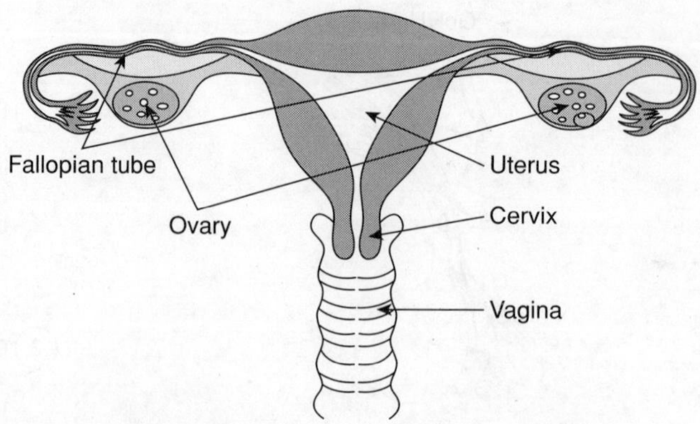

Fig. 13.11 (a) Female reproductive system

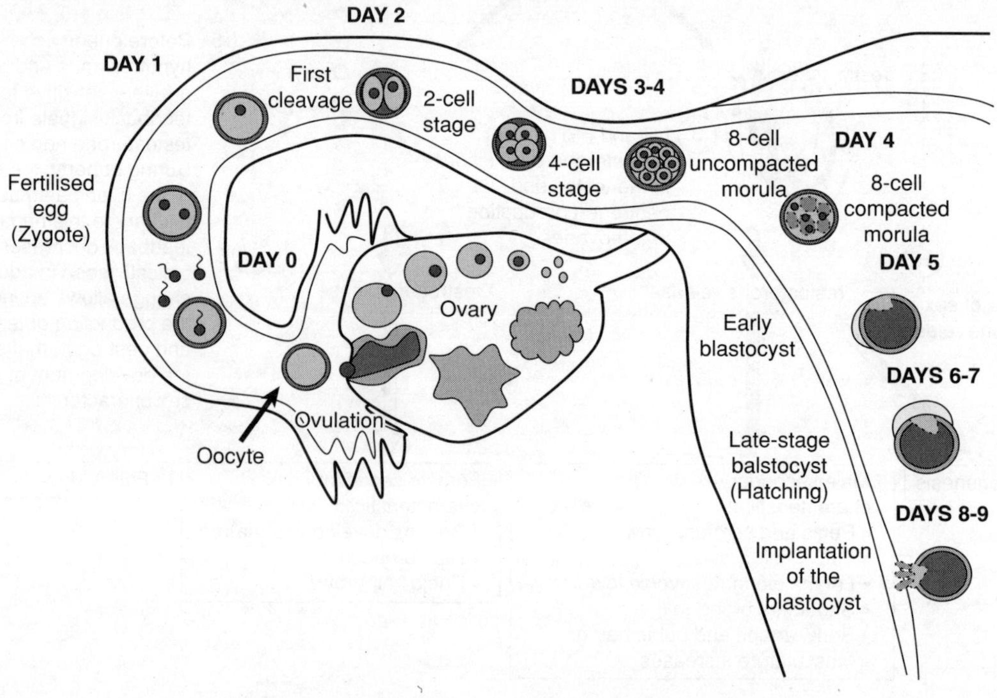

Fig. 13.11 (b) Movement of the fertilised egg

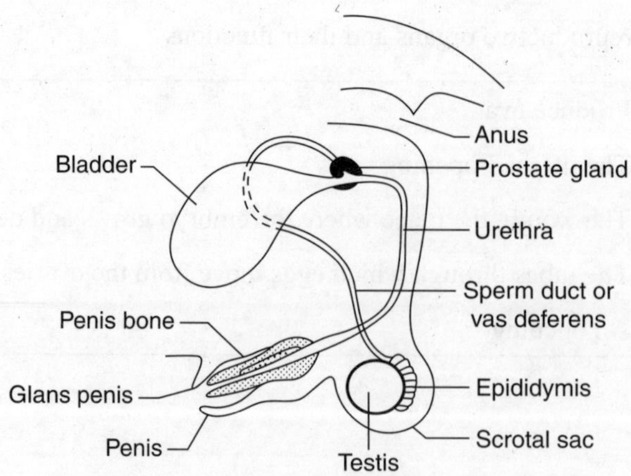

Fig. 13.12 Male reproductive system

Table 13.3 Reproductive organs of the human male

A pair of testes	Produce sperm
Vasa differentia (two sperm ducts)	Each originates in the testis and travels up into the body to unite and form the ejaculatory duct
One ejaculatory duct	A common duct that allows urine and sperm to pass through
One penis	A muscular organ that aids in the transport of sperm into the female body

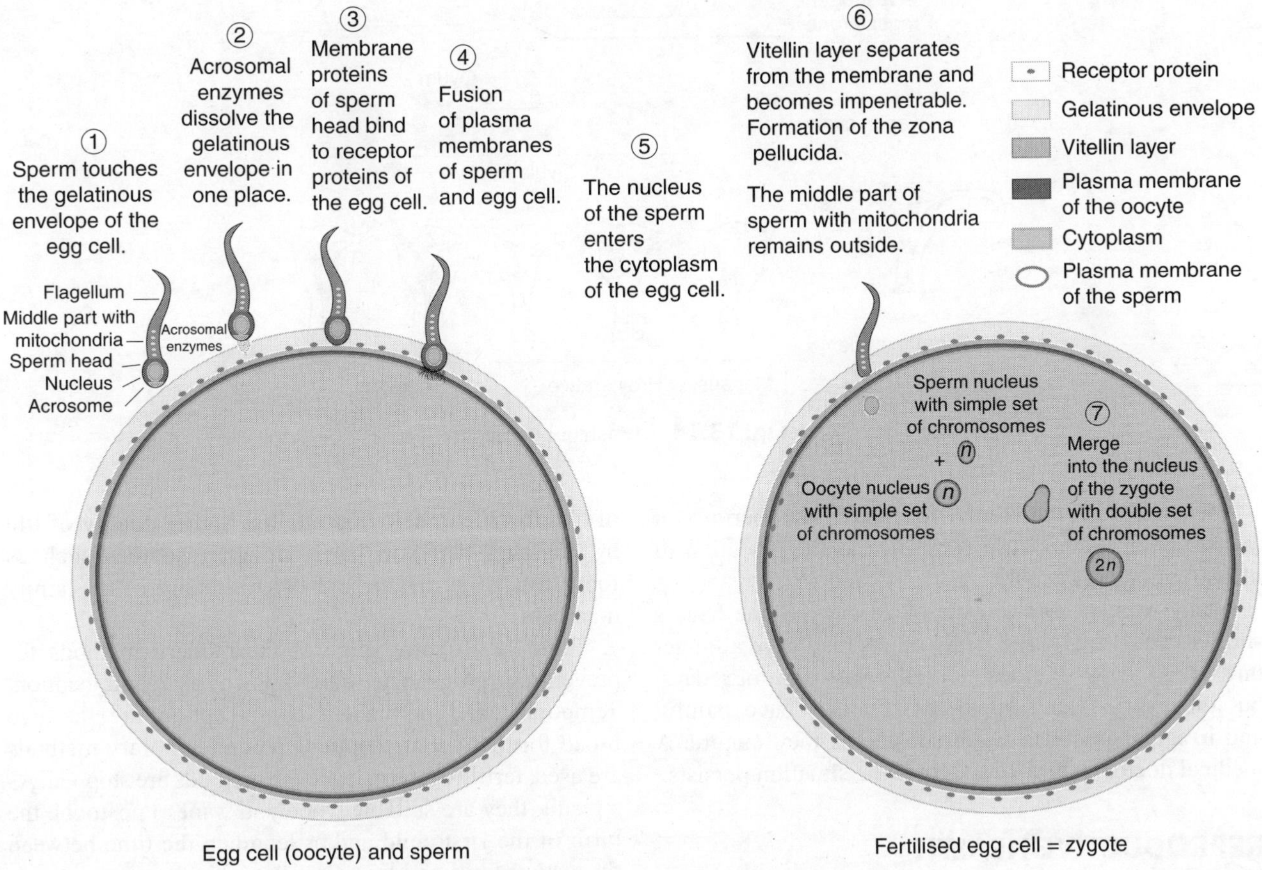

Fig. 13.13 Process of fertilisation

The fertilised egg or zygote begins to divide repeatedly and becomes embedded in the thicker uterine wall in which many blood capillaries have grown when it reaches a stage containing cells and a cavity termed blastocoel. If the egg is not fertilised, it will disintegrate.

The embryo, now known as a foetus, develops into a fully formed individual inside the uterus over the course of 280 days and is born under the influence of oxytocin, a hormone produced by the posterior pituitary.

Menstrual Cycle

A female has a set number of ova when she is born (eggs). These do not mature until adolescence. One ovum matures at a time, and the ovary releases it once every 28–30 days. This occurs as a result of FSH, or the follicle stimulating hormone, produced by the anterior pituitary. Every month, one ovum (egg) is shed alternatively from each ovary.

At the same time, another hormone from the anterior pituitary called LH or lutenising hormones thickens the uterus wall to receive the fertilised egg when the egg (ovum) travels down the fallopian tube. If there is no fertilisation, the uterus' thicker lining and blood vessels are lost, resulting in bleeding. This is referred to as menstruation or periods (Fig. 13.14).

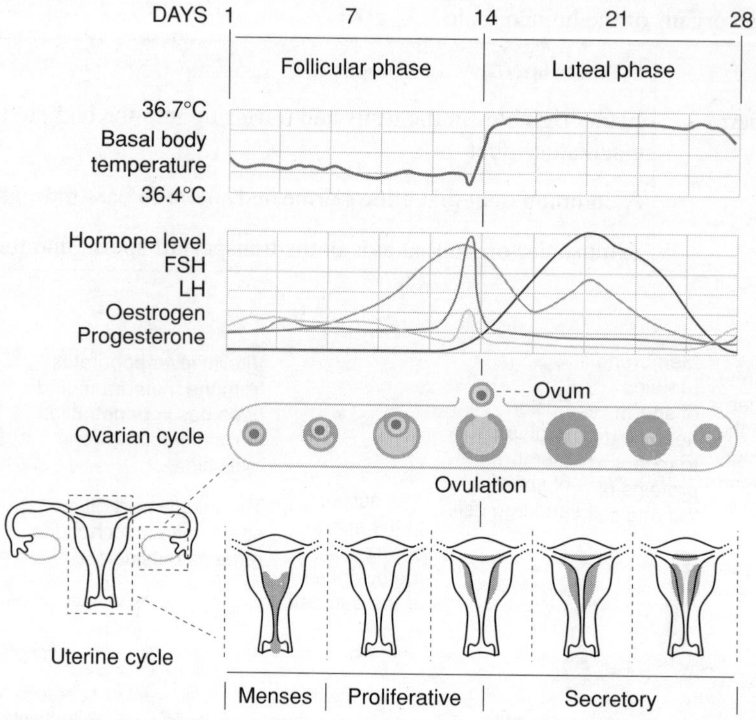

Fig. 13.14 Menstrual cycle

Menarche is the term for the first period of menstruation. Menopause is defined as the cessation of menstruation between the ages of 45 and 55.

Many women have a cycle of 28 days. Some have a shorter cycle of 21 days, while others may have a longer one of 35 days. Periods normally last for four days, but they vary. Many adolescent females have painful and irregular periods, which subside as they mature. A medical doctor should be sought if the situation persists.

REPRODUCTIVE HEALTH

Individual decisions on the size of families help keep the country's population in check. The goals and resources of people drive decisions about family size. Young couples, particularly females, are under pressure to show their fertility and bear male children. Children are unquestionably an important part of any family. Unplanned pregnancy, on the other hand, may threaten the health and wellbeing of both the mother and the child. Furthermore, if the parents are unable to provide for their children, both financially and emotionally, the child may be deprived of opportunities for holistic development.

We know that planning is an important component of decision-making, but it is also critical to have accurate and relevant information in order to make well-informed decisions. Planning the size of the family and the time of childbirth can help you attain a higher quality of life by ensuring that you have enough resources such as food, education, health and the wellbeing of all family members.

Table 13.4 shows some of the modern methods for preventing pregnancy, also known as contraception. Temporary and permanent contraception are the two broad forms of contraception. When temporary methods are used, fertility returns once the methods are stopped. As a result, they are suitable when you want to postpone the birth of the first child and/or lengthen the time between the births of two children. For all practical considerations, permanent techniques are acceptable for couples who have completed their family and do not desire any more children. Contraceptive methods should be used based on the needs of the couple and after consulting a medical professional.

The government has developed a variety of health service delivery organisations at various levels where contraceptive treatments are provided for free or at significantly subsidised costs.

Infections of the reproductive organs are known as reproductive tract infections (RTIs). Poor genital hygiene, such as poor menstrual hygiene in females, can cause several disorders. RTIs, also known as sexually transmitted diseases (STDs), are illnesses that are transmitted from one person to another by sexual intercourse.

Table 13.4 Common methods of contraception

Device name	Function
Condom in males/Diaphragm in females	Physical barrier that prevents the sperm from meeting the egg.
Oral contraceptive pills	Pills inhibit eggs from being released from the ovaries and interfere with ovulation. As a result, fertilisation is impossible. These should be started under the supervision of a qualified medical professional.
Intra Uterine Contraceptive Device (IUCD), for example, Copper-T	Inserted in the female body by a medical practitioner to prevent implantation of the growing embryo.
Vasectomy in males /Tubectomy in females	Surgical treatments for tying the vas deferens /fallopian tube, thereby preventing fertilisation.

STDs

These infections can spread through vaginal or anal intercourse, as well as genital skin contact. Common STDs include gonorrhoea, syphilis, herpes, chlamydia, warts and chancroid. Sexual contact can also transmit the Human Immunodeficiency Virus (HIV).

STD symptoms include:

- Genital itching or pain or anus blisters
- Blisters, lumps and rash in the uro-genital area
- Discoloured discharge from the vagina in females and the penis in males
- Pain while urinating

Most STDs can be cured with prompt and thorough treatment. Infertility can result from untreated STDs. The infected person's sexual partner should likewise seek medical advice and treatment. Until the diseased person is cured, he or she should avoid sexual activity. STDs can be avoided by having only one sex partner and using condoms correctly and consistently.

SUMMARY

- The cells involved in reproduction create a copy of the DNA as well as new cellular components.
- Depending on their bodily composition, various species adopt various means of reproduction.
- Many bacteria and protozoa simply divide into two or more daughter cells when they undergo fission.
- If an organism is split into parts, it can regenerate, as with *Hydra*. They can also produce buds that develop into new organisms.
- Some plants reproduce vegetatively, growing new plants from their roots, stems and leaves.
- In some instances, new generations are produced from a single individual through asexual reproduction.
- Two people must be involved in sexual reproduction in order to create a new individual.
- Variations produced by the DNA copying processes are important for guaranteeing the survival of the species. Sexual reproduction techniques enable the generation of more variety.
- Pollination is the process by which pollen grains are moved from the anther to the stigma to initiate reproduction in flowering plants. It is followed by fertilisation.
- Puberty-related physical changes, such as enlarging breasts in girls and the appearance of new facial hair in boys, signal the development of sexual maturity.
- The testes, which generate the sperm, vas deferens, seminal vesicles, prostate gland, urethra and penis make up the male reproductive system in humans.
- In humans, the female reproductive system consists of the ovaries, fallopian tubes, uterus and vagina.

PRACTICE QUESTIONS

Analyse Your Concepts (School Exam Based)

Multiple Choice Questions

1. Fruits develop out of the:
   (a) Stamen (b) Stigma
   (c) Ovary (d) Ovule
2. Living things require reproduction in order to:
   (a) Keep the individual organism alive
   (b) Fulfil their energy requirements
   (c) Maintain growth
   (d) Continue the species generation after generation
3. The male reproductive parts of a flower, the stamens, are:
   (a) Androecium (b) Filament
   (c) Anther (d) Gynoecium
4. In the figure, parts A, B and C are, sequentially:

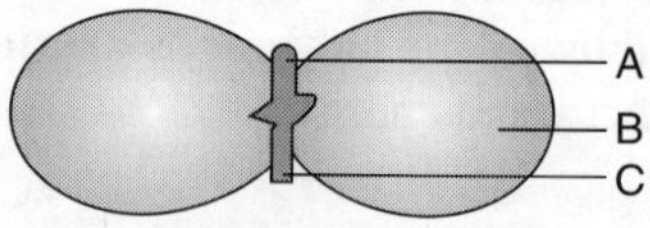

   (a) Cotyledon, plumule and radicle
   (b) Plumule, radicle and cotyledon
   (c) Plumule, cotyledon and radicle
   (d) Radicle, cotyledon and plumule
5. Which of the following is a sexually transmitted disease?
   (a) Kala azar (b) Jaundice
   (c) Cholera (d) Syphilis
6. IUCD is for:
   (a) Vegetative propagation
   (b) Contraception
   (c) Increasing fertility
   (d) Avoiding miscarriage
7. Which among the following is not the function of testes at puberty
   (i) Formation of germ cells
   (ii) Secretion of testosterone
   (iii) Development of placenta
   (iv) Secretion of oestrogen
   (a) (i) and (ii)
   (b) (ii) and (iii)
   (c) (iii) and (iv)
   (d) (i) and (iv)
8. Which is a contraceptive?
   (a) Copper-T
   (b) Condom
   (c) Diaphragm
   (d) All of these
9. The male reproductive system's organs should be placed in the following order to transfer sperm:
   (a) Testis → vas deferens → urethra
   (b) Testis → ureter → urethra
   (c) Testis → urethra → ureter
   (d) Testis → vas deferens → ureter
10. During adolescence, several changes occur in the human body. Mark one change from the following associated with sexual maturation in boys:
    (a) Increase in height
    (b) Cracking of voice
    (c) Weight gain
    (d) Loss of milk teeth

Match the Following

Match the organisms given in Column I with the methods of reproduction/propagation given in Column II:

Column I	Column II
1. *Plasmodium*	a. Spore formation
2. *Spirogyra*	b. Leaves
3. Jasmine	c. Regeneration
4. Apple tree	d. Budding
5. *Bryophyllum*	e. Binary fission
6. Potatoes	f. Layering
7. *Rhizopus*	g. Fragmentation
8. *Hydra*	h. Tubers
9. *Planaria*	i. Cuttings
10. *Leishmania*	j. Multiple fission
11. Sugarcane	k. Grafting
12. Rose	

True or False

Read the following statements and write your answer as true or false.

1. AIDS is caused by Human Immunodeficiency Virus (HIV).
2. Thyroxine is an animal hormone.
3. The narrow opening between the uterus and the vagina is the urethra.
4. The process of fusion of gametes is called fertilisation.
5. The transfer of pollen grains from the anther of a stamen to the stigma of a carpel is called germination.

Fill in the Blanks

Complete the following statements with an appropriate word/term to be filled in the blank spaces.

1. The process of reproduction involving only one cell or one parent is called __________.
2. __________ is a duct from the urinary bladder which carries sperms.
3. The process of fertilisation takes place in the __________ tube in humans.
4. __________ is the ability of an organism to replace its lost body parts.
5. __________ is called the production of new plants from stems, roots or leaves.
6. __________ is the term used to refer to the commencement of menstruation at puberty.
7. __________ is the virus that causes AIDS.
8. Pollen grains contain__________ gametes of a plant.
9. Ovules contain__________ gametes of a plant.
10. The ovary of a flower becomes __________ after fertilisation.
11. The ovule becomes a__________ after fertilisation.
12. Flowering plants reproduce by__________ method of reproduction.
13. The female organ of reproduction in the flower is the __________.
14. The male organ of reproduction in the flower is the __________.
15. The name of the structure in the flower in which the male gamete is formed is __________
16. The __________ at the base of the carpel contains egg cells.
17. The term used to refer to the transfer of pollen from the stamen of one flower to the carpel of another flower of the same species is__________.
18. The cells involved in sexual reproduction are called__________.
19. Fusion of gametes gives rise to a single cell called __________.
20. The process of fusion of gametes is called __________.
21. A multicellular animal starts its life from a __________ through sexual reproduction.
22. The union of a sperm nucleus with an egg nucleus is known as __________ and results in a __________ egg.
23. The menstrual cycle is controlled by __________.

Assertion–Reason Questions

The following questions consists of two statements: Assertion (A) and Reason (R). Answer these questions selecting the appropriate option given below:
For each question, choose the best option based on whether the Assertion and Reason are true and whether the Reason correctly explains the Assertion:

(a) Both A and R are true and R is the correct explanation of A.
(b) Both A and R are true but R is not the correct explanation of A.
(c) A is true but R is false.
(d) A is false but R is true.

1. Assertion (A): In flowering plants, the ovule develops into a seed after fertilisation by a sperm cell.
   Reason (R): The sperm cell enters the ovule through the micropyle and fuses with the egg cell.
2. Assertion (A): Pollination is the transfer of pollen from the stigma to the ovule.
   Reason (R): Pollination can occur through wind, water, or animal pollinators.
3. Assertion (A): In sexual reproduction, the offspring inherit half of their genetic information from each parent.
   Reason (R): This is because during gamete formation, each gamete receives a random assortment of genetic information from the parent cell.

4. Assertion (A): The process of ovulation involves the release of a mature egg cell from the ovary.
   Reason (R): Ovulation is controlled by hormones such as luteinising hormone (LH) and follicle-stimulating hormone (FSH).

5. Assertion (A): Sperm cells are produced in the testes and stored in the epididymis before ejaculation.
   Reason (R): During ejaculation, millions of sperm cells are released into the female reproductive tract.

One-word Answers

1. What is the acrosome is made up of?
2. Fertilisation begins in which part of humans?
3. What is the future shoot hidden in a seed called?
4. Fission is also known as?
5. Name one components of the male reproductive system in humans.

Define the Term

1. Replication
2. Mitosis
3. Fertilisation
4. Micropyle
5. Hormones

Groups of Terms

Given below are certain groups of terms. In each group the first pair indicates a relationship between the two terms. Rewrite and complete the second pair on a similar basis.

1. Hormone Production: Gonads :: (Gamete Formation) : _____
2. Ovulation: Release of an Egg :: (Spermatogenesis) : _____
3. Fertilisation: Fusion of Egg and Sperm :: (Pollination) : _____
4. Mitosis: Cell Division for Growth and Repair :: (Meiosis) : _____
5. Asexual Reproduction: Offspring Produced without Gametes :: (Sexual Reproduction) : _____

Identify the Odd One Out

1. Meiosis, mitosis, gamete, fertilisation
2. Binary fission, budding, regeneration, conjugation
3. Spermatogenesis, oogenesis, parthenogenesis, zygote
4. Sexual reproduction, asexual reproduction, fragmentation, regeneration
5. Embryo, zygote, ovary, testis

Image-Based Questions

1. Name the parts labelled A, B, C, D and E.

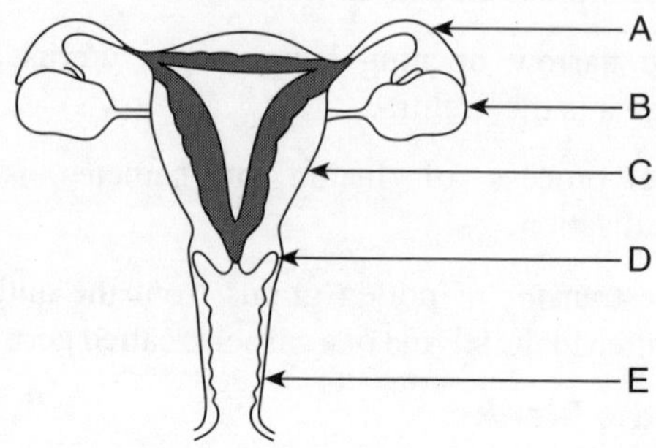

2. Draw a neat diagram of a flower showing its various parts.
   (a) In this diagram, mark the stem, receptacle, sepals, petals, stamen and carpel.
   (b) What name is given to (i) all the petals of a flower and (ii) all the sepals of a flower?
   (c) What are (i) stamen and (ii) carpel in a flower?
   (d) What is another name for the carpel of a flower?
   (e) What is the name of the yellow powdery substance present in the anther of a flower?

3. Show the direction of the pollen tube from the pollen on the stigma in the embryo sac in the given diagram.

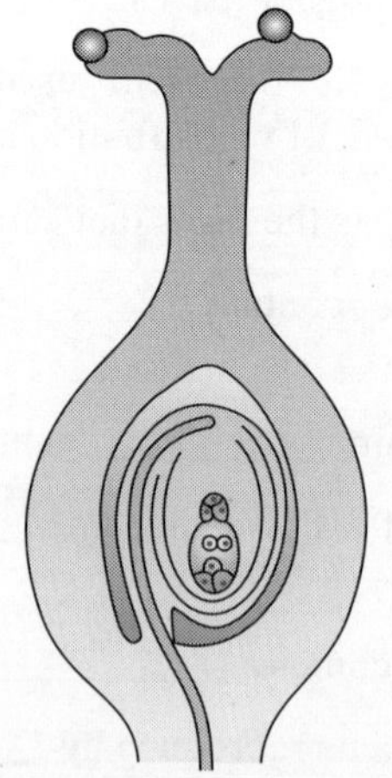

4. Look at the below picture. It shows a mode of reproduction. Answer the following questions based on this picture.

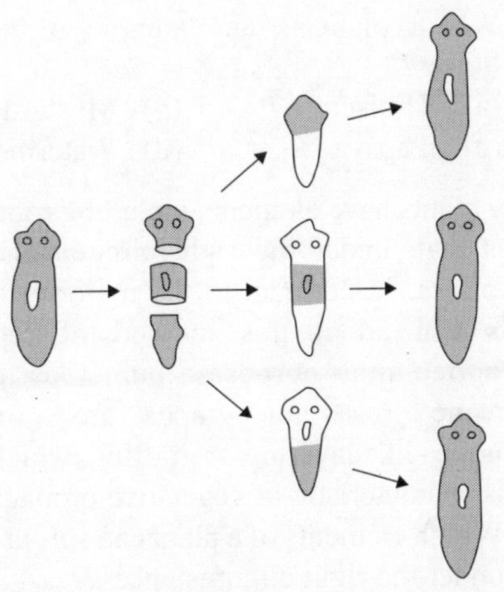

1. What does this picture sho? (i) Budding (ii) Regeneration (iii) Tissue culture (iv) Vegetative propagation
2. It is a kind of: (i) Sexual reproduction (ii) Asexual reproduction
3. Which organism is shown in the picture above? (i) *Planaria* (ii) *Plasmodium* (iii) *Paramecium* (iv) *Spirogyra*
4. *Planaria* and *Hydra* are ______ organisms. (i) Simple (ii) Complex
5. Regeneration is not the same as reproduction. True or false? (i) True (ii) False

5. Label the diagram.

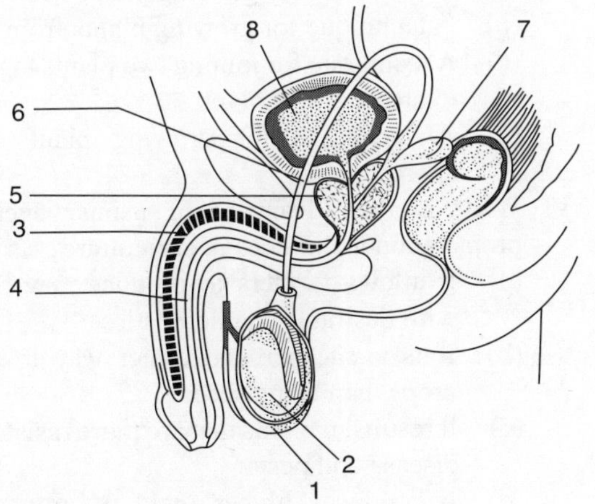

Very Short Answer Questions

1. When a cell reproduces, what happens to its DNA?

2. What is DNA?

3. Name the life process of an organism that helps in the growth of its population.

4. What is cell division?

5. What is germination?

Short Answer Questions

1. How do organisms reproduce?

2. What are the different modes of asexual reproduction?

3. How does sexual reproduction occur in plants?

4. What is budding?

5. What is common in the functions performed by the nucellus and cotyledon?

Long Answer Questions

1. Explain triple fusion in plants and how it occurs.

2. Explain the female reproductive system and its parts.

3. Explain the male reproductive system and its parts.

4. Write down the basic differences between asexual and sexual reproduction.

5. Why are population control measures important?

6. What changes occur in girls and boys in the age group of 10–14 years?

7. Describe sexually transmitted diseases (STDs) and mention the ways of preventing them.

8. Name the surgical methods of birth control in human males and females, respectively.

9. What is grafting? Why is it used in horticulture?

10. What are the different methods of contraception?

11. Write the reasons for adopting contraceptive methods.

Case Study-Based Questions

1. Thousands of immature eggs are present in the ovaries when a girl is born. Some of these start maturing once she reaches puberty. One of the ovaries releases one egg each month. Through a delicate oviduct or fallopian tube, the egg is transported from the ovary to the womb. The uterus is an elastic, bag-like structure formed by the union of the two oviducts.

(i) How many eggs are typically released by one ovary each month?

(a) None (b) One
(c) Two (d) Three

(ii) Which structure transports the egg from the ovary to the womb?
(a) Uterus (b) Vagina
(c) Fallopian tube (d) Cervix

(iii) When do some of the immature eggs start to mature?
(a) At birth
(b) During childhood
(c) During adolescence
(d) During menopause

(iv) What is the uterus?
(a) A structure that releases an egg each month
(b) A structure that transports the egg from the ovary to the womb
(c) A structure that houses the developing foteus during pregnancy
(d) A structure that produces the hormones necessary for pregnancy

(v) How many oviducts unite to form the uterus?
(a) One (b) Two
(c) Three (d) Four

2. Angiosperms' reproductive organs are found in the flower. The sepals, petals, stamens and pistil are the various components of a flower. The reproductive components of a flower, which also contain the germ cells, are the stamens and the pistil. When a flower has stamens or anthers, it can be unisexual (papaya, watermelon). When it has stamens and a pistil, it is said to be bisexual (hibiscus, mustard).

(i) In which part of the flower are the reproductive organs found?
(a) Sepals
(b) Petals
(c) Stamens and pistil
(d) Stamen and anthers

(ii) Which components of the flower contain the germ cells?
(a) Sepals and petals
(b) Stamens and pistil
(c) Sepals and pistil
(d) Stamens and anthers

(iii) What is a unisexual flower?
(a) A flower with stamens and a pistil
(b) A flower with stamens or anthers
(c) A flower with only sepals and petals
(d) A flower that does not contain germ cells

(iv) What is a bisexual flower?
(a) A flower with stamens and a pistil
(b) A flower with only sepals and petals
(c) A flower that does not contain germ cells
(d) A flower with stamens or anthers

(v) Which plant is an example of a unisexual flower?
(a) Hibiscus (b) Mustard
(c) Papaya (d) Watermelon

3. Many plants have elements including root, stem, and leaves that, under the right circumstances, sprout new plants. In contrast to the majority of animals, plants can adopt this method of reproduction. For agricultural purposes, numerous plants like sugarcane, roses or grapes are grown using techniques like layering or grafting, which make use of this characteristic of vegetative propagation.

(i) Which elements of a plant can sprout new plants under the right circumstances?
(a) Roots, stems, and leaves
(b) Flowers and fruits
(c) Seeds and spores
(d) All of the above

(ii) What is the name of the agricultural technique that uses the characteristic of vegetative propagation to grow new plants?
(a) Irrigation (b) Pollination
(c) Layering (d) Photosynthesis

(iii) Which plant is an example of a crop that can be grown using vegetative propagation techniques?
(a) Wheat (b) Sugarcane
(c) Carrots (d) Broccoli

(iv) What is grafting?
(a) A technique for growing plants from seeds
(b) A technique for growing plants from roots
(c) A technique for joining two plants together to create a new plant
(d) A technique for growing plants from spores

(v) What is the advantage of using vegetative propagation techniques in agriculture?
(a) It allows farmers to produce new plants with desirable characteristics
(b) It is a cheaper and easier way to grow crops than using seeds
(c) It results in plants that are more resistant to disease and pests
(d) It produces higher yield of crops than traditional methods

4. In the form of DNA molecules, chromosomes in the nucleus of a cell store information for the inheritance of traits from parents to the next generation. The information source for constructing proteins is the DNA found in the cell nucleus. If the data is modified, different proteins will be produced. Eventually, different proteins will change how our bodies look.

Therefore, the act of making a DNA duplicate is a fundamental aspect of reproduction. Cells create copies of their DNA through chemical processes.

(i) Where is genetic information stored in a cell?
  (a) Mitochondria
  (b) Ribosomes
  (c) Nucleus
  (d) Golgi apparatus

(ii) What is the role of DNA in protein synthesis?
  (a) DNA is the source of energy for protein synthesis
  (b) DNA provides the instructions for protein synthesis
  (c) DNA carries the finished proteins to their destination in the cell
  (d) DNA breaks down proteins into their component amino acids

(iii) What happens if DNA is modified?
  (a) It does not affect protein synthesis
  (b) It leads to the production of different proteins
  (c) It results in the production of fewer proteins
  (d) It causes the cell to die

(iv) What is the significance of DNA duplication in reproduction?
  (a) It ensures that the offspring inherit the same genetic information as the parent
  (b) It allows the cell to repair damaged DNA
  (c) It produces new genetic information for the offspring
  (d) It ensures that the DNA remains stable and does not change over time

(v) How do cells create copies of their DNA?
  (a) Through photosynthesis
  (b) Through respiration
  (c) Through chemical processes
  (d) Through osmosis

5. The ovulation cycle is the process by which a woman's body releases an egg from the ovary, allowing for the possibility of pregnancy. This cycle is controlled by various hormones and involves several stages, including follicular development, ovulation, and luteal phase.

(i) Which hormone is responsible for the initial development of a follicle in the ovary?
  (a) Follicle-stimulating hormone (FSH)
  (b) Luteinising hormone (LH)
  (c) Oestrogen
  (d) Progesterone

(ii) When does ovulation typically occur in the menstrual cycle?
  (a) On the first day of the cycle
  (b) Around day 7 of the cycle
  (c) Around day 14 of the cycle
  (d) Around day 21 of the cycle

(iii) What happens during ovulation?
  (a) The uterus sheds its lining
  (b) An egg is released from the ovary
  (c) A fertilised egg implants in the uterus
  (d) The cervix opens to allow sperm to enter

(iv) Which hormone is responsible for the development of the corpus luteum?
  (a) Follicle-stimulating hormone (FSH)
  (b) Luteinising hormone (LH)
  (c) Oestrogen
  (d) Progesterone

(v) What is the role of progesterone in the menstrual cycle?
  (a) It triggers ovulation
  (b) It causes the uterus to shed its lining
  (c) It prepares the uterus for pregnancy
  (d) It regulates the production of FSH and LH

NCERT Corner

1. (a) Describe in detail the terms 'self pollination' and 'cross-pollination'.
  (b) How do insects support cross-pollination?

2. How is the process of pollination different from fertilisation?

3. Name one sex hormone.

4. (a) What is the name of surgical method of birth control in human males in which the sperm ducts are cut and ligated (tied) at both ends?
  (b) What is the name of surgical method of birth control in human females in which the oviducts are cut and ligated (tied) at both ends?
  (c) Name the contraceptive device used by the human males which acts as a sheath over the male organ and traps the sperms in it.
  (d) Name the contraceptive device used by human females which is placed over the cervix.

5. Give an example of a plant hormone that promotes growth.

COMPETITION WINDOW (OBJECTIVE TYPE)

1. In favourable conditions, *Amoeba* reproduces by:
   (a) Binary fission (b) Multiple fission
   (c) Budding (d) Fragmentation
2. A feature of reproduction that is common to yeast, *Amoeba* and *Spirogyra* is that:
   (a) They are all multicellular
   (b) They are all unicellular
   (c) They reproduce only sexually
   (d) They reproduce asexually
3. The ability of a cell to divide into several cells during reproduction in *Plasmodium* is called:
   (a) Budding
   (b) Binary fission
   (c) Multiple fission
   (d) Reduction division
4. *Bryophyllum* is propagated vegetatively by the process known as:
   (a) Bud (b) Leaf
   (c) Root (d) Stem
5. *Spirogyra* reproduce by the process known as:
   (a) Budding (b) Regeneration
   (c) Fragmentation (d) Fission
6. In *Rhizopus*, tubular thread-like structures bearing sporangia at their tips are called:
   (a) Hyphae (b) Filaments
   (c) Rhizoids (d) Roots
7. Plants like banana, rose, jasmine and orange have lost the capacity to produce:
   (a) Roots (b) Buds
   (c) Flowers (d) Seeds
8. The flower of the *Hibiscus* plant is:
   (a) Bisexual (b) Unisexual
   (c) Neuter (d) Very small
9. The part of the flower which is present at the centre of the flower is:
   (a) Carpels (b) Petals
   (c) Sepals (d) Stamens
10. The seed that contains the future plant is called the:
    (a) Cotyledon
    (b) Seed coat
    (c) Embryo
    (d) Germ cell
11. The period of pregnancy is called:
    (a) Incubation period
    (b) Gestation period
    (c) Ovulation
    (d) Menstruation period
12. The process of release of eggs from the ovary is called:
    (a) Menstruation (b) Ovulation
    (c) Insemination (d) Reproduction
13. The period during adolescence when the reproductive tissues begin to mature is called:
    (a) Ovulation (b) Puberty
    (c) Germination (d) Propagation
14. In human beings, fertilisation occurs in the:
    (a) Uterus (b) Ovaries
    (c) Vagina (d) Fallopian tubes
15. Along the path of the vas deferens, the secretions of which gland provide nutrition to the sperms?
    (a) Seminal vesicles
    (b) Urinary bladder
    (c) Scrotum
    (d) Prostate glands
16. The embryo in humans gets nutrition from the mother's blood with the help of a special tissue called:
    (a) Villi (b) Placenta
    (c) Uterus (d) Womb
17. Which among the following diseases is not sexually transmitted?
    (a) Hepatitis (b) Syphilis
    (c) HIV-AIDS (d) Gonorrhoea
18. Which of the following methods of contraception protects us from acquiring sexually transmitted diseases?
    (a) Copper-T (b) Condoms
    (c) Oral pills (d) Surgery
19. In human males, the testes lie in the scrotum, because it helps in the:
    (a) Process of mating
    (b) Formation of sperms
    (c) Easy transfer of gametes
    (d) Secretion of oestrogen

20. In vegetative reproduction, the new individuals are genetically:
(a) Similar (b) Dissimilar
(c) Abnormal (d) None of these

21. An organism divides into several parts and each component grows into a separate individual. This is called:
(a) Budding (b) Binary fission
(c) Regeneration (d) Spore formation

22. Fertilisation of the ovum in females takes place in the:
(a) Vagina (b) Ovary
(c) Uterus (d) Fallopian tubes

23. The portion of plant that is grafted is called:
(a) Stock
(b) Scion
(c) Stalk
(d) Stem

24. What portion of the bloom gives rise to the fruit?
(a) The whole flower
(b) Only the stamens and carpel
(c) Only the ovary
(d) Only the carpel

25. The agent of anemophilic pollination is:
(a) Birds (b) Rain
(c) Insects (d) Wind

Answer Keys

Multiple Choice Questions

1. (c) 2. (d) 3. (a) 4. (c) 5. (d)
6. (b) 7. (c) 8. (d) 9. (a) 10. (b)

Match the Following

Column I	Column II
1. *Plasmodium*	j. Multiple fission
2. *Spirogyra*	g. Fragmentation
3. Jasmine	f. Layering
4. Apple tree	k. Grafting
5. *Bryophyllum*	b. Leaves
6. Potatoes	h. Tubers
7. *Rhizopus*	a. Spore formation
8. *Hydra*	d. Budding
9. *Planaria*	c. Regeneration
10. *Leishmania*	e. Binary fission
11. Sugarcane	i. Cuttings
12. Rose	i. Cuttings

True or False

1. T 2. T 3. F 4. T 5. F

Fill in the Blanks

1. Uniparental/ Asexual reproduction
2. Vas deferens
3. Fallopian
4. Regeneration
5. Vegetative propagation
6. Menarch
7. HIV or Human Immunodeficiency Virus
8. Male
9. Female
10. Fruit
11. Seed
12. Sexual
13. Carpel
14. Stamen
15. Anther
16. Ovary
17. Cross pollination
18. Gametes
19. Zygote
20. Fertilisation
21. Single cell (zygote)
22. Fertilisation; fertilised
23. Hormones

Assertion–Reason Questions

1. (a) 2. (c) 3. (a) 4. (a) 5. (a)

One-word Answers

1. Golgi bodies 2. Oviduct
3. Plumule 4. Binary fission
5. testes

Define the Term

1. Replication: The process of copying genetic information (DNA) to create an exact replica of the original molecule.
2. Mitosis: The process of cell division in which a single cell divides into two identical daughter cells, each having the same number of chromosomes as the parent cell.
3. Fertilisation: The fusion of a sperm cell and an egg cell to form a zygote, which is the first cell of a new organism.
4. Micropyle: A small opening in the outer covering of a plant ovule through which the pollen tube can enter to fertilize the egg.
5. Hormones: Chemical messengers produced by endocrine glands in the body that regulate various physiological processes such as growth, development, metabolism, and reproduction.

Groups of Terms

1. (Reproductive Organs)
2. (Production of Sperm)
3. (Transfer of Pollen)
4. (Cell Division for Gamete Production)
5. (Offspring Produced from Union of Gametes)

Identify the Odd One Out

1. Gamete
2. Conjugation
3. Parthenogenesis
4. Regeneration
5. Ovary

Image-Based Questions

1. A – Oviduct or Fallopian tube, B – Ovary, C – Uterus, D – Cervix, E – Vagina.
2. (a)

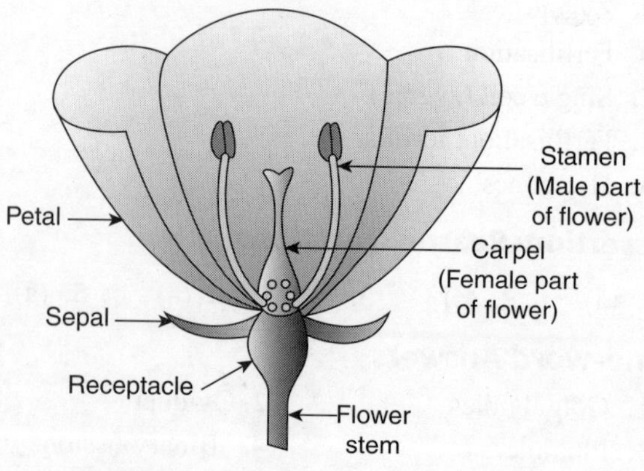

(b) (i) The corolla, (ii) The calyx

(c) (i) Stamen is the male reproductive organ in a flower. (ii) The carpel is the female reproductive organ in a flower.

(d) Another name for carpel is pistil.

(e) The yellow powdery substance present in the anther of a flower is pollen grains.

3.

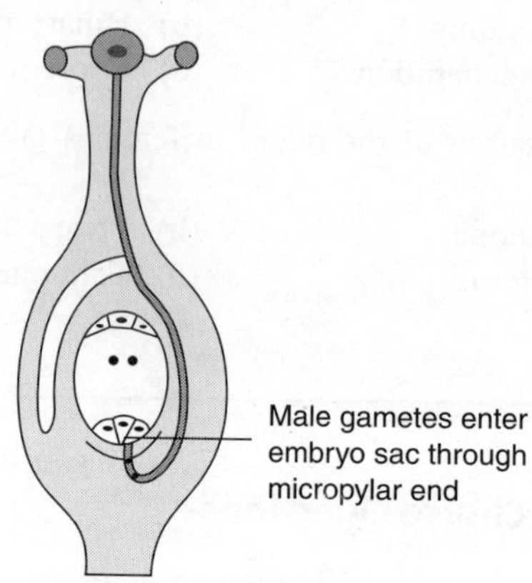

Path of the pollen tube into the embryo sac:

- The pollen tube grows through the tissues of the stigma and the style.
- It enters through the micropylar end into the ovule.
- The synergid cell of the egg apparatus secretes chemicals and directs the movement of the pollen tube inside the synergid through the filiform apparatus.
- The pollen tube penetrates and releases two male gametes into the synergid cells.

4. 1. (ii), 2. (ii) 3. (i) 4. (i) 5. (i)
5. 1-scrotum, 2-testis, 3-penis, 4-urethra, 5-prostrate gland, 6-vas deferns, 7-seminal vesicles, 8-urinary bladder

Very Short Answer Questions

1. When a cell reproduces, DNA replication occurs which forms two similar copies of DNA.
2. DNA (deoxyribonucleic acid) is a polymer made up of large number of nucleotide units. It carries genetic information from generation to generation.
3. Reproduction is a life process that helps in multiplication of an organism and growth of its population.
4. Cell division is a fundamental process in which a parent cell divides to produce two or more daughter cells. It is essential for growth, development, and the replacement of damaged or old cells.
5. Germination is the process by which a seed sprouts and begins to grow into a new plant. It involves the activation of the dormant embryo inside the seed

and the emergence of the radicle (root) and plumule (shoot).

Short Answer Questions

1. Organisms reproduce in two ways: sexually, where the fusion of male and female gametes takes place, and asexually, where the fusion of male and female gametes does not take place.
2. The different modes of asexual reproduction include:
    - Fission
    - Budding
    - Fragmentation
    - Spore-formation
    - Vegetative propagation
3. Plants reproduce sexually through pollination. In this process, the pollen from the anther of the flower is transferred to the stigma. The fusion of the male and female gametes takes place. A zygote is then formed which gives rise to an embryo. This leads to the formation of seed which gradually turns into a fruit.
4. In this method of asexual reproduction, cell division is used to create a new organism from an outgrowth. It stays tethered to the mother and feeds off of her. When it reaches maturity, it separates from the mother and develops into a distinct organism. In *Hydra*, this kind of reproduction is seen.
5. The common function performed by nucellus and cotyledon is to provide nutrition to the developing embryo. In gymnosperms, the nucellus acts as a food source for the growing embryo while in angiosperms, the cotyledons store and provide nutrients to the developing embryo until it is able to perform photosynthesis on its own.

Long Answer Questions

1. Hint: Triple fusion is a process that occurs during double fertilisation in plants. It involves the fusion of two polar nuclei with a sperm cell. Think about the specific structures involved and the resulting products. Elaborate on this.
2. Hint: Focus on the major organs and structures involved in the female reproductive system, such as the ovaries, fallopian tubes, uterus, and vagina. Consider their functions and roles in reproduction. Elaborate on this.
3. Hint: Focus on the major organs and structures involved in the male reproductive system, such as the testes, epididymis, vas deferens, and penis. Consider their functions and roles in reproduction. Elaborate on this.
4. Hint: Explain the mode of reproduction, involvement of gametes, genetic variation, and the number of parents involved in each type. Elaborate on this.
5. Hint: Consider the impact of population growth on resources, environment, and social well-being. Give the reasons why controlling population growth is necessary. Elaborate on this.
6. Hint: Focus on the onset of puberty and the associated physical changes that occur in both girls and boys during this period. Elaborate on this.
7. Hint: Explain the types of STDs, their modes of transmission, common symptoms and methods of prevention such as practicing safe sex and getting regular check-ups. Elaborate on this.
8. Hint: Consider the common surgical methods used for male and female sterilisation as forms of permanent birth control. Elaborate on this.
9. Hint: Grafting is a horticultural technique that involves joining two plant parts to create a new plant. Think about the reasons why grafting is used, such as creating desired characteristics, improving plant health, and enhancing growth. Elaborate on this.
10. Hint: Consider various methods of contraception such as barrier methods, hormonal methods, intrauterine devices (IUDs), fertility awareness methods, and surgical methods. Elaborate on this.
11. Hint: Explain the benefits of using contraceptive methods, such as preventing unwanted pregnancies, spacing births, promoting family planning, and protecting against sexually transmitted infections. Elaborate on this.

Case Study-Based Questions

1. (i) (b), (ii) (c), (iii) (c), (iv) (c), (v) (b)
2. (i) (c), (ii) (b), (iii) (b), (iv) (a), (v) (c)
3. (i) (a), (ii) (c), (iii) (b), (iv) (c), (v) (a)
4. (i) (c), (ii) (b), (iii) (b), (iv) (a), (v) (c)
5. (i) (a), (ii) (c), (iii) (b), (iv) (b), (v) (c)

NCERT Corner

1. (a) (i) Self pollination: When the pollen grains from the anther of a flower are transferred to the stigma of the same flower (or another flower on the same plant), it is called self pollination.

   (ii) When the pollen grains from the anther of a flower on one plant are transferred to the stigma of a flower on another similar plant, it is called cross pollination.

(b) When an insect sits on the flower of a plant for sucking nectar, then the pollen grains from the anther of this flower sticks to its body. And when this insect sits on another flower of another similar plant, then the pollen grains sticking to its body are transferred to the stigma of this second flower. In this way, the insect transfers the pollen grains from the anther of flower in one plant to the stigma of flower in another plant and causes cross pollination.

2.

Pollination	Fertilisation
The pollen grains are transferred from the anther to the stigma of the carpel	The male gamete fuses with the female gamete present in the ovule inside the ovary of the flower to form a fertilised egg or zygote

3. An example of a sex hormone is testosterone.
4. (a) Vasectomy
   (b) Tubectomy
   (c) Condom
   (d) Diaphragm
5. A plant hormone that promotes the growth of cells is auxin.

Competition Window

1. (a)	**2.** (d)	**3.** (c)	**4.** (a)	**5.** (c)
6. (a)	**7.** (d)	**8.** (a)	**9.** (a)	**10.** (c)
11. (a)	**12.** (b)	**13.** (b)	**14.** (d)	**15.** (a)
16. (b)	**17.** (a)	**18.** (b)	**19.** (b)	**20.** (a)
21. (c)	**22.** (d)	**23.** (b)	**24.** (c)	**25.** (d)

CHAPTER

14 The Reproductive System

PREREQUISITES

You must have knowledge of:

- Human anatomy and physiology
- The concept of sexual reproduction and its importance in the continuation of a species
- Hormones involved in the reproductive system
- Various reproductive health issues and their prevention
- Importance of reproductive health and its impact on overall health and well-being

LEARNING OBJECTIVES

At the end of this chapter, you will be able to:

- Describe the male reproductive system
- Describe the female reproductive system
- Explain the menstrual cycle, fertilisation and pregnancy

INTRODUCTION

Reproduction is the process of creating new individuals by sexual or asexual methods. Numerous smaller creatures, such as bacteria and yeast, reproduce by fission, budding, etc. Numerous plants use a variety of means to reproduce vegetatively in a non-sexual way. However, sexual reproduction is the only type of reproduction in humans.

In asexual reproduction, any part of the parent body may be used to make the reproductive units. The units can be of any size, from the entire parent body to a tiny fragment. An example would be a bud that emerges from a *Hydra's* body, separates and develops into a full *Hydra*.

In sexual reproduction, the gametes or sexual cells (egg and sperm) are extremely tiny (microscopic in human beings). In order to create a zygote, which develops into a child through the process of development, the gametes must combine (fertilise).

Typically, different sexes create different types of gametes (male and female). Secondary sexual qualities refer to the visible characteristics that these sexes frequently display. Animals use secondary sexual characteristics to identify and entice mates. Such cues include body odour, calling noises, skin colour, body shape and behavioural tendencies. In humans, the secondary sexual characteristics include distribution of hair on the body and face (beard and moustaches in males), breasts in females, muscularity or stronger build in males, skeletal structure, psyche and behaviour and deeper voice in males.

REPRODUCTION IN HUMANS

The organs of the reproductive system are divided into primary and accessory parts:

- The gonads, or testes and ovaries, in males and females, are the main reproductive organs and are responsible for producing sperms and eggs, also known as gametes.
- The structures that aid in the movement and interaction of the two types of sex cells that results in fertilisation as well as the growth and development of the egg up until the birth of the child are referred to as accessory reproductive components.

Male Reproductive System

The male reproductive system (Fig. 14.1) consists of the following organs:

- Testes (one pair) to produce sperms
- Sperm duct (vas deferens) from each testis to carry the sperms
- Accessory glands (seminal vesicles, prostate and bulbo-urethral glands) to contribute to the seminal fluid
- A penis for transferring the sperms into the female
- Urethra contained inside the penis, to convey the sperm received from the sperm ducts

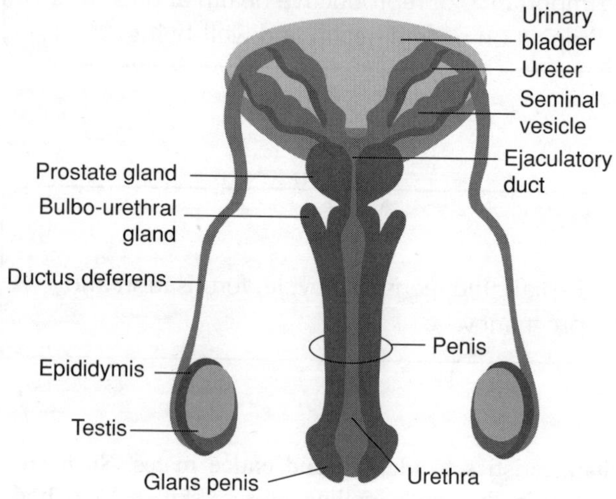

Fig. 14.1 Male reproductive system

Testes

The scrotum is a thin-walled bag of skin that houses the two oval testes (Fig. 14.2), often known as testicles (or scrotal sacs). In the embryonic stage, the testicles are housed inside the abdomen. Shortly before birth, they enter the scrotum. Sterility, or the inability to create sperms, is the result of the testes not descending, which is an aberrant state. Sperm maturation cannot occur because of the body's high temperature. However, in their natural state, the testes lose some body heat because they are suspended from the body in a separate sac.

A temperature that is 2–3°C below body temperature is used to generate sperms in the testes. The movements of the scrotum wall control the temperature. The skin of the scrotum loosens with extreme heat, causing the testes to dangle down and away from the body. The testes are drawn closer to the body for warmth and the skin constricts when it is chilly.

Each testis is enclosed in a capsule with 15–20 lobules on the inside (segments). Every lobule includes the following:

- Sperm, which are produced in the seminiferous tubules. Spermatogenesis is the name of the process.
- Interstitial cells, which act as packing tissues between the coils of the seminiferous tubules. Testosterone is created by the interstitial cells, commonly known as Leydig cells.

The seminiferous tubules' sperm-producing cells continue to divide and create sperms. The fully developed sperms enter a compact system of tubes. About 12–14 ducts (efferent ducts) emerge from the network and join

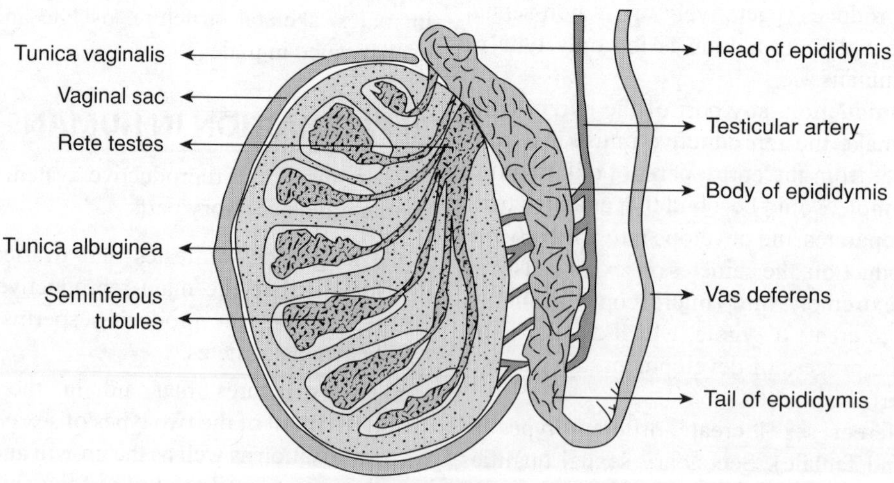

Fig. 14.2 Longitudinal section of testis

a little tubular knot, with the epididymis fitting like a cap on the top of the testis. The epididymis continues down the side of the testis all the way to the back, where it gives way to the separate tubular sperm duct known as the vas deferens. Internal to the epididymis is a single, coiling tube that extends from the top of the testis to its rear before continuing into the sperm duct. This tube is actually around 6 m long. For a few days, the sperms mature and become motile while being stored in the epididymis (capable of moving).

Sperm Ducts

Each testis' sperm duct, or vas deferens, travels through the inguinal canal and up into the abdomen. The testes, along with their ducts, blood vessels, nerves and other structures, descend through the inguinal canal.

The most typical type of hernia occurs when the intestines protrude into the scrotum through the inguinal canal as a result of pressure on the abdomen. The median duct, or urethra, is a blood vessel that runs along the rear of the urinary bladder and connects the two sperm ducts, which loop around the ureters on either side.

Accessory Glands

The three male accessory glands are:

- **Seminal vesicles:** A pair of lobulated glands situated between the rectum and the urinary bladder's posterior side. Just before it connects with the urethra, a duct from each seminal vesicle joins the associated sperm duct. It creates a secretion that acts as a vehicle for sperm movement. The sperms and this fluid combine to form semen, a milky fluid. The inclusion of this secretion makes the sperms, which are passive in the sperm duct, active.
- **Prostate gland:** A bilobed structure that encircles the urethra near the origin of the bladder. As the semen travels through the urethra, it secretes an alkaline fluid. This balances the acid in a woman's vagina.
- **Bulbo-urethral glands:** These two tiny ovoid glands are located just before the urethra reaches the penis. The exudate acts as a lubricant.

Penis

The penis, which is cylindrical in shape and is located in front of the scrotum, is used for the discharge of both urine and semen. With erectile tissues and vascular spaces, it is a highly vascular organ. The penis becomes hard and erect when sexual excitement causes a substantial amount of blood to flow into it and enter the vascular spaces or sinuses. These areas' expansion compresses the veins, preventing blood from draining. Erection is the name for such a penile condition.

Sperms and secretions from the seminal vesicle, prostate, Cowper's and urethral glands combine to form semen. The liquid is milky. A single ejaculation typically comprises between 200,000,000 and 400,000,000 sperms and contains 2–3 ml on average.

Female Reproductive System

The female reproductive system (Fig. 14.3) consists of the following reproductive organs:

- A pair of ovaries
- A pair of oviducts (or fallopian tubes) to convey the egg released from the ovary
- A sac-like or pear-shaped uterus for the growth and development of the embryo developed from the egg
- A vagina
- A vulva, the outermost part

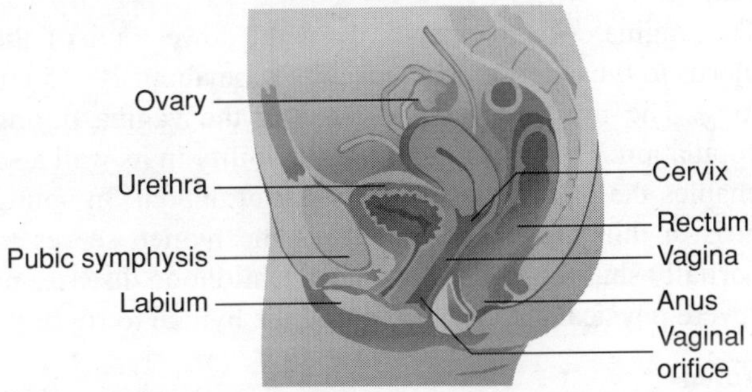

Fig. 14.3 Female reproductive system

Ovaries

Small, ovoid bodies make up the two ovaries. The ova or eggs are produced by their periphery. Each ovary typically produces one mature egg every other month. The follicle is a cellular sac that houses a maturing egg. The follicle, which is known as the Graafian follicle, develops along with the egg and becomes filled with fluid. It protrudes from the ovary's surface when it is mature. The process through which the ova-producing cells develop the mature ovum is known as oogenesis. The follicle ruptures during ovulation, releasing the egg. The cilia of the oviduct's oviducal funnel take up the discharged egg. Oestrogen is secreted by the follicle prior to ovulation. Progesterone prepares the uterus for receiving the embryo.

Ovum and Egg

While all ova are eggs, not all eggs are ova. For instance, the ovum (the central yellow sphere) of a hen's egg is joined by the nourishing albumen (the white sphere) and the shell as it travels down the oviduct. Simply put, an ovum is a female gamete or a healthy egg cell (without any extra parts).

Oviducts

The two oviducts, commonly known as uterine or fallopian tubes, measure around 12 cm in length. Each oviduct contains an aperture that resembles a funnel, called the oviducal funnel, that is close to the associated ovary. The discharged ovum is picked up and pushed into the oviduct by cilia that line the funnel. The egg is then pushed into the uterus by waves of muscle contractions (also known as peristalsis) of the oviduct wall.

Uterus

The uterus (Fig. 14.4), a hollow, muscular organ with a pear shape (7 × 5 cm), is located in the pelvic cavity between the rectum and the urinary bladder. It has two regions: a small, constricted zone at the bottom called the cervix or neck and an upper, broader region that receives the two oviducts.

Vagina

The vagina, which extends from the lower end of the uterus to the exterior, is a muscular tube about 10–15 cm long. The male penis is inserted in the vagina during copulation. The large amount of flexibility in its wall also enables the passage of the baby during labour. In young girls, a thin membrane known as the hymen serves to partially shut the vaginal entrance. Childhood diseases or severe physical activity may cause the hymen to rupture.

Vulva

The term vulva refers to the external female genitalia. The urethra and vagina have separate entrances in it. The lesser lips (labia minora), which are two little fleshy folds on the sides of the vulva, are concealed by the labia majora, which are two bigger hairy folds (greater lips). The male scrotum's analogues are these folds. There is a small, extremely sensitive erectile clitoris, which is similar to the male penis, in the topmost angle of the vulva in front of the urethral opening.

Puberty

During puberty, the undeveloped reproductive systems of boys and girls mature and become ready to bear children. The earliest indication in girls is the growth of breasts around the age of 10 years. The earliest indication in boys is the growth of the testicles at the age of 11 years.

Menstrual Cycle

The human female reproductive cycle lasts from around the age of 13 years to 45–50 years. The menstrual flow, a distinctive phenomenon that occurs roughly every month (28 days with minimal fluctuation), characterises this time. Only pregnancy has the potential to temporarily stop it.

Menarche, which occurs around the age of 13 years in females, is the start of menstruation. Menstruation ends permanently with menopause at around the age of 45.

A menstrual cycle's duration (Fig. 14.5) is measured from the first day of flow to the next onset, which occurs 28 days later. There are four key phases to this time period:

- **Menstrual phase:** Blood is discharged for 3–5 days. The blood vessels are damaged when the uterine lining sheds, which results in bleeding. The ovary starts to develop a fresh egg in a follicle as soon as the menstrual flow starts.

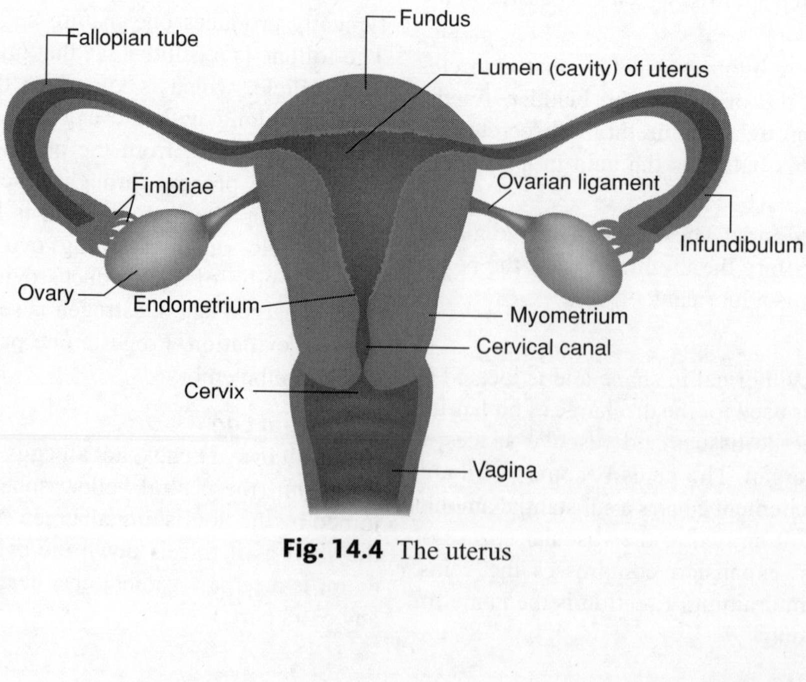

Fig. 14.4 The uterus

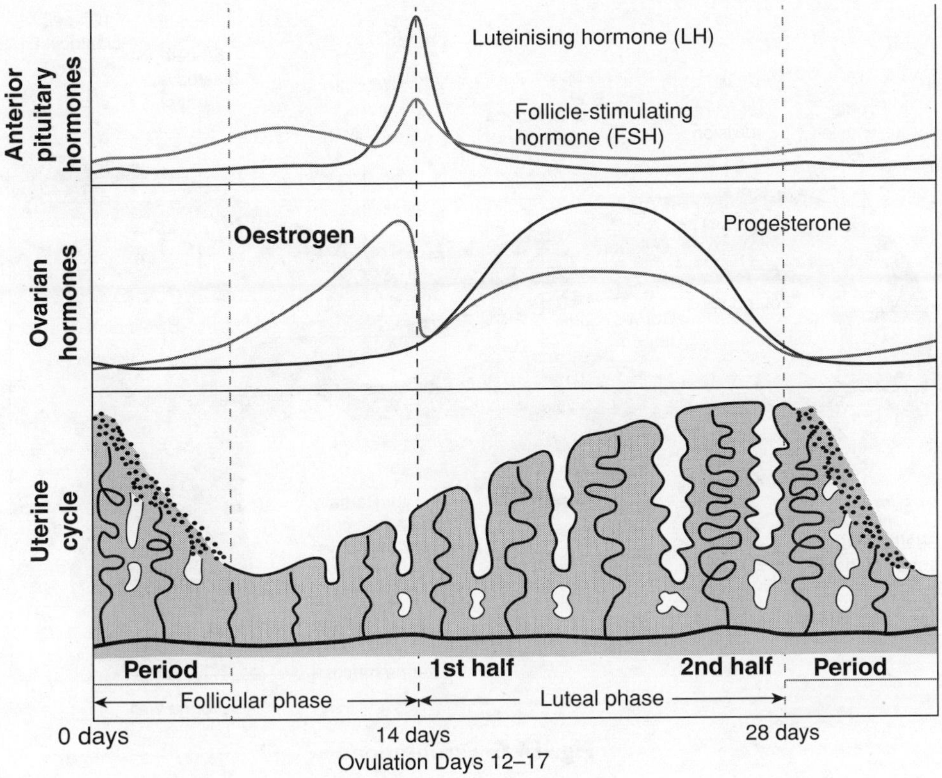

Fig. 14.5 The menstrual cycle

- **Follicular phase:** As the follicle grows, the uterine wall thickens and receives a lot of blood as a result of specific hormones that the body produces. The uterus is changed in a way that will allow it to receive the fertilised egg and support the growth of the future embryo.
- **Ovulatory phase:** The follicle bursts (ovulation) on the 13th or 14th day, and the liberated egg travels down the oviduct.
- **Luteal phase:** This may last for 15–28 days. The uterus lining becomes thicker. An ovary's empty follicle develops into the corpus luteum, a hormone-producing tissue. The egg disintegrates if there is no fertilisation, and on the 28th day, the uterine lining starts to shed again. In contrast, if the egg is fertilised, it is fixed or implanted in the uterine wall and there is no menstrual flow. This occurs because the persistent corpus luteum secretes more progesterone, which raises levels and inhibits the development of further follicles. In the absence of implantation, the corpus luteum degenerates and the thickened state of the uterine wall cannot be preserved.

From Fertilisation to Birth

The sperms are discharged in the vagina near the uterine cervix during copulation (mating or coitus). The average human male ejaculates roughly 400 million sperms in one go. These sperm actively swim through the cervix into the uterine canal with the aid of their tails, moving forward at a speed of about 1.5 mm per minute. They ascend from the uterus to the oviducts. One sperm is all that is needed to fertilise an ovum (egg) if it happens to be present in the oviduct.

Fertilisation (Fig. 14.6) is the process through which the male gamete (sperm) and female gamete (ovum) combine to generate a zygote. The best window for conception is a short time after ovulation (12th or 13th day following the start of menstruation). Few sperms, out of the millions released into the vagina, ascend to the upper oviducts; the majority perish on the journey and are absorbed.

The sperm's acrosome at the top of its head secretes an enzyme called hyaluronidase that dissolves the ovum's wall to make it easier for the sperm to enter the egg. During fertilisation, the nucleus of the egg having

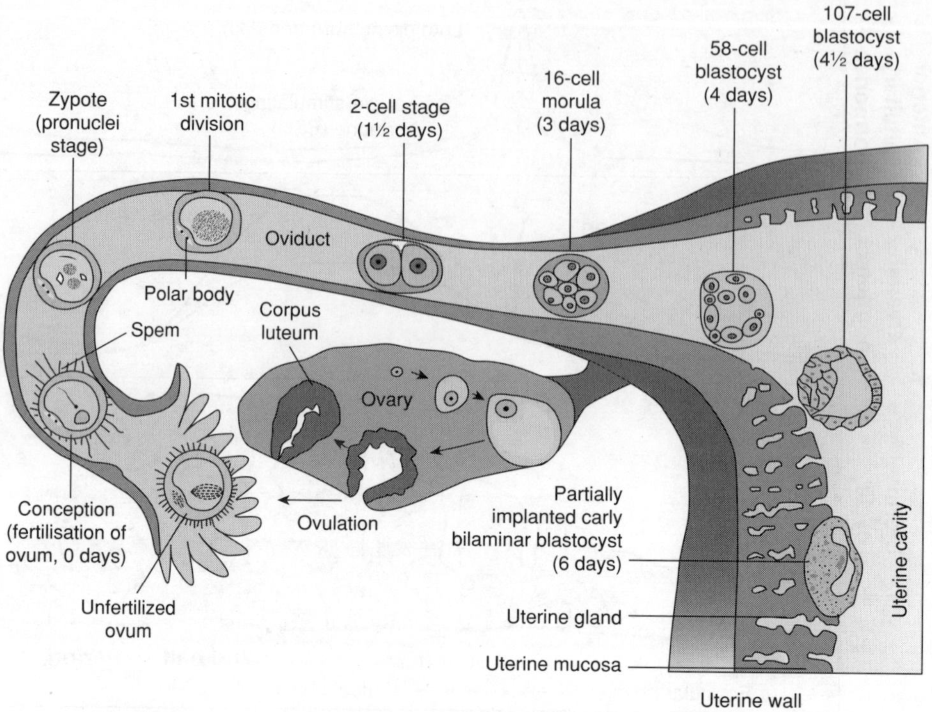

Fig. 14.6 Fertilisation

22 + X chromosomes joins with the nucleus of the egg holding 22 + Y chromosomes, restoring the usual number of 23 pairs.

The mitochondria found in the centre piece supply energy (ATP), thereby enabling the sperm to swim. For propulsion, the tail creates lashing motions. Sperms may live for a few days in the female genital tract, but their ability to fertilise only lasts for one or two days. Only one sperm enters an egg, even though numerous sperms may surround it during fertilisation. The tail is left behind as a result, and only the head portion, which contains the nucleus, enters the egg.

The nucleus of the sperm and the nucleus of the egg fuse together. Once one sperm enters the egg (ovum), some chemical barriers prevent the entry of additional sperms. As soon as the egg (zygote) is fertilised, it begins to divide (cleave) into 2, 4, 8, 16 cells, and so forth. It forms a little hollow ball composed of many cells by the time it enters the uterus (blastocyst). This type of embryo forms a pit in the uterine wall (endometrium) and becomes stuck in it a week (or 5–7 days) following ovulation. This procedure, known as implantation, results in the pregnant state. To care for the growing embryo, the uterus continues to grow. The embryo is quite developed at the end of five weeks of pregnancy. The circulatory system and heart develop. Arms and legs also develop after two months.

Placenta

The developing embryo, often known as the foetus (Fig. 14.7) (8 weeks old and resembling a human), is a living thing. It demands nourishment and air. It excretes carbon dioxide and nitrogenous waste, both of which must be regularly eliminated. The placenta, a bodily organ, carries out these unique tasks. The placenta is a disc-shaped organ joined to the uterine wall. The placenta and the foetus are joined by an umbilical cord, which has blood vessels. The villi, which are two pairs of little finger-like structures, make up the placenta. The uterine wall releases one set of villi and an extension (allantois) from the embryo releases the other set. Although they become intertwined, the two sets of villi never meet.

Here, the mother's blood and the embryo's blood are in close proximity but never mix. The blood of the foetus is brought to it through an artery, and it contains expelled substances like urea and carbon dioxide. The placenta's very small capillaries allow this blood to pass through. Close to these are the capillaries and blood sinuses of the mother.

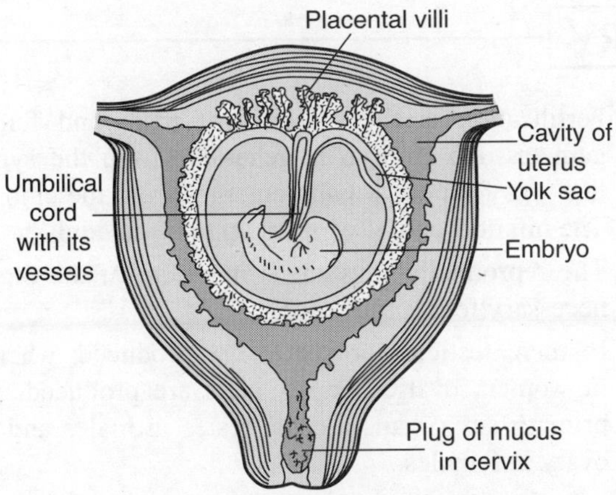

Fig. 14.7 The foetus in a uterus

The placenta permits diffusion of the following material from the mother to the foetus and from the foetus to the mother:

- **From mother to foetus (for utilisation):** Oxygen, nutrients such as glucose, amino acids, vitamins, mineral ions (Na, Cl, K, Ca, Fe, etc.)
- **From foetus to mother (for elimination):** Carbon dioxide, urea and other waste

The placenta has the ability to be permeable. It is permeable to nutrients, antibodies and breathing gases, but is not permeable to germs that travel from the mother to the foetus. If the mother is already infected, some viruses, such HIV that causes AIDS, can travel through the placenta. Secondly, if there was no placenta and there was a direct line of blood flow between the mother and the foetus, the comparatively high pressure of the mother's blood would severely harm the foetus' sensitive and delicate tissues. The placenta serves as an endocrine gland as well. It creates the hormones progesterone and **oestrogens.** Certain tests for pregnancy are made possible by the presence of progesterone in urine.

Amnion

Even before the allantois develops, a sac called the amnion forms around the embryo. The space between the amnion and the embryo is filled with a liquid (amniotic fluid). Amniotic fluid's functions include:

- Protecting the embryo from physical damage through jerks or mechanical shocks. For example, when the mother falls over.
- Keeping even pressure all around the embryo.
- Allowing the foetus some restricted movement.
- Preventing sticking of the foetus to the amnion.

Parturition (Birth)

Gestation refers to the entire period during which an embryo develops in the uterus. It lasts roughly 280 days in humans. When a baby is born, the uterus's strong muscles force the infant out of the womb head first. The blood arteries in the umbilical cord constrict and can be knotted and cut in a matter of minutes. The placenta separates from the uterus and is discharged outside the body as after-birth after about 15 minutes. In a few days, the uterus contracts and gradually returns to normal. The newborn 'announces' its birth by crying loudly. This cry is the baby's body's natural way of cleaning its airways.

- **Fraternal twins** (produced from two eggs): Two eggs may occasionally be discharged from the ovaries at once, and if both are fertilised, they may result in the birth of two children. Fraternal twins are those born from two eggs. A boy and a girl, or both males and both females, can be fraternal twins. Fraternal twins are identical to typical sisters and brothers in every way, with the exception that they share the same age because they were born just a few minutes apart. This type makes up 75% of all twins. The terms diovular (meaning generated from two ova) and dizygotic (meaning produced from two zygotes or two fertilised eggs) are used in biology to describe fraternal twins.
- **Identical twins** (produced from one egg): In the early phases of cell division, a single fertilised egg may occasionally split and separate into two parts. Each of these two split portions then functions like an independent egg and gives birth to a single full human. Both boys and girls can develop into identical twins from a single egg, and for the same reason, they are remarkably similar to one another. However, not everything about identical twins is the same; for instance, there are differences in the handwriting, birthmarks and even fingerprints.
- **More than two:** Sometimes, a single pregnancy can result in the birth of three, four or even more children. Such births can occur for a number of reasons. Triplets (three born at once) can result from a single zygote splitting into two and one of the split parts splitting into two again, or they can result from two eggs in which one zygote develops normally into one baby while the other produces a pair of identical twins. The birth of four children or quadruplets at once can occur as two sets of identical twins from two eggs or in different ways that combine identical and fraternal twins. As the number of children in multiple births increases, their chances of survival decrease. Rare cases of six or seven babies produced are known but only a fewer number survive.

SUMMARY

- Reproduction is the process through which an organism creates offspring in order to maintain its existence. Asexual and sexual reproduction are the two forms.
- Asexual reproduction refers to the process where a new organism is created from a single parent. Asexual reproduction results in offspring that are genetically identical to the parent.
- Sexual reproduction is when male and female sex cells combine to produce a zygote, which then grows into a new organism.
- Fertilisation is the merging of male and female gametes. An embryo is generated from the zygote that was formed. A baby emerges from the embryo. The infant eventually grows up to be an adult.
- The reproductive system is made up of the organs necessary for reproduction.
- In men, testicles or sperm are produced, whereas in women, ovaries or sex cells are produced. The primary sex organs are the testes in males and the ovary in females.

PRACTICE QUESTIONS

Analyse Your Concepts (School Exam Based)

Multiple Choice Questions

1. How soon after fertilisation does the fertilised egg in a female get implanted in the uterine wall?
   (a) A few months (b) One month
   (c) Three weeks (d) About seven days
2. The process of fertilisation occurs in humans in the:
   (a) Uterus (b) Oviduct funnel
   (c) Fallopian tube (d) Vagina
3. The sperm's middle fragment offers:
   (a) Energy (b) Food
   (c) Gene (d) Chromosomes
4. In humans, the typical gestational period is:
   (a) 270 days (b) 290 days
   (c) 280 days (d) 295 days
5. Which of the following best describes the path a sperm takes after leaving a mammal's testis?
   (a) Vas deferens → epididymis → urethra
   (b) Urethra → epididymis → vas deferens
   (c) Epididymis → urethra vas deferens
   (d) Epididymis deferens-urethra
6. When there is no pregnancy, the corpus luteum lives for around:
   (a) 4 days (b) 10 days
   (c) 14 days (d) 28 days

Match the Following

Each question contains statements given in two columns which have to be matched. Statements in column I have to be matched with statements in column II.

Column I	Column II
1. Oestrogen	(i) The shedding of the uterine lining when fertilisation does not occur.
2. Testosterone	(ii) Stimulates ovulation and the production of progesterone in females.
3. Luteinising hormone	(iii) Stimulates the development of male secondary sex characteristics.
4. Fertilisation	(iv) Stimulates the development of female secondary sex characteristics.
5. Menstruation	(v) The fusion of the egg and sperm to form a zygote.

True or False

Read the following statements and write your answer as true or false.

1. Oviducts and fallopian tubes are one and the same thing.
2. One egg is released from each of the two ovaries every month.
3. The enlarged mature follicle bursts to release the egg.
4. The egg is passed down through the oviduct by muscular contractions.
5. The clitoris is equivalent to the male penis.
6. The vagina is a muscular tube.

Fill in the Blanks

Complete the following statements with an appropriate word/term to be filled in the blank spaces.

1. Implantation is the process of fixing of the __________ in a pit in the wall of __________.
2. Heart and circulatory system are formed in the embryo by the end of __________ months of pregnancy.
3. A cord containing __________ vessels connect the __________ with the foetus.
4. List the substances which the foetus receives from the mother through the placenta: __________.
5. __________ is the process of fusion of gametes resulting in the formation of a zygote.

Assertion–Reason Questions

The following questions consist of two statements: Assertion (A) and Reason (R). Answer these questions selecting the appropriate option given below:

(a) Both A and R are true and R is the correct explanation of A
(b) Both A and R are true but R is not the correct explanation of A
(c) A is true but R is false
(d) A is false but R is true

1. Assertion (A): A bisexual bloom makes ova in addition to pollen.
   Reason (R): The carpel is where the ova and pollen are made.
2. Assertion (A): Females create two kinds of gametes.
   Reason (R): There are two X chromosomes in females.
3. Assertion (A): Efficiency of enzymes regulates pea plant height, which is genetically controlled.
   Reason (R): The information needed to make proteins in the cell comes from cellular DNA.
4. Assertion (A): The primary function of the male reproductive system is to produce and deliver sperm.
   Reason (R): Sperm production occurs in the testes and is regulated by hormones such as testosterone.
5. Assertion (A): Fertilisation in humans occurs in the uterus.
   Reason (R): The sperm and egg meet and fuse together in the fallopian tube, which then moves towards the uterus for implantation.

One-word Answers

1. What is the male sex hormone?
2. What is the primary female sex organ
3. What is the process by which a mature egg is released from the ovary?
4. What is the tube that carries sperm from the testes to the urethra?
5. What is the name for the union of sperm and egg?

Define the Term

1. Gamete
2. Zygote
3. Embryo
4. Ovulation
5. Fertilisation
6. Menstruation
7. Contraception

Groups of Terms

Given below are certain groups of terms. In each group the first pair indicates a relationship between the two terms. Rewrite and complete the second pair on a similar basis.

1. Ovary : Oogenesis :: Testis : ____
2. Uterus : Implantation :: Cervix :____
3. Ovulation : Release of an egg :: Ejaculation : ____
4. Follicular phase : Development of ovarian follicles :: Luteal phase : ____
5. Menstruation : Shedding of the uterine lining :: Ejaculation : ____

Identify the Odd One Out

1. Ovaries, Testes, Uterus, Oviduct.
2. Oogenesis, Spermatogenesis, Fertilisation, Meiosis.
3. Fallopian tubes, Vas deferens, Uterus, Cervix.
4. Ovulation, Menstruation, Implantation, Gestation.
5. Sperm, Ovum, Zygote, Placenta.

Image-Based Questions

1. Given below is a diagram of two systems together in the human body.

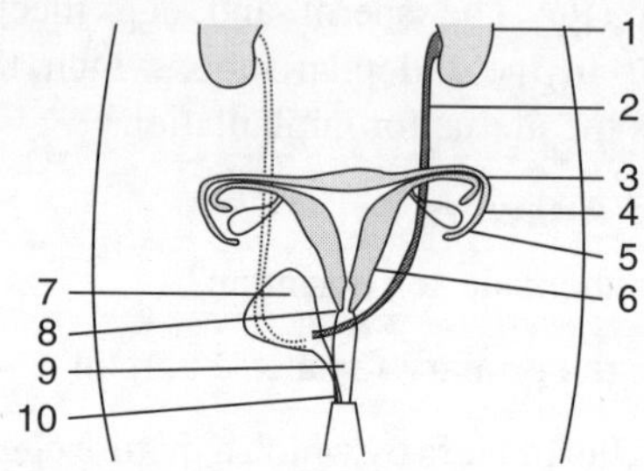

(a) Name the systems.
(b) Name the parts numbered 1–10.
(c) Describe the functions of the parts 8, 9, 4 and 3.

Very Short Answer Questions

1. What is the function of the uterus in the reproductive system?
2. What is the primary function of the fallopian tubes?
3. What is the role of the prostate gland in the male reproductive system?
4. What is the purpose of the epididymis in the male reproductive system?
5. What is the function of the seminal vesicles in the male reproductive system?

Short Answer Questions

1. What is sexual reproduction? What are its key characteristics?
2. Name two of the human reproductive system's key distinguishing characteristics.
3. What are the various organs of male reproductive system?
4. Write the functions of the testes.
5. What are the locations that a mature sperm passes through as it moves from the seminiferous tubules to the urethral opening, in order?
6. What variations exist between the male and female gametes in terms of number, make-up and activity?
7. Explain the different stages in the uterine cycle.
8. (i) How does fertilisation take place?
   (ii) Identify the two key components of fertilisation.
   (iii) After fertilisation, what happens to the zygote?

Long Answer Questions

1. Describe the process of fertilisation in humans.
2. What are the male and female reproductive systems, and what are their primary functions?
3. Discuss the hormonal control of the menstrual cycle in females.
4. What are some common methods of contraception, and how do they work?
5. What is infertility, and what are some common causes?

Case Study-Based Questions

1. Thousands of immature eggs are already present in the ovaries when a girl is born. Some of these start maturing once they reach puberty. One of the ovaries releases one egg each month. Through a delicate oviduct or fallopian tube, the egg is transported from the ovary to the womb. The uterus is an elastic bag-like structure formed by the union of the two oviducts. Through the cervix, the uterus enters the vagina.

   I. What is fertilisation?
   (a) The process of releasing eggs from the ovaries.
   (b) The process of combining male and female gametes to form a zygote.
   (c) The process of developing a fetus in the uterus.
   (d) The process of delivering a baby.

   II. Where does fertilisation occur?
   (a) In the uterus
   (b) In the vagina
   (c) In the cervix
   (d) In the oviduct or the Fallopian tube

   III. What is the placenta?
   (a) The tube that connects the ovary and the uterus.
   (b) The tissue that feeds the developing embryo with nutrients from the mother's blood.
   (c) The organ that produces sperm in males.
   (d) The structure that holds the developing fetus in the uterus.

   IV. Which components make up the female reproductive system?

(a) Fallopian tube, ovary, uterus, vagina and cervix.
(b) Testes, epididymis, vas deferens, urethra, prostate gland and seminal vesicles.
(c) Ovary, uterus, cervix, penis and scrotum.
(d) Fallopian tube, cervix, uterus, urethra and vagina.

V. What happens when an egg is not fertilised?
(a) The thick lining of the uterus breaks down and comes out as blood and vaginal discharge.
(b) The thick lining of the uterus is maintained for the next menstrual cycle.
(c) The egg is reabsorbed by the body.
(d) The egg travels back up to the ovary.

2. In the form of DNA (deoxyribonucleic acid) molecules, chromosomes in the nucleus of a cell store information for the inheritance of traits from parents to the following generation. The information source for constructing proteins is the DNA found in the cell nucleus. Proteins will be produced differently if the information is altered. Eventually, different proteins will change how our bodies look. Therefore, the act of making a DNA duplicate is a fundamental aspect of reproduction. Chemical processes are used by cells to replicate their DNA.

I. What is the importance of variation?
(a) Variation increases the likelihood of species survival
(b) Variation makes a species genetically identical to its parent
(c) Variation reduces the likelihood of species survival
(d) Variation has no impact on species survival

II. Which method of reproduction exhibits less variety?
(a) Asexual reproduction
(b) Sexual reproduction
(c) Both methods exhibit similar variety
(d) None of the above

III. What are the different ways in which a single cell organism reproduces?
(a) Sexual reproduction and asexual reproduction
(b) Fission, fragmentation, budding, and regeneration
(c) Only fragmentation and fission
(d) Budding and regeneration

IV. What characteristics define sexual reproduction?
(a) It requires two parents to produce offspring
(b) It involves the transfer of genetic material from one parent to another
(c) It does not involve the transfer of genetic material
(d) It only occurs in plants

3. Pregnancy is always a possibility with sexual activity. If a woman is not prepared, pregnancy will place significant demands on her body and mind and negatively impact her health. As a result, numerous methods have been developed to prevent pregnancy.

I. What are the different methods to prevent pregnancy?
(a) Condoms, pills, copper-T and surgical methods
(b) Surgery, meditation and exercise
(c) Home remedies, diet and exercise
(d) None of the above

II. What are the surgical methods to prevent pregnancy?
(a) Vasectomy in females and tubectomy in males
(b) Tubectomy in females and vasectomy in males
(c) Both methods are the same
(d) None of the above

III. Why is the female sex ratio declining day by day?
(a) Due to illegal sex-selective abortion of male foetuses
(b) Due to illegal sex-selective abortion of female foetuses
(c) Due to natural causes
(d) None of the above

IV. Which methods of contraception have significant adverse effects?
(a) Condoms and copper-T
(b) Surgical methods only
(c) Oral medications and hormonal methods
(d) All methods have significant adverse effects

V. Why is it important to understand sexual activity and the diseases it causes?
(a) To be healthy and free from sexually transmitted diseases
(b) To avoid pregnancy
(c) To have a better sex life
(d) None of the above

NCERT Corner

1. What is the role of the seminal vesicles and the prostate gland?
2. What are the changes seen in girls at the time of puberty?
3. How does the embryo get nourishment inside the mother's body?
4. If a woman is using Copper-T, will it help in protecting her from sexually transmitted diseases?

COMPETITION WINDOW (OBJECTIVE TYPE)

1. Which of the following is true of the path a sperm takes after leaving a mammal's testicles?
   (a) Vas deferens → epididyrnis → urethra
   (b) Urethra → epididymis → vas deferens
   (c) Epididyrnis → urethra → vas deferens
   (d) Epididymis → vas deferens → urethra
2. If a man has his vas deferens surgically detached or removed:
   (a) Semen will be without sperm
   (b) Sperms in the semen will be without nuclei
   (c) Spermatogenesis will not take place
   (d) Sperms in semen will be non-motile
3. What will happen if the scrotal sacs' temperature momentarily drops to around 2°C below body temperature?
   (a) Sperm formation will be stopped
   (b) Testes will start rotation movement
   (c) Scrotal sacs will contract and come close to the body
   (d) Scrotal sacs will relax and go down away from the body
4. Fertilisation means:
   (a) The transfer of male gamete to the female gamete
   (b) The adhesion of male and female reproductive organs
   (c) The fusion of nuclei of male and female gametes
   (d) The shedding of gametes from a reproductive organ
5. The location where human sperm matures is the:
   (a) Seminiferous tubule (b) Interstitial cells
   (c) Epididymis (d) Prostate gland
6. The beginning of a woman's menstrual cycle is known as:
   (a) Ovulation (b) Menarche
   (c) Menopause (d) Parthenogenesis
7. In mammals, ovulation is a process that involves:
   (a) The egg is released from the Graffian follide
   (b) The egg is fertilised in the fallopian tube
   (c) The egg passes through fimbriated funnel of the fallopian tube
   (d) The unfertilised egg passes out of female's body
8. Female sterilisation entails cuffing and binding the:
   (a) Ureter (b) Uterus
   (c) Urethra (d) Oviduct
9. Mammals have corpus luteum in:
   (a) Brain and connects the two cerebral hemispheres
   (b) Ovaries and produces progesterone hormone
   (c) Heart and initiates atrial contraction
   (d) Skin and acts as a pain receptor

Answer Keys

Multiple Choice Questions

1. (d) **2.** (c) **3.** (a) **4.** (c) **5.** (d) **6.** (b)

Match the Following

Column I	Column II
1. Oestrogen	(iv) Stimulates the development of female secondary sex characteristics.
2. Testosterone	(iii) Stimulates the development of male secondary sex characteristics.
3. Luteinising hormone	(ii) Stimulates ovulation and the production of progesterone in females.
4. Fertilisation	(v) The fusion of the egg and sperm to form a zygote.
5. Menstruation	(i) The shedding of the uterine lining when fertilisation does not occur.

True or False

1. T **2.** T **3.** T **4.** T **5.** T **6.** T

Fill in the Blanks

1. Embryo, uterus
2. Five
3. Blood, placenta
4. Oxygen, glucose and other nutrients
5. Fertilisation

Assertion–Reason Questions

1. (c) **2.** (d) **3.** (a) **4.** (a) **5.** (d)

One-word Answers

1. Testosterone
2. Ovary
3. Ovulation
4. Vas deferens
5. Fertilisation

Define the Term

1. Gamete is a haploid reproductive cell (sperm or egg) that fuses with another gamete during fertilisation to form a zygote.

2. Zygote is a diploid cell that is formed when a sperm cell fertilises an egg cell, resulting in the fusion of their nuclei and the combination of their genetic material.

3. Embryo is the early stage of development of a multicellular organism, specifically after fertilisation and prior to the fetal stage.

4. Ovulation is the process in which a mature egg is released from the ovary into the fallopian tube, ready for fertilisation.

5. Fertilisation is the process in which a sperm cell fuses with an egg cell to form a zygote.

6. Menstruation is the shedding of the lining of the uterus that occurs when a fertilised egg is not implanted, typically occurring on a monthly basis in females.

7. Contraception is the use of methods or devices to prevent pregnancy, such as condoms, birth control pills, and intrauterine devices (IUDs).

Groups of Terms

1. Spermatogenesis
2. Dilation
3. Release of sperm
4. Maintenance of the corpus luteum
5. Release of semen

Identify the Odd One Out

1. Uterus
2. Fertilisation
3. Uterus
4. Gestation
5. Placenta

Image-Based Questions

1. (a) Excretory system and female reproductive system

   (b) 1. Kidney
   2. Ureter
   3. Fallopian Tube
   4. Infundibulum
   5. Ovary
   6. Uterus
   7. Urinary Bladder
   8. Cervix
   9. Vagina
   10. Vulva

   (c) Function of the fallopian tube (part 3): Carries the ovum released from the ovary to the uterus.
   Function of the infundibulum (part 4): Picks up the released ovum and pushes it further on its passage into the fallopian tube.
   Function of the ovary (part 5): Produces female gametes or ova.
   Function of the uterus (part 6): Enables the growth and development of the embryo.

Very Short Answer Questions

1. The uterus provides a site for implantation of a fertilized embryo and supports its development during pregnancy.
2. The fallopian tubes are responsible for transporting eggs from the ovaries to the uterus and providing a site for fertilisation to occur.
3. The prostate gland produces and secretes seminal fluid, which nourishes and protects sperm during ejaculation.
4. The epididymis is responsible for storing and maturing sperm cells as they develop the ability to swim and fertilize an egg.
5. The seminal vesicles secrete fluids that contribute to the formation of semen and provide energy for the sperm's movement.

Short Answer Questions

1. It is a type of reproduction usually found in higher animals and plants where male and female sexes produce male and female heterogametes separately, which on fertilisation produce a zygote.
2. (i) Its reproductive phase is the longest.
   (ii) There is no set period of time for breeding. Any time of the year is suitable for it.
3. 1. A pair of testes
   2. A pair of epididymis
   3. A pair of vas deferentia
   4. A pair of seminal vesicle
   5. Urethra
   6. Penis
   7. Male reproductive glands (Cowper's and prostate gland).
4. The creation or development of sperms is the primary purpose of the testes. Additionally, they release testosterone, a hormone associated with male sex, which helps males develop secondary sexual traits.
5. Seminiferous tubules
   (Produce sperms)
   ↓
   Network of tubules
   ↓
   Efferent ducts
   ↓
   Upper part (head) of epididymis
   ↓
   Middle part (body) of epididymis
   ↓
   Hind part (tail) of epididymis
   ↓
   Sperm duct (vas deferens)
   ↓
   Urethra (in penis)

6.

Male Gametes	Female Gametes
Number: Very large.	Normally one at a time.
Structure: A sperm is about 50–55 microns long with head, acrosome neck and lashing tail.	It is spherical and non-motile.
Activity: With the help of lashing tail, it moves actively towards the oviducts.	Passively carried from ovary to oviduct.

7. The uterine cycle consists of the following four phases:
   (i) The menstrual cycle: It typically lasts five days.
   (ii) Follicle stage: This stage lasts 10–14 days from the end of menstruation until the release of the ovum.
   (iii) Ovulation is the process through which an ovary releases an egg.
   (iv) The corpus luteum stage, which lasts from 10 to 14 days from ovulation to menstruation.
8. (i) Fallopian tube.
   (ii) During fertilisation, the sperm and egg combine to create the zygote.
   (iii) The zygote divides after fertilisation to create a multicellular embryo, which is then deposited in the uterus for continued development.

Long Answer Type Questions

1. Hint: Explain the steps by which fertilisation takes place. Elaborate on this.
2. Hint: Male reproductive system: Includes testes, penis, epididymis, vas deferens, prostate gland, and seminal vesicles. Functions to produce and deliver sperm for fertilisation.
   Female reproductive system: Includes ovaries, fallopian tubes, uterus, cervix, and vagina. Functions to produce eggs, provide a site for fertilisation and embryo development, and support pregnancy. Elaborate on this.
3. Hint: Explain the role of oestrogen, progesterone, follicle-stimulating hormone (FSH) and luteinising hormone (LH).
4. Hint: Explain the common methods of contraception and how they work, such as barrier methods, hormonal methods, intrauterine devices (IUDs) and surgical methods.
5. Hint: Explain infertility and causes such as hormonal imbalances, ovulation disorders, tubal blockages, sperm abnormalities, uterine abnormalities, age-related decline in fertility, and certain medical conditions.

Case Study-Based Questions

1. (i) b, (ii) d, (iii) b, (iv) a, (v) a
2. (i) a, (ii) a, (iii) b, (iv) a
3. (i) a, (ii) b, (iii) b, (iv) c, (v) a

NCERT Corner

1. Lubrication of sperms and providing of a fluid medium for the easy transportation of sperms takes place with the help of secretions from the seminal vesicles and the prostate gland. These secretions also provide nutrients in the form of fructose, calcium and some enzymes.
2. Following are the changes seen in girls at the time of puberty:
   - Hair growth appears in genital area.
   - Hair growth in other areas like underarms, face, hands and legs.
   - The size of the uterus and ovary increases.
   - The size of the breast increases followed by darkening of the nipple skin that is present at the tip of the breast.
   - Beginning of menstrual cycle.
   - Appearance of pimples, as there is more oil secretion from the skin.
3. The lining of the uterus thickens after fertilisation. The blood flow is good so as to nourish the growing embryo. Placenta is a special tissue which is embedded in the uterine wall and helps the embryo get the nourishment from the mother's tissue. Placenta has villi on the embryo side and blood space on the

mother's side. This spacing provides a large area between the mother and the embryo and also for waste removal.

4. No, the use of copper-T cannot stop the contact of body fluids. Hence, it cannot protect her from getting sexually transmitted diseases.

Competition Window

1. (d) **2.** (a) **3.** (c) **4.** (c) **5.** (c) **6.** (b) **7.** (a) **8.** (d) **9.** (b)

CHAPTER 15

Population – The Increasing Numbers and Rising Problems

PREREQUISITES

You must have knowledge of:

- Basic mathematical concepts such as percentage, ratio and proportion
- Basic biology concepts such as reproduction, heredity and genetics
- The concept of population growth, including factors that contribute to it, such as birth rate, death rate, migration and life expectancy
- Measures taken by the government to control population growth, such as family planning, education and healthcare

LEARNING OBJECTIVES

At the end of this chapter, you will be able to:

- Explain the meaning of population, population density, birth rate, death rate and migration
- Differentiate between population growth and population change
- List the causes of low sex ratio in India
- Compare and contrast the age composition of India with Japan

RISING POPULATION - A GLOBAL THREAT

The population issue in emerging nations like India is receiving a lot of attention. Radio, television, Internet, press and other forms of mass communication have made every effort to inform the public about the serious risks posed by population growth. This is evolving into a significant threat for the entire world, and not just for any one nation.

Around 50,000 years ago, the world's population was thought to have been about one million people (1,000,000). Back then, people led an extremely primitive lifestyle. Humans began using tools created from stones, sticks and bones. They were nomads who sought refuge in caves. Humans killed animals and roasted them over a fire or ate them raw. They fished in the seas. They gathered fruits, roots and leaves from wild plants or eggs from bird nests. In the meantime, they learned to live in groups for greater safety and to use their collective intelligence in a variety of ways, including when hunting enormous animals like mammoths. Slowly, humans improved the tools which were still mainly made of stones (Stone Age). This tool-making revolution thus helped humans to improve their lives.

The world's population rose to about 5.3 million around 10,000 years ago. The way people lived at the time underwent a significant transformation, almost a revolution. People began residing in simple huts in settlements. Humans learned to sow crops and store food. Domestication of animals began, with the dog being the first. Agriculture had its beginnings at this time. Early stone tools and weapons were increasingly supplanted by bronze and iron ones (Fig. 15.1). Humans therefore entered the era of metal. As a result of this, the human population began to grow faster.

Fig. 15.1 Iron Age family
(Image from page 307 of Ridpath's history of the world …; John Clark Ridpath; 1897; https://www.flickr.com/photos/internetarchivebookimages/14760471766. Licensed under: CC0)

> **Fun Fact!**
>
> **World Population: Some Important Dates and Figures**
>
> - World population crossed the 5.0 billion mark on July 11, 1987.
> - World population was nearly 5.6 billion in 1995.
> - The number of babies born every day is more than 1,00,000.
> - World population crossed the 6 billion figure on October 12, 1999. The six billionth human was born in Sarajevo (Yugoslavia).
> - According to one estimate, the total world population will reach 10 billion around the year 2050.
> - In 2023, India overtook China to become the world's most populous country.
> - July 11 is observed as World Population Day.

The Industrial Revolution, which began in the seventeenth century, marked the beginning of a significant era in the expansion of the human population. Rapidly expanding industries improved the quality of human life by increasing work opportunities and food production. All this favoured population growth. There was one problem, though. As the number of community organisations grew, there were more widespread outbreaks of contagious diseases as a result of intimate human contact. Poor diet also led to the emergence of nutritional disorders.

Medical science made significant advancements in the twentieth century, particularly with the development of antibiotics and preventative immunisations against a variety of diseases ('prophylactic' means 'guarding beforehand'). The number of deaths across all age groups, particularly those of newborns and the elderly, significantly decreased as a result. As more children per household reached reproductive age and started having children, the population growth rate accelerated dramatically.

Figure 15.2 depicts a diagrammatic curve of the growth of human population where the main revolutions in human culture have been indicated.

Following each cultural revolution, the population grew quickly until it stabilised or rose extremely slowly (flattened curve) over an extended period of time (shown by wavy gaps). The succeeding revolutions brought

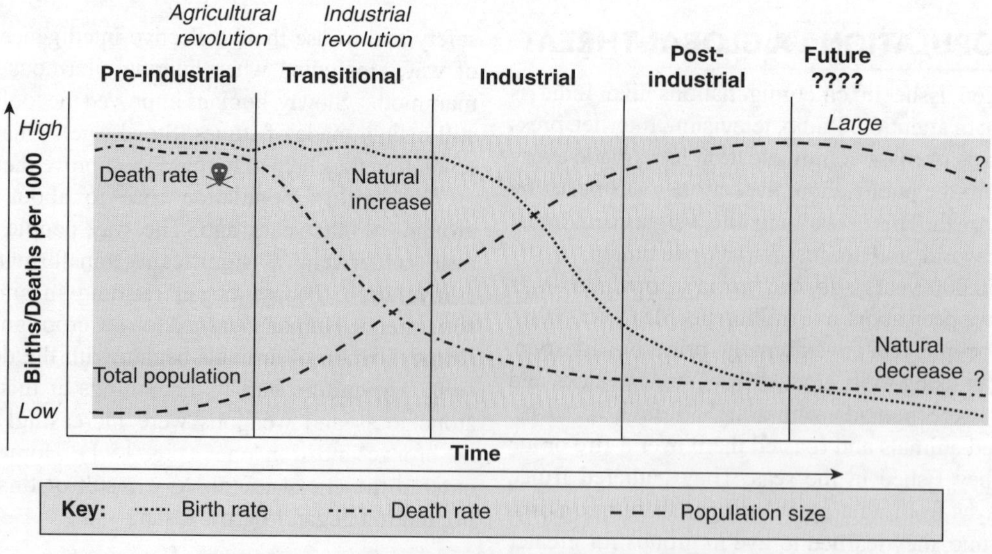

Fig. 15.2 Human population curve of revolutions

about even better means of subsistence, and the human population once more increased quickly.

The end of the solid curve denotes the beginning of the modern era; the two large question marks refer to our uncertainty about how quickly or slowly the population would increase going forward or whether it would eventually tend to stabilise.

It is important to note that **population growth** refers to the increase in the number of individuals in a population over a specific period of time. It is typically measured as a percentage or an absolute increase in the population size. Population growth can occur through natural means, such as births and immigration, or through artificial means, such as the addition of new individuals through policies or programmes.

On the other hand, **population change** refers to any alteration in the size or composition of a population. It encompasses both population growth and population decline. Population change can be positive or negative, indicating an increase or decrease, respectively, in the population size. Population change can also refer to shifts in the demographic characteristics of a population, such as changes in age structure, gender distribution or ethnic composition.

POPULATION EXPLOSION: A SERIOUS GLOBAL CONCERN

Demographers believe that the population expansion that started in the middle of the nineteenth century is the reason why around one-third of all humans who have ever lived are still alive today. Over half of the population of the globe today lives in developing countries, which make up two-thirds of the total population. China and India together account for one-third of global population. The profile of human population expansion throughout history, especially over the past 12,000 years, is seen in Fig. 15.3.

The human population was relatively steady in times past. Severe epidemics, particularly the Black Death, were responsible for the abrupt decline that occurred shortly after 1000 ACE. Later, the scientific and industrial era enabled the population to increase with a new impetus.

At the current rate of population growth, the population will more than double in the next 35 years. On October 31, 2011, there were 7 billion people on the planet. By the end of this century, the population is predicted to exceed 50 billion people (more than seven times the current number).

The main reasons for the sharp rise in world population are:

- Better healthcare for people of all ages. There are more hospitals, clinics and practicing physicians to provide assistance.
- Fewer deaths owing to improved medical care. This is due to several factors: first, there are ongoing immunisation campaigns; second, more people are being treated and living longer; and third, numerous diseases have been controlled or even eradicated.
- The Green Revolution has reduced food shortages. Food production and storage has increased. Deaths from malnutrition are quite rare.
- Better nutrition, especially for growing youngsters, as a result of awareness. They maintain good health, experience fewer illnesses, and live longer.
- Widespread vaccination against deadly diseases.
- Fewer infant deaths. Earlier, an average family included about 4–6 children, of which, one or more would not

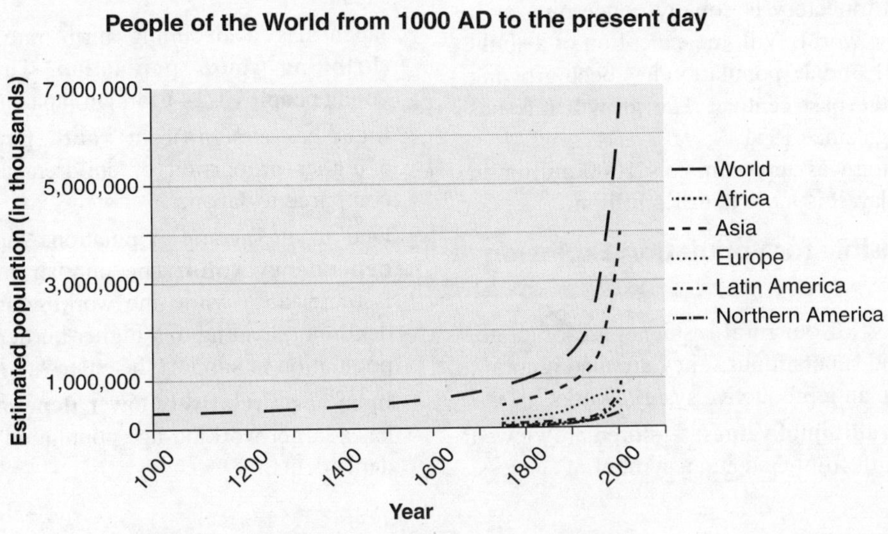

Fig. 15.3 World population growth

survive due to diseases. Today, most newborns survive and thrive due to better healthcare. Maternity homes provide safety for the child as well as the mother. More and more children are reaching the reproductive age and they contribute to population growth.

Fun Fact!

1 square foot of earth per person!
If there are no further checks or controls, the population in the next 700 years would become so high that only 1 sq. foot of earth per person would be available. Can this really happen? No one can answer at present.

Need for Adopting Control Measures

With the fast-increasing world population, there are numerous problems which are serious threats; for example:

- Decreasing open spaces
- Shrinking of forests
- Increasing industrialisation, causing air, water and soil pollution
- Rapid and intense shortage of drinking water and other resources
- Increasing pollution, which is a health hazard

In order to reduce the above threats, we need to control population growth. This can be achieved only by taking measures primarily in the form of contraception.

POPULATION IN INDIA

India's population trajectory is equally concerning as it is for the rest of the world. With the exception of a slight decline in 1911–21, India's population has been growing continuously for the past century. The growth rate has been extremely high since 1951.

India's population was approximately 1000 million in 2001 ACE, but today, it is more than 1.4 billion.

Factors Responsible for Population Explosion in India

- **Illiteracy:** Most of our rural residents are illiterate, uneducated and superstitious. They are also ignorant of how the human reproductive system works.
- **Customs or traditional values**: Children are viewed by many as a gift from God and a symbol of success. As a result, people do not try to avoid becoming pregnant.
- **Rate of mortality:** People from the economically disadvantaged segment believe it is safer to have more children because of the high infant death rate in our country so that at least some may live.
- **Financial factors**: Children are thought of as earning potential boosters for the family.
- **Social and religious customs:** The majority of people in India do not adopt family planning standards due to religious and social customs.
- **Wanting a boy as a child:** Many Indian families still believe that having a boy is crucial to upholding the family name. Additionally, a male child is considered a huge aid to elderly parents. These two factors frequently result in couples having multiple children before having a boy, or occasionally not even that.
- **Lack of leisure activities:** Poor living conditions and poverty result in lack of leisure activities.

POPULATION IN INDIA AND JAPAN

The age composition of India and Japan differs significantly due to variations in population dynamics, historical factors and socioeconomic development. Here is a comparison of the age composition of the two countries:

- Japan has one of the **oldest populations** in the world. It is experiencing rapid population ageing, with a large proportion of elderly individuals aged 65 and above.

  India, on the other hand, has a **relatively younger population**. It has a larger proportion of individuals in the working-age group (15–64 years) compared to Japan.
- Japan has a declining birth rate, resulting in a **shrinking youth population**. The proportion of young people (0–14 years) in Japan is relatively small.

  India has a significant **youth population**. It has a higher proportion of children and young adults compared to Japan.
- Due to its ageing population, Japan has a **high dependency ratio**. The number of elderly people is increasing, while the working-age population is declining, leading to a higher burden on the working population to support the elderly.

  India has a **relatively lower dependency ratio** as it has a larger working-age population that can support dependents.

- Japan has a **declining population** due to low birth rates, low fertility rates and limited immigration. This demographic trend has contributed to population decline and a smaller overall population size.

  India has a **rapidly growing population**. It has a higher birth rate, relatively higher fertility rates and significant population growth fuelled by a larger number of births.
- The ageing population in Japan poses **challenges** such as increased healthcare and pension costs, labour shortage, and a shrinking workforce. It also has implications for economic growth and social welfare systems.

  India's younger population offers **potential advantages** in terms of a larger workforce, demographic dividend and potential for economic growth. However, it also brings challenges in providing education, employment and infrastructure to support the growing population.

Demographic Transition

Demographic transition refers to the historical shift of a population from high birth and death rates to low birth and death rates, accompanied by changes in population growth and age composition. It typically occurs as countries undergo economic and social development. There are four main stages in the demographic transition model:

- **Stage 1:** Pre-transition

  High birth rates and high death rates. Population growth is slow and fluctuates due to factors such as disease, famine and limited medical care. Historically, both India and Japan were in this stage.
- **Stage 2:** Transition

  High birth rates but declining death rates. Improved healthcare, sanitation and food supply lead to reduced mortality rates. Rapid population growth due to a significant gap between birth and death rates. India experienced rapid population growth during this stage, while Japan transitioned to Stage 3.
- **Stage 3:** Post-transition

  Birth rates start to decline. Death rates remain low or continue to decline. Population growth slows down as birth and death rates become more balanced. Japan is currently in this stage, with a declining birth rate and an ageing population.
- **Stage 4:** Post-industrial

  Low birth rates and low death rates. Stable or slightly declining population size. Birth and death rates are both relatively low, resulting in negligible population growth. Some developed countries, such as Japan, have reached this stage. India is still in the midst of the demographic transition, with a high birth rate but experiencing a gradual decline. It is moving from Stage 2 to Stage 3, facing the challenges of managing population growth and providing resources for its growing population. In contrast, Japan has completed the demographic transition and is now facing the consequences of an ageing population, including a declining workforce, strain on healthcare and social security systems, and implications for economic productivity. It is important to note that while these stages provide a general framework, each country's demographic transition can vary based on unique factors, including culture, policy and socioeconomic conditions.

RISING POPULATION: PRESSURE ON NATURAL RESOURCES

A resource is any substance (natural or artificial), energy or organism used by humans for their welfare. In the context of rising human population, the six main resources under pressure are as follows:

- **Food:** Food is a necessity for all living things. It is true that people have been developing more efficient ways to increase food production, but this has not matched the increase in population. If left uncontrolled, the population will grow in geometric progression, roughly in the pattern of 1, 2, 4, 8, 16 …; that is, the numbers will multiply at each step. Food production increases in arithmetic progression, in a pattern of 1, 2, 3, 4, 5 …; that is, by adding up at each stage, by cultivating more land and employing better farming techniques. This suggests that there will be a shortage of food for an unchecked rising population.
- **Water:** With a rise in population, it will become increasingly difficult to find clean, germ-free water for drinking; this is primarily due to water contamination in rivers, ponds, lakes, etc.
- **Land:** Humans are cultivating crops in more land while also utilising more land for the construction of industrial facilities, residential colonies and factories. Thus, the amount of usable land would diminish over time.
- **Forests:** In a sense, forests are the most significant natural resources. They provide wood, fuel, resins, medicines and wood pulp (for making paper). The main source of the oxygen needed for life is forests (natural air purifiers by photosynthesis). They deliver rain (adding moisture through transpiration)

and prevent floods at the same time (by holding soil particles firmly). By absorbing CO_2, they stabilise weather conditions, including the temperature of the atmosphere. An enormous variety of large and small wild animals use forests as their natural habitats.

One of the most harmful effects of a growing population is deforestation, which refers to the destruction of forests in order to obtain wood and timber, cultivate land and build homes. Numerous other issues, such as flash floods, droughts, soil erosion, extinction of wild creatures, the threat of extinction for hundreds of species and global warming, have been brought on by deforestation. Forests are one of our renewable resources. By careful management and by afforestation (planting more trees), we can maintain and even improve the existing forests. Shrinking of forest cover in the world needs to be stopped without delay to maintain the habitability of our planet.

- **Energy:** Energy is required for transportation, factory operations, home cooking and other uses. Coal and petroleum are currently the two main energy sources; both are classified as fossil fuels, and their subsurface deposits are quickly running out. The reserves of petroleum may run out within 150 years at the current consumption rates, which are continually rising, whereas coal may last a little longer. Coal and petroleum can never be created again once they are used up. Due to their lengthy history of use, coal and petroleum are categorised as traditional sources of energy. A growing population would require increasing amounts of energy. New energy sources include solar, nuclear, tidal and wind power.
- **Minerals:** Higher population means greater requirements, industries and means of transportation – all of which utilise minerals such as iron, copper, etc.

Population Growth and Urbanisation Causing Serious Pressure on Resources

Worldwide, the human population is expanding at a rapid rate. Villages are becoming towns, towns are becoming cities, and cities are becoming megacities, or what is known as the metropolis process. Some of the major changes occurring in the process of urbanisation are as follows:

- More residential buildings, markets, schools, hospitals, etc.
- More streets and roads, highways, railway lines and airports
- More industrial installations, godowns and reservoirs

All of these construction projects are encroaching on both the surrounding agricultural land and open spaces. Even woodlands are being overrun. Overall, natural vegetation is declining, which is harming the ecosystem, especially climate.

The living standards of people are rising very fast:

- Increasing amounts of domestic furnishing, dinnerware, technology, accessories and other items.
- There are more clothes: simply consider how many shirts and suits, sarees, pairs of shoes each individual needs.
- Some modes of personal transportation, including bicycles, motorbikes, cars (some people may have many cars for a single family), taxis and others.

Need to Check Exploitative Use of Resources

As you now know, a rapid depletion of natural resources is being brought on by the rising population and, more importantly, rising standards of modern living. Some of these resources cannot be replenished, meaning that they are gone forever and cannot be found again; others take a very long period to be replenished. Therefore, it is imperative that each individual adopts awareness and makes a contribution towards the sustainable use of natural resources.

Sustainable methods can be carried out indefinitely without negatively affecting the supply of the necessary goods. Development that satisfies current demands without jeopardising the ability of future generations to meet their own needs is referred to as sustainable development.

In order to achieve sustainable development, it is primarily necessary to:

- Reduce the excessive use of natural resources
- Recycle and reuse resources whenever possible
- Increase the use of renewable resources, such as wind and solar energy

The population growth rate needs to be reduced for the benefit of all life. As a result, there will be less strain on finite resources, better healthcare, better education, and other benefits. Governments and social organisations are paying close attention to how this can be achieved.

Let us first become familiar with a few terms and some of their applications in the study of human population.

- **Demography** is the measurement of populations, particularly in relation to size and density, distribution and other important statistics. For example, more people live in towns and major cities than in rural areas.

- The number of people per square kilometre (km^2) at any given period is known as **population** density. Netherlands has the highest population density of about 407 persons/km^2, India has 392 persons/km^2, Japan about 348 persons/km^2 and US only about 33 persons/km^2.
- **Birth rate** or natality is the number of live births per 1000 people per year. For example, if there were 440 live births in a population of 20,000, the birth rate would be $440 \div 20 = 22$.
- **Death rate** or mortality is the number of deaths per 1000 people per year.
- The **growth rate** of a population is the difference between the birth rate and the death rate. As long as birth rate exceeds death rate, the population will grow. If the birth rate is lower than the death rate, the population will decline.

Between 1975 and 2010, the population doubled to 1.2 billion, reaching the billion mark in 2000. India overtook China to become the most populous country at the end of April 2023. It is expected to become the first country to be home to more than 1.5 billion people by 2030, and its population is set to reach 1.7 billion by 2050. In 2017, its population growth rate was 0.98%, ranking 112th in the world; in contrast, from 1972 to 1983, India's population grew by an annual rate of 2.3%.

Need for Adopting Control Measures

A nation's per capita income declines as population density rises beyond its capacity, and natural resources like land, minerals, timber and fuel are depleted. Additionally, general health declines. Huge families generally result in a nation having a large population, which therefore lowers overall quality of life. Therefore, implementing population control measures is necessary.

It is necessary that people should be educated about the need to limit the population and about the steps which can be taken in this direction.

- The population problem should be brought to the attention of as many individuals as possible, especially those who reside in extremely distant places.
- People should be made aware of the benefits of having small families as well as the drawbacks of having a large family.
- China has reached the norm of 1 child per family, whereas India has to strive hard to reach the 2 children per family norm.
- Education should change the traditional belief that it is best to have at least one son, especially in Indian society.
- The legal marriageable age should to be raised.
- Married couples should be taught to delay having their first child and to space the second child's birth with enough time for proper rearing.
- People should be encouraged to use family planning strategies to avoid getting pregnant (contraception). These include contraceptives (such as Nirodh), intrauterine devices (such as Copper-T) and oral medications (such as Mala-D). Several forms of contraception are being used nowadays.

Family Welfare

In India, the inverted red triangle has gained popularity as a symbol for family stability. All offices and hospitals where any assistance or advice about family planning is offered without charge display this symbol publicly.

The term family welfare has three aspects:

- Planning for a modest family.
- The general wellbeing of the family, including the mother's nutrition while she is pregnant.
- Continued care for the child, such as immunisation and oral rehydration therapy, to ensure its survival.

Methods of Contraception

- **Hormonal methods (Pills):** Different hormonal treatments are available as tablets or pills (commonly called contraceptive pills). These hormones stop the egg from being released by the ovary.
- **Barrier methods:** These are as follows:
  - **Condoms (such as Nirodh):** These are used by men. It is made of latex (rubber sheath). It prevents the sperms from being deposited in the vagina.
  - **Diaphragms:** These are coil-spring-equipped circular latex caps. These caps can be placed on the uterus' mouth, in the vagina (cervix). These covers stop the sperm from entering the uterus.
  - **Sperm-killing (spermicidal) agents:** These are chemicals placed in the vagina near the cervix, which kill the sperms if present.
  - **Intrauterine devices (IUDs):** Copper-T and Lippe's loop are two gadgets that are frequently utilised in India. They are placed in the uterus. They do not stop fertilisation but prevent the blastocyst from being implanted (embryo).
- **Surgical methods:** These are as follows:
  - **Tubectomy (for females):** Here, the Fallopian tubes (oviducts) are tied with nylon thread to close the passage of the egg (Fig. 15.4).

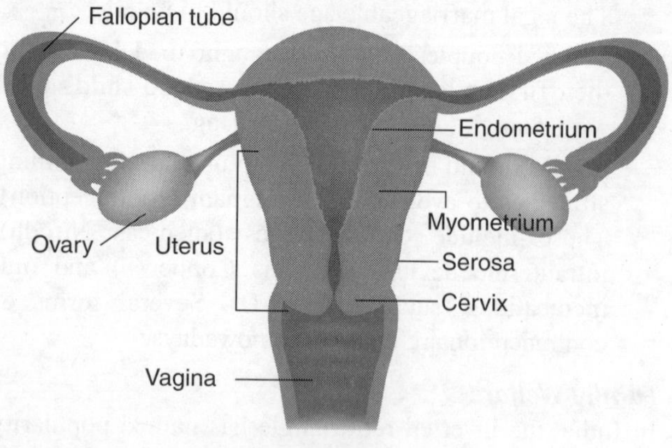

Fig. 15.4 Tubectomy

- **Vasectomy (for males):** Here, a little cut is made in the scrotum, the vas deferens (sperm ducts) of both testicles are tied together, and a tiny piece is taken out from between the two ligatures (Fig. 15.5). This procedure is simpler, faster and safer. Therefore, it is advised that the husband undergo surgery rather than the wife. This procedure does not negatively impact the man in any way, nor does it lessen the enjoyment of sex or libido.

Medical Termination or Induced Abortion of Pregnancy (MTP): The foetus may be removed if the woman has become pregnant and the couple does not wish to have the child or if a particular test reveals conclusive proof of any significant genetic disorder in the embryo. This approach should only be used as a last resort rather than as a contraceptive strategy. Only a licensed physician working in a hospital should carry out this procedure (forced abortion). Only if the pregnancy is less than five months is abortion lawful and it may be requested for free at any government hospital by any willing female. Even the husband's approval is not required for this.

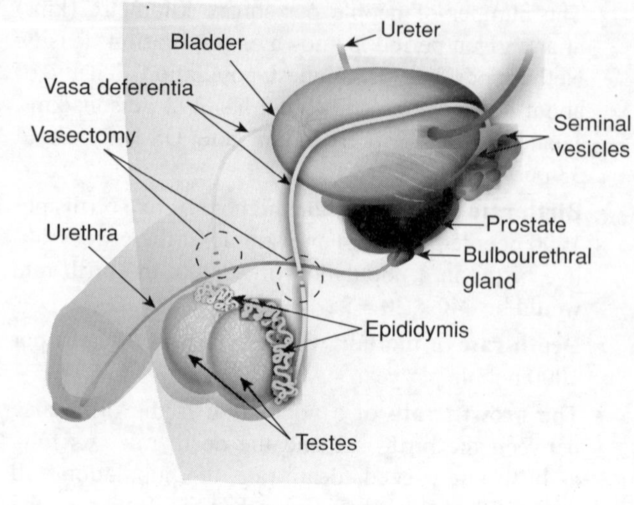

Fig. 15.5 Vasectomy

SUMMARY

- A population is defined as a group of individuals of the same species living and interbreeding within a given area.
- Members of a population often rely on the same resources, are subject to similar environmental constraints, and depend on the availability of other members to persist over time.
- Overpopulation affects all of us as scarcity of food will increase, pollution will increase and global warming becomes even more of a problem.
- With more and more people putting a strain on the planet's already limited resources, measures must be taken to curb population growth.
- The concept of population growth includes factors that contribute to it, such as birth rate, death rate, migration and life expectancy.
- The impact of population growth on the environment includes issues such as deforestation, depletion of natural resources and pollution.
- The impact of population growth on the economy includes issues such as unemployment, poverty and inflation.
- The impact of population growth on society includes issues such as overcrowding, urbanisation and social inequality.
- The measures taken by the government to control population growth are family planning, education and healthcare.

PRACTICE QUESTIONS

Analyse Your Concepts (School Exam Based)

Multiple Choice Questions

1. What was directly responsible for the rapid rise in world population in the twentieth century?
   (a) Increased food production
   (b) Better transport facilities
   (c) Better education and job prospects
   (d) Use of antibiotics and prophylactic vaccinations

2. Birth rate is the number of live births:
   (a) Per 1000 people per year
   (b) Per 100 people per decade
   (c) Per 1000 people per decade
   (d) Per 100 people per year

3. Which of the following is a factor that contributes to population growth?
   (a) Death rate
   (b) Migration
   (c) Low birth rate
   (d) Short life expectancy

4. Which of the following is an impact of population growth on the environment?
   (a) Decrease in pollution
   (b) Increase in natural resources
   (c) Deforestation
   (d) Reduction in greenhouse gas emissions

5. What is demographic transition?
   (a) The process of demographic change that occurs as a country transitions from high birth and death rates to low birth and death rates.
   (b) The process of migration from rural to urban areas.
   (c) The process of achieving zero population growth.
   (d) The process of industrialisation and economic development.

6. What measures can be taken by the government to control population growth?
   (a) Providing free electricity to households
   (b) Encouraging early marriage
   (c) Providing access to family planning services
   (d) Limiting access to education

7. Which of the following is an impact of population growth on society?
   (a) Increased social equality
   (b) Decreased urbanisation
   (c) Overcrowding
   (d) Decrease in crime rate

Match the Following

Each question contains statements given in two columns which have to be matched. Statements in column I have to be matched with statements in column II.

Column I	Column II
(a) Stage 1	(i) Providing opportunities for skill development and awareness
(b) Stage 2	(ii) Providing access to medical facilities and services
(c) Environmental	(iii) High birth rate and high death rate
(d) Education	(iv) High birth rate and declining death rate
(e) Healthcare	(v) Issues such as deforestation and pollution

True or False

Read the following statements and write your answer as true or false.

1. Urbanisation is the enlargement of towns covering a very large area together with modern facilities.
2. Rising population is leading to increased use of land for growing forests.
3. Coal and iron are inexhaustible resources.
4. Cow was the first domesticated animal.
5. Rapidly growing industries favoured a population rise.
6. Present human population growth is following arithmetic progression.
7. Vasectomy is the surgical method of sterilization in human males.

8. Tubectomy is the placing of a diaphragm on the cervix.

9. Birth rate (natality) is the number of live births per 100 people of population per decade.

10. Tubectomy is a popular surgical method of contraception in human males.

Fill in the Blanks

Complete the following statements with an appropriate word/term to be filled in the blank spaces.

1. Public conveyances like buses and railways are ___________.

2. Many sets of garments per member in a family are ___________.

3. Petroleum is a resource of the same category as that of___________.

4. ___________ is a measure of the number of births per 1000 individuals in a population in a year.

5. ___________ is the study of the size, composition and distribution of human populations.

6. The process of demographic change that occurs as a country transitions from high birth and death rates to low birth and death rates is known as ___________.

7. The impact of population growth on the environment includes issues such as ___________, depletion of natural resources, and pollution.

8. The measures taken by the government to control population growth include providing access to ___________, education, and healthcare.

9. Overcrowding, urbanisation and social inequality are some of the impacts of population growth on ___________.

10. ___________ is a measure of the number of deaths per 1000 individuals in a population in a year.

Assertion–Reason Questions

The following questions consist of two statements: Assertion (A) and Reason (R). Answer these questions selecting the appropriate option given below:

(a) Both A and R are true and R is the correct explanation of A
(b) Both A and R are true but R is not the correct explanation of A
(c) A is true but R is false
(d) A is false but R is true

1. Assertion (A): Rapid population growth can put pressure on limited resources.
   Reason (R): As the population increases, the demand for resources such as food, water, and energy also increases.

2. Assertion (A): The carrying capacity of an ecosystem determines the maximum population size it can sustain.
   Reason (R): Carrying capacity is influenced by factors like resource availability, environmental conditions, and interactions between species.

3. Assertion (A): Age pyramid provides information about the distribution of different age groups in a population.
   Reason (R): Age pyramid reflects the birth rate, death rate, and life expectancy of a population.

4. Assertion (A): High population density can lead to increased competition for resources.
   Reason (R): When there are more individuals in a given area, there is a higher demand for limited resources, which can lead to competition.

5. Assertion (A): Demographic transition refers to the change in the birth and death rates of a population over time.
   Reason (R): Demographic transition is influenced by factors like social and economic development, healthcare, and education.

One-word Answers

1. What is the term for the number of individuals per unit area?
2. What is the term for the movement of individuals into a population?
3. What is the term for the maximum population size an environment can sustain?
4. What is the term for the total number of individuals of a species in a specific area at a given time?
5. What is the term for the average number of children born per woman during her reproductive years?

Define the Terms

1. Birth rate
2. Death rate
3. Population density
4. Exhaustible resource
5. Rate of growth of population

Groups of Terms

Given below are certain groups of terms. In each group the first pair indicates a relationship between the two terms. Rewrite and complete the second pair on a similar basis.

1. Birth rate : Natality :: Death rate : ____
2. Population density : Number of individuals per unit area :: ____
3. Immigration : Incoming population movement :: Emigration : ____
4. Limiting factors : Environmental factors that restrict population growth :: ____

Identify the Odd One Out

1. (a) Birth rate (b) Death rate (c) Infant mortality rate (d) Population density
2. (a) Family planning (b) Urbanization (c) Migration (d) Poverty
3. (a) Life expectancy (b) Healthcare (c) Education (d) Corruption
4. (a) Overcrowding (b) Deforestation (c) Pollution (d) Industrialization
5. (a) Stage 1 (b) Stage 2 (c) Stage 3 (d) Stage 5

Image-Based Questions

1.

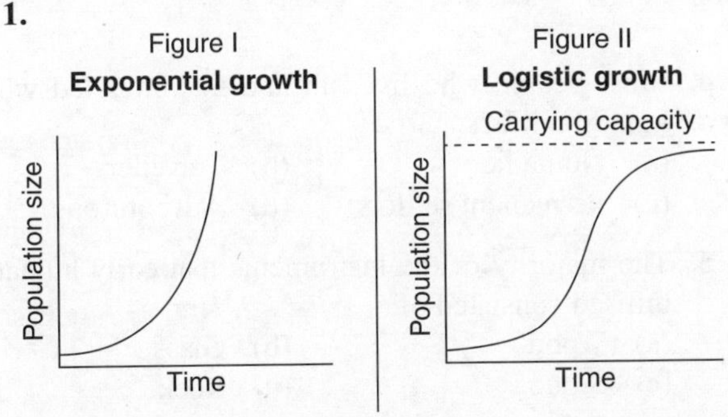

Which of the following statements is true in relation to the above population growth curves given under Figure I and Figure II?

(a) Both the curves show the population growth under unlimited resources.
(b) Population growth curve of Figure I grows in geometric fashion.
(c) Presence of natural checks causes the growth curve of Figure I.
(d) $dN = rN$ is applicable to growth curve of Figure II.

Very Short Answer Questions

1. Give the technical term for the statistical study of the human population of a region.
2. Name two surgical techniques (one for the human male and another for the human female) that can be used to prevent pregnancy.
3. What is the effect of population growth on the environment?
4. What are the different stages of demographic transition?
5. What are the adverse effects of poverty on population growth?

Short Answer Questions

1. What is the present rate of growth of population of the world and of our own country? Give two advantages of a small family.
2. Mention some reasons for the rapid increase of population in India.
3. Someone has said that a population explosion is far more dangerous than an atomic explosion. Justify this statement.
4. Explain briefly the relationship between poverty and population and how one affects the other.
5. List three major landmarks in human history which contributed to the sudden rise in population of the world.
6. What are the age restrictions for marriage by law for boys and girls in India? What is meant by family welfare centre? What is the symbol of family welfare in our country?
7. List the advantages of having small families.

Long Answer Questions

1. Our resources cannot keep pace with the rising population. Give three examples in support of this statement.
2. How can knowledge of processes of reproduction help people in limiting the size of their families? Give two concrete examples.
3. What is the idea behind the phrase "population explosion"?
4. What do you mean when you refer to the 3 Rs in terms of sustainable development?
5. How can people reduce the number of children they have by understanding how reproduction works? Give two specific instances.

Case Study-Based Questions

1. Given below are hypothetical figures in regard to population (in crores) of two countries A and B during the last three decades.

Country	1971	1981	1991
A	6.4	9.6	10.6
B	15.7	15.7	15.3

I. Which country had a declining population growth rate between 1981 and 1991?
(a) Country A (b) Country B
(c) Country C

II. Which country had negative population growth between 1981 and 1991?
(a) Country A (b) Country B
(c) Country C

III. Between which years did Country B show zero population growth?
(a) 1971 and 1981 (b) 1981 and 1991
(c) 1991 and 2001

NCERT Corner

1. Why is the rate of population growth in India declining since 1981?
2. Discuss the major components of population growth.
3. Define age structure, death rate and birth rate.
4. How is migration a determinant factor of population change?
5. Distinguish between population growth and population change.
6. What is the relation between occupational structure and development?
7. What are the advantages of having a healthy population?

COMPETITION WINDOW (OBJECTIVE TYPE)

1. Which of the following is thought to be the most serious problem that, if properly addressed, could pose a threat to one's later years?
(a) Rapid increase in human population
(b) Rapid decrease in human population
(c) Rapid change in human population
(d) Stagnation in human population

2. When the population grows significantly, it serves as a:
(a) Threat on a temporary basis
(b) Threat on a global basis
(c) Threat on a limited basis
(d) Threat on a minimal basis

3. Multiple forms of the media are trying extremely hard to highlight which negative impact in order to raise public awareness?
(a) Declining population
(b) Stagnating population
(c) Rising population
(d) All of the above

4. Most people who lived in ancient times led what kind of a life?
(a) Nomadic (b) Traveller
(c) Permanent settlers (d) Adventurous

5. The majority of the instruments that early humans utilised consisted of:
(a) Wood (b) Glass
(c) Stone (d) Rock

6. What was the overall population of the earth nearly ten thousand years ago?
(a) 4.5 million (b) 5.3 million
(c) 4.8 million (d) 5.5 million

7. When people became aware of bronze and iron, that age was classified as:
(a) Metal age (b) Traditional age
(c) Stone age (d) Modern age

8. When is "World Population Day" celebrated?
(a) 12th May (b) 14th June
(c) 5th July (d) 11th July

Answer Keys

Multiple Choice Questions

1. (d) 2. (a) 3. (b) 4. (c) 5. (a)
6. (c) 7. (c)

Match the Following

(a) - (iii); (b) - (iv); (c) - (v); (d) - (i); (e) - (ii)

True or False

1. T 2. F 3. F 4. F 5. T
6. F 7. T 8. F 9. F 10. F

Fill in the Blanks

1. Mass transportation
2. High clothing consumption
3. Fossil fuels
4. Birth rate
5. Demography
6. Demographic transition
7. Deforestation
8. Family planning services
9. Society
10. Death rate

Assertion-Reason Questions

1. (a) 2. (a) 3. (a) 4. (a) 5. (a)

One-word Answers

1. Population density
2. Immigration
3. Carrying capacity
4. Population
5. Fertility rate

Define the Term

1. Birth rate is the ratio between the number of live-born births in the year and the average total population of that year.
2. Death rate refers to the number of deaths that occur in a population within a specific time period, usually per 1,000 individuals.
3. Population density is the concentration of individuals within a species in a specific geographic locale.
4. The phrase exhaust refers to the complete utilisation of something or finishing up of something. Similarly, resources that will be depleted after a few hundred years are referred to as exhaustible resources.
5. It is the difference between the birth rate and the death rate. As long as the birth rate exceeds the death rate, the population grows. If the birth rate is lower than the death rate, the population declines.

Groups of Terms

1. Mortality
2. Age structure : Distribution of individuals among different age groups
3. Outgoing population movement
4. Density-independent factors : Factors unaffected by population density

Identify the Odd One Out

1. Population density
2. Poverty
3. Corruption
4. Industrialisation
5. Stage 5

Image-Based Questions

1. (b)

Very Short Answer Questions

1. Demography is the technical term for the statistical study of human population of a region.
2. Surgical procedures for family planning include tubal ligation for female sterilisation and vasectomy for male sterilisation.
3. Population growth has adverse effects on the environment such as deforestation, pollution and depletion of natural resources.
4. The different stages of demographic transition are Stage 1 (high birth and death rates), Stage 2 (high birth rates and declining death rates), Stage 3 (declining birth rates and low death rates), and Stage 4 (low birth and death rates).
5. Poverty can lead to a lack of education, healthcare and family planning, which can result in high birth rates and population growth.

Short Answer Questions

1. The global population is increasing at a rate of roughly 1.12% per year, which translates to a net gain of 83 million people every year. India's population is growing at a rate of 1.2% annually.
2.
   - Illiteracy and lack of awareness as far as the lower class is concerned.
   - Lack of sex education in schools.
   - Poor access to contraceptives and early marriages resulting in early pregnancies, particularly in the rural areas.
   - Social ills like child marriage.

3. A population explosion is even more harmful than an atomic explosion since it causes many serious issues that raise doubts about whether humans will survive. The earth and its resources are being mortally taxed by the fastest expansion in the human population
4. Lack of literacy leads to poverty. Also, the majority of the rural residents who make up the majority of our society are still uneducated, ignorant and superstitious. Additionally, they lack knowledge of how the human reproductive system works. This causes excessive copulation, which fuels an even greater increase in population.

   Poor people also believe that the children's employment will enable them to support the family financially. Likewise, the income will increase if there are more children. Due to these poverty-related causes, more people are having big families, which raises the population as a whole.
5. (i) Tool making revolution.
   (ii) Agricultural revolution.
   (iii) Scientific and industrial revolution.
6. According to the 1981 population census, the Indian population was 685 million and as per the 1991 population census, it was 846 million.

   In India, the age restriction for marriage for boys is 21 and for girls, it is 18 years. The age restriction for marriage means the age before which a person cannot get married. If he or she does so, the marriage would not be considered legal.

7. Each child receives more parental attention and educational advantages, which generally raise their self-esteem. Children in small families, especially first and only children, tend to have higher school and personal achievement levels than do children of larger families.

Long Answer Questions

1. Hint: Give examples such as food demand greater than agricultural resources, limited arable land, water and fertilisers.
2. Hint: Knowledge of reproduction processes helps in limiting family size. Elaborate on this.
3. Hint: Elaborate on rapid population growth and exponential increase in population.
4. Hint: Reduce, Reuse, Recycle. Minimise waste and resource consumption.
5. Hint: Explain the family planning methods and contraception techniques.

Case Study-Based Questions

1. (i) b, (ii) b, (iii) a

NCERT Corner

1. The decline is due to greater awareness and usage of birth control measures.
2. Birth rate, Death rate, Migration
3. Age structure: Refers to the number of people in different age groups in that population.

   Death rate: The number of deaths per thousand persons in a year.

   Birth rate: The number of live births per thousand persons in a year.
4. It changes the demographics of a country.
5. Population growth:
   - Increase in the number of inhabitants of a region during a specific time period.
   - Birth rate and migration are the major causes of population growth.

   Population change:
   - During a specific time period, it is the change in the distribution, composition or size of the population.
   - Birth rate, migration and emigration are the major causes of population change.
6. The higher percentage of the population involved in primary occupations like agriculture, animal husbandry, forestry and fishing implies a less developed country.

   We see more development in countries where people move into secondary occupations like manufacturing.

   A high percentage of the population involved in a tertiary occupation like banking, commerce, transport and administration implies a highly developed country.
7. - It will make the nation strong in all spheres.
   - It will help in creating a developed and prosperous nation.
   - It will help in economic growth.
   - The nation can compete globally with all other nations in any sphere of requirement.

Competition Window

1. (a) 2. (b) 3. (c) 4. (a) 5. (c)
6. (b) 7. (a) 8. (d)

CHAPTER 16

Aids to Health

PREREQUISITES

You must have knowledge of:

- Basic biology, including human anatomy and physiology
- Human immune system
- Understanding of infectious diseases and how they are spread
- Signs and symptoms of HIV/AIDS and how it can be diagnosed

LEARNING OBJECTIVES

At the end of this chapter, you will be able to:

- Explain the concept of health and its importance in maintaining a balanced and fulfilling life
- Explore the concept of immunisation and the role of vaccines in preventing infectious diseases
- Describe the HIV/AIDS pandemic, its causes, modes of transmission and methods of prevention
- Explain first aid and its use in emergency situations

INTRODUCTION

Everyone wants to be in good health and free from illness. This requires a number of actions, some of which are personal and others which are communal. On a personal level, one must maintain bodily cleanliness by cleaning, bathing, brushing teeth, engaging in regular exercise, eating a healthy diet, and getting enough relaxation and sleep. To prevent the growth of disease-transmitting insects and pests, our surroundings should be kept clean and devoid of accumulations of trash, stagnant water, etc.

April 7 is celebrated every year as World Health Day. This day helps to create consciousness about healthy living and disseminate the message at all levels.

IMMUNITY

Immunity is the body's line of defence against illness. Germs and other hazardous contaminants, such as pollutants (poisonous compounds), constantly assault our bodies. They infiltrate our body in the following ways:

- Directly through the skin
- Through the mucous membranes of the eyes, nose, urinary or genital tracts
- Through food or water intake
- Through the air we breathe in

Our body's defence mechanism usually prevents them from entering the body. Even if they do enter, the body fights them off so they do not cause any damage (rendered harmless). Thus, the defence system works at two levels:

- **Local defence system (prevention of entry of germs):** This type of barrier system combats germs at their potential points of entry. It includes:
  - protective mechanical barriers
  - expelling them, by sneezing
  - germ-killing secretions
  - germ-fighting WBCs
- **Immune system:** It deals with the germs after they have entered the body tissues.

Local Defence System

- Protective mechanical barriers are as follows:
  - **Skin:** The protein keratin makes up the strong outer layer of the skin, which is nearly impervious to microorganisms. Numerous bacteria that enter the body through the air or through direct contact with infected substances are constantly present on the skin's outer layer. Even a handshake or a kiss on a baby's cheeks spreads viruses from one person to another. The bacteria are eliminated by washing with soap and water. Germs can enter the body through any wound or scratch on the skin. If the skin is injured in any way, a blood clot covers the cut ends of the blood vessels at the wound to keep germs out.
  - **Hair:** The path of the germ up to the skin is blocked by hair. The nostril hairs catch bacteria-filled dust.
  - **Mucus:** The slimy discharge from the epithelial lining of different organs is known as mucus. For example, the mucus secreted in the nose and windpipe captures a lot of microorganisms and keeps them from entering other tissues. The mucus-trapped bacteria are spat out by the cilia of the windpipe.
- **Coughing, sneezing and vomiting** are direct ways to remove any bacteria or foreign organisms that enter the digestive and respiratory systems. If the illness is present in the digestive tract, diarrhoea (loose motions) aids in the removal of the germs.
- **Germ-killing secretions:**
  - Saliva, sweat, tears and nasal secretions contain germ-killing substances.
  - Hydrochloric acid secreted by the stomach kills germs that gain entry along with food.
- **Germ-fighting white blood cells (WBCs):** The white blood cells (phagocytes) attack microbes when they infiltrate bodily systems. They extrude from the blood capillary walls (through diapedesis), envelop the bacteria or germs, and then phagocytose (kill) them (Figs 16.1 and 16.2). For instance, pus in a wound is a combination of slain WBCs, destroyed pathogens and injured tissue cells.

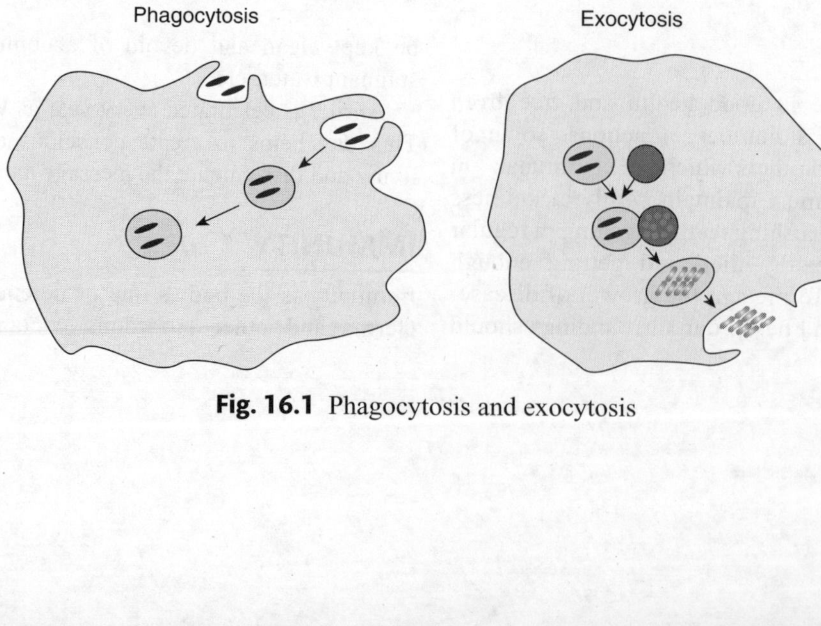

Fig. 16.1 Phagocytosis and exocytosis

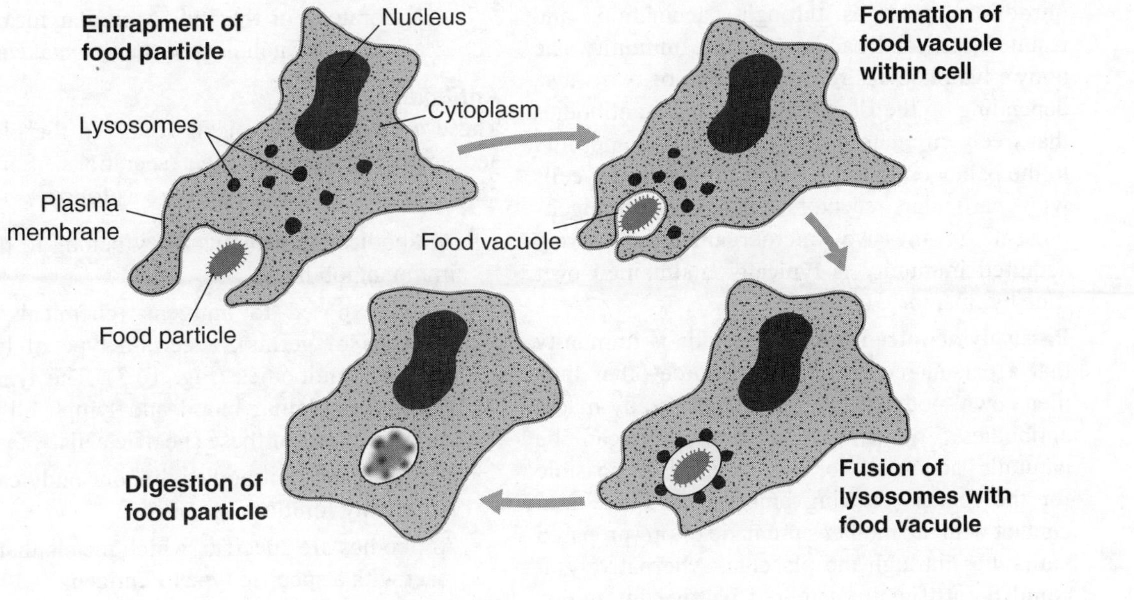

Fig. 16.2 WBCs engulfing bacteria

Merits of local defence systems:

- They immediately set to work.
- They are independent of prior infection exposure.
- They work well against many different potentially contagious agents.

Immune System

In some situations and settings, the local defence mechanisms are ineffective. By their unique means of entrance or through any breaks in the protective barriers, some microorganisms (germs) or their toxic secretions (toxins) permeate the deeper tissues and different organ systems.

At this point, blood and other bodily fluids begin to combat any bacteria or other undesirable external substances. The body fluids include specialised proteins called antibodies that respond with invasive bacteria and antitoxins that react with their toxins to eliminate them, providing protection against disease (immunity). Immunity is the body's defence against the development of a disease following infection by dangerous microbes. It is the capacity of our body to deal with foreign substances, such as bacteria, viruses and toxins, that enter our body and to render them harmless, or simply, it provides resistance against disease-causing germs.

Categories of Immunity

Immunity can be classified into two main categories, innate and acquired, and their different subcategories are as follows:

- **Native or natural immunity**, commonly known as internal immunity, results from the genetic constitution. Without any external stimulus or prior infection, it is already present in the body.
  - **General innate immunity** shows generalised innate resistance to all illnesses. For instance, neither the highly contagious illnesses of plants nor even some animal diseases affect people.
  - **Particular innate immunity** is a built-in defence mechanism against a certain type of microbe. Specific infectious diseases do not affect certain races or individuals. For instance, humans are immune to canine distemper, a highly contagious disease that kills nearly half of all infected canines.
- **Acquired immunity** is the resistance to an illness that a person develops over the course of their lifetime. This may be a previous illness (actively acquired immunity); for example, a person who has previously experienced measles will typically not experience it again. Alternatively, it could result from 'ready-made' antibodies provided from the outside (passively acquired immunity); such as an injection of antivenin given to a person bitten by a deadly snake.
  - **Actively acquired immunity:** This is the resistance that a person develops in response to a prior infection or antigen (chemical found on the surface of the disease-causing germ cell), which naturally enters the body and results in naturally acquired active immunity or is artificially

introduced, such as through vaccination, and results in artificially acquired active immunity. The body's lymphocytes respond in one of two ways, depending on the situation: they create antibodies that freely circulate in blood and lymph and bind to the pathogen to kill it or they create killer cells with particular receptors for foreign antigens present on invasive microorganisms. Actively acquired immunity is typically maintained over time by 'memory' cells.

- **Passively acquired immunity:** This is immunity that a person receives from a source other than their own body in the form of 'ready-made' antibodies. Again, passive immunity can be naturally acquired; for instance, it is possible for the foetus (growing embryo) to come into contact with the mother's premade or pre-prepared antibodies through the placenta. Alternatively, it could be artificially acquired passive immunity. By injecting germs into the body of a horse or another animal, antibodies are created in the blood. Injections of antiserum are made from the serum (which contains antibodies) of the blood of such animals and are injected into the patient. Several of these anti-sera are being prepared at Haffkine Institute in Mumbai and at the Central Drugs Laboratory in Kasauli. A similar idea underpins antivenin, which is used to treat snakebites.

Antibodies

These are special chemicals found in the blood which act against germs or their secretions. Some of the **characteristics** of antibodies are as follows:

- Antibodies are proteins (they belong to the class of immunoglobulins).
- When exposed to antigens (chemicals found on the cells of germs), a certain type of lymphocyte produces antibodies (Fig. 16.3). The lymph nodes, spleen, circulating blood and lymph all have high concentrations of these specific cells.
- The variety of antibodies that our body can produce is virtually limitless.
- Antibodies are specific, which means that they only react with a specific type of antigen.
- A specific antigen is recognised by an antibody, which then binds to it and renders it harmless. The harmless antigen is then destroyed and removed by the body.
- Some people have certain antibodies in their blood since birth. Even if the microorganisms that cause these diseases have got past the barriers and escaped

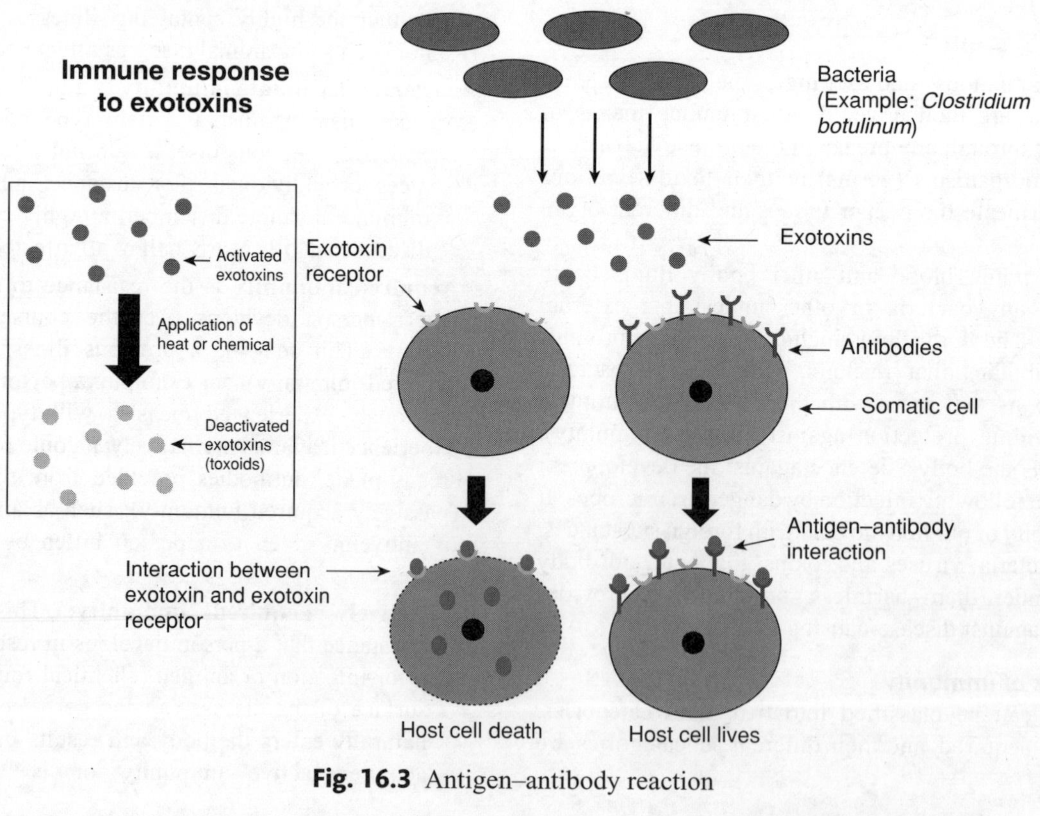

Fig. 16.3 Antigen–antibody reaction

the phagocytes, those persons who have these specific antibodies do not get those diseases.

- Immunity generated by antibodies in response to antigen exposure may last a short time (as in the case of the common cold or cholera) or a longer time (as in smallpox and measles).

AIDS: A Condition in which the Immune System Fails

Acquired Immunodeficiency Syndrome is a dreaded illness for which there is currently no cure. It is brought on by HIV (Human Immunodeficiency Virus) infection. It can be spread through blood transfusion, contaminated syringes and sexual contact. The virus assaults the immune system cells and significantly reduces the number of T-cells in the thymus. These are the cells that stimulate the immune system-producing lymphocytes. In full-blown AIDS, immunity collapses, allowing other serious illnesses to proliferate unchecked and eventually lead to death. December 1 is marked every year as World AIDS Day.

VACCINATION AND IMMUNISATION

The method of injecting a synthetic germ substance into the body to increase resistance to specific diseases is known as vaccination. Scientifically speaking, this procedure is known as prophylaxis, and the substance that is ingested is known as the vaccine. The vaccine or germ material is injected or, less frequently, taken orally into the body (for example, polio drops). The vaccine induces the WBCs to create antibodies against the specific disease's pathogens inside the body.

Originally used to refer to vaccinations against smallpox, the terms vaccine and vaccination are now used more broadly.

Categories of Vaccines

- **Killing germs**, such as the typhoid TAB vaccine, poliomyelitis Salk vaccine and rabies vaccine (dog-bite).
- **Living, weakened pathogens**, such as the BCG vaccine for tuberculosis and the measles vaccine, both of which are freeze-dried. The names of two French scientists who created this strain for vaccination are the full form of BCG, which is called Bacillus of Calmette and Guerin.
- **Fully poisonous living organisms**, such as smallpox. An individual receives an injection with the cowpox virus, which is remarkably similar to the smallpox virus. Instead of the smallpox virus causing several pustules to appear all over the body, the cowpox virus only causes one pustule to appear. The cowpox vaccine also offers protection against smallpox. Smallpox vaccinations are no longer offered since, according to authorities, the illness has been completely eradicated.
- **Toxoids**, such as those for diphtheria and tetanus (bacteria-secreted toxins). By adding diluted formalin, these toxins (poisons) are rendered harmless while still having the ability to create antibodies (antitoxins). Additionally, efforts are being made to create an AIDS vaccine.

Antitoxins

Any dangerous compound produced by an animal, plant or bacterium is referred to as a toxin in general usage. As an illustration, consider dangerous substances generated by infections that are developing inside the body, such as snake venom, scorpion, insect and other sting toxins. Any molecule produced by the body in response to the ingestion of a foreign poison was referred to as an antitoxin. Animals like horses create antivenins to treat snake bites. Instead of antitoxin, the term antibody is now more commonly used.

A blood serum protein known as an antibody is created in reaction to the injected antigens. For example, if someone has a disease like diphtheria, injecting ready-made antibodies from a different source is beneficial. For this, blood from horses, rabbits, etc. is used to create an antibody-containing serum (often known as antiserum), in which a minor form of the disease is artificially created. Passive immunisation is the term for a treatment involving such antibodies.

Antiseptics

These are gentle chemicals that kill bacteria when applied to the body. These compounds have such a low concentration that they have no negative effects on the body or skin. Good antiseptics include carbolic acid, iodine, benzoic acid, mercurochrome and boric acid in diluted solutions. The same function is also accomplished by several antibiotic creams.

Antibacterials

These are potent chemicals used on areas and locations where germs thrive and proliferate. Cresol, phenol, Lysol, 40% formalin, lime, Bordeaux combination and DDT are examples of commonly used disinfectants. All disinfectants should be used with caution because they are potent and should not be applied to human skin.

Boiling and high heat may kill bacteria and are referred to as physical disinfectants. Deodorants are neither disinfectants nor antibiotics. They only cover up unpleasant odour.

Antibiotics

Antibiotics are chemicals that some microbes make; they have the ability to kill or stop the growth of other germs. Although the first antibiotic penicillin was discovered in 1929, it was not until the 1940s that it was actually used to treat a human disease. Selman Waksman came up with the name antibiotic for this class of medicines in 1942.

In a London hospital, Alexander Fleming (1881–1955) began cultivating *Staphylococcus* laboratory cultures (round bacteria growing in clusters). He discovered in 1928 that one of the cultures on a petri dish had developed a specific unwanted mould called *Penicillium*, which had prevented the bacteria from growing. It was obvious that some material that was released by the mould must have killed the germs there. In 1929, Fleming gave the bacterial-killing chemical that had been freed from the mould the name penicillin. This drug was tested on lab animals in the hope that it would eliminate disease-causing bacteria effectively, and it did so without endangering the animals.

When penicillin was originally tested on people in the early 1940s, it was discovered to be quite efficient against a number of infections. It was particularly effective in treating the sexually transmitted disease gonorrhoea (STD). In those early days, doctors occasionally advised administering pentidsulph, a medication cocktail made up of both penicillin and sulpha. After that, many antibiotics were discovered and are now being used to treat a variety of infectious ailments. These antibiotics include streptomycin, chloromycetin, aureomycin and ampicillin. In addition, antibiotics are used to avoid infections following surgical procedures.

Penicillin interferes with cell wall construction; therefore, bacteria are unable to develop and reproduce. Streptomycin attaches to bacterial ribosomes, preventing them from synthesizing proteins, which prevents the bacteria from proliferating and growing.

In Fleming's research, *Pencillium notatum* was identified as the mould that produced penicillin. This mould resembles that which frequently develops on oranges, lemons and other citrus fruits. Penicillin is mostly manufactured commercially from the related species *P. chrysogenum,* while it is also made synthetically. One of the most often used antibiotics is streptomycin. The source of this is the bacterium *Streptomyces*. Antibiotics are increasingly frequently produced synthetically rather than by cultivating microorganisms.

Uses

- Antibiotics are mostly used in medicine to treat infections.
- Some antibiotics are used to preserve food, particularly fresh fish and meat.
- Some are used in treating animal feed to prevent infections.
- Some antibiotics are used for controlling plant pathogens.

Sulphonamide Group of Medicines

We have been searching for ever-newer treatments for illnesses since the dawn of time. We have experimented with many different plant and animal products, many of which were successful. We have also experimented with several chemicals, some of which had positive outcomes. Chemotherapy, a form of treatment involving the use of chemicals, was created in this manner.

In 1910, a drug called Salvarsan (based on an arsenic compound) was produced. It killed the germs of syphilis as well as of sleeping sickness. However, it proved poisonous to the patient as well.

Sulphonamides are a class of compounds that were found to be beneficial against many different bacterial illnesses in the 1930s. The two sulpha medications are sulphadiazine and sulphanilamide. These medications are synthetic, and they stop the metabolism of bacteria, killing them. Sulphonamides are currently only occasionally used, and only in conjunction with antibiotics, to treat some conditions.

FIRST AID

Health emergencies may suddenly arise in numerous ways. Some such situations could be as follows:

- Serious cuts on the body
- Breaking of a bone
- A particle falling in the eye
- Sudden unconsciousness
- A heart attack
- Burns of various kinds
- Swallowing a poisonous substance
- Drowning accident
- Bite or sting by poisonous animals or insects
- An electric shock

In all such cases, whatever is needed to be performed before the doctor arrives or the patient is taken to the nearest available hospital or a dispensary comes under First Aid. Some such treatments are as follows:

- **Bleeding:** In the event of bleeding, elevate the area to reduce blood loss due to gravity. If feasible, apply some mild antiseptic after cleaning the wounded surface with clean water and pressing it with some clean cotton wool.
- **Fractures:** In the event of a fracture, lay the victim down comfortably and take the damaged clothing off. Move the broken part(s) at your own risk. Tie a sling around the fractured arm to provide support.
- **Eyes:** Do not rub your eyes if something gets in them. Sprinkle some clean water into the eye to gently wash it.
- **Confusion:** Lay the unconscious person down comfortably on a bed as soon as possible. Loosen the clothing.
- **Heart attack:** In the event of a heart attack, lay the patient straight down, let fresh air in, and urgently seek medical attention.
- **Burns:** If the burn is slight, wash the affected area immediately with plenty of cold water and let it air dry. Do not scratch the area. Apply a cream or ointment that has been specifically advised for burns, if any are on hand.
- **Ingestion of poison:** If a poisonous substance has been ingested, try to induce vomiting and force the patient to drink as much salt water as they can.
- **Snake bite:** In the event of a snake bite, apply pressure to the wound and tie a tourniquet above it to stop the spread of poison throughout the body.
- **Artificial breathing:** Accidents like drowning, electric shock or head injuries can cause someone to cease breathing. To save lives, breathing needs to be restarted right away. This is accomplished by artificially generating breathing. Figure 16.4 displays one such approach.

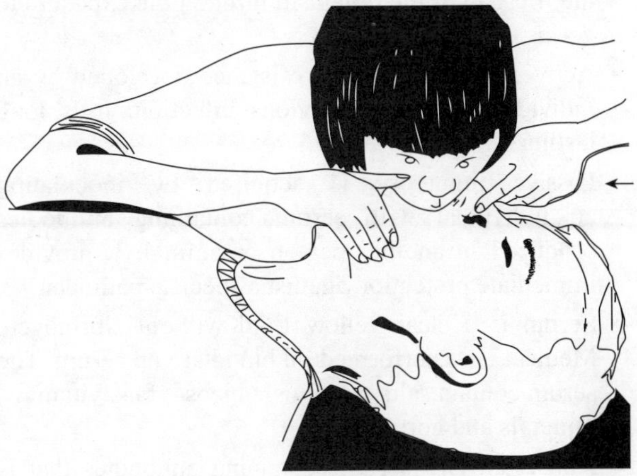

Fig. 16.4 Artificial respiration

If the victim of a drowning incident has consumed a lot of water, some of which may have entered the lungs, immediately lay him down with his back facing up and his head turned to one side. To force the water out, repeatedly press the back. As soon as the water has completely dried up, begin artificial breathing.

The subject should be laid flat on his back with a pillow or folded towel placed under his shoulders so that his chest is elevated and his head is thrown back while artificial respiration is administered.

Hold the victim's arms while sitting close to his head. Pull the arms backward and upward. His chest will enlarge as a result, drawing air in. The victim's arms should then be folded and pressed against their ribs. Now the air will be released. At a rate of around 15 times per minute, repeat the two stages. Continue until the victim is able to breathe on their own or until the doctor arrives.

SUMMARY

- Health is a state of the complete physical, mental and social wellbeing of an individual. Any disturbance in this state may lead to disease.
- Disease is a disorder or a disturbance in the structure or function of any organ or part of the body.
- Hygiene is the practice of good and healthy habits. Personal hygiene involves keeping clean, eating a balanced diet, obtaining sufficient exercise and sleep, and regular medical check-up. Public hygiene means keeping the surroundings clean.
- Immunity is the ability of an organism to resist the infection of a disease-causing organism or pathogen. There are two types of immunity, natural and acquired.
- Natural immunity is the inherited ability of an organism to resist the attack of pathogen. The human body has natural immunity to many bacteria and viruses. The immunity depends on the antibodies, the protective substances, which are produced by lymphocytes.

- Acquired immunity is the ability of an organism to resist the attack of a pathogen which develops during the lifespan of the organism through the exposure to germs.
- Active immunity is the resistance developed by an individual due to a previous infection. It is long lasting.
- Passive immunity is acquired by inoculating an individual with serum containing antibodies produced in another person or animal. It provides immediate protection against a specific pathogen.
- Serum is a clear yellow fluid without fibrinogen. Medical tests performed on blood are on serum. The serum contain salts, proteins, glucose, fats, vitamins, minerals and hormones.
- Antiserum is a serum containing antibodies that is taken from a person or animal and injected to a patient.
- Vaccination is the practice of injecting mild or weakened germs into the body to develop total resistance against a particular disease.
- Vaccine is a preparation of dead or weakened (attenuated) germs which when introduced into the body of an individual stimulates the production of antibodies against that particular germs. Examples: BCG (Bacillus Calmette–Guerin), TAB (typhoid–paratyphoid A and B vaccine), DPT (diphtheria, pertussis and tetanus), Salk's, etc.
- A toxin is a product of pathogen metabolism which is harmful to the host and which stimulates the production of antitoxins or antibodies specific to it.
- A toxoid is a treated or weakened toxin which has lost its harmful property but which stimulates the production of antibodies or antitoxins when injected in the body.

PRACTICE QUESTIONS

Analyse Your Concepts (School Exam Based)

Multiple Choice Questions

1. Penicillin is:
   (a) An antiseptic (b) A disinfectant
   (c) An antibiotic (d) An antitoxin

2. 'T' in DTP vaccine stands for:
   (a) Tuberculosis (b) Typhoid
   (c) Tetanus (d) Tonsilitis

3. World Health Day is celebrated on:
   (a) January 15 (b) February 21
   (c) April 7 (d) October 10

4. The cells in the _____ are infected and colonised by the retrovirus known as HIV:
   (a) Immune system
   (b) Immune system and central nervous system
   (c) Endocrine system
   (d) Immune system and endocrine system

5. The prevalence of HIV infection worldwide in 2015 was:
   (a) 0.2% (b) 0.4%
   (c) 0.6% (d) 0.8%

6. The most typical reason for HIV-positive patients to be admitted as inpatients to a hospital is:
   (a) Tuberculosis and other mycobacterial infections
   (b) Cardiovascular disorders
   (c) Bacterial infections
   (d) None of these

Match the Following

Each question contains statements given in two columns which have to be matched. Statements in column I have to be matched with statements in column II.

Column I	Column II
1. Gonorrhoea	(i) HIV
2. Syphilis	(ii) *Neisseria*
3. Genital warts	(iii) *Treponema*
4. AIDS	(iv) Human papilloma virus
5. *Plasmodium*	(v) Malaria

True or False

Read the following statements and write your answer as true or false.

1. Lysol is an antibiotic.
2. Sweat and tears contain germ-killing substances.
3. Our body can make only a limited variety of different antibodies.
4. Salk vaccine is used against tuberculosis.
5. Treatment by the use of chemicals is known as allopathy.
6. Alexander Fleming coined the term 'antibiotic' for substances like penicillin.

Fill in the Blanks

Complete the following statements with an appropriate word/term to be filled in the blank spaces.

1. __________ is a disease caused by *Mycobacterium tuberculosis* that primarily affects the lungs and can spread to other parts of the body.
2. The most common mode of transmission of __________ is through the bite of infected female *Anopheles* mosquitoes.
3. The use of __________ is an effective measure to prevent the spread of many infectious diseases, including COVID-19.
4. The Human Immunodeficiency Virus (HIV) attacks the __________ cells of the immune system, which weakens the body's defence against infections and diseases.
5. __________ is an acute intestinal infection caused by the bacterium *Vibrio cholerae*, which spreads through contaminated water and food.

Assertion–Reason Questions

The following questions consist of two statements: Assertion (A) and Reason (R). Answer these questions selecting the appropriate option given below:

(a) Both A and R are true and R is the correct explanation of A
(b) Both A and R are true but R is not the correct explanation of A
(c) A is true but R is false
(d) A is false but R is true

1. Assertion (A): Droplets are used to propagate several infectious diseases.
Reason (R): Saliva and nasopharyngeal secretions may include millions of germs and viruses in their tiny droplets.
2. Assertion (A): Stomach bleeding and pain associated with acidity are symptoms of peptic ulcers.
Reason (R): Peptic ulcers are a result of stress.
3. Assertion (A): All forms of pathogens require the same type of therapy.
Reason (R): Many infectious diseases can be prevented by vaccination.
4. Assertion (A): The malarial parasite in the human body is killed by chloroquine.
Reason (R): Quinine is obtained from the *Cinchona* tree's bark.
5. Assertion (A): Tuberculosis is a bacterial disease.
Reason (R): Most cases of tuberculosis are spread through droplet infection.

One-word Answers

1. Syphilis germs were eliminated by a medication based on an arsenic compound created in 1910.
2. The first antibiotic to be found.
3. The type of immunity required for treating snake bites.
4. Support tied to the arm in case of fracture.
5. A substance to neutralise the effect of poison.
6. Chemical substance which destroys microorganisms and prevents their further growth.

Define the Term

1. Pathogen
2. Immunity
3. Vaccination
4. Antibiotic
5. Epidemic

Groups of Terms

Given below are certain groups of terms. In each group the first pair indicates a relationship between the two terms. Rewrite and complete the second pair on a similar basis.

1. Antibody : Immunity :: Antigen :
2. HIV : Retrovirus :: AIDS :
3. Blood transfusion : Direct contact :: Sexual intercourse :

4. ELISA : Antibody detection :: PCR :
5. Condoms : Barrier method :: Vaccines :

Identify the Odd One Out

1. Category: Diseases
   Terms: (a) Tuberculosis (b) Cholera (c) Polio (d) Malaria
2. Category: Prevention Measures
   Terms: (a) Handwashing (b) Vaccination (c) Antibiotic (d) Use of mosquito nets
3. Category: Pathogens
   Terms: (a) HIV (b) *Salmonella* (c) *Plasmodium* (d) Cancer
4. Category: Symptoms
   Terms: (a) High fever, chills and sweating (b) Fatigue, fever and recurrent infections (c) Sudden fever, diarrhoea and vomiting (d) Itching, redness and inflammation
5. Category: Modes of Transmission
   Terms: (a) Airborne (b) Waterborne (c) Foodborne (d) Sexual

Image-Based Questions

1.

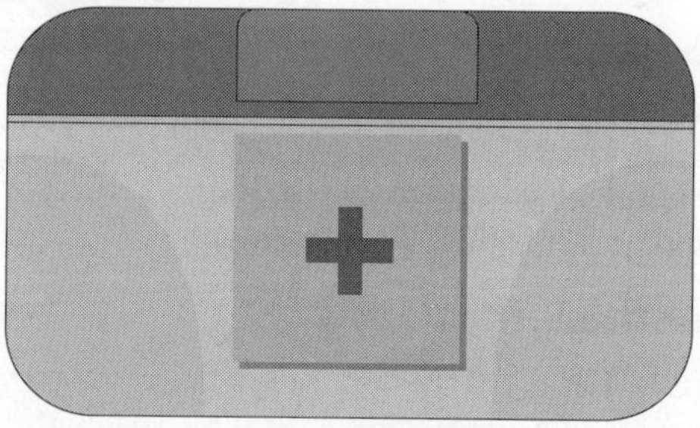

(a) Identify the object in the image.
(b) What is the purpose of having a first aid box in a workplace?
(c) What are the basic items that should be present in a first aid box?
(d) What is the recommended size for a first aid box?
(e) Why is it important to regularly check the contents of a first aid box?

2.

(a) Identify the correct head-tilt and chin-lift technique for performing artificial respiration on an adult.
(b) What is the recommended method for performing artificial respiration on an infant?
(c) Name the device used for artificial respiration.
(d) What is the recommended frequency for performing rescue breaths during artificial respiration?
(e) Identify the position of the hands and the recommended depth for chest compressions when performing CPR on an adult.

Very Short Answer Questions

1. Give the full form of the following:
   (i) DPT vaccine (ii) AIDS
   (iii) BCG (iv) WHO
2. Give the technical term for the kind of proteins produced in blood to fight and destroy harmful microbes.
3. What is the name of the virus that causes AIDS?
4. What are the three main modes of transmission of HIV?
5. What are some common symptoms of HIV?

Short Answer Questions

1. What first aid steps would you take in the following cases of accident?
   (a) If someone is injured.
   (b) If someone is unconscious and not breathing.
   (c) Burns and scalds.
2. Differentiate between:
   (a) Antiseptic and antibiotic
   (b) Antiseptic and disinfectant
   (c) Disinfectant and deodorant
   (d) Vaccination and sterilisation
   (e) Active immunity and passive immunity
   (f) Innate immunity and acquired immunity
3. Name any three vaccines and the diseases for which they provide immunity.
4. Describe any four applications of antibiotics.
5. How can HIV be prevented?

Long Answer Questions

1. The goal of vaccination is to build up one's resistance to a disease. Explain.
2. Explain briefly, the role of the following health aids: antiseptics, disinfectants and vaccines.

3. Explain the three stages of HIV infection.
4. What are some common misconceptions about HIV/AIDS, and how can they be addressed?
5. What are some challenges in preventing the spread of HIV, and how can they be addressed?

Case Study-Based Questions

1. A group of friends went on a camping trip. Upon returning, one of them started experiencing symptoms like high fever, headache, and muscle pain. After a few days, the symptoms worsened, and the person developed a rash. Based on the provided information, answer the following questions:
   (i) What type of infectious disease is the person most likely experiencing?
   (a) Bacterial infection
   (b) Viral infection
   (c) Fungal infection
   (d) Parasitic infection
   (ii) Which of the following is a common vector for transmitting the disease mentioned?
   (a) Mosquito (b) Tick
   (c) Flea (d) All of the above
   (iii) What is the most appropriate next step to diagnose the disease?
   (a) Perform a blood test
   (b) Conduct a urine test
   (c) Examine a stool sample
   (d) Take a throat swab
2. A student was exposed to a contagious respiratory infection at school. Shortly afterward, the student started experiencing symptoms like a sore throat, cough, and runny nose. The student's friend is worried about contracting the infection. Based on the provided information, answer the following questions:
   (i) Which of the following is the most likely mode of transmission for the respiratory infection?
   (a) Direct contact
   (b) Airborne droplets
   (c) Contaminated food or water
   (d) Indirect contact with fomites
   (ii) What precautionary measure should the friend take to minimize the risk of infection?
   (a) Wear a mask
   (b) Wash hands frequently
   (c) Avoid close contact with the infected person
   (d) All of the above
   (iii) What is the primary defence mechanism of the body to prevent respiratory infections?
   (a) Antibodies
   (b) Cough reflex
   (c) Mucus secretion
   (d) White blood cells
3. A group of students was studying different infectious diseases caused by bacteria. They were discussing the symptoms and possible treatments for each disease. Answer the following questions based on the provided information:
   (i) Which of the following diseases is caused by the *Mycobacterium tuberculosis*?
   (a) Tuberculosis (b) Pneumonia
   (c) Meningitis (d) Tetanus
   (ii) Which of the following is an appropriate treatment for bacterial infections?
   (a) Antiviral medications
   (b) Rest and hydration
   (c) Antibiotics
   (d) Antifungal creams
   (iii) What is the primary mode of transmission for bacterial infections?
   (a) Inhalation of contaminated droplets
   (b) Direct contact with bodily fluids
   (c) Consumption of contaminated food or water
   (d) All of the above
4. A student was preparing a biology experiment involving the handling of microorganisms. While performing the experiment, the student accidentally pricked their finger with a contaminated needle. Based on the provided information, answer the following questions:
   (i) What type of infection is the student at risk of developing?
   (a) Viral infection
   (b) Fungal infection
   (c) Bacterial infection
   (d) Bloodborne infection
   (ii) What is the first immediate action the student should take after the needlestick injury?
   (a) Wash the wound with soap and water
   (b) Apply an antiseptic solution
   (c) Seek immediate medical attention
   (d) All of the above

(iii) What preventive measure should be considered to avoid such incidents in the future?
(a) Wearing gloves and appropriate protective equipment
(b) Proper disposal of needles and other sharp objects
(c) Following laboratory safety guidelines
(d) All of the above

5. During the flu season, a student's entire family falls ill with symptoms such as fever, cough, and body aches. The student wants to take preventive measures to avoid getting sick. Based on the provided information, answer the following question:
(i) What is the most effective preventive measure the student can take to reduce the risk of contracting the flu?
(a) Regularly washing hands with soap and water
(b) Getting vaccinated against the flu
(c) Avoiding close contact with sick family members
(d) All of the above

NCERT Corner

1. List any four ways in which the antibiotics are being used.
2. List the merits of local defence system.
3. Suppose a person develops the disease diphtheria. Comment upon the principle of the treatment he should receive.
4. Abnormally large number of WBCs in the blood are usually an indication of some infection in our body. Comment on the statement.
5. Explain briefly, the role of the following health aids:
   - Antiseptics
   - Disinfectants
   - Vaccines

COMPETITION WINDOW (OBJECTIVE TYPE)

1. Vaccines are:
(a) An extract of toxins secreted by bacteria
(b) An extract of dead and weakened microbes
(c) Strong chemicals obtained from fungi
(d) Chemicals that are applied on the skin to kill bacteria.

2. The letter 'B' in BCG vaccination stands for:
(a) Brief (b) Beri-beri
(c) Bacteria (d) Bacillus

3. Which of the following modes of transmission is NOT a way in which HIV spreads?
(a) Sharing of needles (b) Blood transfusion
(c) Mosquito bites (d) Sexual contact

4. Which of the following is NOT a symptom of AIDS?
(a) Persistent cough
(b) Diarrhoea
(c) Skin rashes
(d) None of the above

5. What is the role of NGOs in the management of AIDS patients?
(a) Diagnosis of the disease
(b) Providing financial assistance to patients
(c) Providing emotional support to patients and their families
(d) None of the above

6. Which of the following is a common opportunistic infection seen in AIDS patients?
(a) Malaria
(b) Tuberculosis
(c) Typhoid
(d) Cholera

7. How can the spread of HIV be prevented?
(a) By using condoms during sexual contact
(b) By avoiding sharing of needles
(c) By getting tested for HIV
(d) All of the above

Answer Keys

Multiple Choice Questions

1. (c) 2. (c) 3. (c) 4. (b) 5. (d)
6. (a)

Match the Following

1. 1-(ii) 2-(iii) 3-(iv) 4-(i) 5-(v)

True or False

1. F 2. T 3. F 4. F 5. F
6. F

Fill in the Blanks

1. Tuberculosis 2. Malaria
3. Face masks 4. CD4
5. Cholera

Groups of Terms

1. Response 2. Disease
3. Direct contact 4. DNA amplification
5. Preventive method

Define the Term

1. A pathogen is a microorganism such as a virus, bacterium, fungus or parasite that causes disease in the host organism.
2. Immunity refers to the ability of the body to resist infections and diseases by recognising and fighting off harmful pathogens.
3. Vaccination is the process of administering a vaccine to stimulate the body's immune system to produce antibodies against specific pathogens, which helps to prevent or control the spread of infectious diseases.
4. An antibiotic is a type of medicine that is used to treat bacterial infections by killing or inhibiting the growth of bacteria.
5. An epidemic is the occurrence of a disease that affects a large number of people in a community or region at the same time.

Identify the Odd One Out

1. Polio 2. Antibiotic
3. Cancer
4. Itching, redness, and inflammation
5. Sexual

Assertion-Reason Questions

1. (a) 2. (c) 3. (d) 4. (b) 5. (b)

One-word Answers

1. Salvarsan 2. Penicillin
3. Passive immunity 4. Splint
5. Antidote 6. Antiseptic

Image-Based Questions

1. (a) First aid box
   (b) The purpose of having a first aid box in a workplace is to provide immediate medical assistance and basic first aid treatment to injured or ill employees until professional medical help arrives.
   (c) The basic items that should be present in a first aid box include:
   - Adhesive bandages (various sizes)
   - Sterile gauze pads
   - Adhesive tape
   - Antiseptic solution or wipes
   - Scissors
   - Tweezers
   - Disposable gloves
   - Pain relievers (such as acetaminophen or ibuprofen)
   - Antihistamines (for allergic reactions)
   - CPR mask or face shield
   - Instant cold packs
   - First aid manual or instructions

   These are some essential items, but the specific contents may vary based on the intended use and regulations in different workplaces or environments.
   (d) The recommended size for a first aid box depends on the number of employees or individuals present in the workplace or the specific environment. As a general guideline, the first aid box should be adequately stocked and large enough to accommodate all necessary supplies for the expected number of people.
   (e) It is important to regularly check the contents of a first aid box to ensure that all items are in good condition, within their expiration dates, and fully stocked.
2. (a) The correct head-tilt and chin-lift technique for performing artificial respiration on an adult is as follows:
   - Place one hand on the victim's forehead and gently tilt their head backward.

- With the fingertips of your other hand, lift the chin to open the airway.
- Maintain the head tilt and chin lift while performing rescue breaths.

(b) The recommended method for performing artificial respiration on an infant is the modified head-tilt and chin-lift technique. Place the infant on their back on a firm surface.

- Place your hand on the infant's forehead, keeping their head in a slightly extended position.
- Use your fingers to gently lift the chin, keeping the mouth and nose uncovered.
- Be cautious not to tilt the head too far back, as it may obstruct the airway in infants.

(c) CPR mask

(d) The recommended frequency for performing rescue breaths during artificial respiration depends on the situation and the individual's condition. However, a common guideline is to deliver rescue breaths at a rate of approximately 10 to 12 breaths per minute. This translates to one breath every 5 to 6 seconds.

(e) When performing CPR on an adult, the recommended hand position for chest compressions is as follows:

- Place the heel of one hand on the center of the chest, between the nipples.
- Place the other hand on top of the first hand, interlocking the fingers.
- Keep your arms straight and position your shoulders directly above your hands.
- Compress the chest downward, using your upper body weight to achieve a depth of at least 2 inches (5 centimeters) but not exceeding 2.4 inches (6 centimeters).
- Allow for complete chest recoil between compressions.

Very Short Answer Questions

1. (i) DPT vaccine: Diphtheria Tetanus Pertussis vaccine
   (ii) AIDS: Acquired Immunodeficiency Syndrome
   (iii) BCG: Bacille Calmette–Guerin
   (iv) WHO: World Health Organization
2. Antibodies or immunoglobulins are produced in blood to fight and destroy harmful microbes.
3. The virus that causes AIDS is called Human Immunodeficiency Virus (HIV).
4. The three main modes of transmission of HIV are sexual transmission, transmission through blood, and mother-to-child transmission during pregnancy, delivery, or breastfeeding.
5. Some common symptoms of HIV include fever, fatigue, swollen lymph nodes, night sweats, weight loss, and recurrent infections. However, it is important to note that many people with HIV do not experience any symptoms for years.

Short Answer Questions

1. (a) (i) Assess the situation and take necessary precautions to avoid any additional harm.
   (ii) Call for help.
   (iii) Assess the person's condition.
   (iv) Carry out basic first aid.

   (b) (i) Ensure safety.
   (ii) Check responsiveness.
   (iii) Call for help.
   (iv) Position the person on their back on a firm surface, such as ground.
   (v) Tilt the person's head back gently by lifting their chin.

   (c) (i) Ensure safety.
   (ii) Check responsiveness.
   (iii) Call for help.
   (iv) Cool the burn.
   (v) Protect the burn area.
   (vi) Monitor breathing and vital signs
   (vii) Prevent hypothermia: If the person's body temperature drops due to prolonged exposure to cool water during burn treatment, cover them with a clean, dry sheet or blanket to prevent hypothermia.

2. (a) Antiseptic and antibiotic

Antiseptics	Antibiotics
Antiseptics are the chemicals applied on the body to kill the microorganisms.	Antibiotics are chemical substances produced by microorganisms, which inhibit the growth of another microorganism.

(b) Antiseptic and disinfectant

Disinfectant	Antiseptic
Chemical substances which destroy some bacteria and prevent the growth of others are called disinfectants.	Chemical substances which are used to destroy microbes in spots and places and for the sterlisation of instruments are called antiseptics.

(c) Disinfectant and deodorant

Disinfectant	Deodorant
It is a strong chemical used to sanitise inert surfaces.	They are neither antiseptics not disinfectants. They are aerosols that can mask bad odour.

(d) Vaccination and sterilisation

Vaccination	Sterilisation
It is the introduction of any kind of weakened/ dead germs into the body of a living beings to develop resistance against diseases.	Process of elimination of microbes from a surface contained in a fluid, in a compound such as biological culture media or in medication.

(e) Active immunity and passive immunity

Active Immunity	Passive Immunity
It refers to the protective immunity that occurs when the immune system of an individual is stimulated to produce antibodies and lymphocytes.	It is a type of immunity in which a person obtains antibodies or lymphocytes from another person's immune system.

(f) Innate immunity and acquired immunity

Innate Immunity	Acquired Immunity
It is the immunity obtained by virtue of genetic constitutional make-up; that is, it is inherited from parents. It is present in the body without any external stimulation or a previous infection.	Acquired immunity is the resistance to a disease which an individual acquires during a lifetime. It may be the result of either a previous infection or from readymade antibodies supplied from outside.

3. The TAB vaccine provides immunity against typhoid. It contains the killed germ. Salk's vaccine provides immunity against poliomyelitis. It contains the killed germ. The BCG vaccine provides immunity against tuberculosis. It contains living weakened germs.

4. They are widely used to fight infections. They are used as food preservatives, especially in fish and meat. They are used in animal feed to prevent internal infections. They are used for controlling plant pathogens.

5. HIV can be prevented through various ways such as practicing safe sex, avoiding sharing needles, using sterile medical equipment, getting tested regularly, and taking pre-exposure prophylaxis (PrEP) medication. Additionally, education and awareness campaigns can also play a crucial role in preventing the spread of HIV.

Long Answer Questions

1. Hint: Vaccination stimulates the immune system to recognize and remember specific pathogens. If the vaccinated individual later encounters the actual pathogen, the immune system recognizes it quickly and mounts a rapid and effective response, preventing or minimizing the severity of the disease. Elaborate on this.

2. Hints: Antiseptics: Used on living tissues to prevent or reduce the growth of microorganisms and prevent infection. Examples include alcohol-based hand sanitizers and iodine solutions.

   Disinfectants: Used on surfaces or objects to destroy or inactivate microorganisms. They are not typically safe for use on living tissues. Examples include bleach and hydrogen peroxide.

   Vaccines: Prepared from weakened or inactivated pathogens or their components, vaccines stimulate the immune system to produce an immune response without causing the disease. They provide protection against specific infectious diseases by building immunity and preventing or reducing the severity of future infections.

   Elaborate on these points.

3. Hint:

   Stage 1: Acute HIV infection (initial infection stage) - This stage occurs shortly after contracting HIV and is often accompanied by flu-like symptoms.

Stage 2: Clinical latency (chronic or asymptomatic stage) - During this stage, the virus replicates at lower levels, and there may be no noticeable symptoms.

Stage 3: Acquired Immunodeficiency Syndrome (AIDS) - In this stage, the immune system is severely damaged, and the individual becomes susceptible to opportunistic infections and certain cancers.

Elaborate on these points.

4. Hint: Common misconceptions about HIV/AIDS include:
   (i) HIV can be transmitted through casual contact (hugging, sharing utensils):
   (ii) HIV can be cured by herbal remedies or alternative therapies: Address by promoting accurate information about the effectiveness of antiretroviral therapy (ART) in managing HIV and the importance of medical treatment.
   (iii) Only certain groups of people are at risk for HIV.

   Elaborate on these points.

5. Hint: Challenges in preventing the spread of HIV and their potential solutions:
   (i) Lack of awareness and education: Address by promoting comprehensive sex education, HIV awareness campaigns, and community outreach programs to increase knowledge about transmission and prevention.
   (ii) Stigma and discrimination: Address by advocating for acceptance, understanding, and inclusion of people living with HIV, challenging stereotypes, and promoting empathy and support.
   (iii) Limited access to healthcare and prevention services: Address by improving healthcare infrastructure, expanding access to HIV testing, treatment, and prevention methods, and reducing barriers such as cost and stigma.
   (iv) High-risk behaviours and misconceptions: Address by promoting harm reduction strategies, encouraging condom use, promoting regular testing, and addressing misconceptions through targeted educational campaigns.

   Elaborate on these points.

Case Study-Based Questions

1. (i) (b), (ii) (d), (iii) (a)
2. (i) (b), (ii) (d), (iii) (c)
3. (i) (a), (ii) (c), (iii) (d)
4. (i) (d), (ii) (c), (iii) (d)
5. (i) (d)

NCERT Corner

1. Listed below are the four applications of antibiotics:
   - They are widely used to fight infections
   - They are used as food preservatives, especially in fish and fresh meat
   - Used in treating animal feed to prevent internal infections
   - Used for controlling plant pathogens
2. Merits of the local defence system are as follows:
   - They instantaneously start working
   - They are not dependent on previous exposure to infections
   - Are effective against a wide range of potentially infectious agents.
3. Diphtheria is a bacterial disease and is infectious. It causes cold, coughing and sneezing; if left untreated in severe cases, it may lead to paralysis or heart failure.

   Treatment:
   - Combination of supportive care and medication
   - Prompt intravenous administration of diphtheria toxoid which is made harmless
   - Once this is administered into the body of the patient, it stimulates antibody production against pathogens that cause diphtheria
4. Whenever there is an infection in the body, the immune system receives a signal to synthesize particular antibodies. In response to the quantity of germ-multiplication in the body, white blood cells multiply rapidly which enables it to produce more antibodies, thereby stopping an infection on time. Consequently, abnormal numbers of WBCs are an indication of some infection in the body.
5. (a) Antiseptic: It is a mild chemical substance applied to the body to prevent its growth and kill germs. Examples are iodine and Lysol.
   (b) Disinfectant: A chemical that kills microbes once it come in contact with it; disinfectants are too strong to be used on the body. Examples are phenol and cresol.
   (c) Vaccine: A substance that is administered in the body to provide passive immunity. These are the materials that are germs or secreted by germs. Examples are DTP and OPV.

Competition Window

1. (b) 2. (d) 3. (c) 4. (b) 5. (c)
6. (b) 7. (d)

CHAPTER
17 Health Organisations

PREREQUISITES

You must have knowledge of:

- The overall wellbeing of an individual, including physical, mental and social aspects
- Healthcare
- Public health and health policies
- Health systems

LEARNING OBJECTIVES

At the end of this chapter, you will be able to:

- Define public health
- Monitor the dangers to public health
- Explain the need to ensure that everyone has access to healthcare
- Explain how to plan for medical crises and to advance human health and wellbeing
- Describe worldwide health standards and recommendations

INTRODUCTION

Health is one of the primary concerns of all the governments of the world. They strive to take steps individually at the national level as well as collectively at the global level to safeguard the health interests of their people. The Red Cross and the World Health Organisation (WHO) are two organisations that work towards better health for all at an international level.

HEATH IN INDIA

India is a vast country. It has highly diverse geographical and climatic conditions. Its population is distributed in broadly five types of habitations:

- **Large cities:** These have several multi-story buildings, an excessive number of automobiles, overworked sewage and water systems and numerous industries, some of which are located inside and outside the municipality.
- **Minor cities:** These are urban areas that are smaller in size and population compared to major cities or metropolitan areas.
- **Settlements:** These have a small population and are heavily dependent on agriculture; the primary sources of income are dairies, poultry farming, cottage businesses, etc.
- **Remote locations:** Here there are tribal people, who mostly depend on forests and forest products for survival; neither clean water nor any sort of medical facility is available.
- **Homes in slums:** These are seen everywhere, but especially on the fringes of large cities. They are characterised by unhealthy, unhygienic environments.

Each type of habitation has a different set of health issues. We can generalise them as follows:

- **Water and food-borne illnesses:** Diseases like diarrhoea, gastroenteritis, typhoid and dysentery are more prevalent in places with inadequate water supply. Water from hand pumps and other sources is frequently polluted. In some locations, water derived from wells or hand pumps contains toxic minerals that might lead to health issues. People who live nearby are harmed by untreated sewage or industrial effluents discharged into rivers and other bodies of water.
- **Insect and air-borne diseases:** People may be unaware that flies contaminate exposed food. Lack of general hygiene encourages the growth of mosquitoes, houseflies and other disease-carrying insects.
- **Lack of medical facilities:** Although preventive immunisation is gaining popularity, it requires better implementation. Lack of medical resources, particularly in rural areas, frequently causes preventable deaths and health harm. Due to ignorance and superstitious beliefs, timely treatment is delayed, which has catastrophic implications.
- **Professional hazards:** Effective controls help prevent accidents, illnesses and incidents at work, reduce or eliminate safety and health risks, and assist employers in providing employees with safe and healthy working environments. To successfully manage and prevent dangers, employers should engage employees, as they frequently have the best knowledge of the circumstances that lead to hazards and knowledge on how to control them. They can also utilise a ladder of controls to identify and assess potential solutions for hazard control.

CATEGORIES OF HEALTH ORGANISATIONS

Every nation cares about the health and welfare of its citizens. Various types of organisations have been working towards this objective at various levels. In general, we can divide them into three categories based on the level of labour:

Local Bodies

Every city, town and even bigger villages have some sort of local organisation to look after the welfare of the populace's health and cleanliness. Municipal corporations, municipalities, town area committees, etc. are a few examples. They have regular personnel and tools for taking care of the following:

1. **Sanitation:** Elimination of fly and mosquito breeding grounds, removal and correct disposal of garbage, sewage and other pollutants.
2. Availability of clean drinking water
3. **Vaccinations:** Immunisation (building up disease resistance) of children and others to protect them from a number of contagious illnesses as tetanus, pertussis (whooping cough), diphtheria, TB, poliomyelitis, measles, etc. Whenever there is a chance for an outbreak to spread, these programmes are even more aggressive.
4. **Keeping statistical records:** Apart from compulsory registration of births and deaths, other information about the health and diseases of the people in their area is to be regularly maintained.

National Bodies

In India, several agencies, institutes and research centres are actively engaged in studying the health problems of the country. These centres focus on the following diseases:

- Malaria
- Dengue
- Tuberculosis
- Leprosy
- Cholera

For each of these diseases, they conduct a routine survey, identify the transmitting agent and its breeding grounds, set up preventive immunisation, and take any other necessary actions to combat the ailment.

The Delhi-based National Institute of Communicable Diseases (NICD), formerly known as the Malaria Institute, is actively conducting research on infectious diseases. To completely eradicate poliomyelitis, a comprehensive national campaign called Pulse Polio has been started.

International Bodies

The two most important international bodies concerned with people's health are the Red Cross and WHO (World Health Organisation).

Red Cross

The Red Cross is both a national and an international organisation. The primary goal of the Red Cross Society is to carry out initiatives that lessen or end human suffering both during times of peace and during times of war. It was formally established in 1864. A red cross painted on a white background serves as the Red Cross Society's logo.

Red Cross members are permitted to visit battlefields and provide care for injured soldiers, whether they are friends or foes. The main activities of Red Cross Societies are as follows:

- To extend relief and help to the victims of any calamity such as flood, fire, famine and earthquake

- To procure and supply blood for the needy victims of war or other calamities
- To extend all possible first-aid in any accident
- To educate people in accident prevention
- To arrange for ambulance service in all emergencies
- To look after maternal and child welfare centres
- The Indian Red Cross Society has also been engaged in training midwives

May 8 is celebrated as Red Cross Day. The Red Cross emblem is very often used by hospitals, ambulances, doctors and nurses for quick identification, but legally, only the Red Cross Society units can use it.

WHO

The World Health Organisation (WHO), established in 1948, is a specialised agency of the United Nations Organisation (UNO).

The main activities of WHO are as follows:

- To gather and provide data about the occurrence of epidemic diseases (such as cholera, plague, typhus, yellow fever and smallpox)
- To fund and encourage initiatives for disease research
- To provide information on the most recent advancements in the use of vaccines, cancer research, nutritional discoveries, the management of drug addiction and nuclear radiation health risks
- To recommend quarantine measures (patient isolation) to stop the disease from spreading to others
- To establish pharmaceutical standards to assure the purity and dosage of vital medications
- To plan programmes to combat endemic (local) and epidemic (widespread) diseases

SUMMARY

- Health organisations are those that look after people's wellbeing.
- The two principal international organisations are the Red Cross and the World Health Organisation.
- The World Health Organisation (WHO) was established as a specialised organisation of the United Nations in 1948. Geneva serves as its administrative centre.
- Its six regional hubs are located in the United States, Denmark, Congo, India, Egypt and the Philippines. In India, the WHO headquarters are in New Delhi.
- A worldwide and national voluntary health organisation called the Red Cross was established in 1864. Switzerland is where it has its main office. A red cross on a white backdrop is the logo.

PRACTICE QUESTIONS

Analyse Your Concepts (School Exam Based)

Multiple Choice Questions

1. Red Cross Day is celebrated on:
   (a) May 8 (b) June 23
   (c) September 17 (d) November 21
2. The WHO headquarters are located in:
   (a) Paris (b) Berlin
   (c) Geneva (d) New York
3. Which of the following is an example of an international health organisation?
   (a) Centers for Disease Control and Prevention (CDC)
   (b) World Health Organisation (WHO)
   (c) National Institutes of Health (NIH)
   (d) American Red Cross
4. What is the primary goal of a non-governmental organisation (NGO) focused on health?
   (a) To provide healthcare services to individuals in need
   (b) To conduct research on health issues
   (c) To advocate for policies and programmes that promote health
   (d) All of the above

5. Which of the following is an example of a government agency responsible for regulating and overseeing healthcare in the United States?
   (a) World Health Organisation (WHO)
   (b) Centers for Disease Control and Prevention (CDC)
   (c) National Institutes of Health (NIH)
   (d) Food and Drug Administration (FDA)

Match the Following

Each question contains statements given in two columns which have to be matched. Statements in column I have to be matched with statements in column II.

Column I	Column II
1. UNESCO	(a) Promotes health education among the masses
2. UNICEF	(b) Helps in exchange of currencies
3. WHO	(c) Tries to educate millions of people
4. ILO	(d) Provides help to children in the member-states
5. IMF	(e) Improves the working conditions of workers

True or False

Read the following statements and write your answer as true or false.

1. As defined by the World Health Organisation, health is the absence of disease or infirmity.
2. To improve health in their community, managers of healthcare organisations often do more than just provide healthcare.
3. All healthcare organisations provide healthcare services to people.
4. A healthcare organisation (HCO) must be concerned about stakeholders because stakeholders have an interest in what the HCO does.
5. The primary goal of health organisations is to provide healthcare services to individuals in need.

Fill in the Blanks

Complete the following statements with an appropriate word/term to be filled in the blank spaces.

1. The ____________ is a United Nations agency that is responsible for promoting health and providing technical assistance and leadership in global public health issues.
2. A ____________ is an organisation that is not affiliated with the government and is focused on promoting health through various activities such as providing healthcare services, conducting research, and advocating for health policies.
3. A ____________ is a type of health organisation that provides healthcare services to individuals in a specific geographic area, such as a city or county.
4. The ____________ is a federal agency in the United States that is responsible for conducting research and providing funding for research related to health and medical issues.
5. The ____________ is a global alliance of organisations and governments that is focused on the goal of ending the AIDS epidemic.

Assertion–Reason Questions

The following questions consist of two statements: Assertion (A) and Reason (R). Answer these questions selecting the appropriate option given below:
   (a) Both A and R are true and R is the correct explanation of A
   (b) Both A and R are true but R is not the correct explanation of A
   (c) A is true but R is false
   (d) A is false but R is true

1. Assertion (A): Both private and cooperative sectors are significant in the healthcare industry.
   Reason (R): Numerous multispecialty hospitals are run by the commercial and cooperative sectors.
2. Assertion (A): Let us say that a state has a literacy rate of 78% and a secondary net attendance ratio of 47%.
   Reason (R): More than half the students attend elementary schools in other states.
3. Assertion (A): Sustainable development is a crucial step for the development of a country.
   Reason (R): Sustainability is the ability to use resources in a judicious way and maintain the ecological balance.
4. Assertion (A): The availability of resources is a prerequisite for any region's development, but in the absence of equivalent advances in technology and institutions, development may be hampered by the availability of resources alone.
   Reason (R): Although there are numerous resource-rich places in our country, they are also located in economically depressed areas.

5. Assertion (A): A country at present has very high national income due to surplus oil resource. However, it is essential to look for alternative sources of income also.
   Reason (R): Oil is a natural resource that is non-renewable in nature. If it is not used judiciously, it will become exhausted soon.

One-word Answers

Name the organisation responsible for the following:

1. To extend relief to victims of earthquakes.
2. To lay down pharmaceutical standards for important drugs.
3. Arranging ambulances in emergencies.
4. To suggest quarantine measures
5. Training of midwives
6. Procuring and supplying blood for transfusion.

Define the Term

1. Health system
2. Public health agency
3. Healthcare provider
4. Non-profit healthcare organisation
5. Community health centre

Groups of Terms

Given below are certain groups of terms. In each group the first pair indicates a relationship between the two terms. Rewrite and complete the second pair on a similar basis.

1. Pathogen : Disease :: Carcinogen :
2. Vaccine : Immunity :: Antibiotic :
3. Diagnosis : Symptoms :: Prescription :
4. Health promotion : Prevention :: Disease management :
5. Epidemiology : Outbreak :: Surveillance :

Identify the Odd One Out

1. Out-of-pocket payment, Health savings account, Medicaid, Health insurance
2. Swachh Bharat Abhiyan, Ayushman Bharat, Mid-day meal scheme, National Rural Health Mission (NRHM)
3. American Medical Association (AMA), World Health Organisation (WHO), Indian Medical Association (IMA), American Hospital Association (AHA)
4. Immunisation, Disease surveillance, Health education, Clinical interventions
5. Ambulatory care centre, Long-term care facility, Psychiatric hospital, Community health centre

Image-Based Questions

1.

(a) Identify the given image.
(b) When is World Red Cross Day celebrated?
(c) What is the theme for World Red Cross Day 2022?
(d) Who founded the Red Cross and when?
(e) What is the significance of Red Cross emblem?

2.

(Public health officials from..., Flickr: Logo of the World Health Organization, United States Mission Geneva, 2011, https://commons.wikimedia.org/wiki/File:Logo_of_the_World_Health_Organization.jpg. Licensed under: CC-BY-2.0)

(a) Identify the logo.
(b) Who is the current Director-General of the World Health Organisation?
(c) Which year was the World Health Organisation established?
(d) What is the mission statement of the World Health Organisation?
(e) What are some of the global health issues that the World Health Organisation works to address?

3.

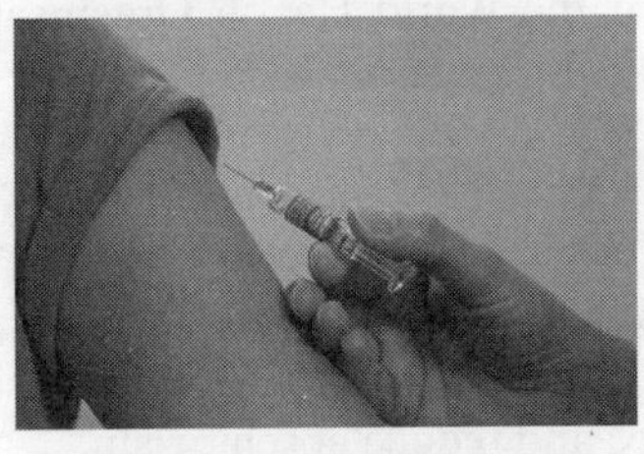

(Courtesy: Katja Fuhlert; https://pixabay.com/photos/vaccination-doctor-injection-1215279/)

(a) What is the process shown in this image called?
(b) What is the purpose of the tool shown in this image?
(c) What is the name of the scientist who developed the first vaccine?
(d) What is the name of the organisation responsible for developing this vaccine?

Very Short Answer Questions

1. What do the following abbreviations stand for?
   (a) WHO: World Health Organisation
   (b) UNO: United Nation Organisation
2. What is the role of World Health Organisation (WHO)?
3. What is the aim of the National Health Policy (NHP) of India?
4. What is the difference between a primary healthcare centre and a community healthcare centre?
5. What is the purpose of the Health Information System (HIS)?

Short Answer Questions

1. Name any two habitation-based health problems.
2. Name three duties that city municipalities have in relation to people's health.
3. Name three Indian health issues.
4. List the key responsibilities of the Red Cross.
5. What is the role of the National Rural Health Mission (NRHM)?

Long Answer Questions

1. Describe the reasons for the formation of WHO.
2. Mention any three health problems in India.
3. Give two functions of WHO.
4. Explain the functions and importance of the Primary Health Centre (PHC) in the Indian healthcare system.
5. Explain the role of the World Health Organisation (WHO) in global health governance.

Case Study-Based Questions

1. BC Company has noticed an increase in the number of sick leaves taken by its employees. As an HR manager, suggest some measures to improve the health of employees and reduce absenteeism.
   (a) Encourage employees to take frequent breaks and exercise during work hours.
   (b) Conduct regular health check-ups and provide counseling services.
   (c) Promote healthy eating habits by providing healthy snacks in the office.
   (d) All of the above.
2. DEF Ltd. recently introduced a work-from-homepolicy for its employees due to the pandemic. However, the company is noticing a decline in productivity and employee morale. As a consultant, suggest some measures to improve the situation.
   (a) Offer regular training and development opportunities to keep employees engaged.
   (b) Set clear expectations and provide regular feedback to employees.
   (c) Host virtual team-building activities to promote collaboration and communication.
   (d) All of the above.
3. GHI Hospital is facing an issue of employee burnout and high turnover among its nursing staff. As a hospital manager, suggest some strategies to address this issue.
   (a) Provide opportunities for career development and growth to encourage employee retention.
   (b) Offer flexible work schedules to reduce stress and burnout.
   (c) Provide counselling services to help employees cope with the demands of the job.
   (d) All of the above.
4. JKL Company wants to introduce a wellness program for its employees. As an HR consultant, suggest some components of the wellness program.
   (a) Conduct regular health check-ups and provide access to counseling services.
   (b) Provide healthy meal options and promote healthy eating habits.
   (c) Offer fitness classes and encourage employees to exercise.
   (d) All of the above.
5. MNO School is noticing an increase in student absenteeism due to health issues. As a school administrator, suggest some measures to improve the health of students and reduce absenteeism.
   (a) Conduct regular health check-ups and provide access to medical facilities.
   (b) Promote healthy eating habits by providing nutritious meals in the school canteen.
   (c) Encourage physical activity by introducing sports and fitness programmes.
   (d) All of the above.

NCERT Corner

1. List any three health problems in India.
2. State two functions of WHO.
3. State the main functions of Red Cross.
4. State the main functions of World Health Organisation.
5. Describe the reasons for the formation of WHO.

COMPETITION WINDOW (OBJECTIVE TYPE)

1. Which international health organisation is responsible for providing global leadership in public health and coordinating responses to health emergencies?
   (a) World Health Organization (WHO)
   (b) Centers for Disease Control and Prevention (CDC)
   (c) United Nations Children's Fund (UNICEF)
   (d) Red Cross
2. Which organisation focuses on immunization campaigns, disease surveillance, and promoting child and maternal health worldwide?
   (a) World Health Organization (WHO)
   (b) United Nations Children's Fund (UNICEF)
   (c) Centers for Disease Control and Prevention (CDC)
   (d) World Food Programme (WFP)
3. What is the primary mission of the Centers for Disease Control and Prevention (CDC)?
   (a) Promoting global health security
   (b) Conducting scientific research on diseases
   (c) Providing healthcare services to underserved populations
   (d) Ensuring access to clean water and sanitation
4. Which organisation is responsible for providing humanitarian aid and assistance during and after natural disasters, conflicts, and health emergencies?
   (a) World Health Organization (WHO)
   (b) United Nations Children's Fund (UNICEF)
   (c) Red Cross
   (d) World Food Programme (WFP)
5. What role does the World Food Programme (WFP) play in global health?
   (a) Providing emergency medical supplies
   (b) Ensuring access to nutritious food
   (c) Promoting vaccination campaigns
   (d) Conducting research on infectious diseases

Answer Keys

Multiple Choice Questions

1. (a) 2. (c) 3. (b) 4. (d) 5. (d)

Match the Following

1. 1-c 2-d 3-a 4-e 5-b

True or False

1. F 2. F 3. F 4. T 5. F

Fill in the Blanks

1. World Health Organisation (WHO)
2. Non-governmental organisation (NGO)
3. Local health department
4. National Institutes of Health (NIH)
5. Joint United Nations Programme on HIV/AIDS (UNAIDS)

Assertion-Reason Questions

1. (c) 2. (b) 3. (b) 4. (a) 5. (a)

One-word Answers

1. National Disaster Management Authority (NDMA)
2. WHO
3. Red Cross
4. WHO
5. Red Cross
6. Red Cross

Define the Term

1. A health system refers to the complex network of people, institutions and resources involved in providing healthcare services to individuals and communities. It includes healthcare providers, public health agencies, insurers, policymakers and others who work together to promote health and provide healthcare services.

2. A public health agency is a government or non-governmental organisation that is focused on promoting and protecting the health of communities through disease prevention, health education and other measures. Public health agencies may conduct research, provide healthcare services and develop policies and programmes that promote health and prevent disease.
3. A healthcare provider is a person or organisation that is involved in providing healthcare services to individuals. This can include doctors, nurses, therapists, hospitals, clinics and other healthcare facilities. Healthcare providers may provide a range of services such as medical treatment, diagnostic testing and preventive care.
4. A non-profit healthcare organisation is one that provides healthcare services, but its main goal is not to generate profits for its owners or shareholders. Instead, it is typically focused on a specific mission, such as providing care to underserved populations or advancing medical research. Non-profit healthcare organisations may receive funding from donors, grants or government sources to support their work.
5. A community health centre is a type of healthcare facility that provides primary care services to individuals in underserved communities. Community health centres may be located in rural or urban areas and may serve low-income or uninsured patients. They are typically funded through a combination of public and private sources and may offer a range of services such as medical care, dental care and mental health services.

Groups of Terms

1. Cancer
2. Infection
3. Treatment
4. Control
5. Monitoring

Identify the Odd One Out

1. Health insurance
2. National Rural Health Mission (NRHM)
3. American Medical Association (AMA)
4. Health education
5. Community health centre

Image-Based Questions

1. (a) World Red Cross Day
   (b) World Red Cross Day is celebrated on May 8 every year.
   (c) The theme for World Red Cross Day 2022 was BeHumanKind.
   (d) The Red Cross was founded by Swiss businessman and social activist, Henry Dunant, in 1863.
   (e) The Red Cross emblem, a red cross on a white background, is a symbol of protection for medical and humanitarian workers, as well as a symbol of neutrality and impartiality in times of conflict and crisis.
2. (a) World Health Organisation
   (b) Dr. Tedros Adhanom Ghebreyesus.
   (c) April 7, 1948.
   (d) The mission statement of the World Health Organisation (WHO) is: "The WHO's primary role is to direct and coordinate international health within the United Nations' system. WHO main areas of work include: health systems, promoting health through the life-course, noncommunicable diseases, communicable diseases, corporate services, preparedness, surveillance, and response."
   (e) Communicable diseases: WHO works to prevent the spread of infectious diseases such as HIV/AIDS, malaria, tuberculosis and Ebola.

   Non-communicable diseases: WHO works to reduce the burden of chronic diseases such as heart disease, cancer and diabetes.

   Maternal and child health: WHO works to improve maternal and child health through programmes that promote safe childbirth, immunisation and nutrition.

   Environmental health: WHO works to prevent environmental risks such as air pollution, unsafe water, and toxic chemicals.

   Health emergencies: WHO works to prevent and respond to health emergencies such as outbreaks of infectious diseases and natural disasters.

   Universal health coverage: WHO works to promote access to quality health services for everyone, regardless of their ability to pay.

   Health systems strengthening: WHO works to strengthen health systems to ensure they are able to provide high-quality, accessible health services to populations.
3. (a) Vaccines
   (b) The tool shown in the image (syringe vaccine) is a medical device used to administer vaccines. The purpose of the syringe is to inject a specific

amount of the vaccine into the body of an individual in order to help them develop immunity to a particular disease. Vaccines are a critical tool in preventing the spread of infectious diseases and protecting individuals from illness.

(c) Edward Jenner

(d) Academic institutions, government agencies, non-governmental organisations and pharmaceutical companies.

Very Short Answer Questions

1. (a) WHO: World Health Organisation

   (b) UNO: United Nations Organisation
2. The World Health Organisation (WHO) is a specialised agency of the United Nations that is responsible for coordinating international public health efforts and providing technical support to countries.
3. The National Health Policy (NHP) of India aims to achieve universal health coverage and to provide affordable, accessible and quality healthcare services to all citizens.
4. A primary healthcare centre is a basic level of healthcare that provides basic medical and diagnostic services, while a community healthcare centre provides a broader range of services including specialist care and hospitalisation.
5. The Health Information System (HIS) is used to collect, process and analyse health-related data to inform decision-making and improve the delivery of healthcare services.

Short Answer Questions

1. The two habitation-based health problems are:

   Food-borne and water-borne diseases:
   - Contaminated food and water cause several diseases especially in rural/slum areas.
   - Many pathogens breed in areas where there is stagnant water.
   - Water-borne diseases spread due to contaminated water from mixing of untreated sewage with river water.

   Lack of proper medical facilities:
   - Lack of medical facilities, especially in many rural areas, leads to unavoidable deaths and damage to health.
   - Lack of knowledge on the availability and use of the medical facilities in rural areas.
   - Superstitious beliefs of people also delay timely treatment which may result in serious consequences.
2. (a) The first duty is to ensure good sanitation by maintaining a clean environment. It should be in charge of eliminating fly and mosquito breeding grounds, as well as the removal and correct disposal of rubbish, sewage, and other wastes.

   (b) Supplying the locality with clean drinking water.

   (c) Maintaining population statistics for the city. It should keep track of information regarding the health and illnesses of the local populace in addition to the records of births and deaths.
3. Three Indian health issues:

   Obesity: 15%

   Cancer: 14%

   Diabetes: 2%
4. The main function of the Red Cross is to extend relief and help to the victims of any calamity – fire, flood, famine, earthquakes, etc.
5. The National Rural Health Mission (NRHM) is a flagship programme of the Indian government that aims to provide accessible, affordable and quality healthcare services to the rural population of India. The NRHM is focused on improving the health status of the rural population by strengthening the healthcare system, including infrastructure, human resources and health financing mechanisms. The NRHM also focuses on reducing maternal and child mortality rates, preventing and controlling communicable and non-communicable diseases and promoting healthy lifestyles.

Long Answer Questions

1. HInt: Reasons for the formation of WHO were to establish a global health authority to address international health issues and coordinate responses to health emergencies; to promote collaboration among nations in improving health outcomes and ensuring access to healthcare for all; and to provide technical assistance, guidance, and support to countries in addressing health challenges and achieving better health standards.
2. Hint: Three health problems in India are communicable diseased, non-communicable diseases and maternal and child health issues. Elaborate on these points.
3. Hint: Functions of WHO:

   Providing leadership in global health by setting norms and standards, formulating health policies, and providing technical expertise and guidance.

   Coordinating international responses to health emergencies, conducting research, and promoting the development and distribution of essential medicines and vaccines.

Elaborate on these points.

4. Hint: Explain all the functions such as providing basic healthcare services, offering immunisation and conducting health education and awareness programs, as well as the importance of Primary Health Centres (PHCs). Elaborate on these points.
5. Hint: Setting global health policies and standards, providing technical guidance, and promoting international cooperation in addressing health challenges. Facilitating the sharing of scientific research, expertise, and best practices among member countries. Collaborating with governments, NGOs, and other stakeholders to strengthen health systems, promote universal health coverage, and achieve the Sustainable Development Goals related to health. Monitoring global health trends, conducting research, and providing timely information and guidance during health emergencies. Advocating for health equity, addressing health inequalities, and promoting access to essential medicines and vaccines for all. Elaborate on these points.

Case Study-Based Questions

1. (d) **2.** (d) **3.** (d) **4.** (d) **5.** (d)

NCERT Questions

1. Health problems differ with different habitations. Listed below are three common health problems in India:
   - Food and water-borne diseases – certain diseases are common in water-deficient areas and areas where water is obtained from wells or hand-pumps wherein water is contaminated. Few of these diseases are diarrhoea, dysentery and typhoid. Few other causes are toxic levels of mineral content found in drinking water, untreated sewage or effluents from industries poured into water bodies.
   - Insect and air-borne diseases – lack of awareness amongst people in the remote areas about the wide range of factors causing food contamination has affected and caused diseases. Factors such as lack of general cleanliness can lead to breeding of houseflies, mosquitoes and other insects which can potentially cause diseases has often been overlooked.
   - Unavailability of proper medical facilities and nutrition due to a combination of financial distress and medical help in remote areas has increased the mortality rate and is one of the major health concerns.
2. The two functions of the WHO (World Health Organisation) are as follows:
   - To collect and supply information about the occurrence of epidemic diseases such as cholera and typhoid.
   - To promote and support projects for research on diseases.
3. The main functions of the Red Cross are:
   - To extend relief and help to the victims of any calamity – fire, flood, famine, earthquake, etc.
   - To procure and supply blood for the victims of war and other calamities
   - To extend all possible first-aid in an accident
   - To educate people in accident prevention
   - To arrange for ambulance service in all emergencies
   - To look after maternal and child welfare centres
   - To train midwives
4. The main functions of the World Health Organisation are:
   - To collect and supply information about the occurrence of epidemic diseases such as cholera and typhoid
   - To promote and support projects for research on diseases
   - To supply information on the latest developments about vaccine use, cancer research, control of drug addiction, health hazards of nuclear radiation
   - To suggest quarantine measures to avert the spreading of diseases
   - To lay pharmaceutical standards for important drugs
   - Organisation of campaigns to control the epidemic and endemic diseases
5. The World Health Organisation (WHO) established in 1948, is a specialised agency of the United Nations Organisation (UNO). The reasons for the formation of WHO are as follows:
   - Member countries of the UNO focused on the need for creating an international body to look after the health problems of people of the world.
   - The need for having an organisation was felt in fields of research on the causes and cures of diseases.
   - These measures were attempted in order to provide better, faster results.
   - Most importantly, it aimed at proving to be beneficial for poor and developing countries.

Competition Window

1. (a) **2.** (b) **3.** (b) **4.** (c) **5.** (b)

CHAPTER

18 Our Environment

PREREQUISITES

You must have knowledge of:

- Environmental science
- Ecosystem
- Biodiversity
- Pollution
- Climate change
- Sustainable development
- Conservation

LEARNING OBJECTIVES

At the end of this chapter, you will be able to:

- Define the ecosystem and offer examples of interdependence between aquatic and terrestrial ecosystems
- Build a link between an ecosystem's biotic and abiotic components
- Describe how autotrophs, heterotrophs, and saprophytes in a biotic community are interdependent
- Describe the role of the food chain, food web, trophic levels and energy flow in an ecosystem
- Emphasise the importance of mineral cycling, mentioning examples of carbon, nitrogen and water, as well as human involvement in maintaining the cycles

INTRODUCTION

Everything that surrounds us – our house, garden, town, shops, hills, rivers, ocean, air, soil, sunlight and other plants and animals – is referred to as the environment. The natural environment is made up of all the living and non-living objects that occur naturally on this planet (Fig. 18.1). We also know that all living organisms rely on their surroundings for survival. Their lives are governed by the environment, and they, in turn, have an impact on it. As a result, the ecosystem as a whole is a vibrant and complex organism. We must understand our environment in order to nurture it.

Fig. 18.1 Natural environment
(Nature capture from a top of the mountain…; Chethiyanidushan; 2020; https://commons.m.wikimedia.org/wiki/File:Beauty_Of_Natural_Environment.jpg. Licensed under: CC-BY-SA-4.0)

THE ECOSYSTEM AND ITS COMPONENTS

The environment is from where living organisms obtain their sustenance and oxygen. Plants and animals interact with one other as well as their physical environment during this process. As a result, an ecosystem can be defined as, "a biological environment that includes all the species residing in a certain area, as well as the non-living physical components of the environment with which the organisms interact." Tansley proposed the notion of ecosystem in 1935. The word ecosystem comes from the Greek '*oikos*' meaning 'house' and '*systema*' meaning 'system'.

An ecosystem may be natural or man-made. Natural ecosystems are those that exist in nature. They can be terrestrial or aquatic. The terrestrial environment includes grasslands and deserts, whereas the aquatic ecology includes rivers, ponds and oceans. Man-made or human-designed ecosystems, on the other hand, are artificial ecosystems such as gardens, aquariums and crop fields.

Relationship Between the Components of an Ecosystem

An ecosystem is made up of both non-living and living things. They are referred to as abiotic and biotic components, respectively (Table 18.1).

- **Abiotic:** The non-living physical and chemical factors in an ecosystem's environment are referred to as abiotic components.
- **Biotic:** In an ecosystem, the biotic components are organisms such as plants, animals and microorganisms.

Table 18. 1 Components of the ecosystem

Abiotic components	Biotic components
Sunlight	Primary producers
Temperature	Herbivores
Precipitation	Carnivores
Water or moisture	Omnivores
Soil	Detritivores
Air	

All of these elements change across time and space. You may have noticed that the vegetation of coastal regions, hilly areas and deserts are all rather different. Do you have any idea why? It is because each has its own set of abiotic factors, such as temperature, soil and moisture. As a result, we can examine how abiotic elements affect various creatures in the ecosystem (Fig. 18.2).

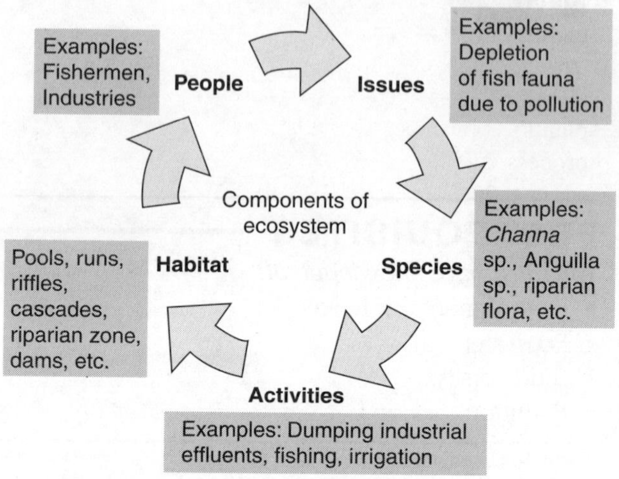

Fig.18.2 Components of ecosystems and their interrelationship

SCIENCE IN DAILY LIFE

List some biotic and abiotic elements in your neighbourhood.

The biotic and abiotic elements play a vital role in all ecosystems. Can you see how all living things are completely reliant on the environment's abiotic components? With the help of sunlight, CO_2 and chlorophyll, green plants make their own nourishment. Plant-eating herbivores and carnivores, including humans, rely on plant-produced food. Carbon dioxide, oxygen, water and other nutrients are returned to the environment by plants, animals and other species. This not only benefits the soil, but also the environment. As a result, we may conclude that we all consume a small amount of sunlight on a daily basis.

Biotic Community

The term biotic community refers to a group of organisms that live together and share the same habitat (Fig. 18.3). An ecosystem is made up of multiple biotic communities that interact with one another. On a tree, for example, one can see communities of various birds, insects and a variety of other species living together in the same environment, mutually maintaining and interdependent. A biotic community is a collection of diverse creatures. Members of a biotic community are classified as autotrophs, heterotrophs or saprotrophs based on their mode of nourishment.

- **Autotrophs:** These are organisms that feed on themselves (in Greek, '*auto*' means 'self and '*trophos*' means 'feeder'). You probably know that all plants (with the exception of a few parasitic plants) can make their own food through the process of photosynthesis, but did you know that some organisms can make their own food through the process of chemosynthesis? This category includes a number of bacteria. Plants are also known as producers since they supply food to all creatures, either directly or indirectly. Autotrophs are the foundation of any biotic system because they capture solar energy and use it to produce food for everyone.

  Autotrophs in terrestrial habitats are primarily rooted plants, but in aquatic ecosystems, they include floating phytoplankton and shallow-water rooted plants known as macrophytes.
- **Heterotrophs:** These are consumers that eat both plants and animals (in Greek, '*heteros'* means 'other' and '*trophos*' means 'feeder'). Herbivores (animals that eat plants) and carnivores (animals that eat meat) are examples of consumers.
- **Saprotrophs:** Decomposers, also known as saprotrophs, feed on dead and decaying substances (in Greek, '*sapros*' means 'rotten' and '*trophos*' means 'feeder'). They decompose the complex organic molecules of deceased plants and animals into simpler forms before reintroducing them into the ecosystem. Decomposers serve as a vital link between the ecosystem's living and non-living components. This category includes bacteria and fungi.

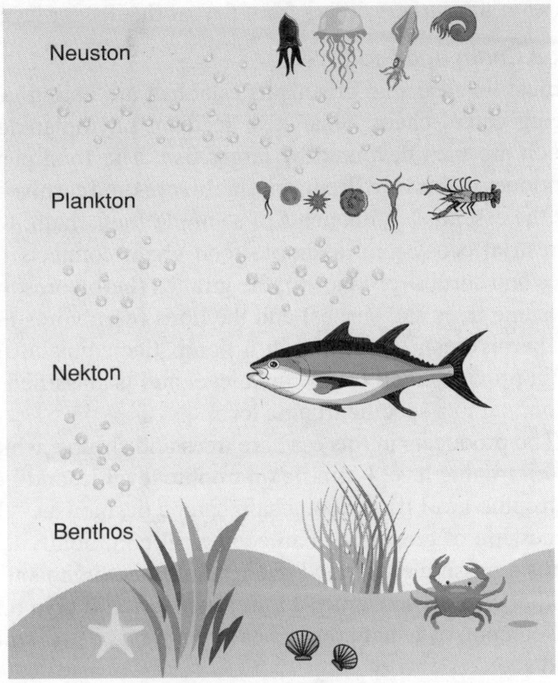

Fig. 18.4 Pond ecosystem

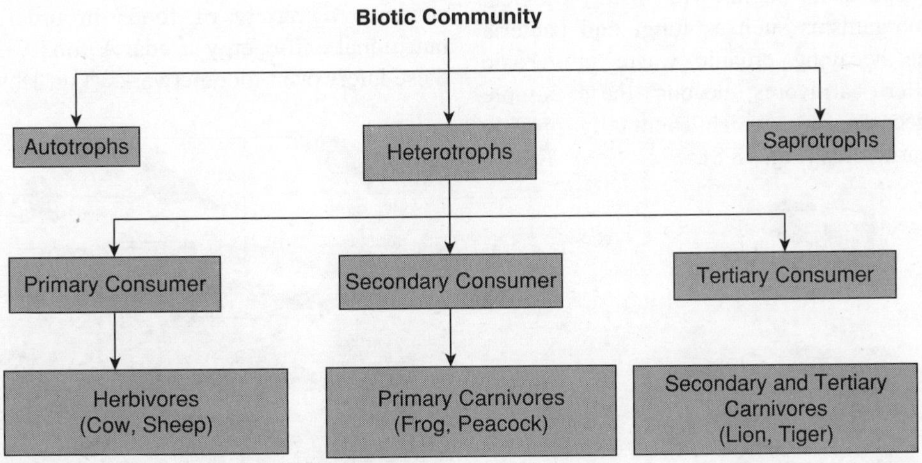

Fig. 18.3 Biotic community

Pond is a wonderful example of an ecosystem to learn about the link between abiotic and biotic components. There are three layers to a pond; top, middle and bottom. Temperature, light conditions, oxygen concentration and other elements that affect the lives of the biotic components residing in each stratum varies substantially from one another. If you have ever taken a dip in a pond, you have probably noticed that the temperature of the water at the surface differs from that of the deeper layers.

Food Chain and Food Web

It must be clear that organisms establish an 'eating' and 'being eaten' chain. Small fish feed on phytoplankton, which are then devoured by larger fish. The food chain continues in this way. Producers, herbivores and carnivores are the essential components of a simple food chain. In a terrestrial ecosystem, a simple food chain connects the trees and shrubs (producers), the giraffes (herbivores that consume trees and shrubs) and the lions (carnivores that eat herbivores), much like in a pond. Each link in the chain provides food for the next level and is classified as belonging to a specific trophic level.

The producers in this case are trees and shrubs, which are on trophic level I, giraffe on trophic level II, and lion on trophic level III. A food chain can be defined as, "the succession of creatures in an ecological community that forms a passing on of food energy from one organism to another as each consumes a lower member and is in turn preyed upon by a higher member of the food chain," since food supplies energy.

In an ecosystem, different types of food chains can exist. A grazing food chain can be found in both aquatic and grassland ecosystems. In terrestrial ecology, it is the most common food chain (Figs 18.5, 18.6 and 18.7).

The detritus food chain begins with dead biological materials. Microorganisms such as fungi and bacteria break down the decaying organic waste into basic nutrients. Smaller carnivores devour these simple nutrients and decomposers, which eventually provide food for larger carnivores (Fig. 18.8).

Fig. 18.5 Food chain in a forest

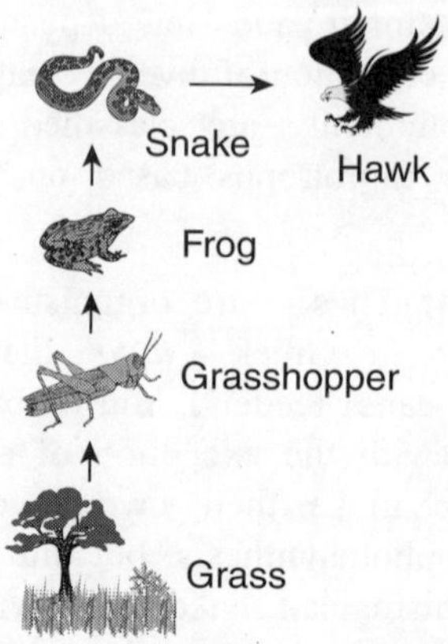

Fig. 18.6 Food chain in a grassland ecosystem

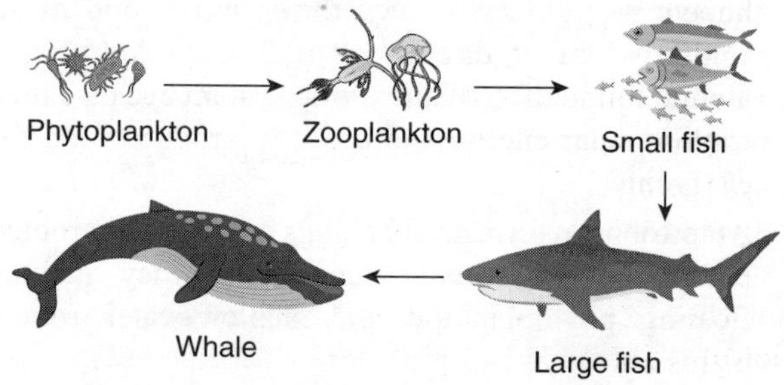

Fig. 18.7 Food chain in an aquatic ecosystem

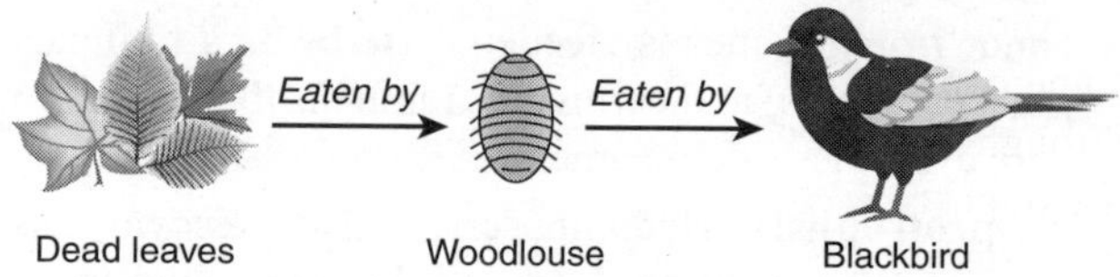

Fig. 18.8 Detritus food chain

Most animals are members of many food chains and consume a variety of foods in order to meet their nutritional and energy needs. A food web is formed by these interwoven food networks (Fig. 18.9).

Fig. 18.9 Food web, consisting of many food chains

The importance of the food web and food chain is stated below:

- They aid in the preservation of ecological balance.
- They aid in our understanding of organism-to-organism feeding relationships.
- They are the conduits for energy flow and the nutrition cycle.

Energy Flow in an Ecosystem

We know that food provides energy, and that energy is transmitted from one link to the next in a food chain. The energy that is transported from one trophic level to the next is unidirectional, meaning it does not return. When a herbivore consumes plant food, only a small portion of the energy is converted into new body mass; the remainder is lost as heat or utilised by the herbivore to carry out its life functions (such as movement, digestion, respiration, reproduction). As a result, when a carnivore eats a herbivore, the carnivore only receives a small portion of the total energy. Some energy will be lost as heat or 'used up' by the carnivore during the transfer. To gain enough energy to expand, the carnivore must eat a lot of herbivores. As we move up the food chain, the amount of energy transferred decreases as a result of the huge amount of energy wasted at each link (Fig. 18.10).

Another fascinating feature of the food chain is how some dangerous chemicals reach our bodies unwittingly through it. You have learned about water pollution. The use of pesticides and other chemicals to protect our crops from diseases and pests is one of the causes of water pollution. These pollutants are either washed into the soil or end up in bodies of water. These are absorbed by plants together with water and minerals from the soil, while aquatic plants and animals take them up from water bodies. These are some of the ways by which they enter the food chain. As these compounds are not biodegradable, they build up at each trophic level.

Since humans are at the top of any food chain, we accumulate the highest concentration of these substances in our body. Biological magnification is the term for this process. This is why pesticide residues can be found in food grains like wheat and rice, as well as vegetables and fruits, and even meat. They cannot be got rid of by washing or other methods.

What Effect Do Our Actions Have on the Environment?

We are a vital element of the natural world. Changes in the environment have an impact on humans, and our actions have an impact on the environment. Here, we will look at two environmental issues in depth: ozone depletion and trash disposal.

The Ozone Layer and How it is being Depleted

Ozone (O_3) is a molecule made up of three oxygen atoms. While O_2, commonly known as oxygen, is required for all aerobic forms of life, ozone is a poisonous gas. It plays an important role at higher altitudes in the atmosphere. It protects the earth's surface from the Sun's ultraviolet (UV) radiation. This radiation is extremely harmful; for example, it has been linked to skin cancer in humans.

UV radiation acting on the oxygen (O_2) molecule produces ozone at higher levels of the atmosphere. UV radiation with a greater energy splits some molecular oxygen (O_2) into free oxygen (O) atoms.

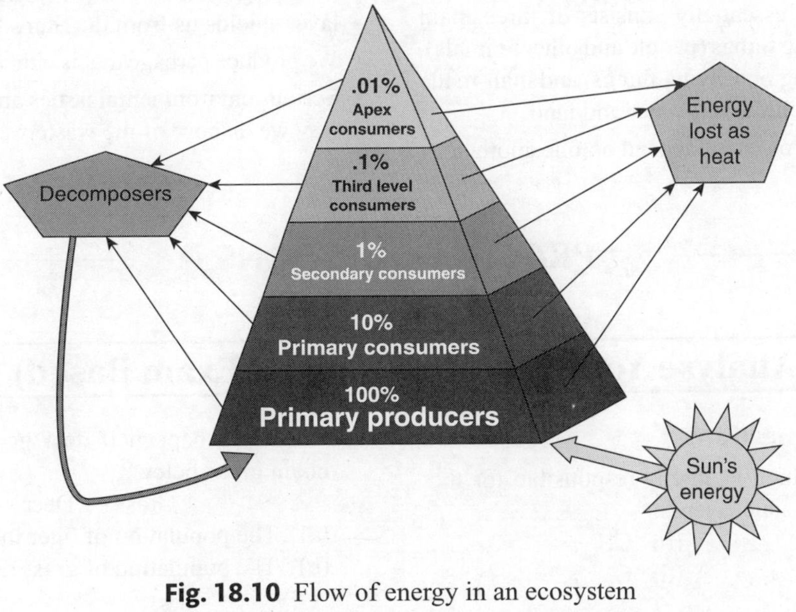

Fig. 18.10 Flow of energy in an ecosystem

As depicted, these atoms interact with molecular oxygen to produce ozone:

$$O_2 \xrightarrow{UV} O + O$$
$$O + O_2 \longrightarrow \underset{\text{(Ozone)}}{O_3}$$

In the 1980s, the amount of ozone in the atmosphere began to fall. Synthetic chemicals such as chlorofluorocarbons (CFCs), which are used as refrigerants and in fire extinguishers, have been linked to this decline. The United Nations Environment Programme (UNEP) was successful in negotiating a deal in 1987 to limit CFC production at 1986 levels.

Managing the Waste We Create

Any town or city we visit will undoubtedly have mountains of waste strewn about. We can be sure that any tourist attraction will be cluttered with empty food wrappers. Increases in trash output have resulted from improvements in our way of life. Changes in mentality are also important, since more and more of the objects we use become disposable. Due to changes in packaging, much of our waste is now non-biodegradable.

Biogeochemical or Nutrient Cycles

The biotic population is always in need of nutrients to survive, and they obtain these from the environment. The environment contains a certain amount of nutrients in the form of oxygen, carbon dioxide, nitrogen, phosphorus, sulphur and water. However, the amount of these nutrients fluctuates throughout time in different sections of an ecosystem. These elements, on the other hand, are never lost, and nature has its own periodic mechanism of renewing them. The biogeochemical cycles are the cyclic flow of these nutrients through the environment. It can be defined as "the cycle in which nitrogen, carbon and other inorganic elements of the soil and atmosphere of a region are changed into the organic compounds of animals or plants and released back into the environment."

SUMMARY

- Both living and non-living things coexist naturally in the environment. It involves the dynamic interaction of environmental elements, biological systems and natural resources. These factors all directly affect human survival and economic activity.
- Life on Earth is entirely a product of the environment. Factors in our environment are biotic and abiotic.
- Abiotic factors refer to things that are not alive, whereas biotic factors pertain to living things.
- The environment essentially consists of three main components: living things (people and other animals), naturally occurring non-living things, and man-made objects. It also includes air, water and land.
- The parts of an ecosystem depend on one another.
- Producers enable the rest of the ecosystem to access the energy from sunshine.
- The number of trophic levels in a food chain is constrained by the energy loss that occurs as we move from one trophic level to the next.
- The environment is impacted by human activity.
- The ozone layer is in risk due to the use of chemicals like CFCs.
- This might harm the ecosystem because the ozone layer shields us from the Sun's UV rays.
- We produce garbage that is either biodegradable or not.
- Serious environmental issues are being created by the way we dispose of the waste we produce.

PRACTICE QUESTIONS

Analyse Your Concepts (School Exam Based)

Multiple Choice Questions

1. The substance which is chiefly responsible for the depletion of the ozone layer is:
   (a) CFCs (b) CH_4
   (c) DDT (d) O_3

2. What will happen if deer go missing in the food chain given below?
   Grass → Deer → Tiger
   (a) The population of tiger increases
   (b) The population of grass decreases

(c) Tiger will start eating grass
(d) The population of tiger decreases and the population of grass increases

3. Which of the following substances will not be converted to compost when added in a composting pit?
(a) Waste paper
(b) Fruit and vegetable peels
(c) Human and animal excreta
(d) Plastic bags

4. Global warming is a phenomenon related to:
(a) Evaporation (b) Ecological balance
(c) Greenhouse effect (d) Desertification

5. Which of the following is a greenhouse gas that contributes to climate change?
(a) Oxygen (b) Nitrogen
(c) Carbon dioxide (d) Argon

Match the Following

Each question contains statements given in two columns which have to be matched. Statements in column I have to be matched with statements in column II.

1. Match the 'environment words' with their meanings.

Words	Meanings
Ecology	A person who advocates preservation and careful management of natural resources.
Habitat	The variety and variability among living organisms and the ecological complexes in which they occur.
Herpetologist	The relationship of living things to one another and their environment, or the study of such relationships.
Extinct	Tropical evergreen trees or shrubs forming dense thickets along tidal shores.
Aquifer	A place where a population lives and its surroundings, both living and non-living.
Naturalist	A zoologist who studies reptiles and amphibians.
Biodiversity	Having no living representative.
Mangrove	An underground geological formation, containing water.
Conservationist	A person who studies plants and animals, especially outdoors.

True or False

Read the following statements and write your answer as true or false.

1. The substance which is chiefly responsible for the depletion of the ozone layer is ozone.
2. Global warming is a phenomenon related to Greenhouse Effect.
3. World Environment Day is celebrated on 3 June of every year.
4. Biodegradable wastes are those substances which are broken down into simpler, harmless substances in nature in due course of time by biological processes such as the action of microorganisms like certain bacteria.
5. Grass and plants belong to the third trophic level of a food chain.

Fill in the Blanks

Complete the following statements with an appropriate word/term to be filled in the blank spaces.

1. Those waste materials which can be broken down to non-poisonous substances in nature in due course of time by the action of microorganisms are called _________ wastes.
2. The waste materials which cannot be broken down into harmless substances in nature are called _________.
3. _________ is the ultimate source of energy.
4. In 1987, _________ succeeded in forging an agreement to freeze CFC production at 1986 levels.
5. Ozone at the higher levels of the atmosphere is a product of UV radiation acting on _________ molecules.
6. _________ can be classified as herbivores, carnivores, omnivores and parasites.

Assertion-Reason Questions

The following questions consists of two statements: Assertion (A) and Reason (R). Answer these questions selecting the appropriate option given below:
(a) Both A and R are true and R is the correct explanation of A
(b) Both A and R are true but R is not the correct explanation of A
(c) A is true but R is false
(d) A is false but R is true

1. Assertion (A): The food chain present in water bodies is called the aquatic food chain.
   Reason (R): An example of an aquatic food chain is Phytoplankton → Zooplankton → Fish → Shark
2. Assertion (A): Decomposers decompose organic matter and keep the environment clean.
   Reason (R): They recycle matter by breaking down the organic remains of dead plants and animals.
3. Assertion (A): Ozone is both good and bad.
   Reason (R): Chlorofluorocarbons damage the ozone layer.
4. Assertion (A): An example of an artificial ecosystem is a garden
   Reason (R): Biotic and abiotic components are altered by humans.
5. Assertion (A): Humans live in the troposphere.
   Reason (R): The atmosphere extends further than the troposphere.

One-word Answers

1. What is the process by which water evaporates from plants called?
2. What is the term for the variety of different species in a given area?
3. What is the process by which soil and rock are removed from the Earth's surface called?
4. What is the term for the study of the relationships between living organisms and their environment?
5. What is the term for the process by which carbon dioxide is released from burning fossil fuels and contributes to climate change?

Define the Term

1. Biodiversity
2. Pollution
3. Sustainability
4. Greenhouse gas
5. Ecosystem

Groups of Terms

Given below are certain groups of terms. In each group the first pair indicates a relationship between the two terms. Rewrite and complete the second pair on a similar basis.

1. Producers : Decomposers :: Grass : ___________
2. Evaporation : Condensation :: Sun : ___________
3. Birth Rate : Death Rate :: 10 : ___________
4. Primary : Secondary :: Bare Rock : ___________
5. Forest: Desert :: Decidious Trees : ___________

Identify Odd One Out

1. Oxygen, Nitrogen, Carbon dioxide, Water vapour.
2. Plastic bags, Glass bottles, Biodegradable containers, Styrofoam cups.
3. Renewable energy, Nuclear energy, Fossil fuels, Hydroelectric power.
4. Acid rain, Greenhouse effect, Ozone depletion, Soil erosion.
5. Ecosystem, Endangered species, Urbanisation, Climate change.

Image-Based Questions

1. A teacher draws a pyramid of energy on the board and writes A, B, C and D, in each trophic level as shown in the diagram given below. Which level represents the herbivores?

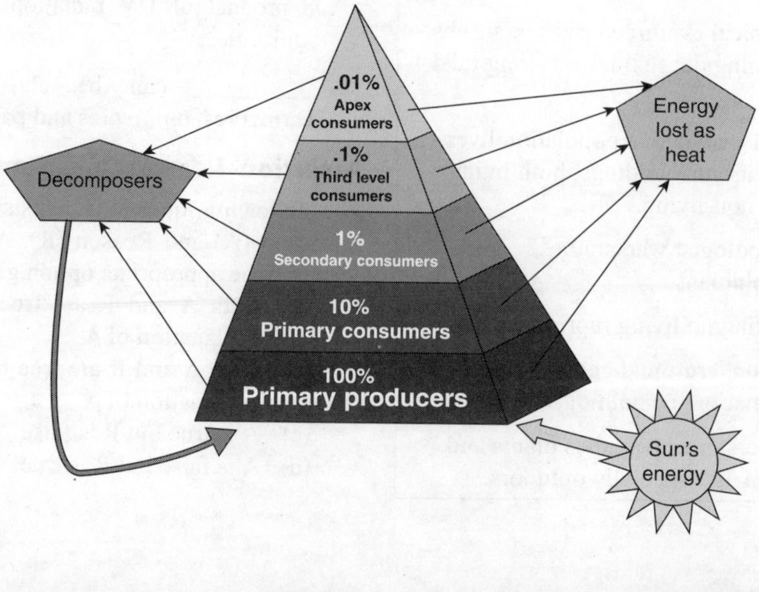

(a) A (b) B
(c) C (d) D

2. An ecosystem is represented in the figure given below.

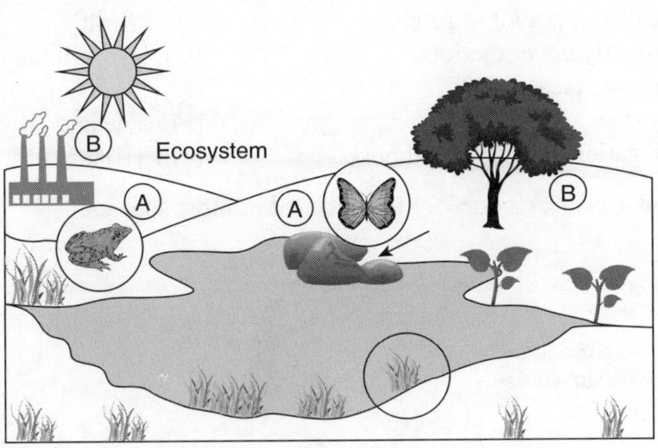

This ecosystem will be self-sustaining if:

(a) The type of organisms represented by B are eliminated.
(b) Materials cycle between the organisms labelled A and the organisms labelled B.
(c) The organisms labelled A outnumber the organisms labelled B.
(d) The organisms labelled A are equal in number to the organisms labelled B.

3. In class, the teacher explained the concept of food chain and energy flow. She made a diagram as given below and asked the students to identify the producer organism in the chain. What do you think will be the student's answer?

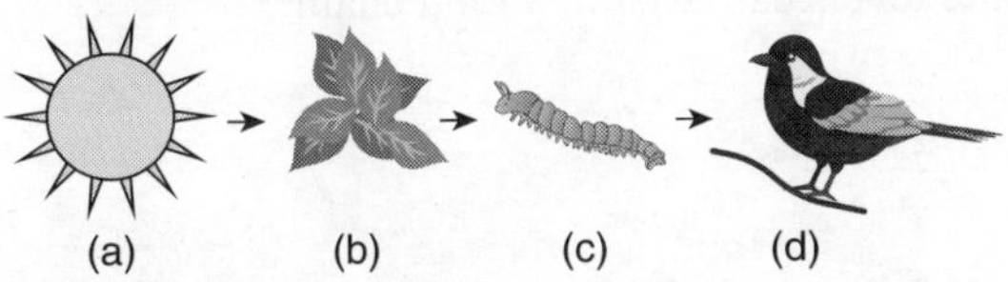

4. Why do producers always occupy the first trophic level in every food chain?

5. We often use the word environment. What does it mean?

Very Short Answer Questions

1. What is renewable energy?
2. What is the main cause of climate change?
3. What is the 3Rs approach to waste management?
4. What is sustainable development?
5. What is an ecological footprint?

Short Answer Questions

1. What is an ecosystem?
2. Why is a lake considered to be a natural ecosystem?
3. In the following food chain, plants provide 500 J of energy to rats. How much energy will be available to hawks from snakes?

Plants → Rats → Snakes → Hawks

Long Answer Questions

1. What are the main causes and consequences of air pollution?
2. How does climate change impact biodiversity?
3. What are the benefits and drawbacks of using renewable energy sources?
4. How can we promote sustainable agriculture practices?
5. What are the major impacts of deforestation?

Case Study-Based Questions

1. In the modern world, effective waste management is crucial. Population growth is increasing trash creation, which is doubling daily. Additionally, the rise in trash is having an impact on a lot of people's lives. Trash management is the control of waste through recycling and disposal. Additionally, effective waste management methods must be used while keeping environmental conditions in mind. For instance, there are a number of methods and strategies used to dispose of the garbage. You must be familiar with the five Rs of environmental

conservation: refuse, reduce, reuse, repurpose, and recycle.

(i) Although recycling paper is an excellent idea, why should it not be used for food packaging?
  (a) Recycled paper takes up a lot of room
  (b) They cannot properly cover the food
  (c) They can spread disease
  (d) They are expensive

(ii) Effective waste segregation is crucial. Choose the phrases that best convey the value of waste segregation.
  (A) Less garbage is dumped in landfills
  (B) Better for the environment and public health
  (C) Contribute to waste reduction
  (D) Worsen a waste picker's health
  (a) (A) and (B) (b) (A) and (C)
  (c) (B) and (C) (d) (A) and (D)

2. The food chain is a crucial aspect of nature for the survival of most species.

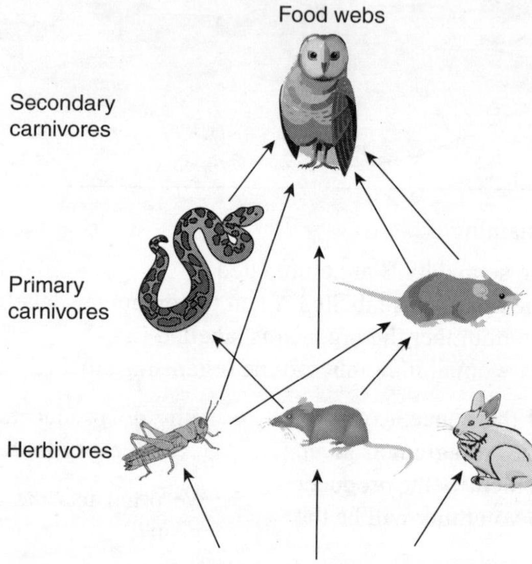

(i) Which of the following factors restricts how many trophic levels can exist in a food chain?
  (a) Less energy is available at higher trophic levels
  (b) There is less food available
  (c) Air pollution
  (d) Water pollution

(ii) If Sophia is eating curd, which trophic level would that be considered as?
  (a) 1st (b) 2nd
  (c) 3rd (d) 4th

(iii) An ecosystem requires two basic inputs: matter and energy.
  (a) Matter is continuously circulating and energy is bidirectional
  (b) Energy is continuously circulating and matter is unidirectional.
  (c) Both energy and matter are continuously moving in one direction.
  (d) Matter is bidirectional and energy is multidirectional.

3. Aditya Aggarwal and his elder brother Amit Aggarwal are the owners of Tee Cee Industries, a steel mill founded by their ancestors in 1984 in Kunjpura village, Karnal district, Haryana. They also manage a gaushala, a facility for 1,200 cows that are no longer able to produce milk. The cow shelter was manageable, but the steel plant was becoming expensive to run because they paid Rs 5 lakh in energy every month. The brothers had an idea. Why not use the biogas generated from cow manure from the shelter and other gaushalas to power the factory? This thought inspired Aditya and Amit to launch Amrit Fertilizers, a biogas initiative, in 2014 without receiving any funding from the government.

(i) What ingredients are used in biogas plant?
(a) Animal dung (b) Crop waste
(c) Food residue (d) All of the above

(ii) Why is biogas better than animal dung cake?
(A) The calorific value of biogas is lower.
(B) The calorific value of animal dung cake is higher.
(C) Biogas has a large capability for heating.
(D) Biogas burns smokelessly.
(a) (A) only (b) (B) only
(c) (C) and (D) (d) (A) and (B)

(iii) How is biogas formed?
(a) In the presence of air only.
(b) In the presence of water only.
(c) In the absence of air only.
(d) In the presence of water and absence of air

4. The environment is a vital component of who we are. We are affected by environmental changes, yet we also cause environmental changes through our actions. Three oxygen atoms combine to produce a molecule of ozone (O_3). While all aerobic forms of life require O_2, or oxygen as we often refer to it, ozone is a poison that can kill. However, ozone serves a crucial purpose at greater atmospheric altitudes. It protects the earth's surface from the Sun's ultraviolet (UV) radiation.

(i) How is ozone formed?
(a) By burning fossil fuels
(b) By photosynthesis
(c) By dividing molecular oxygen into nascent oxygen and then uniting to form O_3
(d) By volcanic activity

(ii) In which layer does ozone exist?
(a) Troposphere (b) Mesosphere
(c) Thermosphere (d) Stratosphere

(iii) What do UV radiations do?
(a) Provide energy for photosynthesis
(b) Help in the breakdown of ozone
(c) Cause genetic mutations and may lead to different types of cancers in humans
(d) All of the above

5. Air pollution in Delhi:

(i) What are the primary sources of air pollution in Delhi?
(a) Industrial emissions
(b) Vehicle exhaust
(c) Crop burning
(d) All of the above

(ii) What are the health impacts of air pollution on the residents of Delhi?
(a) Respiratory diseases
(b) Heart disease
(c) Cancer
(d) All of the above

(iii) What measures have been taken to reduce air pollution in Delhi, and what are the results?
(a) Implementing odd-even vehicle restrictions
(b) Closing thermal power plants
(c) Encouraging public transportation
(d) All of the above

NCERT Corner

1. Why are some substances biodegradable and others non-biodegradable?

2. Give any two processes in which biodegradable substances affect the environment.

3. How does ozone affect the ecosystem?

4. What are the problems caused by the non-biodegradable wastes that we generate?

5. If all the waste we generate is biodegradable, will this have no impact on the environment?

COMPETITION WINDOW (OBJECTIVE TYPE)

1. Which of the following groups contains only biodegradable items?
   (a) Grass, flowers and leather
   (b) Grass, wood and plastic
   (c) Fruit peels, cake and lime juice
   (d) Cake, wood and grass

2. Which of the following constitutes a food chain?
   (a) Grass, wheat and mango
   (b) Grass, goat and human
   (c) Goat, cow and elephant
   (d) Grass, fish and goat

3. Identify the original source of the energy which flows through a food chain?
   (a) Carbon dioxide (b) Glucose
   (c) Oxygen (d) Sunlight

4. Which of the following statements about food chains are correct?
   (a) It includes repeated eating, that is, each group eatstheotherandissubsequentlyeatenbysomeother group of organisms.
   (b) It shows a series of branching lines and unidirectional flow of energy.
   (c) It shows the unidirectional flow of energy and proceeds in a progressive straight line.
   (d) Both (a) and (c)

5. What is the role of decomposers in the environment?
   (a) Transform organic materials into inorganic ones.
   (b) Transform inorganic material into more basic forms.
   (b) Create the organic chemical from the inorganic ingredient.
   (d) Avoid dissolving the organic ingredient.

6. The amount of energy transferred from one trophic level to the next is:
   (a) 5% (b) 10%
   (c) 20% (d) 15%

7. Who are known as the primary consumers?
   (a) Herbivores (b) Carnivores
   (c) Omnivores (d) Producers

8. Which one of the following causes the ozone layer to thin out?
   (a) CFCs (b) Oxygen
   (c) SO_2 (d) CO_2

9. We should reduce the use of the plastic bags, bottles etc. because:
   (a) They are not durable
   (b) They are non-biodegradable
   (c) They are made of toxic materials
   (d) They react with the atmospheric gases

10. Among the following, choose the correct option which contains only biodegradable items?
    (i) Wood, paper, PVC
    (ii) Paper, seeds, detergent,
    (iii) Paper, animal excreta, wood
    (iv) Wool, leaves, paper
    (a) (i), (ii) and (iii)
    (b) (i) and (iii)
    (c) (ii), (iii) and (iv)
    (d) (iii) and (iv)

11. Which among the following statements is incorrect in view of plants?
    (a) They convert solar energy into mechanical energy
    (b) They prepare their food from organic compounds
    (c) They are also called producers
    (d) They are the initial source of energy in a food chain

12. In a food chain, the second trophic level is occupied by:
    (a) Carnivores (b) Autotrophs
    (c) Herbivores (d) Producers

13. Green plants utilise _________ percent of the sun's energy to prepare their food by the process of photosynthesis?
    (a) 1 (b) 10
    (c) 20 (d) 99

14. The process of accumulation of harmful chemical substances like pesticides, in the body of living organisms at each trophic level of a food chain is known as:
    (a) Biological magnification
    (b) Biological accumulation
    (c) Chemical magnification
    (d) Chemical accumulation

15. Which of the following may be a conclusion of the excessive exposure of humans to sun's ultraviolet rays?

(i) Peptic ulcers
(ii) Eye disease like cataract
(iii) Damage to lungs
(iv) Skin cancer
(a) (i) and (iv)
(b) (ii), (iii) and (iv)
(c) (ii) and (iv)
(d) Only (iv)

16. Which among the following is the full form of DDT?
(a) Dichlorodiphenyltrichloroethane
(b) Dichlorodiphenyltetrachloroethane
(c) Dichlorodecaphenyltrichloroethane
(d) Dichlorodiethyltrichloroethane

17. If 100 J of energy is available at the producer level in a food chain, then the energy available to the secondary consumer will be:
(a) 10 J (b) 0.1 J
(c) 1 J (d) 0.01 J

18. Which of the following radiations is responsible for the conversion of atmospheric oxygen to ozone?
(a) Gamma radiations
(b) Cosmic radiations
(c) Infrared radiations
(d) Ultraviolet radiations

Answer Keys

Multiple Choice Questions

1. (a) 2. (d) 3. (d) 4. (c) 5. (c)

Match the Following

Words	Meanings
Ecology	The relationship of living things to one another and their environment, or the study of such relationships.
Habitat	A place where a population lives and its surroundings, both living and non-living.
Herpetologist	A zoologist who studies reptiles and amphibians.
Extinct	Having no living representative.
Aquifer	An underground geological formation, containing water.
Naturalist	A person who studies plants and animals, especially outdoors.
Biodiversity	The variety and variability among living organisms and the ecological complexes in which they occur.
Mangrove	Tropical evergreen trees or shrubs forming dense thickets along tidal shores.
Conservationist	A person who advocates preservation and careful management of natural resources.

True or False

1. F 2. T 3. F 4. T 5. F

Fill in the Blanks

1. Biodegradable
2. Non-biodegradable wastes
3. Sun
4. United Nations Environment Programme (UNEP)
5. Oxygen (O_2)
6. Consumers

Assertion-Reason Questions

1. (a) 2. (a) 3. (b) 4. (a) 5. (b)

One-word Answers

1. Transpiration
2. Biodiversity
3. Erosion
4. Ecology
5. Combustion

Define the Term

1. Biodiversity is the variety of life on Earth, including the diversity of species, ecosystems and genetic diversity within species.
2. Climate change is a long-term shift in global weather patterns caused by changes in temperature, precipitation and other factors, primarily as a result of human activities such as burning fossil fuels and deforestation.
3. Ecological footprint is a measure of the impact that human activities have on the environment, expressed as the amount of land and resources required to sustain a given population or activity.
4. Greenhouse gases are gases that trap heat in the Earth's atmosphere and contribute to global warming, such as carbon dioxide, methane and water vapour.
5. Renewable energy are energy sources that are replenished naturally and can be used indefinitely, such as solar, wind and hydro power.

Groups of Terms

1. Bacteria
2. Clouds
3. 5
4. Wildfire
5. Cacti

Identify the Odd One Out

1. Water vapour
2. Glass bottles
3. Nuclear energy
4. Soil erosion
5. Urbanisation

Image-Based Questions

1. (c)
2. (b)
3. (b)

Very Short Answer Questions

1. Renewable energy refers to energy sources that are replenished naturally and can be used indefinitely, such as solar, wind and hydro power.
2. The main cause of climate change is the increase in greenhouse gas emissions, primarily carbon dioxide, resulting from human activities such as burning fossil fuels and deforestation.
3. The 3Rs approach to waste management refers to the principles of reduce, reuse and recycle, which aim to minimise the amount of waste that is generated and promote sustainable use of resources.
4. Sustainable development is a concept that aims to balance economic, social and environmental considerations in development activities, in order to meet the needs of the present generation without compromising the ability of future generations to meet their own needs.
5. An ecological footprint is a measure of the impact that human activities have on the environment, expressed as the amount of land and resources required to sustain a given population or activity.

Short Answer Questions

1. An ecosystem is defined as a structural and functional unit of the biosphere. It comprises living organisms and their non-living environment that interact by means of food chains and biogeochemical cycles resulting in energy flow, biotic diversity and material cycling to form a stable self-supporting system.
2. Lake is an ecosystem where living organisms grow, reproduce and interact among each other as well as with abiotic components and carry out other activities in nature by themselves without any human interference. Therefore, it is referred to as a natural ecosystem.
3. In an ecosystem, only 10% of energy is transferred from one trophic level to the next, that is, the ten percent law, and the rest is dissipated into the environment. Therefore, if plants (being producers at the first trophic level) transfer 500 J of energy to rats (second trophic level), then rats would transfer 50 J of energy to snakes (third trophic level), which in turn would transfer only 5 J of energy to hawks (fourth or last trophic level) in a food chain.
4. Producers are the green plants that can manufacture food using CO_2 and H_2O in the presence of sunlight; they are autotrophs. They serve as a source of food for all non-producers or consumers directly or indirectly. Hence, producers occupy the first trophic level in a food chain.
5. Environment can be defined as the physical or biological world where an organism lives. Literally speaking, an organism's immediate surrounding constitutes its environment which includes both biotic and abiotic components.

Long Answer Questions

1. Hint: Causes: Emissions from vehicles, industrial activities, power generation, burning of fossil fuels, agricultural practices, and indoor pollutants.

   Consequences: Respiratory diseases, cardiovascular problems, reduced lung function, allergies, smog formation, acid rain, global warming, ozone depletion, and adverse effects on ecosystems and biodiversity.

   Elaborate on these points.
2. Hint: Climate change can disrupt ecosystems and negatively impact biodiversity in several ways.

   Shifts in temperature and precipitation patterns can alter habitats, affecting the distribution and abundance of species.

   Rising temperatures can lead to changes in the timing of biological events such as flowering, migration, and breeding, causing mismatches with food sources and impacting species' survival.

   Sea-level rise and increased ocean acidification can harm marine ecosystems and coral reefs.

   Extreme weather events can cause habitat destruction, population declines, and increased vulnerability to diseases.

   Elaborate on these points.
3. Hint: Benefits: Renewable energy sources such as solar, wind, hydro, and geothermal power offer several advantages. They are sustainable, do not deplete natural resources, reduce greenhouse gas emissions, improve air quality, and mitigate climate change. They also promote energy independence, create jobs, and support economic growth.

   Drawbacks: Renewable energy sources have some limitations. They may require substantial upfront investments, depend on weather conditions (solar and wind energy), and have intermittency issues. Some renewable technologies may have visual and noise impacts, and there can be concerns related to land use and wildlife impacts (large-scale hydroelectric projects).

   Elaborate on these points.
4. Hint: Encourage organic farming methods that minimize the use of synthetic fertilizers and pesticides.

Support crop diversification and agroforestry practices to improve soil health, reduce erosion, and enhance biodiversity.

Promote efficient water management techniques such as drip irrigation and rainwater harvesting.

Advocate for sustainable land management practices like conservation tillage and crop rotation.

Promote integrated pest management strategies to minimize the use of harmful chemicals.

Encourage the use of renewable energy in farming operations and promote sustainable waste management practices in agriculture.

Elaborate on these points.

5. Hint: Loss of biodiversity, climate change, soil erosion and degradation, disruption of water cycles and socioeconomic impact. Elaborate on these points.

Case Study-Based Questions

1. (i) (c), (ii) (a)
2. (i) (a), (ii) (c), (iii) (c)
3. (i) (d), (ii) (c), (iii) (d)
4. (i) (c), (ii) (d), (iii) (c)
5. (i) (d), (ii) (d), (iii) (d)

NCERT Corner

1. Biodegradable substances are those that can be broken down by biological processes. The environment contains a wide variety of compounds that decomposers (bacteria and fungus) can break down quickly because they have the necessary enzymes. There are additional compounds, referred to as non-biodegradable substances, that do not break down in this way. These compounds last a long time because bacteria and fungus do not break them down. Physical processes like heat and pressure will have an impact on these non-biodegradable materials.
2. During decomposition, a foul smell produced. Harmful gases such as ammonia, methane, carbon dioxide, etc., are produced which can further cause global warming.
3. Ozone (O_3) is a molecule made from three atoms of oxygen. At the higher levels of the atmosphere, it shields the surface of the earth from ultraviolet (UV) radiation from the Sun. It may affect any ecosystem in the following ways:
   - At the surface of the earth, it is a deadly poison for all lower forms of life.
   - If this layer gets depleted, it may cause cancer in humans.
4. (i) Non-biodegradable wastes persist in the environment for a long time and cause greater harm to the various members of the ecosystem by causing biological magnification.

   (ii) Non-biodegradable waste such as fertilisers, pesticides, weedicides, etc., changes the soil chemistry, and in turn affects the fertility of soil and subsequently reduces the crop yield.
5. Biodegradable waste can be easily recycled by bacterial and fungal decomposers. The only negative environmental effect will be that many of the gases emitted during the decomposition process may cause global warming.

Competition Window

1. (c)	2. (b)	3. (d)	4. (d)	5. (a)
6. (b)	7. (a)	8. (a)	9. (b)	10. (d)
11. (b)	12. (c)	13. (a)	14. (a)	15. (c)
16. (a)	17. (c)	18. (d)		

CHAPTER 19

Pollution – A Rising Environmental Problem

PREREQUISITES

You must have knowledge of:

- Industrialisation
- Urbanisation
- Population growth
- Agricultural practices
- Transportation
- Improper waste disposal

LEARNING OBJECTIVES

At the end of this chapter, you will be able to:

- Define the terms pollution and pollutants
- Describe the types of pollution, their sources and the detrimental consequences on human health
- Explain how to control air pollution
- Describe water pollution and its sources
- Describe thermal pollution
- List soil pollution's causes, effects and prevention methods
- Explain radiation pollution's sources and dangers
- Describe noise pollution

INTRODUCTION

Pollution is a serious problem of modern times due to rapid advancement in several areas such as industries, transport, agriculture, residential set-ups and thermal and nuclear power generation plants. It causes deterioration and even poisoning of the environment including air, water and soil. It poses a serious threat to human health as well as to almost all living beings and our surroundings.

Any substance that is added to the air, water or land and which degrades the environment's natural quality is considered a pollutant. A pollutant is any constituent which causes pollution.

There are five main types of pollution: air, water, soil, radiation and noise.

AIR POLLUTION

Air pollution is the term used to describe deterioration in air quality that harms both living things and certain goods. It is of two types:

- **Gaseous pollution:** The release of hazardous gases from various sources.
- **Particulate pollution:** Particles from mining, stone drilling, smoking, rubber tyre wear in automobiles and dust.

The main sources of air pollution are as follows:

- **Sources from nature:** Examples include dust from storms, forest fires and volcanic ash. These sources cannot be controlled by humans and will keep occurring.

- **Artificial sources:** Examples include automobiles, businesses, dumpsters and brick kilns. These are the four main sources, all of which can be managed by humans and for which solutions are being developed.

Vehicular Air Pollution

There are many different types of vehicles, including motor vehicles like automobiles, trucks, buses, scooters, tractors, motorcycles and locomotive engines. They mostly run on gasoline and diesel. Three main gaseous pollutants, CO_2, SO_2 and CO, are present in the exhaust that is released. In addition to raising road dust particles into the air, automobiles' worn rubber tyres leak rubber particles into the environment. You may have also noticed a lot of dust being created as the morning road sweepers sweep the roadways.

This type of pollution can be controlled by better use of compressed natural gas, lead-free gasoline, high-quality car fuels and efficient engines (CNG).

Industrial Air Pollution

The Industrial Revolution has swept the entire planet. It includes all industries, from little potteries and carpet-weaving businesses to huge sugar, clothing and metal mills as well as companies that make cars, medicines and other goods. Pollutants of all types are released by them.

The main industrial gaseous pollutants are oxides of nitrogen, CO_2, SO_2 and CO. There are numerous particle contaminants in the smoke emitted into the atmosphere by industry chimneys (Fig. 19.1). Fine particles produced during stone drilling and mining (of coal, metals, etc.) pose substantial health risks to the workers involved.

Fig. 19.1 Smoke from factories chimney
(Smoke pollution; Adina Voicu; 2016; https://commons.wikimedia.org/wiki/File:Smoke_Pollution.jpg. Licensed under: CC0)

Smog is a byproduct of the mixing of smoke emitted from numerous sources with dust and tiny drops of fog. Smog harms plants and can lead to asthma and allergies in people.

Burning Garbage

The majority of rubbish is made up of things like peelings from fruits and vegetables, shells from dry fruits, washed rice, lentils and other foods, as well as any leftovers from meals. Such garbage is either burned in designated areas or composted to create manure for use in agriculture. Rubbish that is burnt emits CO_2 as well as other dangerous pollutants. The smoke released increases the number of airborne particulates.

Brick Kilns

Bricks are made in fire-heated structures called brick kilns. To contain the flames inside, raw, moist clay bricks are piled up with firewood in between. This arrangement is then covered in a coating of a mud mixture. This set-up is left undisturbed for a few days. A hole allows the internal smoke to be expelled. At the end of this process, the original mud-coloured bricks become red, and there is a significant amount of ash and fragments of broken brick produced as trash. All of them are pollutants that are either discarded or reused in some way. The contaminants come from the heat and smoke that are released out.

WATER POLLUTION

Any alteration in the water's quality that renders it unfit for consumption by humans and other living things is referred to as water pollution.

The main sources of water pollution are discussed below:

- **Household detergents:** Every home uses some detergents (cleansing agents) to wash and clean soiled or worn garments, crockery, utensils, etc. The dirty water flows into the drains.
- **Sewage:** The liquid waste from residential activities is known as sewage. It is made up of toilet and other domestic waste water, as well as culinary trash. Most cities have sewage treatment facilities that remove the undesirable material before discharging the clean water into neighbouring bodies of water or rivers. In some locations, people still practice disposal of dead bodies into rivers. Sometimes, the cremated ashes are thrown into rivers like the Ganga at numerous locations. These actions contribute to water contamination.
- **Industrial waste:** Waste water from several enterprises, both small- and large-scale, contains different kinds of chemical contaminants. These wastes are frequently dumped into waterways.

Industries that treat fish close to the coast discharge hazardous wastes into the water, causing pollution.

- **Oil spills:** Accidental release of petroleum into estuaries or oceans are known as oil spills. Overturned oil tankers, offshore oil mining, and oil refineries are the origins of spills. Many marine animals die as a result of oil pollution (fish, birds, etc.). The duck in Fig. 19.2 is covered in oil.

Fig. 19.2 Duck affected by oil spill
(An oiled surf scoter (a duck) after the 2007 oil spill in San Francisco Bay…; Brocken Inaglory; 2007; https://commons.wikimedia.org/wiki/File:Oiled_bird_3.jpg. Licensed under: CC-BY-SA-3.0)

- **Thermal pollution:** Water is used by many sectors, including nuclear power plants, thermal power plants and oil refineries, to cool their equipment. This hot waste water, which is dumped into neighbouring streams, rivers or the sea and causes warming; it may be 8–10°C hotter than the intake water. The creatures and plants living there are negatively impacted by the warming of the water, which has even been known to kill fish.

SOIL POLLUTION

In contrast to air and water pollution, which may travel great distances, soil pollution is largely localised. The main sources of soil pollution are discussed below:

- **Industrial waste:** In addition to releasing gaseous air pollutants and chemical-laden water, industries also give out much solid waste. This is in the form of chemical residues, flyash, metallic ash, etc.
- **Urban commercial and domestic waste:** There are numerous marketplaces selling fruits, vegetables and food grains in cities and towns. Banquet halls, restaurants and tailors produce a lot of solid trash of various kinds. Plastic bags, glass bottles, electric bulbs, kitchen waste, paper packaging, etc. are only a few examples of the waste we make at home. This is all solid garbage. Municipal employees collect it and dump it for proper disposal; some of it can be turned into manure, and the remainder can be disposed of in sanitary landfills. These are places where waste is dumped in a ground depression and covered with dirt every day.
- **Chemical fertilisers:** Nitrates, phosphates and ammonium salts are examples of chemical fertilisers that aid in faster and higher crop output. However, using too much of them can be dangerous. Rainwater gradually washes these fertilisers away until they reach lakes and ponds. This causes germs to proliferate more quickly, which depletes the water's oxygen supply and kills fish and other aquatic life.
- **Biomedical waste:** There are numerous items under this category:
    - Needles, syringes, dirty dressings, etc.
    - Unused discarded medicinal tablets and powders
    - Discarded biological research materials carelessly disposed of in the municipal garbage.

  Such waste often reaches the soil and may cause harm to soil organisms, and also to humans indirectly through some agents.
- **Pesticides:** DDT and other pesticides are frequently employed to eradicate pests in houses, godowns and agricultural crops. Most of these remain (undegraded) as contaminants in the environment and significantly harm life in indirect ways. Even the milder pesticides that are now in use are harmful to the environment. Pesticides like DDT change the fundamental makeup of the soil, eradicate the microorganisms that live there, and may even make their way into human bodies through food produced in such soils. The majority of the microorganisms that are killed are those that reuse soil nutrients.

The wastes generated from the various sources described above can be categorised into two types:

- **Biodegradable waste:** Material which can be degraded (broken down) by microorganisms into relatively harmless compounds. Examples are paper, vegetable peels, food left-overs, sewage, cow dung, dried leaves, hay, etc.
- **Non-biodegradable waste:** Material which cannot be degraded. Examples are plastic, glass, styrofoam (a kind of packing material) and pesticides like DDT.

RADIATION

High-energy particles make up the energy in radiation. It is widely used in the domains of medicine (X-rays, etc.) and in nuclear power plants to produce energy.

X-rays have a lot of energy in them (measured in photons, the units of energy). The cells in the body can be harmed by these rays. X-ray exposure that is too frequent and prolonged poses major health risks (damaged chromosomes, genetic variations, bone cancer, etc.). Operators of X-ray machines must exercise extreme caution to minimise exposure.

Nuclear radiation leaks from nuclear power plants may result through careless waste disposal or an unintentional explosion at the facility. Both Chernobyl (Ukraine) in 1986 and Fukushima Daiichi Nuclear Plant (Japan) on March 11, 2011, experienced very significant nuclear plant explosions. Numerous people were exposed to radiation, and many more developed health issues like bleeding and possibly cancer. Iodine-131 was the main radioactive contaminant; it increases the risk of thyroid cancer. Radiation even seeps into the ocean and passively enters humans through seafood.

Radioactive material that is carelessly disposed of has the potential to do serious harm, as was the case in Mayapuri, Delhi, in 2010. Through 'kabaris', radioactive waste (Cobalt-60) from a research facility found its way to the market streets. Many affected individuals are still in pain and some have even passed away. Due to carelessness, a resident of Delhi who was exposed to radioactive radiation stored Cobalt-60 pencils in his back pocket, which resulted in severe burns.

Fig. 19.3 Noise pollution
(A Qantas Boeing 747-400…; Adrian Pingstone; 2004; https://commons.wikimedia.org/wiki/File:Qantas_b747_over_houses_arp.jpg. Licensed under: PD-user)

NOISE POLLUTION

Any unpleasant or loud unwanted sound that interferes with hearing or concentration is referred to as noise.

Industrial machinery, workshops, trains, cars on the road, jet aircrafts landing and taking off in the air, loud conversations and the radio or television inside homes, loudspeakers and musical bands in public spaces, and so on are major sources of noise (Fig. 19.3).

Even a quiet discussion in a room becomes noisy if you are reading something carefully or working on a maths problem. However, you might not be disturbed if you are reading casually in a train.

The harmful effects of noise pollution are:

- Interferes with communication
- Interrupts concentration of thought and disturbs peace of mind
- Lowers efficiency of work
- Disturbs sleep and leads to nervous irritability
- Can damage the ear drum; prolonged noise can even lead to deafness
- Bird life is disturbed by aircrafts landing or taking off from airports

Measures to minimise noise pollution:

- Prohibiting blowing of horns.
- Restriction on loud speakers, especially at night.
- Planting trees by the road and keeping doors closed in houses.
- Not setting off fire crackers.

EFFECTS OF VARIOUS TYPES OF POLLUTION

Pollution affects the environment, humans and other life in numerous ways. The more significant ones are as discussed below.

Effects on Human Health

- Lung and respiratory illnesses are brought on by air pollution. A lack of blood cells could be caused by lead particles. Asthma and reduced visibility are caused by smog (smoke and fog). The lungs' airways are severely harmed by the exhaust gas SO_2 (the disease bronchitis).
- Cholera, jaundice and typhoid are a few diseases that can be brought on by water contamination.
- Human health may be indirectly impacted by soil contamination. Agricultural farming pesticides may have negative health effects.

Effects on Climate and Environment

Pollution affects the climate and environment in several ways but the three most significant ones are:

- **Acid rain:** Normally, clean rain is only slightly acidic (pH of about 6) because of some CO_2 naturally present in the atmosphere. Due to increasing industrialisation, there is lot of emission of CO_2, and sulphur dioxide (SO_2) and oxides of nitrogen which are dissolved in the raindrops falling on earth as rain. This is acid rain.

  Harmful effects of acid rain:

  - Damage to vegetation by pollution of the soil (Fig. 19.4).

Fig. 19.4 Damage to vegetation (tomato leaf potassium deficiency; Goldlocki; 2005; https://commons.wikimedia.org/wiki/File:Tomate_Blatt_Kalimangel.jpg. Licensed under: CC-BY-SA=3.0-migrated)

  - Decay of building material and paints.
  - The reaction between the acid and calcium causes antique sculptures, monuments and sculpture to erode.
  - Sulphates and nitrates, which are small particles, reduce visibility and are harmful to human health.
  - Due to the increasing acidity of the water in such lakes and rivers, fish and other aquatic species suffer.

- **Greenhouse effect and global warming:** Small glass structures called greenhouses are made specifically to cultivate plants, especially in the chilly winter months. While allowing light to enter from all sides, the glass panel prevents heat from escaping. A similar thing occurs when an automobile is left in the sun for a few hours. Certain gases, in particular CO_2 and methane (CH_4), build up in the atmosphere, preventing heat from escaping and warming the air (Greenhouse Effect). More CO_2 is released as a result of increased industrialisation, domestic fuel combustion and population growth. Conversely, the loss of flora and deforestation is reducing the amount of CO_2 used in photosynthesis. Thus, overall rise in CO_2 is leading to greater retention of the solar radiation in the atmosphere, causing global warming. During the last three decades, the temperature of the earth's atmosphere has increased by about 0–5°C. This is causing melting of snow caps, rise in sea level, as well as decline in food production in agricultural fields and in fishery resources in the oceans.

- **Ozone layer depletion:** Ozone is an allotropic form of oxygen. In the upper atmosphere, it resembles a sunscreen. It stops the sun's damaging UV-B rays from entering the atmosphere. Some gaseous substances, such as CFCs (chlorofluorocarbons), ascend into the atmosphere when they are emitted by appliances like refrigerators, aerosol sprayers and packing materials like styrofoam. They disintegrate into chlorine atoms there, which then disintegrate ozone (O_3) into oxygen (O_2) and O. As ozone levels decline, dangerous UV-B rays that cause skin cancer, genetic diseases, sunburn, and decreased ocean and forest productivity can still reach the planet. September 16 is marked as International Ozone Day.

ABATEMENT OF POLLUTION

Although pollution cannot entirely be eliminated, it can be reduced in a number of ways.:

- Using CNG (compressed natural gas) and unleaded gasoline in vehicles
- Switching off the car's engines while it is not in use and at red lights
- Installing tall chimneys with filters and electrostatic precipitators in industrial settings
- Treating industrial waste before releasing them into bodies of water
- Not disposing of food trash in drains or on the ground
- Installing septic tanks in homes and sewage treatment units in bigger settlements
- Using more compost (organic waste) in place of chemical fertilisers and using pesticides sparingly
- Recycling plastic, metal and glass as well as burning (incineration) of non-recyclable garbage
- Planting trees along the roadside, waste land and mountainous slopes to reduce soil erosion as and

improve the air quality (reduction of the amount of CO_2 in the atmosphere)

Vehicular standards: The term Euro/Bharat norm refers to a set of standards that apply to all automobiles and require tougher emission controls to be upheld in the most major cities. The Indian government has implemented an auto fuel strategy to significantly reduce sulphur and nitrogen oxides in vehicle exhausts.

SUMMARY

- Pollution is a complex issue that involves the introduction of harmful substances or products into the environment that cause adverse effects on living beings and the natural world. Pollution can take different forms, and each type of pollution has unique characteristics and causes.
- Any substance that is added to the air, water or land and which degrades the environment's natural quality is considered a pollutant. Air pollution is the term used to describe the undesired elements that contaminate the air and have a negative impact on both living things and non-living things.
- Industrial source includes the noise from various industries and big machines woring at a very high speed and high noise intensity.
- Sulfur dioxide, nitrogen oxides, carbon dioxide, fly ash and suspended particle matter are the main air pollutants (SPM).
- Human activities burning of fossil fuel increase the carbon dioxide content in the atmosphere. The increased concentration of carbon dioxide may bring about drastic changes in the world climate, in the near future, in the form of increased temperature or global warming.
- Carbon monoxide is a harmful gas that is produced by vehicles, along with carbon dioxide, nitrogen oxides, sulphur oxides and smoke. It decreases the blood's ability to carry oxygen.
- Acid rain refers to the presence of excessive acids such as sulphuric acid, sulphurous acid and nitric acid.
- Water pollution, on the other hand, occurs when harmful substances, such as chemicals, toxins and waste products, contaminate water bodies such as rivers, lakes and oceans. Water pollution can be caused by different factors, including industrial activities, agricultural practices, sewage and wastewater disposal and oil spills. The contamination of water bodies can lead to the death of aquatic life, affect the quality of water, and cause health problems in humans who consume contaminated water.
- Soil pollution is the contamination of soil with harmful substances, such as pesticides, chemicals and industrial waste. Soil pollution can have severe impacts on plant growth, leading to the death of vegetation, and can also contaminate groundwater and surface water.
- Noise pollution is the presence of unwanted and excessive sound that can cause discomfort, hearing loss and other health problems. Noise pollution can be caused by different sources, including traffic, construction activities and industrial processes. Noise pollution can affect the behaviour and communication of animals, leading to their displacement or death.
- Thermal pollution occurs when the temperature of water bodies or air increases due to the discharge of heated water or air by industries and other sources. Thermal pollution can have severe impacts on aquatic life, leading to changes in the ecosystem and the displacement of fish species.
- Controlling pollution is essential to ensure a sustainable future for all. Governments, industries and individuals can take steps to reduce pollution levels through sustainable practices such as reducing waste generation, using eco-friendly products, promoting clean energy and implementing policies that encourage waste reduction, recycling and responsible waste disposal. It is crucial to address pollution as it has severe impacts on human health, wildlife and the natural world, and the effects can be long-lasting and irreversible.

PRACTICE QUESTIONS

Analyse Your Concepts (School Exam Based)

Multiple Choice Questions

1. Which one of the following is an example of a pollutant?
   (a) A rat entering your kitchen.
   (b) CO_2 given out by the respiring organism.
   (c) The heat given out while cooking food.
   (d) The dust raised during road-cleaning.
2. Which one of the following is a 'dirty' practice contributing to pollution and needs to be stopped by educating the public?
   (a) Very loud marriage musical bands
   (b) Disposing of corpses in rivers
   (c) Floating lighted earthen lamps in rivers
   (d) Wearing scented/perfumed clothes
3. One of the examples of radiation pollutants is
   (a) Sulphur dioxide
   (b) Ozone
   (c) Iodine-131
   (d) Discarded fused electric bulbs
4. The causes of water pollution are:
   (a) Afforestation (b) Oil refineries
   (c) Paper factories (d) Both (b) and (c)
5. Chlorofluorocarbon are non-flammable chemicals mainly used in:
   (a) Perfumes (b) Refrigerators
   (c) Air conditioners (d) All of the above
6. Which of the following is not a waterborne disease?
   (a) Measles (b) Typhoid
   (c) Cholera (d) Hepatitis

Match the Following

Each question contains statements given in two columns which have to be matched. Statements in column I have to be matched with statements in column II.

Column I	Column II
A. Air pollution	1. Occurs when the quality of water is adversely affected by human activities.
B. Water pollution	2. Occurs when the air's quality is reduced by pollutants, such as smoke, dust and gases.
C. Soil pollution	3. Occurs when the soil's natural composition is altered by human activities, making it less fertile.
D. Noise pollution	4. Occurs when unwanted sound, such as from traffic, construction or loud music, disrupts the environment.

(Continued)

True or False

Read the following statements and write your answer as true or false.

1. Pollution is defined as the introduction of harmful substances or products into the environment.
2. Carbon dioxide is not a greenhouse gas and does not contribute to global warming.
3. Indoor air pollution can be caused by factors such as tobacco smoke, radon gas and household cleaning products.
4. Acid rain is caused by the emission of sulfur dioxide and nitrogen oxides into the atmosphere.
5. The ozone layer helps to protect the Earth from harmful ultraviolet radiation.

Fill in the Blanks

Complete the following statements with an appropriate word/term to be filled in the blank spaces.

1. Rubber particles and dust raised by running motor vehicles are examples of ________ pollutants.
2. Too frequent exposure to ________ in a medical diagnostic technique may damage chromosomes.
3. Thermal power plants give out a lot of ________ waste water.
4. Sewage is a liquid waste from ________
5. ________ is a harmful gas that is produced by vehicles

Assertion–Reason Questions

The following questions consist of two statements: Assertion (A) and Reason (R). Answer these questions selecting the appropriate option given below:

(a) Both A and R are true and R is the correct explanation of A

(b) Both A and R are true but R is not the correct explanation of A
(c) A is true but R is false
(d) A is false but R is true

1. Assertion (A): Potable water is utilised for routine housekeeping.
   Reason (R): Potable water is readily available and easy to access for most households, and is often used for a variety of purposes including routine housekeeping, drinking and cooking.
2. Assertion (A): In 1985, the Ganga Action Plan, a comprehensive river preservation strategy, was introduced.
   Reason (R): This strategy aims to lower the river's pollutant levels.
3. Assertion (A): Water becomes polluted when dangerous substances, including sewage, poisonous chemicals and toxins, interact with it.
   Reason (R): Water pollutants are compounds that contaminate water.
4. Assertion (A): Global warming is the progressive rise in the earth's atmosphere's average temperature.
   Reason (R): Methane, NO_2 and water vapour are other gases that contribute to the so-called greenhouse effect.
5. Assertion (A): CFCs benefit the atmosphere.
   Reason (R): Aerosol sprays, refrigerators and air conditioners are all used with CFCs to protect the ozone layer.

One-word Answers

1. Pollutants which are mainly responsible for causing acid rain.
2. Any two chemicals leading to the formation of ozone holes.
3. The chemical element which caused Minimata disease in Japan.
4. The release of hazardous gases from various sources.
5. Rain refers to the presence of excessive acids such as sulphuric acid, sulphurous acid and nitric acid.

Define the Term

1. Pollution
2. Greenhouse gas
3. Acid rain
4. Ozone layer
5. Air pollution

Groups of Terms

Given below are certain groups of terms. In each group the first pair indicates a relationship between the two terms. Rewrite and complete the second pair on a similar basis.

1. Air Pollution : Acid Rain :: Water Pollution : ______
2. Thermal Pollution : Increase in Water Temperature :: Light Pollution : ______
3. Noise Pollution : Hearing Loss :: Soil Pollution : ______
4. Plastic Pollution : Microplastics :: Chemical Pollution : ______
5. Radioactive Pollution : Nuclear Accidents :: Land Pollution : ______

Identify the Odd One Out

1. Carbon monoxide, nitrogen oxides, sulfur dioxide, helium
2. Lead, mercury, chromium, oxygen
3. Plastic bags, styrofoam containers, glass bottles, paper towels
4. E-waste, batteries, pesticides, household cleaning products
5. Industrial runoff, sewage, oil spills, radioactive waste

Very Short Answer Questions

1. What is the primary cause of air pollution?
2. What is the main cause of water pollution?
3. What is the primary source of plastic pollution in the oceans?
4. What is the main cause of soil pollution?
5. What is the primary cause of climate change?

Short Answer Questions

1. List the three major constituents of sewage.
2. What are the causes of oil spills and how do they impact marine life?
3. Name two strategies to reduce noise pollution.
4. Write two measures to minimise noise pollution.
5. What is pollution?

Long Answer Questions

1. Discuss the effects of air pollution on human health and the environment.
2. Describe the different types of water pollution and their sources.
3. Discuss the impact of plastic pollution on marine life and the environment.
4. Discuss the causes and effects of soil pollution.

5. Discuss the impacts of noise pollution on human health and the environment.

Case Study-Based Questions

1. What are the two potential causes of numerous fish deaths near a seashore?
   (a) Overfishing and climate change
   (b) Industrial waste and thermal pollution
   (c) Oil spills and plastic pollution
   (d) Eutrophication and acidification

2. Look at the cartoonist's presentation of a kind of pollution given below.

(Noise, Rashtriya Paryavaran, 2023, https://commons.wikimedia.org/wiki/File:Noise_pollution_ghdf_lkji_542.png. Licensed under: CC-BY-SA-4.0)

   What is the pollution type that includes industrial equipment, workshops, trains, cars, jet aircrafts, loud conversations, and loudspeakers?
   (a) Air pollution (b) Water pollution
   (c) Noise pollution (d) Soil pollution

3. What are the pollutant sources for noise pollution?
   (a) Chemical factories and oil refineries
   (b) Agricultural practices and fertilizers
   (c) Industrial equipment and loudspeakers
   (d) Plastic waste and littering

4. What are two negative outcomes of noise pollution?
   (a) Decreased soil fertility and loss of biodiversity
   (b) Increased rates of cancer and other diseases
   (c) Less efficient work and sleep disruption
   (d) Disruption to breeding and feeding habits of wildlife

5. The following picture depicts a certain phenomenon.

   What is the phenomenon in which gaseous substances such as chlorofluorocarbons (CFCs) decompose ozone into oxygen and O?
   (a) Greenhouse effect (b) Acid rain
   (c) Ozone layer depletion (d) Water pollution

6. What are the main factors that contribute to ozone layer depletion?
   (a) Agricultural practices and fertilisers
   (b) Burning of fossil fuels
   (c) Industrial waste and emissions
   (d) Gaseous substances like CFCs

7. What are some of the effects of ozone layer depletion on people?
   (a) Increased rates of heart disease and respiratory problems
   (b) Disruption of ecosystems and loss of biodiversity
   (c) Sunburn, genetic conditions and skin cancer
   (d) Depletion of natural resources and habitat destruction

8. The picture below shows a campaign recently started by the Indian Government.

(Clean India Programme by..., Baranagore Ramakrishna Mission Ashrama High School, 2016, https://commons.wikimedia.org/wiki/File:Swachh_Bharat_Abhiyan,_July_2016(8).jpg. Licensed under: PDMark-owner)

   When was the Swachh Bharat Abhiyan campaign officially launched?
   (a) October 2, 2012 (b) October 2, 2013
   (c) October 2, 2014 (d) October 2, 2015

9. What are the main goals of the Swachh Bharat Abhiyan campaign?
   (a) To promote tourism in India
   (b) To improve public transportation
   (c) To clean up streets and roads only
   (d) To end open defecation, create accountable systems for observing latrine use and develop effective strategies for managing waste

NCERT Corner

1. What are the different ways in which water gets contaminated?
2. At an individual level, how can you help reduce air pollution?
3. Clear, transparent water is always fit for drinking. Comment.
4. You are a member of the municipal body of your town. Make a list of measures that would help your town to ensure the supply of clean water to all its residents.
5. Explain the differences between pure air and polluted air.

COMPETITION WINDOW (OBJECTIVE TYPE)

1. Name the first scientist to coin the term smog and to describe the layers of smog?
   (a) Nikola Tesla
   (b) Stephen Hawking
   (c) Henry Antoine
   (d) Nicolaus Copernicus
2. Pollutants responsible for the cause of SMOG:
   (a) From incinerators
   (b) Emissions from vehicles
   (c) Both incinerators and emissions from vehicles
   (d) None of the above
3. Name the secondary air pollutant:
   (a) PANs (b) Ozone
   (c) Carbon monoxide (d) Nitrogen Dioxide
4. Name the particles is called the particulate pollutants?
   (a) Ozone (b) Radon
   (c) Fly Ash (d) Ethylene
5. Which of the following agents is responsible for turning the Taj Mahal yellow?
   (a) Sulphur (b) Chlorine
   (c) Sulphur dioxide (d) Nitrogen dioxide
6. Which river is the world's most polluted?
   (a) Ganga River (b) Chenab River
   (c) Cauvery River (d) Yamuna River
7. The primary causes of water pollution are:
   (a) Plants (b) Animals
   (c) Human activities (d) None of these
8. Techniques used for reducing the total dissolved solids (TDS) in the water are:
   (a) Osmosis (b) Ion exchange
   (c) Distillation (d) Both (b) and (c)
9. The correct statement about CNG is:
   (a) It is a clean fuel
   (b) It is a harmful fuel
   (c) It is a polluting fuel
   (d) All of the above
10. The main cause of permanent hardness of water?
    (a) Magnesium sulphate
    (b) Magnesium bicarbonate
    (c) Magnesium carbonate
    (d) None of the above
11. Write an example of pollutant?
    (a) A rat entering your kitchen
    (b) CO_2 given out by the respiring organism
    (c) The heat given out while cooking food
    (d) The dust raised during road-cleaning
12. A radiation pollutant is:
    (a) Sulphur dioxide
    (b) Ozone
    (c) Iodine-131
    (d) Discarded fused electric bulbs
13. The BS emission standards that went into effect for the entire nation on April 1, 2017 were:
    (a) BS III (b) BS IV
    (c) BS V (d) BS VI

Answer Keys

Multiple Choice Questions

1. (b) 2. (b) 3. (c) 4. (d) 5. (a)

Match the Following

1. A-2, B-1, C-3, D-4

True or False

1. T 2. F 3. T 4. T 5. T

Fill in the Blanks

1. Particulate pollutants
2. X-rays
3. Thermal
4. Waste
5. Carbon monoxide

Assertion–Reason Questions

1. (a) 2. (a) 3. (a) 4. (a) 5. (b)

One-word Answers

1. Sulphur dioxide and nitrogen oxide
2. CFCs and halons
3. Methylmercury
4. Air pollution
5. Acid rain

Define the Term

1. Pollution is the introduction of harmful substances or products into the environment, which can cause adverse effects on living organisms and their surroundings.
2. Greenhouse gas is a gas that traps heat in the Earth's atmosphere, contributing to the warming of the planet. Examples include carbon dioxide, methane and water vapour.
3. Acid rain is rain that is acidic due to the presence of pollutants such as sulfur dioxide and nitrogen oxides, which can harm ecosystems and human-made structures.
4. Ozone layer is a layer in the Earth's atmosphere that contains high levels of ozone, which helps to protect the planet from harmful ultraviolet radiation from the sun.
5. Air pollution is the presence of harmful substances in the air, which can cause adverse health effects on humans and other living organisms.

Groups of Terms

1. Eutrophication
2. Disruption of natural cycles
3. Reduced crop yields
4. Heavy metals
5. Hazardous waste

Identify the Odd One Out

1. Helium
2. Oxygen
3. Paper towels
4. Pesticides
5. Radioactive waste

Very Short Answer Questions

1. Human activities, such as the burning of fossil fuels and industrial processes.
2. Discharge of pollutants into water bodies from human activities such as industrial processes, agriculture and improper waste disposal.
3. Improper disposal of plastic waste, such as littering or dumping plastic waste into bodies of water.
4. Contamination from hazardous waste, industrial activities and agricultural practices.
5. The release of greenhouse gases into the atmosphere, primarily from human activities such as burning fossil fuels and deforestation.

Short Answer Questions

1.
   - Organic matter
   - Nutrients
   - Pathogen
2. Accidents during oil transportation
   - Offshore drilling
   - Natural disaster
3. Soundproofing: One effective strategy for reducing noise pollution is to soundproof buildings, especially those located near sources of noise such as highways or airports. This can involve adding insulation, double-paned windows, and other materials to help block out noise.

   Noise barriers: Another strategy for reducing noise pollution is to install noise barriers such as walls or fences between sources of noise and nearby residential areas. These barriers can help to reduce the amount of noise that reaches nearby communities and can be particularly effective in reducing noise from highways or railways.
4. Limiting noise levels: One effective measure to minimise noise pollution is to limit the levels of noise produced by various sources. This can be done by setting noise limits for industrial and transportation activities, as well as for construction and demolition

projects. In addition, restrictions on the use of loud equipment or activities during certain times of day can also help to minimize noise pollution.

Promoting public awareness: Another measure to minimise noise pollution is to promote public awareness about the impacts of noise pollution and encourage people to take steps to reduce noise in their daily lives. This can involve education campaigns, community outreach, and public policies that promote quieter living environments. By increasing awareness of the problem of noise pollution and providing people with the tools and resources to reduce noise levels, we can work towards minimizing the harmful impacts of noise on human health and the environment.

5. Pollution is the introduction of harmful substances or products into the environment, which can cause harm to living organisms and ecosystems.

Long Answer Questions

1. Hint: Human health: Respiratory problems, cardiovascular issues, increased risk of lung cancer, allergies. Environment: Acid rain, damage to vegetation, depletion of ozone layer, climate change.
   Elaborate on these points.
2. HInt: Point source pollution: Industrial discharge, sewage treatment plants. Non-point source pollution: Agricultural runoff, urban runoff, mining activities.
   Elaborate on these points.
3. Hint: Marine life: Entanglement, ingestion, disruption of ecosystems.
   Environment: Microplastic pollution, contamination of water bodies, negative impact on wildlife.
   Elaborate on these points.
4. Hint: Causes: Industrial activities, agricultural practices, improper waste disposal.
   Effects: Soil degradation, reduced fertility, contamination of crops, groundwater pollution.
   Elaborate on these points.
5. Hint: Human health: Hearing loss, stress, sleep disturbances, impaired cognitive function. Environment: Disruption of animal behavior, ecological imbalance, interference with communication.
   Elaborate on these points.

Case Study-Based Questions

1. (b)	**2.** (c)	**3.** (c)	**4.** (c)	**5.** (c)
6. (d)	**7.** (c)	**8.** (c)	**9.** (d)	

NCERT Corner

1. The different ways in which the water gets contaminated are listed below:
   (i) **Industrial waste:** The harmful chemical wastes get released by the industries into water sources.
   (ii) **Sewage waste:** Kitchen, laundry and toilets release waste materials, and this pollutes water.
   (iii) **Agricultural chemicals:** Excessive amounts of pesticides and fertilizers are used by farmers to enhance crop production. Rains and floods carry these harmful chemicals to water bodies, which results in water pollution.
2. We can take the following steps to reduce air pollution:
   (i) We should avoid using personal cars and opt for public transport as much as possible.
   (ii) We should properly dispose of the garbage and not burn it.
   (iii) For shorter distances, avoid vehicles.
   (iv) We should control the number of emissions from household chimneys and vehicles.
   (v) We should replace our vehicle fuel, diesel and petrol with LPG and CNG.
3. No, we do not agree with the statement that clear water is always fit for drinking because water might appear clean, but it may contain some disease-causing micro-organisms and several other dissolved impurities. Hence, we advise purifying the water by boiling it or by any purifying system before drinking it.
4. The following are the measures taken by us to ensure the supply of clean water to our town residents:
   (a) We clean the area around the water pipes.
   (b) The main water source has to be built in a clean surroundings and should be maintained properly.
   (c) Chemical methods such as chlorination must be used for purifying water.
5. The composition of pure air contains 78% nitrogen, 21% oxygen and 0.03% carbon dioxide. Besides these, there are some other gases like methane, argon, ozone and water vapour. The air is said to be polluted when the composition of air is altered by the addition of harmful substances or gases such as sulphur dioxide, carbon monoxide, nitrogen dioxide and particulate matter.

Competition Window

1. (c)	**2.** (c)	**3.** (b)	**4.** (c)	**5.** (c)
6. (a)	**7.** (c)	**8.** (d)	**9.** (a)	**10.** (a)
11. (d)	**12.** (c)	**13.** (b)		

CHAPTER 20 Management of Natural Resources

PREREQUISITES

You must have knowledge of:

- Natural resources like forests, water, minerals and wildlife, and their significance in our daily lives
- Ecosystem and its components
- Factors that contribute to the degradation of natural resources, such as deforestation, pollution, overexploitation and climate change
- United Nations' Sustainable Development Goals (SDGs) and their role in promoting sustainable management of natural resources

LEARNING OBJECTIVES

At the end of this chapter, you will be able to:

- Describe natural resources
- Explain resource management
- Define sustainable development
- Explain why it is necessary for us to manage our resources

INTRODUCTION

We are not strangers to living in harmony with nature. Sustainable living has always been an important aspect of India's culture and history. It has been woven into our long-standing traditions and customs, as well as our art and crafts, festivals, food, beliefs, rituals and folklore. The notion that 'the entire natural world should be in harmony' is considered to be represented in the classic Sanskrit phrase '*vasudhaiva kutumbakam*', which means 'the world is one family'. The phrase appears in the Maha Upanishad, which is thought to be a part of the Atharva Veda, an ancient Indian text.

In earlier classes, we learned about natural resources such as the soil, air and water, as well as how various components in nature are recycled over and over again. We have also learned about the pollution of these resources as a result of some of our activities. In this chapter, we will look at some of our resources and how we use them. Perhaps we should also consider how we should preserve resources and the environment. We will examine natural resources, such as forests, wildlife, water, coal and petroleum, to identify what concerns are at stake in determining how these resources should be managed for long-term development.

Environmental issues are frequently discussed or read about. These are frequently global issues, and we feel helpless to make any changes. There are international environmental rules and regulations, as well as national environmental laws and acts. National and international organisations are also working to protect our ecosystem. Our society's awareness of the challenges produced by unthinkingly exploiting our resources is a relatively new

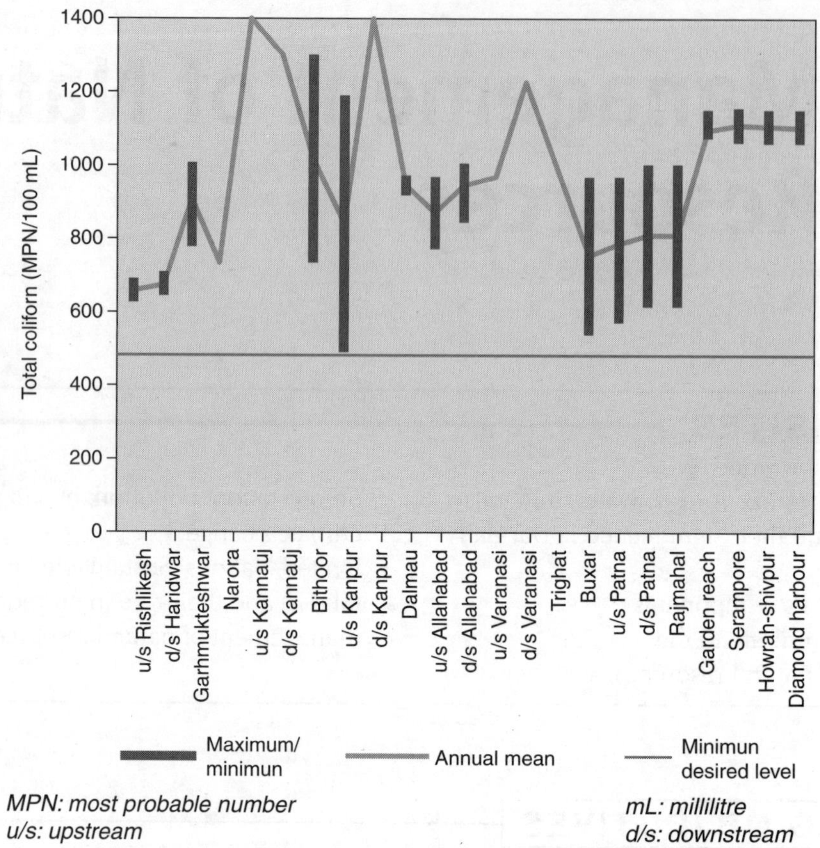

Fig. 20.1 Total coliform count levels in the Ganga, 2012

phenomena. When people become more aware of their surroundings, they are more likely to take action.

You have probably heard of the Ganga Action Plan (GAP). This multi-crore project was started as a result of the poor quality of the Ganga's water. More than 100 cities in Uttar Pradesh, Bihar and West Bengal use the Ganga as a sewage disposal system. The toxicity of the water has been increased by the discharge of untreated sewage, excreta and chemicals from industry. This has made it unusable for the river's vegetation and fauna.

The GAP project was launched in 1985. Coliform bacteria are a type of bacterium prevalent in human intestines, and their presence in water indicates that disease-causing germs have contaminated it (Fig. 20.1).

Some quantitative factors are utilised to quantify pollution or the quality of the water we use for various purposes. Some contaminants are hazardous even in trace amounts, necessitating the use of specialised equipment to detect them. However, as we learned earlier, the pH of water may be simply determined using a universal indicator.

REDUCE, REUSE AND RECYCLE

Every member of our society can participate in the three Rs to conserve the environment:

- **Reduce:** We must reduce our consumption and wasteful practices. For example, not wasting food, turning off lights to save power, replacing broken faucets and minimising the amount of water used for bathing.
- **Reuse:** This refers to the practice of repurposing something rather than discarding them. Reusing plastic utensils and bottles, for example. Many objects cannot be recycled or require a lot of energy to do so; yet, we can put them to better use.
- **Recycle:** This is the process of collecting old plastic, paper, glass or metal objects in order to create products from them rather than producing them from scratch. It is necessary to have a system in place to separate and dispose of each sort of trash.

We can make environmentally beneficial judgments even when making ordinary decisions. To do so, we need to learn more about how our decisions affect the environment, which can be immediate, long-term or far-reaching. The concept of sustainable development promotes growth that meets current basic human requirements while also conserving resources for future generations. Environmental protection and economic development are intrinsically linked. As a result, sustainable growth necessitates a shift in all aspects of life. It is contingent on people's willingness to modify their perceptions of the socioeconomic and environmental realities around them, as well as each individual's willingness to change their current usage of natural resources.

NEED FOR MANAGING NATURAL RESOURCES

- Natural resources are being depleted at an alarming rate as a result of increased population and changing lifestyle demands. Management of resources must be an inherent component of our society in order to maintain a sustainable, equitable distribution of resources and the minimisation of environmental damage. We must guarantee that we make wise use of our natural resources, which are finite and require long-term planning in order to last generations.
- As the world's population grows, so does the demand for more resources, which are depleting at an exponential rate. Changing lifestyles and technological improvements are forcing corporations to abuse our natural resources to meet demand

Wildlife and Forests

Forests are biodiversity hotspots (Fig. 20.2). The variety and range of plant and animal life in a given ecosystem is referred to as biodiversity. Ecological equilibrium may be lost and the ecosystem may be harmed as a result of biodiversity loss.

When it comes to forest conservation, the following stakeholders must be considered:

- People who live in close proximity to forests and rely on forest products.
- The land and resources owned by the government's Forest Department.
- Industrialists or people who use the forest to make items; for example, tendu leaves for bidis and paper mills.
- Wildlife enthusiasts and conservationists who aim to preserve nature

Monoculture is when a single crop is grown in a specific area. Excessive monoculture depletes the area's biodiversity. Various necessities of individuals living in forest areas are ignored, such as fodder leaves, herbs and edible fruits.

Forests are seen as a source of raw resources by industrialists. Industries have greater political influence than ordinary citizens, and they are mainly concerned with meeting their needs. They are unconcerned about the environment's sustainability and will shift from one habitat to the next in pursuit of raw materials.

In the management of forest resources and landscapes, human intervention is required:

- To assure progress while maintaining the environment.
- To ensure that economic progress and conservation occur at the same time.
- To ensure that benefits are distributed to the local population; for example, the Bishnoi community who helps to save Khejri trees in Jodhpur, Rajasthan.

Excessive and illegal logging will degrade forest resources faster than they can be replenished. This will also destroy the ecological balance and threaten the habitats of many flora and animal species.

Fig. 20.2 A view of a forest life
(Pine forest near Klavdievo, Borodianka Raion, Kiev Oblast, Ukraine; Аимаина хикари; 2013; https://commons.wikimedia.org/wiki/File:Klavdievo_forest.JPG. Licensed under: CC0)

Sustainable development necessitates the satisfaction of all forest resource stakeholders. In practice, companies use forest materials at considerably lower prices than the market, resulting in tensions between local residents and industrialists. In the 1970s, the Chipko Andolan ('Hug the Trees Movement') was a conflict between industrialists and rural residents (Fig. 20.3). Reni Garhwal, in the Himalayas, is where it all began. The logging contractor had to withdraw due to a disagreement with the local villagers. Women of the village stopped the felling of trees by hugging them. The movement swiftly gathered traction and media attention, forcing the government to reconsider its approach to forest resource management. Today, it is understood that local people need to be involved in the management of forest resources as well.

Fig. 20.3 Chipko Movement
(Women and men standing around trees, hugging them…; Arnab Chaudhary; 2020; https://commons.wikimedia.org/wiki/File:Big_chipko_movement_1522047126.jpg. Licensed under: CC-BY-SA-4.0)

Water for Everyone

All terrestrial forms of life require water to survive. In India, regions where there is a lack of water are also places where there is a lot of poverty. Despite the monsoon, there is failure to retain groundwater due to loss of vegetation and the discharge of industrial effluents. Due to the loss of the water table and interruption of the water cycle, there is a reduction in fresh drinkable water.

The government's intervention to pursue megaprojects must not ignore local irrigation systems. In Himachal Pradesh, a canal irrigation system known as kulhs channelled flowing stream water to man-made channels that carried it down the slope to communities. The water was first utilised by the community furthest from the kulh's source. This made it easier for water to percolate through the soil. After the government irrigation system took control, it was rendered obsolete.

Strict restrictions on the use of stored water and the construction of dams, tanks and canals must be put in place. Based on water availability, the best cropping patterns must be adopted.

Dams have the ability to store massive amounts of water while also generating electricity. Dam mismanagement, however, leads to exploitation, and the resource is not distributed fairly.

The disadvantages of dams are:

- Displacement of tribals without compensation
- Corruption and consumption of money without generation of benefits
- Environmental problems like deforestation.

In order to preserve and enhance biomass, **water harvesting** encourages soil and water conservation. It increases local revenue while also reducing the effects of droughts and floods. Khadins (Fig. 20.4), tanks and nadis in Rajasthan, bandharas and tals in Maharashtra, bundhis in Madhya Pradesh and Uttar Pradesh, ahars and pynes in Bihar, kulhs in Himachal Pradesh, ponds in the Kandi belt of Jammu region, eris (tanks) in Tamil Nadu, surangams in Kerala, and kattas in Karnataka are just a few examples.

Crescent-shaped earthen embankments/concrete check dams installed in flood-prone locations are known as water harvesting structures. The primary goal is to replenish groundwater.

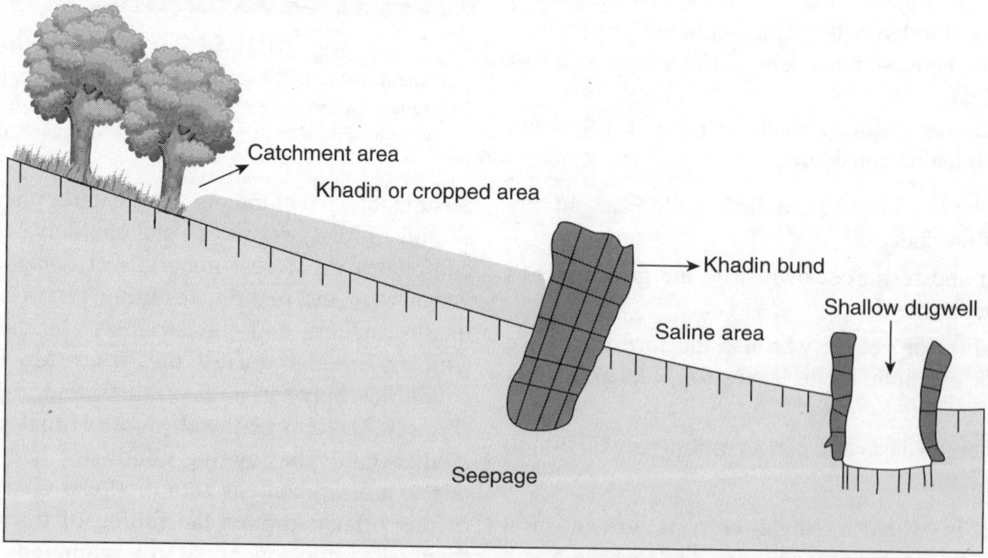

Fig. 20.4 Traditional water harvesting system, an ideal setting of the khadin

The advantages of groundwater are listed below:

- Does not evaporate
- Recharges wells and groundwater when it flows
- Does not allow mosquitos to grow
- Does not come in contact with human contaminants

Coal and Petroleum

Coal and petroleum are derived from non-renewable fossil fuels. They will eventually be depleted. As a result, proper management of fossil fuel consumption is critical. Fossil fuels are made up of a tremendous amount of carbon and are generated through millions of years of decomposing vegetation. When they are burned in an environment with a limited supply of oxygen, they produce toxic gases that damage the atmosphere and contribute to global warming.

The wise use of fossil fuels improves the efficiency of our equipment and protects the long-term viability of our resources.

SUMMARY

- Natural resources should be utilised in a sustainable manner
- By honestly implementing the motto "Reduce, Reuse and Recycle" in our daily lives, we may lessen the strain on the environment.
- The interests of numerous stakeholders must be taken into account when managing forest resources.
- The supply of coal and petroleum, which are fossil fuels, will eventually run out. We must use these resources wisely as their burning pollutes our environment.

PRACTICE QUESTIONS

Analyse Your Concepts (School Exam Based)

Multiple Choice Questions

1. Which of the following canals brought greenery to Rajasthan?
   (a) Indira Gandhi Canal
   (b) Rajiv Gandhi Canal
   (c) Jawaharlal Canal
   (d) Mahatma Gandhi Canal
2. Which of the following is the age-old concept of water harvesting system in Madhya Pradesh?
   (a) Nadis (b) Ponds
   (c) Bandharas (d) Bundhis
3. Surangams are the age-old concept of water harvesting in:
   (a) Karnataka (b) Kerala
   (c) Tamil Nadu (d) Andhra Pradesh
4. Which of the following is not a use of forests?
   (a) Controls floods.
   (b) Used to make paper.
   (c) Causes soil erosion.
   (d) Resin, gum and drugs are obtained.
5. From the list given below select the item that is not a natural resource:
   (a) Soil (b) Water
   (c) Electricity (d) Air

Match the Following

Each question contains statements given in two columns which have to be matched. Statements in column I have to be matched with statements in column II.

1. Match Column I with Column II.

Column I	Column II
(i) Reduce	(A) Reprocess to make desired things
(ii) Recycle	(B) Use again and again.
(iii) Reuse	(C) Sustainable development.
(iv) Safe disposal of waste	(D) Less use of things.

2. Match Column I with Column II.

Column I	Column II
(i) Himachal Pradesh	(A) Surangams
(ii) Rajasthan	(B) Bandharas
(iii) Bihar	(C) Kulhs
(iv) Karnataka	(D) Khadins
	(E) Kattas
	(F) Ahars

True or False

Read the following statements and write your answer as true or false.

1. Control of local people over their own water resources may cause mismanagement and over-exploitation of these resources.
2. River water is said to be polluted with acidic wastes when the pH of the water is above 7.
3. The main cause for abundant coliform bacteria in the river Ganga is disposal of human excreta directly.
4. The GAP project was launched in 1985.
5. Coal and petroleum are derived from renewable fossil fuels.

Fill in the Blanks

Complete the following statements with an appropriate word/term to be filled in the blank spaces.

1. __________ are 'biodiversity hotspots'.
2. The management of __________ resources requires a long-term perspective.
3. The Government of India has recently instituted an 'Amrita Devi Bishnoi National Award' for __________ in the memory of Amrita Devi Bishnoi.
4. 'Khadin' system of rain-water harvesting is practiced in __________.
5. The Chipko Andolan originated from an incident in a remote village called '__________' in Garhwal.

Assertion–Reason Questions

The following questions consists of two statements: Assertion (A) and Reason (R). Answer these questions selecting the appropriate option given below:

(a) Both A and R are true and R is the correct explanation of A

(b) Both A and R are true but R is not the correct explanation of A

(c) A is true but R is false

(d) A is false but R is true

1. Assertion (A): The presence of forests maintains the region's temperature balance.
   Reason (R): By absorbing atmospheric carbon dioxide, forests help to minimise air pollution.
2. Assertion (A): Use of natural resources must be judicious.
   Reason (R): Resources are limited, and the human population is growing rapidly.
3. Assertion (A): In situ conservation is the preservation of biological variety in its native setting.
   Reason (R): One of these examples is the rise in the number of Manipur deer from 17 to 150 in the Calcutta and Delhi zones.
4. Assertion (A): Coal is a naturally occurring fuel that burns.
   Reason (R): It is found in volcanoes.
5. Assertion (A): Poisonous gases are produced when coal and petroleum are burned.
   Reason (R): In high quantities, sulphur, nitrogen and carbon monoxide oxides are toxic.

One-word Answers

1. Who prepared the Ganga Action Plan?
2. What are the three Rs that will help us to conserve natural resources for long-term use?
3. In which year was the Wildlife Protection Act implemented in India?
4. What is the 'Hug the Trees Movement'?
5. What is the term when a single crop is grown in a specific area?

Define the Term

1. Reduce
2. Reuse
3. Recycle
4. Clear-cutting
5. Erosion
6. Greenhouse gas

Groups of Terms

Given below are certain groups of terms. In each group the first pair indicates a relationship between the two terms. Rewrite and complete the second pair on a similar basis.

1. Forests: Timber :: Oceans: Fisheies: ________
2. Solar energy: Sunlight :: Wind energy: ________
3. Recycling: Waste reduction :: Rainwater harvesting: ________
4. Fossil fuels: Non-renewable :: Hydroelectric power: ________
5. Soil conservation: Preventing soil erosion :: Air pollution control: ________

Identify the Odd One Out

1. Renewable energy, fossil fuels, conservation, non-renewable resources.
2. Rainwater harvesting, desalination, water pollution control, water wastage reduction.
3. Clear-cutting, selective logging, reforestation, agroforestry.
4. Strip mining, reclamation, acid mine drainage, underground mining.
5. Organic farming, monoculture, crop rotation, integrated pest management.

Image-Based Questions

1.

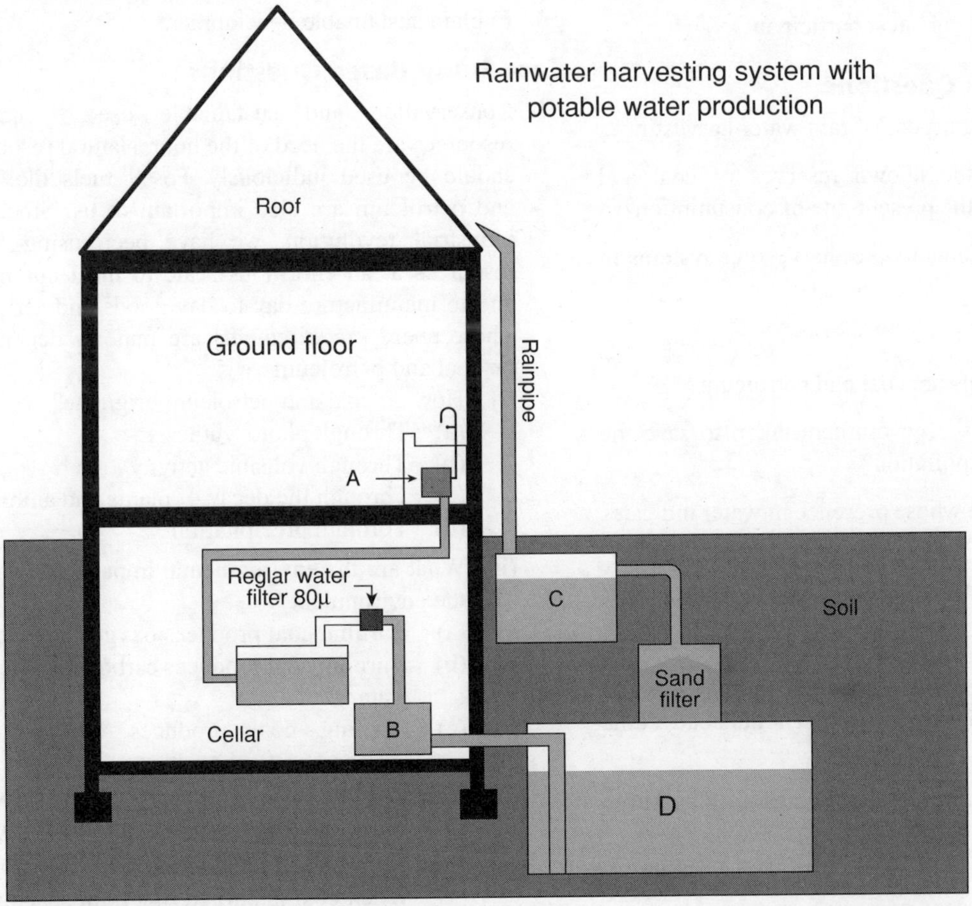

(i) Study an image of a rooftop rainwater harvesting system and label the different components of the system.

(ii) How does this system work to capture and store rainwater for later use, and what are its advantages over traditional water sources?

(iii) How can this system be scaled up for use in larger buildings or communities?

2. 

(i) What is the Chipko movement often called?

(ii) When did the Chipko movement start?

(iii) How many Bishnois sacrificed their lives to save Khejri trees?

(iv) What year was Sarvodaya Movement?

(v) Who were the greatest participant?

Very Short Answer Questions

1. What is the main purpose of rain water harvesting?
2. How long will the known reserves of coal and petroleum last at the present rate of consumption?
3. Name some traditional water harvesting systems in India.
4. State the main purpose of rainwater harvesting.
5. What types of fuels are coal and petroleum?
6. What action has the government taken to save the river Ganga from pollution?
7. Name the bacteria whose presence in water indicates contamination of water.
8. What does coliform presence in Ganga water mean?
9. Name two industries based on forest produce.
10. Name one river which becomes polluted while passing through Haridwar, Kanpur, Allahabad, Patna and Howrah?
11. Name the most common practice used to recharge groundwater.
12. What are the objectives of Namami Gange Programme?
13. When was National Mission for Clean Ganga set up?
14. Who won the Stockholm Water Prize in 2015?

Short Answer Questions

1. Define the term 'conservation of environment'.
2. What are the two kinds of natural resources?
3. A person lives near a forest. Make a list of four items which he can get from the forest to meet his daily needs.
4. Name any two forest products, each of which is the basis for some industry.
5. State an instance where human intervention saved the forests from destruction.

Long Answer Questions

1. Write a brief discussion on water harvesting.
2. Describe two kinds of natural resources.
3. What are the advantages of ground water?
4. Write a short note on wildlife and forest.
5. Explain sustainable development.

Case Study-Based Questions

1. Conservation and sustainable use of natural resources are the need of the hour. Natural resources should be used judiciously. Fossil fuels like coal and petroleum are also important to us. Since the industrial revolution, we have been using these resources at an enormous scale to meet our needs and to manufacture day to day goods and services. These energy requirements are majorly dependent on coal and petroleum.

(i) How do coal and petroleum originate?
   (a) Through photosynthesis
   (b) Through volcanic activity
   (c) Through the decay of plants and animals
   (d) Through precipitation

(ii) What are the environmental impacts of coal on the environment?
   (a) Burning coal produces oxygen and water
   (b) Burning coal produces carbon dioxide and water
   (c) Burning coal produces nitrogen and oxygen
   (d) Burning coal produces carbon dioxide, water, and oxides of sulphur and nitrogen

(iii) How is Carbon Monoxide formed?
   (a) When coal is burned in a tight space with insufficient airflow or oxygen
   (b) When coal is exposed to sunlight
   (c) When coal is mixed with water
   (d) When coal is heated at high pressure

(iv) Which is more lethal CO or CO_2? Give reason.
   (a) CO_2, as it is toxic to humans
   (b) CO, as it can result in serious respiratory issues and even death

(c) Both are equally lethal
(d) None of the above

(v) Why should we use these resources with care?
(a) Because they are unlimited in amount
(b) Because they take millions of years to form and are limited in amount
(c) Because they are harmful to the environment
(d) Because they are expensive

2. In order to increase the production of biomass, watershed management places a strong emphasis on scientific soil and water conservation. The goal is to create secondary resources, such as plants and animals, and to develop primary resources, such as land and water, in a way that prevents ecological imbalance.

(i) What is one importance of watershed management?
(a) It increases the pollution levels in water bodies
(b) It decreases the lifespan of dams and reservoirs
(c) It has no effect on drought and flooding
(d) It lengthens the lifespan of dams and reservoirs and lessens the impact of drought and flooding

(ii) What is rainwater harvesting?
(a) A method of wasting rainwater
(b) A method of gathering rainwater and storing it in ponds or pits for later use
(c) A method of polluting rainwater
(d) A method of using up all the rainwater during the rainy season

(iii) Which of the following is a traditional method used for rainwater harvesting?
(a) Using sprinklers to irrigate fields
(b) Building large dams and reservoirs
(c) Constructing underground tunnels to collect rainwater
(d) Rajasthan's khadins tanks and nadis, Bihar's ahar and pynes, Maharashtra's bandharas and tals, Himachal Pradesh's kulhs, and the Jammu region's Kandi belt's ponds.

(iv) What are the uses of groundwater?
(a) It is used for cooking food
(b) It is used for generating electricity
(c) It doesn't evaporate, gives plants moisture, and keeps out animal and human waste from contaminating the soil
(d) It is used for transportation

(v) Which of the following is true regarding rainwater harvesting?
(a) It is a method of wasting rainwater
(b) It is a method of using up all the rainwater during the rainy season
(c) It is a method of gathering rainwater and storing it for later use
(d) It is a method of polluting rainwater

3. 'Biodiversity hot spots' include forests. The number of species present in a region is one indicator of its biodiversity. The diversity of life forms, such as bacteria, fungus, ferns, flowering plants, nematodes, insects, birds, and reptiles, is also significant. Conserving the biodiversity we have acquired is one of the fundamental goals of conservation.

(i) Who are the stakeholders of forests?
(a) Only government agencies responsible for managing them
(b) Only business owners who utilize forest resources
(c) Only animal and outdoor enthusiasts who visit forests
(d) Those who depend on and live in forests, government agencies responsible for managing them, business owners, and lovers of animals and the outdoors

(ii) What do we get from forests?
(a) Water and minerals
(b) Food, wood, medicines and other things
(c) Fuel for transportation
(d) Electronics and technology components

(iii) What are the repercussions of the disappearance of forests on our environment?
(a) Increased biodiversity
(b) Enhanced ecological balance
(c) Reduced risk of natural disasters
(d) Erosion of soil, ecological imbalance, numerous disasters, threatened biodiversity, potential extinction of species, etc.

(iv) What are the strategies for protecting forests?
(a) Increase deforestation
(b) Overuse forest resources
(c) Lessen deforestation, reforest, and use the resource wisely
(d) None of the above

(v) What are the types of forests?
(a) Rainforests only
(b) Coniferous and deciduous only
(c) Mangrove and thorny only
(d) Rainforests that are evergreen or deciduous, coniferous, mangrove, thorny, Mediterranean and taiga, among others.

NCERT Corner

1. Which fossil fuel is conserved when we use bicycles for covering short distances instead of vehicles?
2. Name a clean fuel other than LPG and natural gas.
3. Name two fossil fuels.
4. Find out about the traditional systems of water harvesting/management in your region.
5. List five things you have done over the last week to (a) conserve our natural resources. (b) increase the pressure on our natural resources.

COMPETITION WINDOW (OBJECTIVE TYPE)

1. Which one of the following is a renewable resource?
   (a) Wildlife (b) Petroleum
   (c) Coal (d) Natural gas
2. Which one of the following is a non-renewable resource?
   (a) Water (b) Coal and minerals
   (c) Wind (d) Vegetation
3. Which of the following can be used for sustainable development?
   (a) Industries (b) Forests
   (c) Crops (d) Resources
4. Which of the following is a 'biodiversity hot spot'?
   (a) Forests (b) Rivers
   (c) Deserts (d) Oceans
5. When we destroy a forest, we are also destroying:
   (a) Trees
   (b) Population of wildlife
   (c) Environment
   (d) Food and shelter of wild animals
6. Which one of the following communities in Rajasthan has a religious tenet of conservation of forest and wildlife?
   (a) Agarwal (b) Jaishwal
   (c) Bishnoi (d) Jain
7. Among the statements given below, select the ones that correctly describe the concept of sustainable development:
   (i) Planned growth with minimum damage to the environment
   (ii) Growth irrespective of the extent of damage caused to the environment
   (iii) Stopping all developmental work to conserve the environment
   (iv) Growth that is acceptable to all the stakeholders
   (a) (i) and (iv)
   (b) (ii) and (iii)
   (c) (ii) and (iv)
   (d) (iii) only
8. In our country, vast tracts of forests are cleared and a single species of plant is cultivated. This practice promotes:
   (a) Biodiversity in the area
   (b) Monoculture in the area
   (c) Growth of natural forest
   (d) The natural ecosystem in the area
9. Sardar Sarovar Dam is constructed on which river?
   (a) Ganga (b) Sutluj
   (c) Narmada (d) Kaveri
10. What is the purpose of rainwater harvesting?
    (a) To hold rain water on the surface of the earth.
    (b) To recharge ground water.
    (c) To use water for the irrigation of crops.
    (d) To rear fish
11. What are the three Rs to save the environment?
    (a) Reduce, Recycle, Reuse
    (b) Replenish, Reduce, Reuse
    (c) Reconstruct, Recycle, Reduce
    (d) Reduce, Recycle, Remove
12. Expand the abbreviation GAP:
    (a) Governmental Agency for Pollution Control
    (b) Gross Assimilation by Photosynthesis
    (c) Ganga Action Plan
    (d) Governmental Agency for Animal Protection
13. Groundwater will not be depleted due to:
    (a) Afforestation
    (b) Thermal power plants
    (c) Loss of forest and decreased rainfall
    (d) Cropping of high-water-demand crops

14. Pick the correct combination of groups which are not fossil fuel:
(a) Wind, ocean and coal
(b) Kerosene, wind and tide
(c) Wind, wood, sun
(d) Petroleum, wood, sun

15. Environment Day falls on:
(a) 28th February (b) 23rd March
(c) 5th June (d) 16th August

16. The pH range most conducive for life of freshwater plants and animals is:
(a) 6.5–7.5 (b) 2.0–3.5
(c) 3.5–7.0 (d) 9.0–10.5

17. Which environmental problem is associated with the construction of high-rise dams?
(a) A large number of human settlements are submerged in the water.
(b) It contributes to deforestation and loss of biodiversity.
(c) It involves the spending of huge amounts of money.
(d) All the above.

Answer Keys

Multiple Choice Questions

1. (a) 2. (d) 3. (b) 4. (c) 5. (c)

Match the Following

1. (i) (D) (ii) (A) (iii) (B) (iv) (C)
2. (i) (C) (ii) (D) (iii) (F) (iv) (E)

True or False

1. F 2. F 3. T 4. T 5. F

Fill in the Blanks

1. Forests
2. Natural
3. Wildlife conservation
4. Rajasthan
5. Reni

Assertion-Reason Questions

1. (a) 2. (a) 3. (c) 4. (c) 5. (a)

One-word Answers

1. Department of Environment
2. Reduce, recycle, reuse
3. 1972

Define the Term

1. Reduce refers to the rational purchase of products. The easiest way to minimize waste is not to generate waste.
2. Reuse refers to the use of goods that are both economical and environmentally sustainable.
3. Recycle refers to the re-usage of goods for creative and by-products. Here are a few ways to recycle:
   - Recycling used paper to produce paper bags and straws.
   - Purchasing various items that are recycled, and those which can be recycled.
4. Clear-cutting or clear-cut harvesting is a forestry method in which most or all trees in the region are cut uniformly for economic growth.
5. Erosion is the activity of surface processes (like water or wind) which remove soil, rock or dissolved materials from one location on the earth's surface and then transport it to another location.
6. Gas whose concentration increases heat retention in the atmosphere is called greenhouse gas.

Groups of Terms

1. Fisheries
2. Wind power
3. Water conservation
4. Renewable
5. Reducing atmospheric pollution

Identify the Odd One Out

1. Conservation
2. Desalination
3. Clear-cutting
4. Reclamation
5. Monoculture

Image-Based Questions

1. (i) A: Water filter; B: Pump; C: rainwater catchment; D: Main tank

   (ii) The rooftop rainwater harvesting system works by collecting rainwater from the roof surface through gutters and downpipes, which then leads it into the storage tank. The first-flush device helps remove pollutants and debris from the initial rainwater flow. The stored rainwater can be used for various purposes like gardening, toilet flushing, and washing.

   Advantages:

   Sustainable water supply: It reduces dependence on traditional water sources by utilizing rainwater, thereby conserving groundwater and surface water.

   Cost-effective: Rainwater is free, and using it for non-potable purposes reduces the need for treated water, resulting in cost savings.

   Environmentally friendly: It helps in reducing stormwater runoff, which can lead to soil erosion and water pollution.

   Self-sufficiency: The system allows individuals or communities to become self-sufficient in meeting their water needs, especially during water scarcity or drought periods.

   (iii) To implement rooftop rainwater harvesting on a larger scale, several steps can be taken:

   Increasing storage capacity: Install larger storage tanks or connect multiple tanks to cater to the water demand of larger buildings or communities.

Proper design and installation: Engage professionals to design and install the system to ensure efficiency and effectiveness.

Community-level implementation: Encourage and educate communities to adopt rooftop rainwater harvesting collectively, considering shared resources and infrastructure.

Government support: Provide incentives, subsidies, and regulations to promote the adoption of rainwater harvesting systems in larger buildings or communities.

Integration with existing water infrastructure: Explore the integration of rainwater harvesting systems with municipal water supply networks to enhance overall water management and distribution.

2. (i) The Chipko movement is often called the "Hug the Trees" movement.
   (ii) The Chipko movement started in the early 1970s.
   (iii) It is reported that 363 Bishnois sacrificed their lives to save Khejri trees.
   (iv) The Sarvodaya Movement was initiated in the year 1951.
   (v) The greatest participants of the Chipko movement were the local villagers, especially women from the rural communities.

Very Short Answer Questions

1. To make rain water percolate under the soil so as to recharge the groundwater.
2. The coal reserve will last for about 200 years and petroleum will last for about 40 years.
3. Traditional water harvesting systems are Khadims, tanks, bandharas, tals, bundhis, pynes etc.
4. It is done to recharge the ground water by percolating water under the soil.
5. Fossil fuels.
6. The government launched the Ganga Action Plan in 1985.
7. Coliform
8. It means contamination by disease-causing microorganisms.
9. Timber industry, paper manufacturing industry
10. Ganga
11. Rainwater harvesting
12. Namami Gange Programme was launched to accomplish twin objectives of effective abatement of pollution and rejuvenation of river Ganga.
13. National Mission for Clean Ganga was set up in October, 2016.
14. Dr. Rajendra Singh won the Stockholm Water Prize in 2015.

Short Answer Questions

1. Conservation of the environment refers to "keeping the environment safe as it is," which involves making wise use of the planet's natural resources to prevent further deterioration and improvement of the ecosystem.
2. Water and forest.
3. Timber, firewood, honey and fruits.
4. (i) Pine wood for matchbox industry (ii) Bamboo for paper industry
5. Human intervention saved the Arabari forest range of West Bengal from destruction with active and willing participation of local community. The Sal forest of Arabari underwent a remarkable recovery.

Long Answer Questions

1. Hint: Water harvesting is the collection and storage of rainwater or runoff for later use. Helps in conserving water, recharging groundwater, and meeting water needs during periods of scarcity.
   Elaborate on these points.
2. Hint: Two kinds of natural resources: renewable (solar energy, wind energy, forests) and non-renewable (fossil fuels, minerals). Elaborate on these points.
3. Hint: Reliable source of water, less susceptible to evaporation and contamination, supports ecosystems and agriculture, helps maintain streamflow during dry periods. Elaborate on these points.
4. Hint: Forests provide habitat for a wide range of wildlife, support biodiversity, regulate climate, and provide resources like timber and medicinal plants. Wildlife plays a crucial role in maintaining ecological balance and offers recreational and aesthetic value. Elaborate on these points.
5. Hint: Balancing economic growth, social well-being, and environmental protection to meet the needs of the present without compromising the ability of future generations to meet their own needs. Focuses on long-term resource management, reducing environmental impact, and promoting social equity. Elaborate on these points.

Case Study-Based Questions

1. (i) (c), (ii) (d), (iii) (a), (iv) (b), (v) (b)
2. (i) (d), (ii) (b), (iii) (d), (iv) (c), (v) (c)
3. (i) (d), (ii) (b), (iii) (d), (iv) (c), (v) (d)

NCERT Corner

1. Petroleum
2. CNG
3. Coal and petroleum

Competition Window

1. (a)	2. (b)	3. (d)	4. (a)	5. (c)
6. (c)	7. (a)	8. (b)	9. (c)	10. (c)
11. (c)	12. (c)	13. (a)	14. (c)	15. (c)
16. (a)	17. (a)			